Maximum Entropy and Bayesian Methods

Fundamental Theories of Physics

An International Book Series on The Fundamental Theories of Physics: Their Clarification, Development and Application

---

Volume 62

Maximum Entropy and Bayesian Methods

Santa Barbara, California, U.S.A., 1993

Proceedings of the Thirteenth International Workshop on Maximum Entropy and Bayesian Methods

edited by

Glenn R. Heidbreder

formerly of the
Department of Electrical and Computer Engineering,
University of California, Santa Barbara,
Santa Barbara, California, U.S.A.

KLUWER ACADEMIC PUBLISHERS

DORDRECHT / BOSTON / LONDON

A C.I.P. Catalogue record for this book is available from the Library of Congress.

ISBN 978-90-481-4407-5

Published by Kluwer Academic Publishers,
P.O. Box 17, 3300 AA Dordrecht, The Netherlands.

Kluwer Academic Publishers incorporates
the publishing programmes of
D. Reidel, Martinus Nijhoff, Dr W. Junk and MTP Press.

Sold and distributed in the U.S.A. and Canada
by Kluwer Academic Publishers,
101 Philip Drive, Norwell, MA 02061, U.S.A.

In all other countries, sold and distributed
by Kluwer Academic Publishers Group,
P.O. Box 322, 3300 AH Dordrecht, The Netherlands.

Printed on acid-free paper

CONTENTS

PREFACE

This volume contains selections from among the presentations at the Thirteenth International Workshop on Maximum Entropy and Bayesian Methods- MAXENT93 for short- held at the University of California, Santa Barbara (UCSB), August 1-5, 1993. This annual workshop is devoted to the theory and practice of Bayesian probability and the use of the maximum entropy principle in assigning prior probabilities. Like its predecessors, MAXENT93 attracted researchers and scholars representing a wide diversity of disciplines and applications. These included physicists, geophysicists, astronomers, statisticians, engineers, and economists, among others. Indeed Bayesian methods increasingly compel the interest of any who would apply scientific inference. The impressive successes, so evident in the proceedings of the past workshops, when adherence to Bayesian principles replaces popular ad hoc approaches in problems of inference, continue. Many are reported in this volume. It is perhaps indicative of the growing acceptance of Bayesian methods that the most prominent controversy at the thirteenth workshop was not a Bayesian- frequentist confrontation but rather a disagreement over the suitability of using an approximation in the Bayesian formalism.

Acknowledgments and thanks are due the several organizations and many individuals who made the workshop possible. The United States Navy Office of Naval Research (ONR) continued its support of the workshop through its grant N00014-93-1-0583, which was further supplemented by the United States Army Research Office (ARO). Kluwer Academic Publishers provided startup funding in the early stages of workshop planning. Special thanks are due Dr. Rabinder Madan of ONR, Dr. William Sander of ARO, and Dr. David Larner of Kluwer for their support. Thanks are also due the University of California, Santa Barbara, which provided its attractive facilities. Support of the UCSB College of Engineering and of the Departments of Physics and of Statistics and Applied Probability is acknowledged with gratitude. I am particularly indebted to my former colleagues in the Department of Electrical and Computer Engineering, Drs. Hua Lee and Glen Wade, for their invaluable assistance as co-organizers and co-hosts.

As has been the practice at recent workshops, MAXENT93 began with a series of tutorials on Bayesian methods and their applications.

Thanks are due to Drs. C, Ray Smith, Anthony Garrett, Ali Mohammad-Djafari, Tom Loredo, Larry Bretthorst, and John Skilling for their efforts in presenting the tutorials and especially to Dr. Bretthorst for detailing his tutorial for presentation as the lead article in this volume.

Finally, I wish to thank the reviewers for their efforts and the many authors for their contributions and their patience with the protracted editorial process. Although many authors supplied camera ready copy, it was necessary, in the interest of consistency of presentation, to re-typeset many contributions. I apologize for any typographical errors or omissions which may have resulted from this process. The help of Drs. Larry Bretthorst and Gary Erickson and of Ms. Mary Sheetz in the re-typesetting is gratefully acknowledged.

Reston, VA, USA — Glenn Heidbreder
November 1995

AN INTRODUCTION TO MODEL SELECTION USING PROBABILITY THEORY AS LOGIC

G. Larry Bretthorst
Washington University
Department of Chemistry
1 Brookings Drive
St. Louis, Missouri 63130

ABSTRACT. Probability theory as logic is founded on three simple *desiderata:* that degrees of belief should be represented by real numbers, that one should reason consistently, and that the theory should reduce to Aristotelian logic when the truth values of the hypotheses are known. Because this theory represents a probability as a state of knowledge, not a state of nature, hypotheses such as "The frequency of oscillation of a sinusoidal signal had value ω when the data were taken," or "Model x is a better description of the data than model y" make perfect sense. Problems of the first type are generally thought of as parameter estimation problems, while problems of the second type are thought of as model selection problems. However, in probability theory there is no essential distinction between these two types of problems. They are both solved by application of the sum and product rules of probability theory. Model selection problems are conceptually more difficult, because the models may have different functional forms. Consequently, conceptual difficulties enter the problem that are not present in parameter estimation. This paper is a tutorial on model selection. The conceptual problems that arise in model selection will be illustrated in such a way as to automatically avoid any difficulties. A simple example is worked in detail. This example, (radar target identification) illustrates all of the points of principle that must be faced in more complex model selection problems, including how to handle nuisance parameters, uninformative prior probabilities, and incomplete sets of models.

Introduction

A basic problem in science and engineering is to determine when a model is adequate to explain a set of observations. Is the model complete? Is a new parameter needed? If the model is changed, how? Given several alternatives, which is best? All are examples of the types of questions that scientists and engineers face daily. A principle or theory is needed that allows one to choose rationally. Ockham's razor [1] is the principle typically used. Essentially, Ockham's razor says that objects should not be multiplied needlessly. This is typically paraphrased: "When two models fit the observations equally well, prefer the simpler model." This principle has proven itself time and time again as a valuable tool of science. From the standpoint of probability theory, the reason that Ockham's razor works is that simpler models are usually more probable. That simpler models are usually more probable was first argued by Jeffreys [2] and later explicitly demonstrated by Jaynes [3], Gull [4], and Bretthorst [5–8]. However, probability theory tempers Ockham's razor and will allow more complex models to be accepted when they fit the data significantly better or when they contain parameters that have higher initial probability.

This paper is a tutorial on model selection. In it the procedures and principles needed to

G. R. Heidbreder (ed.), Maximum Entropy and Bayesian Methods, 1–42.

apply probability theory as extended logic to problems of model selection will be discussed in detail. Primarily these procedures and principles will be illustrated using an example taken from radar target identification. In this example we will illustrate the assignment of probabilities, the use of uninformative prior probabilities, and how to handle hypotheses that are mutually exclusive, but not exhaustive. While we attempt to explain all of the steps in this calculation in detail, some familiarity with higher mathematics and Bayesian probability theory is assumed. For an introduction to probability theory see the works of Tribus [9], Zellner [10], and Jaynes [11]; for a derivation of the rules of probability theory see Jaynes [11,12], and for an introduction to parameter estimation using probability theory see Bretthorst [13]. In this tutorial the sum and product rules of probability theory will be given and no attempt will be made to derive them. However, as indicated in the abstract, if one wishes to represent degrees of belief as real numbers, reason consistently, and have probability theory reduce to Aristotelian logic when the truth of the hypotheses are known, then the sum and product rules are the unique rules for conducting inference. For an extensive discussion of these points and much more, see Jaynes [11].

1 The Rules of Probability Theory

There are two basic rules for manipulating probabilities, the product rule and the sum rule; all other rules may be derived from these. If A, B, and C stand for three arbitrary hypotheses, then the product rule states

$$P(AB|C) = P(A|C)P(B|AC), \tag{1}$$

where $P(AB|C)$ is the joint probability that "A and B are true given that C is true," $P(A|C)$ is the probability that "A is true given C is true," and $P(B|AC)$ is the probability that "B is true given that both A and C are true." The notation "$|C)$" means conditional on the truth of hypothesis C. In probability theory *all* probabilities are conditional. The notation $P(A)$ is not used to stand for the probability for a hypothesis, because it does not make sense until the evidence on which it is based is given. Anyone using such notation either does not understand that all knowledge is conditional, i.e., contextual, or is being extremely careless with notation. In either case, one should be careful when interpreting such material. For more on this point see Jeffreys [2] and Jaynes [11].

In Aristotelian logic, the hypothesis "A and B" is the same as "B and A," so the numerical value assigned to the probabilities for these hypotheses must be the same. The order may be rearranged in the product rule, Eq. (1), to obtain:

$$P(BA|C) = P(B|C)P(A|BC), \tag{2}$$

which may be combined with Eq. (1) to obtain a seemingly trivial result

$$P(A|BC) = \frac{P(A|C)P(B|AC)}{P(B|C)}. \tag{3}$$

This is Bayes' theorem. It is named after Rev. Thomas Bayes, an 18th century mathematician who derived a special case of this theorem. Bayes' calculations [14] were published in 1763, two years after his death. Exactly what Bayes intended to do with the calculation, if anything, still remains a mystery today. However, this theorem, as generalized by

Laplace [15], is the basic starting point for inference problems using probability theory as logic.

The second rule of probability theory, the sum rule, relates to the probability for "A or B." The operation "or" is indicated by a + inside a probability symbol. The sum rule states that given three hypotheses A, B, and C, the probability for "A or B given C" is

$$P(A+B|C) = P(A|C) + P(B|C) - P(AB|C). \tag{4}$$

If the hypotheses A and B are mutually exclusive, that is the probability $P(AB|C)$ is zero, the sum rule becomes:

$$P(A+B|C) = P(A|C) + P(B|C). \tag{5}$$

The sum rule is especially useful because it allows one to investigate an interesting hypothesis while removing an uninteresting or nuisance hypothesis from consideration.

To illustrate how to use the sum rule to eliminate nuisance hypotheses, suppose D stands for the data, ω the hypothesis "the frequency of a sinusoidal oscillation was ω," and B the hypothesis "the amplitude of the sinusoid was B." Now suppose one wishes to compute the probability for the frequency given the data, $P(\omega|D)$, but the amplitude B is present and must be dealt with. The way to proceed is to compute the joint probability for the frequency and the amplitude given the data, and then use the sum rule to eliminate the amplitude from consideration. Suppose, for argument's sake, the amplitude B could take on only one of two mutually exclusive values $B \in \{B_1, B_2\}$. If one computes the probability for the frequency and (B_1 or B_2) given the data one has

$$P(\omega|D) \equiv P(\omega[B_1 + B_2]|D) = P(\omega B_1|D) + P(\omega B_2|D). \tag{6}$$

This probability distribution summarizes all of the information in the data relevant to the estimation of the frequency ω. The probability $P(\omega|D)$ is called the marginal probability for the frequency ω given the data D.

The marginal probability $P(\omega|D)$ does not depend on the amplitudes at all. To see this, the product rule is applied to the right-hand side of Eq. (6) to obtain

$$P(\omega|D) = P(B_1|D)P(\omega|B_1D) + P(B_2|D)P(\omega|B_2D) \tag{7}$$

but

$$P(B_1|D) + P(B_2|D) = 1 \tag{8}$$

because the hypotheses are exhaustive. So the probability for the frequency ω is a weighted average of the probability for the frequency given that one knows the various amplitudes. The weights are just the probability that each of the amplitudes is the correct one. Of course, the amplitude could take on more than two values; for example if $B \in \{B_1, \cdots, B_m\}$, then the marginal probability distribution becomes

$$P(\omega|D) = \sum_{j=1}^{m} P(\omega B_j|D), \tag{9}$$

provided the amplitudes are mutually exclusive and exhaustive. In many problems, the hypotheses B could take on a continuum of values, but *as long as only one value of B is realized when the data were taken* the sum rule becomes

$$P(\omega|D) = \int dB P(\omega B|D). \tag{10}$$

Note that the B inside the probability symbols refers to the hypothesis; while the B appearing outside of the probability symbols is a number or index. A notation could be developed to stress this distinction, but in most cases the meaning is apparent from the context.

The sum and integral appearing in Eqs. (9,10) are over a set of mutually exclusive and exhaustive hypotheses. If the hypotheses are not mutually exclusive, one simply uses Eq. (4). However, if the hypotheses are *not* exhaustive, the sum rule *cannot* be used to eliminate nuisance hypotheses. To illustrate this, suppose the hypotheses, $B \in \{B, \cdots, B_m\}$, are mutually exclusive, but not exhaustive. The hypotheses B could represent various explanations of some experiment, but it is always possible that there is something else operating in the experiment that the hypotheses B do not account for. Let us designate this as

$$\text{SE} \equiv \text{“Something Else not yet thought of.”}$$

The set of hypotheses $\{B, \text{SE}\}$ is now complete, so the sum rule may be applied. Computing the probability for the hypothesis B_i conditional on some data D and the information I, where I stands for the knowledge that amplitudes B are not exhaustive, one obtains

$$P(B_i|DI) = \frac{P(B_i|I)P(D|B_iI)}{P(D|I)} \tag{11}$$

and for SE

$$P(\text{SE}|DI) = \frac{P(\text{SE}|I)P(D|\text{SE}I)}{P(D|I)}. \tag{12}$$

The denominator is the same in both these equation and is given by

$$\begin{aligned} P(D|I) &= \sum_{i=1}^{m} P(DB_i|I) + P(D\text{SE}|I) \\ &= \sum_{i=1}^{m} P(B_i|I)P(D|B_iI) + P(\text{SE}|I)P(D|\text{SE}I). \end{aligned} \tag{13}$$

But this is indeterminate because SE has not been specified, and therefore the likelihood, $P(D|\text{SE}I)$, is indeterminate even if the prior probability $P(\text{SE}|I)$, is known. However, the relative probabilities $P(B_i|DI)/P(B_j|DI)$ are well defined because the indeterminacy cancels out. So there are two choices: either *ignore* SE and thereby assume the hypotheses B are complete or *specify* SE, thereby completing the set of hypotheses. One of the main purposes of this tutorial is to illustrate this last alternative and to show how to apply it in real problems.

2 Assigning Probabilities

The product rule and the sum rule are used to indicate relationships between probabilities. These rules are not sufficient to conduct inference because, ultimately, the "numerical values" of the probabilities must be known. Thus the rules for manipulating probabilities must be supplemented by rules for assigning numerical values to probabilities. The historical lack of these supplementary rules is one of the major reasons why probability theory, as formulated by Laplace, was rejected in the late part of the 19th century. To assign any probability there is ultimately only one way, logical analysis, i.e., non-self-contradictory analysis of the available information. The difficulty is to incorporate only the information one actually possesses without making gratuitous assumptions about things one does not know. A number of procedures have been developed that accomplish this task: Logical analysis may be applied directly to the sum and product rules to yield probabilities (Jaynes [11]). Logical analysis may be used to exploit the group invariances of a problem (Jaynes [16]). Logical analysis may be used to ensure consistency when uninteresting or nuisance parameter are marginalized from probability distributions (Jaynes [21]). And last, logical analysis may be applied in the form of the principle of maximum entropy to yield probabilities (Zellner [10], Jaynes [16,19], and Shore and Johnson [17,18]). Of these techniques the principle of maximum entropy is probably the most powerful, and in this tutorial it will be used to assign all probabilities.

In this tutorial there are three different types of information that must be incorporated into probability assignments: parameter ranges, knowledge of the mean and standard deviation of a probability distribution for several quantities, and some properties of the noise or errors in the data. Their assignment differs only in the types of information available. In the first case, the principle of maximum entropy leads to a bounded uniform prior probability. In the second and third cases, it leads to a Gaussian probability distribution. To understand the principle of maximum entropy and how these probability assignments come about, suppose one must assign a probability distribution for the ith value of a parameter given the "testable information" I. This probability is denoted $P(i|I)$ $(1 \leq i \leq m)$. Information I is testable when, for any proposed probability assignment $P(i|I)$, there exists a procedure by which it can be unambiguously determined whether I agrees with $P(i|I)$. The Shannon entropy, defined as

$$H \equiv -\sum_{i=1}^{m} P(i|I) \log P(i|I), \tag{14}$$

is a measure of the amount of ignorance (uncertainty) in this probability distribution [22]. Shannon's entropy is based on a qualitative requirement, the entropy should be monotonically increasing for increasing ignorance, plus the requirement that the measure be consistent. The principle of maximum entropy then states that if one has some testable information I, one can assign the probability distribution, $P(i|I)$, that contains only the information I by maximizing H subject to the information (constraints) represented by I. Because H measures the amount of ignorance in the probability distribution, assigning a probability distribution that has maximum entropy yields a distribution that is least informative (maximally ignorant) while remaining consistent with the information I: the probability distribution, $P(i|I)$, contains only the information I, and does not contain any additional information not already implicit in I [17,18].

To demonstrate its use, suppose that one must assign $P(i|I)$ and nothing is known except that the set of hypotheses is mutually exclusive and exhaustive. Applying the sum rule one obtains

$$\sum_{i=1}^{m} P(i|I) = 1. \tag{15}$$

This equation may be written

$$\sum_{i=1}^{m} P(i|I) - 1 = 0 \tag{16}$$

and because this equation sums to zero, any multiple of it may be added to the entropy of $P(i|I)$ without changing its value:

$$H = -\sum_{i=1}^{m} P(i|I)\log P(i|I) + \beta\left[1 - \sum_{i=1}^{m} P(i|I)\right]. \tag{17}$$

The constant β is called a Lagrange multiplier. But the probabilities $P(i|I)$ and the Lagrange multiplier β are not known; they must be assigned. To assign them, H is constrained to be a maximum with respect to variations in all the unknown quantities. This maximum is located by differentiating H with respect to both $P(k|I)$ and β, and then setting the derivatives equal to zero. Here there are m unknown probabilities and one unknown Lagrange multiplier. But when the derivatives are taken, there will be $m+1$ equations; thus all of the unknowns may be determined. Taking the derivative with respect to $P(k|I)$, one obtains

$$\log P(k|I) + 1 + \beta = 0, \tag{18}$$

and taking the derivative with respect to β returns the constraint equation

$$1 - \sum_{i=1}^{m} P(i|I) = 0. \tag{19}$$

Solving this system of equations, one finds

$$P(i|I) = \frac{1}{m} \qquad \text{and} \qquad \beta = \log m - 1. \tag{20}$$

When nothing is known except the specification of the hypotheses, the principle of maximum entropy reduces to Laplace's principle of indifference [15]. But the principle of maximum entropy is much more general because it allows one to incorporate any type of testable information.

As noted earlier, in the inference problem addressed in this paper, there are three different types of information to be incorporated into probability assignments. The specification of parameter ranges occurs when the prior probabilities for various location parameters appearing in the calculation must be assigned. (A location parameter is a parameter that appears linearly in the model equation.) For these location parameters, the principle of maximum entropy leads to the assignment of a bounded uniform prior probability. However, care must be taken because most of these parameters are continuous and *the rules and procedures given in this tutorial are strictly valid only for finite, discrete probability distributions.* The concept of a probability for a hypothesis containing a continuous parameter, a probability density function, only makes sense when thought of as a limit. If the preceding calculations are repeated and the number of hypotheses are allowed to grow infinitely, one will automatically arrive at a valid result as long as all probabilities remain finite and normalized. Additionally, the direct introduction of an infinity into any mathematical calculation is ill-advised under any conditions. Such an introduction presupposes the limit already accomplished and this procedure will cause problems whenever any question is asked that depends on how the limit was taken. For more on the types of problems this can cause see Jaynes [21], and for a much more extensive discussion of this point see

Jaynes [11]. As it turns out, continuous parameters are not usually a problem, provided one always uses normalized probabilities. In this tutorial, continuous parameters will be used, but their prior probabilities will be normalized and the prior ranges will never be allowed to go to infinity without taking a limit.

The second type of information that must be incorporated into a probability assignment is knowledge of the mean and standard deviation of a probability distribution. It is a straightforward exercise to show that, in this case, the principle of maximum entropy leads to a Gaussian distribution.

The third type of information that must be incorporated into a probability assignment is information about the true errors or noise in the data. The probability that must be assigned is denoted $P(D|LI)$, the probability for the data given that the signal is L, where the data, D, is a joint hypothesis of the form, $D \equiv \{d_1 \ldots d_N\}$, d_j are the individual data items, and N is the number of data values. If the true signal is known to be $L(r_j)$ at position r_j, then

$$d_j - L(r_j) = n_j \tag{21}$$

assuming that the noise is additive, and n_j is the true noise value. Thus the probability for the data can be assigned if one can assign a probability for the noise.

To assign a probability for the noise the question one must ask is, *what properties of the noise are to be used in the calculations*? For example, should the results of the calculations depend on correlations? If so, which of the many different types of correlations should the results depend on? There are second order correlations of the form

$$\rho'_s = \frac{1}{N-s}\sum_{j=1}^{N-s} n_j n_{j+s}, \tag{22}$$

where s is a measure of the correlation distance, as well as third, fourth, and higher order correlations. In addition to correlations, should the results depend on the moments of the noise? If so, on which moments should they depend? There are many different types of moments. There are power law moments of the form

$$\sigma'_s = \frac{1}{N}\sum_{j=1}^{N} n_j^s, \tag{23}$$

as well as moments of arbitrary functions, and a host of others.

The probability that must be assigned is the probability that one should obtain the data D, but from Eq. (21) this is just the probability for noise $P(e_1 \cdots e_N|I')$, where e_j stands for a hypothesis of the form "the true value of the noise at position r_j was e_j, when the data were taken." The quantity e_j is an index that ranges over all valid values of the noise; while the probability for the noise, $P(e_1 \cdots e_N|I')$, assigns a reasonable degree of belief to a particular set of noise values. For the probability for the noise to be consistent with correlations it must have the property that

$$\rho_s = \langle e_j e_{j+s}\rangle \equiv \frac{1}{N-s}\sum_{j=1}^{N-s}\int de_1 \cdots de_N \; e_j e_{j+s} P(e_1 \cdots e_N|I') \tag{24}$$

and for it to be consistent with the power law moments it must have the additional property that

$$\sigma_s = \langle e^s \rangle \equiv \frac{1}{N} \sum_{j=1}^{N} \int de_1 \cdots de_N \, e_j^s P(e_1 \cdots e_N | I') \tag{25}$$

where the notation $\langle\rangle$ denote mean averages over the probability density function.

In Eq. (22) and Eq. (23), the symbols ρ'_s and σ'_s were used to denote means or averages over the sample noise. These averages are the sample correlation coefficients and moments and they represent states of nature. In Eq. (24) and Eq. (25), the symbols ρ_s and σ_s are used to denote mean averages over the probability for the noise, and they represent states of knowledge. To use information in a maximum entropy calculation, that information must be testable, i.e., the moments and correlation coefficients must be known.

Assuming that none of these quantities are known, how can the principle of maximum entropy be used? Its use requires testable information, and unless at least some of the ρ'_s and σ'_s are known, it would appear that we have no testable information. However, this description of the problem is not what probability theory asks us to do. Probability theory asks us to assign $P(e_1 \cdots e_N | I')$, where I' represents the information on which this probability is based. Suppose for the sake of argument that that information is a mean, ν, and standard deviation, σ, then what probability theory asks us to assign is $P(e_1 \cdots e_N | \nu\sigma)$. This expression should be read as the joint probability for all the errors given that the mean of the errors is ν and the standard deviation of the errors is σ. According to probability theory, in the process of assigning the probability for the errors, we are to assume that both ν and σ are known or given values. This is a very different state of knowledge from knowing that the mean and standard deviation of the sampling distribution are ν and σ. If we happen to actually know these values, then there is less work to do when applying the rules of probability theory. However, if their values are unknown, we still seek the least informative probability density function that is consistent with a fixed or given mean and standard deviation. The rules of probability theory are then used to eliminate these unknown nuisance hypotheses from the final probability density functions.

But which of these constraints should be used? The answer was implied earlier by the way the question was originally posed: what *properties* of the errors are to be used in the calculations? The class of maximum entropy probability distributions is the class of all probability density functions for which sufficient statistics exist. A sufficient statistic is a function of the data that summarizes all of the information in the data relevant to the problem being solved. These sufficient statistics are the sample moments that correspond to the constraints that were used in the maximum entropy calculation. For example, suppose we used the first three correlation coefficients, ρ_1, ρ_2, and ρ_3, as defined by Eq. (24) in a maximum entropy calculation, then the parameter estimates will depend only on the first three correlation coefficients of the data and our uncertainty in those estimates will depend on ρ_1, ρ_2, and ρ_3 if they are known, and on the first three correlation coefficients of the true noise values if ρ_1, ρ_2, and ρ_3 are not known. *All* other properties of the errors have been made irrelevant by the use of maximum entropy. So the real question becomes, what does one know about the errors before seeing the data? If there is information that suggests the errors may be correlated, then by all means a correlation constraint should be included. Additionally, if one has information that suggests the higher moments of the noise can deviate significantly from what one would expect from a Gaussian distribution, then again a constraint on the higher moments should be included. But if one has no information about higher moments and correlations, then one is always better off to leave those constraints out of the maximum entropy calculation, because the resulting probability density function will have higher entropy. Higher entropy distributions are by definition less informative and therefore make more conservative estimates of the parameters. Consequently, these higher entropy probability density functions are applicable under a much wider variety of circumstances, and typically they are simpler and easier to use than distributions having lower entropy.

In assigning the probability density function for the noise, it will be assumed that our parameter estimates are to depend only on the mean and variance of the true errors in the data. The appropriate constraints necessary are on the first and second moments of the probability density function. The constraint on the first moment is given by

$$\nu = \frac{1}{N}\sum_{j=1}^{N}\int de_1\cdots de_N\; e_j P(e_1\cdots e_N|I') \tag{26}$$

and by

$$\sigma^2+\nu^2 = \frac{1}{N}\sum_{j=1}^{N}\int de_1\cdots de_N\; e_j^2 P(e_1\cdots e_N|I') \tag{27}$$

for the second moment, where ν and σ^2 are the fixed or given values of the mean and variance. Note the second moment of the probability distribution, Eq. (27), is written as $\sigma^2+\nu^2$, to make the resulting probability density function come out in standard notation.

We seek the probability density function that has highest entropy for a fixed or given value of σ^2 and ν. To find this distribution Eq. (26) and Eq. (27) are rewritten so they sum to zero:

$$\nu - \frac{1}{N}\sum_{j=1}^{N}\int de_1\cdots de_N\; e_j P(e_1\cdots e_N|I') = 0, \tag{28}$$

and

$$\sigma^2+\nu^2 - \frac{1}{N}\sum_{j=1}^{N}\int de_1\cdots de_N\; e_j^2 P(e_1\cdots e_N|I') = 0. \tag{29}$$

Additionally, the probability for finding the true noise values somewhere in the valid range of values is one:

$$1 - \int de_1\cdots de_N P(e_1\cdots e_N|I') = 0. \tag{30}$$

Because Eq. (28) through Eq. (30), sum to zero, they may each be multiplied by a constant and added to the entropy of this probability density function without changing its value, one obtains

$$\begin{aligned} H &= -\int de_1 \cdots de_N P(e_1 \cdots e_N|I') \log P(e_1 \cdots e_N|I') \\ &+ \beta\left[1 - \int de_1 \cdots de_N P(e_1 \cdots e_N|I')\right] \\ &+ \delta\left[\nu - \tfrac{1}{N}\sum_{j=1}^{N}\int de_1 \cdots de_N\ e_j P(e_1 \cdots e_N|I')\right] \\ &+ \lambda\left[\sigma^2 + \nu^2 - \tfrac{1}{N}\sum_{j=1}^{N}\int de_1 \cdots de_N\ e_j^2 P(e_1 \cdots e_N|I')\right] \end{aligned} \tag{31}$$

where β, δ, and λ are Lagrange multipliers. To obtain the maximum entropy distribution, this expression is maximized with respect to variations in β, δ, λ, and $P(e_1' \cdots e_N'|I')$. After a little algebra, one obtains

$$P(e_1 \cdots e_N|\nu\sigma) = (2\pi\sigma^2)^{-\frac{N}{2}} \exp\left\{-\sum_{j=1}^{N}\frac{(e_j - \nu)^2}{2\sigma^2}\right\}, \tag{32}$$

where

$$\lambda = \frac{N}{2\sigma^2}, \qquad \delta = -\frac{N\nu}{\sigma^2}, \quad \text{and} \quad \beta = \frac{N}{2}\left[\log(2\pi\sigma^2) + \frac{\nu^2}{\sigma^2}\right] - 1 \tag{33}$$

and I' has been replaced by the fixed or given values of the moments.

There are several interesting points to note about this probability density function. First, this is a Gaussian distribution. However, the fact that the prior probability for the errors has been assigned to be a Gaussian makes no statement about the true sampling distribution of the errors; rather it says only that for a fixed value of the mean and variance the probability density function for the errors should be maximally uninformative and that maximally uninformative distribution happens to be a Gaussian. Second, this probability assignment apparently does not contain correlations. The reason for this is that a constraint on correlations must lower the entropy. By definition a probability assignment with lower entropy is more informative, and so must make more precise estimates of the parameters. Instead of saying this probability density function does not contain correlations, it would be more correct to say that this probability density function makes allowances for *every possible correlation* that could be present and so is less informative than correlated distributions. Third, if one computes the expected mean value of the moments, one finds

$$\langle e^s \rangle = \exp\left\{-\frac{\nu^2}{2\sigma^2}\right\}\sigma^{2s}\frac{\partial^s}{\partial\nu^s}\exp\left\{\frac{\nu^2}{2\sigma^2}\right\} \quad (s \geq 0) \tag{34}$$

which reduces to

$$\langle e^0 \rangle = 1, \qquad \langle e^1 \rangle = \nu, \quad \text{and} \quad \langle e^2 \rangle = \sigma^2 + \nu^2 \tag{35}$$

for $s = 0$, $s = 1$, and $s = 2$, just the constraints used to assign the probability density function. Fourth, for a fixed value of the mean and variance this prior probability has highest entropy. Consequently, when parameters are marginalized from probability distributions or when any operation is performed on them that preserves mean and variance while discarding other information, those probability densities necessarily will move closer and closer to this Gaussian distribution regardless of the initial probability assignment. The Central Limit Theorem is one special case of this phenomenon – see Jaynes [11].

Earlier it was asserted that maximum entropy distributions are the only distributions that have sufficient statistics and that these sufficient statistics are the only properties of the data, and therefore the errors, that are used in estimating parameters. We would like to demonstrate this property explicitly for the Gaussian distribution [11]. Suppose the true value of a location parameter is ν_0 and one has a measurement such that

$$d_j = \nu_0 + n_j. \tag{36}$$

The hypothesis about which inferences are to be made is of the form "the true value of the mean is ν given the data, D." Assigning a Gaussian as the prior probability for the errors, the likelihood function is then given by

$$P(D|\nu\sigma I) = (2\pi\sigma^2)^{-\frac{N}{2}} \exp\left\{-\frac{1}{2\sigma^2}\sum_{j=1}^{N}(d_j-\nu)^2\right\}. \tag{37}$$

The posterior probability for ν may be written as

$$P(\nu|D\sigma I) \propto (2\pi\sigma^2)^{-\frac{N}{2}} \exp\left\{-\frac{N}{2\sigma^2}([\overline{d}-\nu]^2 + s^2)\right\} \tag{38}$$

where a uniform prior probability was assigned for ν. The mean data value, $\overline{d}$, is given by

$$\overline{d} = \frac{1}{N}\sum_{j=1}^{N} d_j = \nu_0 + \overline{n} \tag{39}$$

where $\overline{n}$ is the mean value of the true errors. And s^2 is given by

$$s^2 = \overline{d^2} - (\overline{d})^2 = \frac{1}{N}\sum_{j=1}^{N} d_j^2 - \left(\frac{1}{N}\sum_{j=1}^{N} d_j\right)^2 = \overline{n^2} - (\overline{n})^2 \tag{40}$$

where $(\overline{n})^2$ is the mean square of the true noise vales. From which one obtains

$$(\nu)_{est} = \begin{cases} \overline{d} \pm \sigma/\sqrt{N} & \sigma \text{ known} \\ \overline{d} \pm s/\sqrt{N-3} & \sigma \text{ unknown} \end{cases} \tag{41}$$

as the estimate for ν. The actual error, Δ, is given by

$$\Delta = \overline{d} - \nu_0 = \overline{n} \tag{42}$$

which depends only on the *mean of the true noise values*; while our accuracy estimate depends only on σ if the standard deviation of the noise is known, and *only on the mean and mean-square* of the true noise values when the standard deviation of the noise is not known. Thus the underlying sampling distribution of the noise has completely canceled out and the only property of the errors that survives is the actual mean and mean-square of the true noise values. *All* other properties of the errors have been made irrelevant. Exactly the same parameter estimates will result if the underlying sampling distribution of the noise is changed, provided the mean and mean-square of the new sampling distribution is the same, just the properties needed to represent what is actually known about the noise, the mean and mean-square, and to render what is *not* known about it irrelevant.

3 Example – Radar Target Identification

In Section 1 the sum and product rules of probability theory were given. In Section 2 the principle of maximum entropy was used to demonstrate how to assign probabilities that are maximally uninformative while remaining consistent with the given prior information. In this section a nontrivial model selection problem is given. Each step in the calculation is explained in detail. The example is complex enough to illustrate all of the points of principle that must be faced in more complicated model selection problems, yet sufficiently simple that anyone with a background in calculus should be able to follow the mathematics.

Probability theory tells one what to believe about a hypothesis C given all of the available information $E_1 \cdots E_n$. This is done by computing the posterior probability for hypothesis C conditional on all of the evidence $E_1 \cdots E_n$. This posterior probability is represented symbolically by

$$P(C|E_1 \cdots E_n). \tag{43}$$

It is computed from the rules of probability theory by repeated application of the sum and product rules and by assigning the probabilities so indicated. This is a general rule and there are no exceptions to it: *ad hoc devices have no place in probability theory.* Given the statement of a problem, the rules of probability theory take over and will lead every person to the same unique solution, provided each person has exactly the same information.

To someone unfamiliar with probability theory, how this is done is not obvious; nor is it obvious what must be done to obtain a problem that is sufficiently well defined to permit the application of probability theory as logic. Consequently, in what follows all of the steps in computing $P(C|E_1 \cdots E_n)$ will be described in detail. To compute the probability for any hypothesis C given some evidence $E_1 \cdots E_n$, there are five basic steps, which are not necessarily independent:

1. *Define The Problem:* State in nonambiguous terms exactly what hypothesis you wish to make inferences about.

2. *State The Model:* Relate the hypothesis of interest to the available evidence $E_1 \cdots E_n$.

3. *Apply Probability Theory:* The probability for hypothesis C conditional on all the available evidence $E_1 \cdots E_n$ is computed from Bayes theorem. The sum rule is then applied to eliminate nuisance hypotheses. The product rule is then repeatedly applied to factor joint probabilities to obtain terms which cannot be further simplified.

4. *Assign The Probabilities:* Using the appropriate procedures, translate the available evidence into numerical values for the indicated probabilities.

5. *Evaluate The Integrals and Sums:* Evaluate the integrals and sums indicated by probability theory. If the indicated calculations cannot be done analytically then implement the necessary computer codes to evaluate them numerically.

Each of these steps will be systematically illustrated in solving a simplified radar target identification problem. In the last section a numerical simulation is discussed.

3.1 Define The Problem

Probability theory solves specific problems in inference. It does this by summarizing ones state of knowledge about a hypothesis as a probability distribution. Thus, to solve an inference problem, one must first state the hypothesis of interest. Here the identification of radar targets will be used to illustrate how to solve model selection problems using probability. However, the subject of this paper is model selection, not radar target identification. For those interested in a more detailed discussion of the fundamentals of radar target identification using probability theory see Jaynes [23]. The hypothesis about which inferences are to be made is of the form "Target number k is being observed by the radar." The index k will represent a particular type of aircraft, or as the radar target identification community refers to them, a particular type of target. The first $\ell - 2$ of these hypotheses represent real aircraft (the known aircraft) and the last two are "The aircraft is NOT a known target," and "No target is in the data, this is a false alarm." The index k really specifies a series of different hypotheses of the form "Hypothesis k is the best description of this state of knowledge." The probability for the kth hypotheses is written $P(k|DI)$, where D is the data and I stands for all of the assumptions and prior information that go into making this a well defined problem. In this problem, as in all realistic problems, this list will be fairly long.

The kth hypothesis is the quantity about which inferences are to be made. The collection of all of these hypotheses is called a library, $L \equiv \{L_1, \ldots, L_\ell\}$, where ℓ is the total number of the hypothesis to be tested. The library is separated into three types of hypotheses: the "known," the "unknown," and the "no-target" hypotheses. Hypotheses one through $(\ell - 2)$ are the known aircraft. These might include the F15, and 747 and a host of others. When making inferences about the known hypotheses, the hypotheses are all of the form "The aircraft being observed is an F15" or "747," etc. In radar target identification, there are so many different types of aircraft, and the number of them changes so rapidly, that one can never be sure of having a hypothesis for all existing aircraft. That is to say, the set of known targets is *not exhaustive.* As was demonstrated earlier, the sum rule may be used to eliminate uninteresting or nuisance hypotheses, but only if the set of hypotheses is exhaustive. Here the hypotheses are mutually exclusive, but not exhaustive. Thus the sum rule cannot not be used unless the set of hypotheses is completed. The set of hypotheses may be made complete either by assuming the set of hypotheses is complete and there by forcing probability to choose from the given set of targets or by defining a model that completes the set. In the radar target identification problem, there is a requirement to be able to identify a hypothesis of the form "the target is NOT one of the known targets." This hypothesis will be number $(\ell - 1)$ in the library. The third class of hypotheses is the "no-target" hypothesis, i.e., no target is present in the data. This hypothesis will be designated as number ℓ.

The hypotheses about which inferences are to be made have now been defined. The needed probability distribution is given symbolically as $P(k|DI)$. However, the definitions of these hypotheses (k, D, and I) are still vague and could describe a host of different problems. To continue with the analysis of this problem, these hypotheses must be made more specific. The process of identifying the relationships between these hypotheses is a process of model building and it is to this task we now turn.

3.2 State The Model

Probabilities are conditional on evidence. Stating the model is the process of relating the hypotheses to that evidence. All types of evidence could be available. In this problem the evidence will consist of data. information about the orientation angle and range to the target, and information about parameter ranges. All of this evidence enters the calculations in exactly the same way, and it doesn't make any difference whether the evidence is data, parameter ranges, or strong prior information. It is all used to assign probabilities conditional on that evidence. To understand the evidence, one must first understand a little about the radar.

The radar is a fictional two-dimensional radar. Schematically, the radar is located at the origin of a polar coordinate system. These coordinates will be referred to as the radar coordinates; they are shown in Fig. 1. The radar captures three different types of data: range, Doppler velocity, and signature data. Only the signature data will be available to the target identification routines. Information from the range and Doppler velocity data will be available in the form of parameter estimates. Additionally, in the real radar target identification problem, information about the velocity, altitude, and acceleration could be used to help identify targets, because this information would effectively eliminate many different types of aircraft. However, in this tutorial, our attention will be restricted to the signature data and the range and Doppler velocity data will be used only to the degree necessary to locate the target in the signature data.

The range data is the vector position of the target as measured in the radar coordinates. Each measurement consists of three numbers: the vector range to target, R_0, Θ, and the time of the measurement. The radar gathers these range measurements periodically, about one measurement every second or so.

The Doppler velocity data is a scalar and represents the speed of the target as projected along the range vector. That is to say, it represents how fast the target is approaching the radar; it is not the target's velocity vector. These measurements are acquired at the same time as the range measurement.

The information needed by the identification calculation is the true range, R_c and orientation angle, ω, of the target. These are shown schematically in Fig. 2. The radar estimates these quantities from the measured range and Doppler velocity data. These inferred or measured values will be denoted as R_0 and Ω respectively. Inferring these quantities is an extensive calculation using probability theory. The details of these calculations are presented in Bretthorst [24]. The results of these inferences are available to the identification routines in the form of a (mean $\pm$ standard deviation) estimate of these quantities. These estimates are interpreted as probabilities in the form of

$$P(\omega|I_\Omega) = (2\pi\sigma_\Omega^2)^{-\frac{1}{2}} \exp\left\{-\frac{[\Omega-\omega]^2}{2\sigma_\Omega^2}\right\} \tag{44}$$

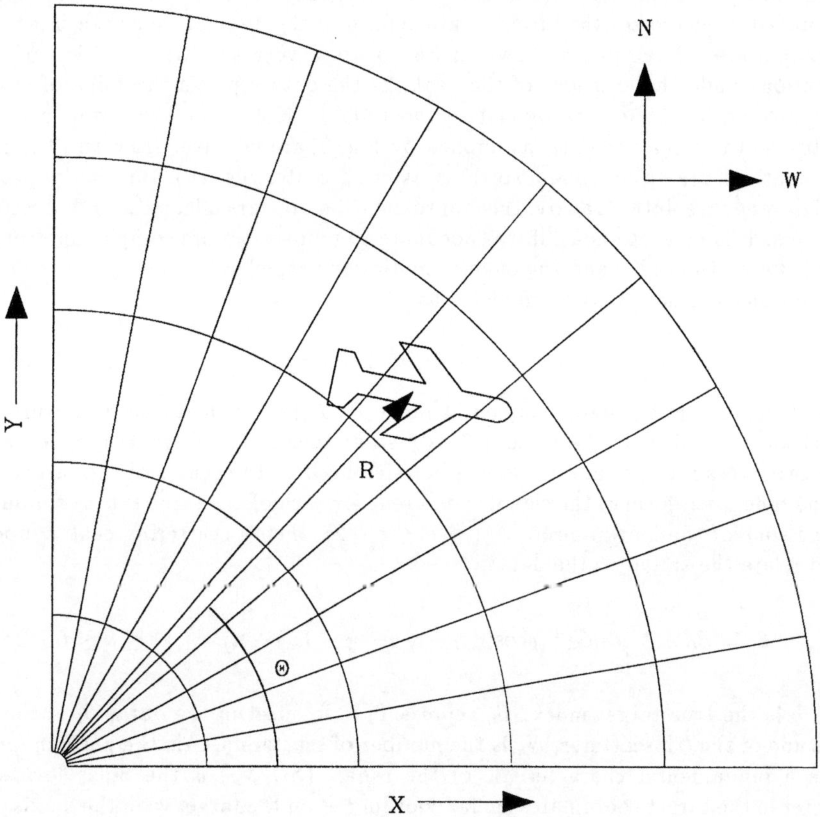

Fig. 1. The radar takes three different types of data: range, Doppler velocity, and signature data. The range data is the vector distance to the center of the target. The Doppler velocity data is the projection of the vector velocity onto the range vector, i.e., it is how fast the target is approaching the radar. Last, the signature data is the envelope of the reflected radar signal, as the signal crosses the target.

where σ_Ω^2 is the uncertainty in this estimate, and I_Ω is the information on which this probability is based. Equation (44) is the probability for a set of hypotheses. The hypotheses are of the form: "The true orientation angle of the target is ω." Similarly for the range to the target one has

$$P(R_c|I_R) = (2\pi\sigma_R^2)^{-\frac{1}{2}} \exp\left\{-\frac{[R_0 - R_c]^2}{2\sigma_R^2}\right\} \tag{45}$$

where σ_R^2 is the uncertainty in the estimated range, and I_R stands for the evidence on which the range estimate is based.

The radar gathers a third type of data, the signature data $D \equiv \{d_1, \cdots, d_N\}$, where N is the number of data values in a signature data set. If the radar were operating in the optical limit, the signature data would be the intensity of the reflected radar signal as the transmitted wave crosses the target. Data typical of this type of radar are shown in Fig. 3. The amplitudes of the peaks shown in Fig. 3 are a very sensitive function of the target orientation, while the locations of the peaks in the data represent the line of site distance to a scatterer (a surface orthogonal to the radar). Note that the radar is an envelope detector, so the signature data, as implied by Fig. 3, are positive. However, the radar does not operate in the optical limit, so the scattering center model is only an approximation. For high range resolution radars, this approximation appears adequate to represent isolated scatterers. It is not yet known if it is adequate to represent more complex interactions, like those between the radar and the engine cavities or propellers.

The signature data may be modeled as

$$d_j = L_k(r_j) + n_j \tag{46}$$

where d_j represents the data sampled at range r_j, $L_k(r_j)$ is the target signature evaluated at position r_j, and n_j is the noise in this measurement. The distances, r_j, correspond to distances across a target and these will be referenced to the center of the target.

The functional form of the signal is different for each of the three types of models. If the target is one of the known aircraft $(1 \leq k \leq \ell - 2)$, then a scattering center model allows one to relate the target to the data:

$$d_j = B_0 + \sum_{l=1}^{N_k} B_l G(S_{kl}\cos(\phi_{kl} - \omega) - r_j + R_c) + n_j \qquad (1 \leq k \leq \ell - 2) \tag{47}$$

where k is the true target index, B_0 represents a dc offset in the data, B_l is the unknown amplitude of the lth scatterer, N_k is the number of scatterers, G is the peak shape function and is a fundamental characteristic of the radar, (S_{kl}, ϕ_{kl}) is the polar location of the scatterer in the target coordinates (polar coordinates on the target with the x axis orientated along the main axis of the aircraft), and (R_c, ω) are the true range and orientation angle of the target. The location of the scatterers (S_{kl}, ϕ_{kl}) and the number of scatterers, N_k, are known quantities and define what is meant by a known target. The constant term may be incorporated into the sum by rewriting the model as

$$d_j = \sum_{l=0}^{N_k} B_l G(S_{kl}\cos(\phi_{kl} - \omega) - r_j + R_c) + n_j \tag{48}$$

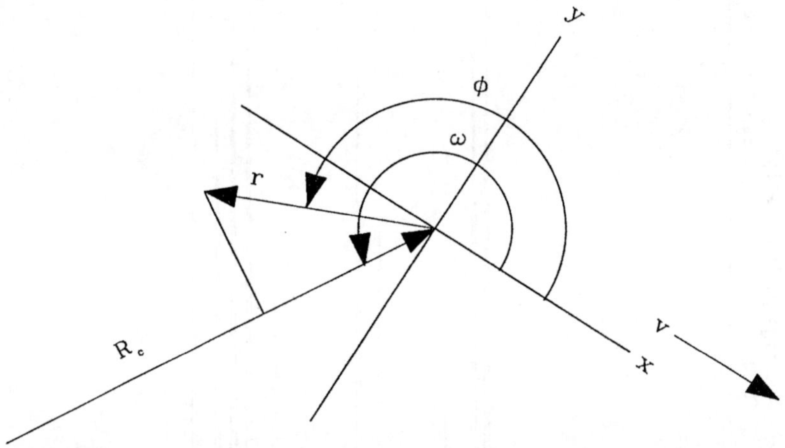

Fig. 2. The observation angle is the difference between the angular location of a scatterer, ϕ, and the orientation angle of the target, ω. These angles are measured in the local target coordinates. The target is orientated along its velocity vector so the observation angle is calculated from the range and velocity vectors of the target.

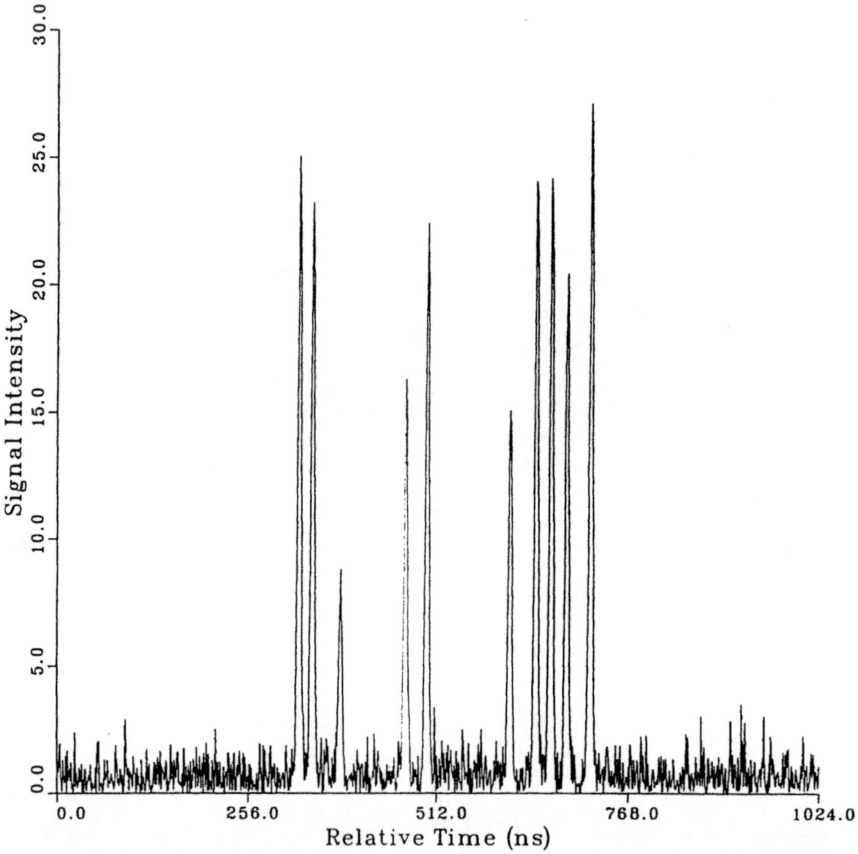

Fig. 3. The signature data represents the intensity of the received signal as the radar signal crossed the target. Locations on the target orthogonal to the radar reflect a large signal, while other locations scatter the radar signal off into space. The peak shape is a characteristic of the radar, while the intensity of the return is a complicated function of range, orientation, and the electromagnetic properties of the target surface.

with the understanding that the signal function G is a constant, for $l = 0$.

The simplest of the three types of models is the no-target model ($k = \ell$). In this model there are no scatterers, only noise. But the noise is positive, the radar is an envelope detector, so the signature data will contain a constant, corresponding to the dc component of the rectified noise. This model may be written as

$$d_j = B_0 + n_j. \tag{49}$$

In addition to the known and no-target hypotheses, the radar must also identify the unknown target. Any type of model that has the ability to expand the data on a complete set is a suitable model for the unknown. However, when expanding the data on a complete set, it is always advisable to choose a basis which captures the essence of the signal. The signal from an unknown aircraft contains an unknown number of scatterers of known peak shape, so an appropriate model would be

$$d_j = \sum_{l=0}^{N_\nu} B_l G(S_{[\ell-1]l} - r_j) + n_j \tag{50}$$

where N_ν is the unknown number of scatterers and the other symbols retain their meaning. From a single data set, there is no information about the angular location of the scatterers, and so no need to include the rotation (the cosine) or to reference the scatterers to the center of the target. Consequently, R_c, ϕ, and ω do not appear in Eq. (50).

The problem now has enough structure to begin the process of applying the rules of probability theory. However, the models are still incomplete in the sense that all of the information available has not yet been supplied. To give just one example, there are a number of amplitude parameters in these models. These amplitudes represent the intensity of the reflected signal. A great deal is known about the possible range of values for these amplitudes. Eventually, probability theory will ask us to supply this information. But supplying it will be delayed until the form in which this information is needed is known.

In one sense delaying this hides some of the beauty of probability theory as logic, because it will appear as if the prior information is being handled differently from the data. In fact this is not the case. For notational convenience, what will be done is that information other than data will be represented as I in all probability symbols. When manipulating probabilities, I must be thought of exactly as if it were any other hypothesis. When prior probabilities are assigned these will typically depend only on I. At that time it must be asked exactly what information is available about the hypothesis and then that information must be used in assigning the probability. If all of the information had been made explicit at the beginning of the calculations this last step would not be necessary because each probability would automatically indicate the evidence on which it is to depend. So by delaying the process of identifying the prior information the notation has been simplified, but at the expense of making prior information seem somehow different from data; which it is not.

3.3 Apply Probability Theory

The problem is to determine which of the hypotheses k is most probable in view of the data and all of one's prior information. This posterior probability is denoted by $P(k|DI)$. To calculate this posterior probability, one applies Bayes' theorem, Eq. (3), to obtain:

$$P(k|DI) = \frac{P(k|I)P(D|kI)}{P(D|I)}. \tag{51}$$

To compute the posterior probability for the target k one must assign three terms. The first term, $P(k|I)$, is the probability for the target given only the information I. This term is referred to as a prior probability, or simply as a "prior" and represents what was known about the presence of this target before obtaining the data D. The second term, $P(D|kI)$, is the probability for the data given that the true hypothesis is k. This term is referred to as the marginal likelihood of the data for reasons that will become apparent shortly. The third term, $P(D|I)$, is the global likelihood for the data, and is a normalization constant.

The prior probability, $P(k|I)$, is sufficiently simplified that it could be assigned. Depending on the location of the radar, there could be either a great deal or very little prior information available. For example, if the radar were located at a civilian airport the types of aircraft one would expect to observe would be very different from what one would expect to observe on an aircraft carrier. Additionally, it is always possible that the radar has just been installed and there is no historical information on which to base a prior. This latter assumption will be used in this tutorial and the principle of maximum entropy will lead us to assign a uniform prior probability to this term.

The global likelihood for the data, $P(D|I)$, is a normalization constant. The way to calculate it is to calculate the joint probability for the data and the model, $P(Dk|I)$, and then apply the sum rule to eliminate k from consideration:

$$P(D|I) = \sum_{k=1}^{\ell} P(Dk|I). \tag{52}$$

This can be factored using the product rule, Eq. (1), to obtain:

$$P(D|I) = \sum_{k=1}^{\ell} P(k|I)P(D|kI). \tag{53}$$

Note, that, as asserted earlier, this term is a sum over all values appearing in the numerator, so it is just the constant needed to ensure the total probability is one. The global likelihood may now be substituted back into the posterior probability for the kth hypothesis, Eq. (51), to obtain

$$P(k|DI) = \frac{P(k|I)P(D|kI)}{\sum_{\eta=1}^{\ell} P(\eta|I)P(D|\eta I)} \tag{54}$$

where the summation index was changed to avoid confusion.

To simplify some of the notation in what follows, the normalization constant will be dropped, and the equal sign will be replaced a proportionality sign. At the end of the calculations the normalization constant will be computed. With this change, the posterior probability for the models becomes

$$P(k|DI) \propto P(k|I)P(D|kI). \tag{55}$$

The only remaining term that must be addressed is the marginal likelihood for the data $P(D|kI)$. The model hypothesis explicitly appears in this term. There are three different types of models each having different parameterizations; consequently there are three distinct applications of the rules of probability theory needed to simplify this term. The no-target model is by far the simplest of the three and it will be dealt with first.

Apply Probability Theory Given The No-Target Model

The marginal likelihood is computed from the joint likelihood of the data and the nuisance hypotheses or parameters. The sum rule is then used to remove the dependence on the nuisance parameters. For the no-target hypothesis there is only a single nuisance parameter, B_0, so the marginal likelihood is given by

$$P(D|\ell I) = \int dB_0 P(DB_0|\ell I) \tag{56}$$

where the integral is over all possible values of the constant B_0, and k has been replaced by ℓ to indicate that it is the marginal likelihood of the no-target model that is being computed. It should now be apparent why $P(D|kI)$ is called a marginal likelihood. It is a likelihood because it is the probability for the data given the model. It is a marginal probability because, to compute it, one must marginalize over all nuisance parameters appearing in the model.

To continue with the calculation, the product rule, Eq. (1), is applied to the right-hand side of the marginal likelihood, Eq. (56), to obtain:

$$P(D|\ell I) = \int dB_0 P(B_0|I)P(D|B_0\ell I) \tag{57}$$

where it has been assumed that the constant dc offset (which is a characteristic of the noise) does not depend on which target is present, and $P(D|B_0\ell I)$ is the direct probability for the data given the hypothesis, or the likelihood function. Substituting the marginal likelihood into the posterior probability, Eq. (55), one obtains

$$P(\ell|DI) \propto P(\ell|I) \int dB_0 P(B_0|I)P(D|B_0\ell I). \tag{58}$$

Given the assumptions made, these probabilities may not be further simplified; the only recourse is to assign them numerical values and perform the indicated integral. These probabilities will be assigned in Section 3.4 and the integrals evaluated in 3.5.

Apply Probability Theory Given The Known Target Model

There are three types of models, so three applications of the rules of probability theory are needed to simplify the marginal likelihoods. The previous subsection dealt with the marginal likelihood for the no-target model; here the marginal likelihood for the known target hypothesis will be simplified. As was indicated previously, the marginal likelihood of the data is computed from the joint likelihood of the data and the nuisance parameters. For the known targets these parameters are the amplitudes, B, the true position R_c, and orientation angle of the target ω. The position of the scatterer (S_{kl}, ϕ_{kl}) and the number of scatterers, N_k, are known. The marginal likelihood for the data given the known target hypothesis is given by

$$P(D|kI) = \int dBd\omega dR_c P(DB\omega R_c|kI) \qquad (1 \leq k \leq \ell - 2) \tag{59}$$

where the range on the integrals will be discussed later. Applying the product rule, to the right-hand side of the marginal likelihood one obtains

$$P(D|kI) = \int dBd\omega dR_c P(B\omega R_c|kI)P(D|B\omega R_c kI) \qquad (1 \leq k \leq \ell - 2) \tag{60}$$

where $P(B\omega R_c|kI)$ is the joint prior probability for the nuisance parameters given the known target hypothesis and the prior information I, and $P(D|B\omega R_c kI)$ is the likelihood of the data given the model parameters.

In the previous example there was only a single nuisance hypothesis or parameter, the dc offset, so after factoring the joint-likelihood the calculation was essentially finished. In this example there are many additional hypotheses which requires many additional applications of the product rule. The process is begun by applying the product rule to the joint prior probability for the parameters:

$$P(B\omega R_c|kI) = P(R_c|kI)P(B\omega|R_c kI) \tag{61}$$

where $P(R_c|kI)$ is the prior probability for the range to the target, and $P(B\omega|R_c kI)$ is the joint prior probability for the amplitudes and the orientation angle given the true target k, and the range R_c. In both these probabilities, the identity of the target is given. However, knowing the target identity may or may not help one in assigning either of these terms. When assigning the prior probability for the range to target, $P(R_c|kI)$, knowing the target index, k, would enable one to limit the range of valid values, because the length of the target k would be known. But compared to the six inch range resolution of the radar, knowing the total length of the target is essentially irrelevant. Consequently, it will be assumed that knowing the target identity does not increase our state of knowledge about the range to target and the reference to hypothesis k will be dropped from $P(R_c|kI)$ giving $P(R_c|I)$.

In the case of the joint prior probability for the amplitudes and the orientation angle, $P(B\omega|R_c kI)$, knowing which target is present does not increase our state of knowledge about either the amplitudes or the orientation angle, because the intensity of a scatterer is determined by constructive and destructive interference of the radar waves in the reflected signal. Because the size of the target is large relative to the wavelength of the transmitted signal, large changes in the amplitudes occur for small changes in the orientation angle. But the orientation angles are known only to about one or two degrees. Consequently, knowing the true hypothesis k does not improve our state of knowledge about the amplitudes. And because our state of knowledge about the amplitudes does not improve, there is no additional information about the orientation angle of the target. So whether or not the true target is known does not change our state of knowledge about the amplitudes or the orientation angle. As a result the reference to true hypothesis k may be dropped from the right-hand side of the prior, giving

$$P(B\omega R_c|kI) = P(R_c|I)P(B\omega|R_c I). \tag{62}$$

The previous discussion is one of deciding the logical independence of two or more hypotheses. It occurs in every problem in probability theory. Sometimes probabilities are logically independent and sometimes they are not; each case must be decided based on what one knows. When hypotheses are logically independent, the independent hypotheses may be dropped from the right-hand side of the appropriate probability. However, if the hypotheses are logically dependent, then one must follow the rules of probability theory to obtain valid results.

To illustrate that nonsense may be obtained if logical dependence is ignored, we give one of E. T. Jaynes' favorite examples: suppose someone polled every person in England about the height of the queen-mother. Then the probability for her height, H, given the responses $d_1, \ldots, d_n$ and the prior information I would be written:

$$P(H|d_1 \ldots d_n I) = P(H|I)P(d_1 \ldots d_n|HI). \tag{63}$$

Assuming logical independence, one obtains

$$P(H|d_1 \ldots d_n I) = P(H|I)P(d_1|HI)P(d_2|HI) \ldots P(d_n|HI) \tag{64}$$

If $N \approx 10^6$ then the square root of N effect would imply that her height may be estimated to roughly a part in a thousand, clearly an absurd result. The reason is because the measurements are correlated. From the product rule one obtains

$$P(H|d_1 \ldots d_n I) = P(H|I)P(d_1|HI)P(Hd_2 \ldots d_n|d_1 I). \tag{65}$$

So only the first data item may be assigned an independent probability. All the others must be assigned assuming the first data item known. But each person's opinion is based on news reports, papers, books, and by discussing her height with other people who all have access to basically the same information. All of the opinions are correlated: the data are not independent. In other words, ten million uninformed opinions are not as good as one expert opinion, a fact many politicians and pollsters have forgotten.

To determine whether one hypotheses is logically independent of another the only relevant question is to ask, would knowing the first hypothesis help to determine the other? If the answer to this is yes, the hypotheses are not logically independent and the rules of probability theory must be followed exactly to obtain a valid result. In this tutorial, logical independence will be assumed in many cases. In each case it will be pointed out when and why it is being used. However, in any given problem logical independence may or may not hold. Each case must be determined on its own merits and failure to resolve the issue correctly can lead to nonsense; not because probability theory is wrong, but because from a false hypothesis all conclusions follow, a simple fact of logic.

If logical independence is assumed, Eq. (62) may be factored to obtain

$$P(B\omega R_c|kI) = P(R_c|I)P(\omega|I)P(B_0|I)P(B_1|I) \cdots P(B_{N_k}|I). \tag{66}$$

Logical independence follows here for all the same reasons given earlier: the scatterers change intensity so rapidly, and in so unpredictable a manner, that knowledge of any one amplitude will not aid one in predicting the amplitudes of the others. Substituting the factored prior back into the posterior probability for the known targets, Eq. (55), one obtains

$$\begin{aligned} P(k|DI) &\propto P(k|I)\int dBd\omega dR_c P(R_c|I)P(\omega|I) \\ &\times P(B_0|I)P(B_1|I)\cdots P(B_{N_k}|I)P(D|B\omega R_c kI) \qquad (1 \le k \le \ell - 2) \end{aligned} \tag{67}$$

as the posterior probability for the known targets. None of the indicated probabilities may be further simplified. The next step in the calculation is to assign these probabilities numerical values and then perform the indicated integrals. These last two steps will be delayed until after the marginal likelihood for the unknown target has been simplified.

Apply Probability Theory Given The Unknown Target Model

Simplifying the marginal likelihood for the unknown target hypothesis is similar to what was done previously. The marginal likelihood given the unknown model is computed from the joint probability for the data and the nuisance parameters. For the unknown hypothesis the nuisance parameters are the amplitudes, B, the locations of the scatterers, $S \equiv \{S_1 \cdots S_{N_\nu}\}$, and the number of scatterers, N_ν. Applying the sum rule, the marginal likelihood is given by

$$P(D|[\ell-1]I) = \sum_{N_\nu=0} \int dBdSP(DBSN_\nu|[\ell-1]I) \tag{68}$$

where the target index k was replaced by $[\ell - 1]$ to indicated that this is the posterior probability for the unknown hypothesis. The upper limit on this sum will be discussed when the prior probability for the number of scatterers is discussed. Also note that scatterer N_0 is the dc offset. Applying the product rule one obtains

$$P(D|[\ell-1]I) = \sum_{N_\nu=0} \int dBdSP(BSN_\nu|[\ell-1]I)P(D|BSN_\nu[\ell-1]I) \tag{69}$$

where $P(BSN_\nu|[\ell-1]I)$ is the joint prior probability for the parameters, and $P(D|BSN_\nu[\ell-1]I)$ is the likelihood of the data given those parameters. Using the logical independence assumption and substituting into the posterior probability for the unknown, one obtains

$$\begin{aligned} P(\ell-1|DI) &\propto P(\ell-1|I)\sum_{N_\nu=0} \int dBdSP(B_0|I)P(B_1R_cI)\cdots P(B_{N_\nu}|R_cI) \\ &\times P(N_\nu|I)P(S_1|I)\cdots P(S_{N_\nu}|I)P(D|BSN_\nu[\ell-1]I). \end{aligned} \tag{70}$$

The discussion on logical independence for the amplitudes given earlier applies equally well to the location of the scatterers. Because, the amplitudes cannot be predicted from first principles, knowing the amplitudes does not help in determining the location of the scatterers and conversely. The point has now been reached where these probabilities may not be further simplified. The next step in the calculation is to assign these probabilities numerical values and it is to this problem that we now turn.

3.4 Assign The Probabilities

The posterior probability for the hypothesis of interest has one of three different functional forms depending on the particular hypothesis, Eqs. (58,67,70). These three equations contain seven prior probabilities and three likelihood functions. The prior probabilities specify what was known about the various hypotheses before obtaining the data; while the likelihood functions tell us what was learned about the hypotheses from the data. These probabilities must be assigned to reflect the information actually available. Earlier, the principle of maximum entropy was used to assign three different probabilities: one when only the number of hypotheses, or range of values, was known. This lead to a uniform probability distribution. The other two cases assumed the first two moments of a probability distribution to be known and led to a Gaussian probability distribution. All three of these calculations will now be used to assign the indicated probabilities.

Assigning The Prior Probabilities

Of the seven prior probabilities that must be assigned, three of them have already been touched on. First, the prior probability for the targets, $P(k|I)$, represents what was known about the target before obtaining the data. In the numerical simulations that follow, the enumeration of possible targets is all that is assumed known. Using this information the principle of maximum entropy will assign a uniform prior probability. Because this prior appears in every target's posterior probability exactly one time, the prior range will cancel when the posterior probability is normalized. The other two prior probabilities discussed were those for the location and the orientation angle of the target, Eqs. (44,45). The remaining four priors that must be assigned are: $P(B_0|I)$, $P(B_l|I)$, $P(N_\nu|I)$, and $P(S_j|I)$. The first step in accomplishing this task is to state the information on which a given prior is to be based. In these four cases, the prior information will consist of the valid range of these parameters. This will result in assigning a uniform prior probability. However, care must be taken in assigning these priors because the three types of models have differing numbers and types of parameters and prior ranges are what sets the scale of comparison between the three types of models.

The prior probability for the constant dc offset, $P(B_0|I)$ is the simplest to address and will be taken first. The dc offset, like the prior probability for the target, occurs in every model exactly one time. Any constants that appear in each posterior probability the same number of times will cancel when the posterior probability is normalized. If a uniform prior probability is assigned, the prior range for B_0 will cancel. But note that it is the prior range that cancels, the integral over B_0 must still be over the valid ranges for this parameter. To specify this prior, the range of valid values must be given. In this calculation an approximation will be used that will simplify the results somewhat while introducing only a small error in the calculation. The approximation is that the integration ranges are wide compared to the expected value of the parameter. Consequently, when the integral over the dc offset is evaluated, the limits on the integral may be extended from minus to plus infinity. This amounts to ignoring a term contributed by an error function. But the error function goes to one so rapidly for large arguments that, for all practical purposes, the approximation is exact. Because the prior ranges cancel, it will not be specified other than saying it is uniform with wide bounds.

The next prior probability to be assigned is $P(B_l|I)$, the prior probability for an amplitude. What is known about the amplitudes? The amplitudes are bounded. The bounds are known based on the transmitted signal power, the distance to the target, the reflectivity of the target surface, the surface area of the scatterer, and the efficiency of the receiver. The amplitude must satisfy

$$0 \leq B_l \leq B_{max} \tag{71}$$

where B_{max} is the maximum signal intensity that one could obtain for any target. Using the principle of maximum entropy results in assigning a uniform prior probability given by

$$P(B_l|I) = \begin{cases} \dfrac{1}{B_{max}} & \text{If } 0 \leq B_l \leq B_{max} \\ 0 & \text{otherwise} \end{cases} . \tag{72}$$

To assign the prior probability for the unknown number of scatterers, $P(N_\nu|I)$, one must again state what is known. In this case, the unknown number of scatterers in a particular data set could range from one (this prior only occurs in models that have at least one scatterer) up to a maximum. But what is the maximum value? There are N data values, and if there were N scatterers, the data could be fit exactly by placing a scatterer at each data value and adjusting its amplitude. Because, no additional information is available about the number of scatterers, N may be taken as an upper bound. Using the principle of maximum entropy, one obtains

$$P(N_k|I) = \begin{cases} \dfrac{1}{N} & \text{If } 1 \leq N_k \leq N \\ 0 & \text{otherwise} \end{cases} \tag{73}$$

as the prior probability for the unknown number of scatterers.

Last, to assign the prior probability for the location of the scatterers, $P(S_l|I)$, one must again state what is actually known about their locations. The location of the target is known to within about 6 inches. The range window (the distance represented by the signature data) is centered on the middle of the target. So the scatterers must be somewhere within the data. If only this is known, then the principle of maximum entropy will again assign a uniform prior probability for the location of the scatterers:

$$P(S_l|I) = \begin{cases} \dfrac{1}{N} & \text{If } 1 \leq S_l \leq N \\ 0 & \text{otherwise} \end{cases} \tag{74}$$

where the range dimensions were taken to be unit steps.

All of the prior probabilities have now been assigned. These priors were uniform priors in the cases where only the valid range of values were known, and they were Gaussians when the prior information consisted of a (mean $\pm$ standard deviation) estimate of a parameter value. The only remaining probabilities that must be assigned are the three likelihoods, and, as it will turn out, these probabilities are also prior probabilities.

Assigning The Likelihoods

In the radar target identification problem there are three likelihoods: the likelihood of the data given the no-target hypothesis, $P(D|\ell B_0 I)$; the likelihood of the data given the known target hypothesis, $P(D|B\omega R_c kI)$; and the likelihood of the data given the unknown target hypothesis $P(D|BSN_k[\ell-1]I)$. To assign them, first note that the data D are not a single hypothesis, rather they represent a joint hypothesis: $D \equiv \{d_1, \ldots, d_N\}$. Applying the product rule and representing all of the given quantities as I', the likelihoods may be factored to obtain

$$P(d_1 \ldots d_N|I') = P(d_1|I')P(d_2 \ldots d_N|d_1 I'). \tag{75}$$

Probability theory tells one to assign the probability for the first data item given the parameters, and then assign the probability for the other data items assuming one knows the first data value. Probability theory automatically guards against the example mentioned earlier where assuming logical independence leads to nonsense. However, the designers of the radar take great care to insure that the errors in the data are independent. Given this is the case, the likelihoods may be factored to obtain

$$P(d_1 \ldots d_N|I') = P(d_1|I') \cdots P(d_N|I'). \tag{76}$$

The probability for the data is just the product of the probabilities for obtaining data items separately. Each of our model equations is of the form

$$n_j = d_j - L_k(r_j) \tag{77}$$

where $L_k(r_j)$ is the kth library model evaluated at position r_j. The probability for obtaining the data is just the probability that one should obtain a particular set of errors given that one knows the true signal $L_k(r_j)$.

Earlier it was shown that the results obtained using a Gaussian noise prior probability depend only on the first and second moments of the true noise values in the data. So if a Gaussian distribution is used for the prior probability for the noise, the results obtained will not depend on the underlying sampling distribution of the errors. But note that assigning a Gaussian noise prior probability in no way says the noise is Gaussian; rather, it says only that our estimates and the uncertainty in those estimates should depend only on the first and second moments of the noise. Notice that the Gaussian probability, Eq. (32), assumes the noise standard deviation is known, so σ must be added to the likelihoods in such a way as to indicate that it is known; this gives

$$P(D|\sigma I') = (2\pi\sigma^2)^{-\frac{N}{2}} \exp\left\{-\sum_{j=1}^{N} \frac{[d_j - L_k(r_j)]^2}{2\sigma^2}\right\}. \tag{78}$$

as the likelihood function. Using this equation as a prototype, the likelihood for the data given the known target hypothesis is given by

$$P(D|B\omega R_c \sigma kI) = (2\pi\sigma^2)^{-\frac{N}{2}}$$
$$\times \exp\left\{-\frac{1}{2\sigma^2}\sum_{j=1}^{N}[d_j - \sum_{l=0}^{N_k} B_l G(S_{kl}\cos(\phi_{kl} - \omega) - r_j + R_c)]^2\right\} \tag{79}$$

where $(1 \leq k \leq \ell - 2)$ and. I' as been replaced by all of the given parameters. Similarly, the likelihood for the data given the unknown target hypothesis is given by

$$P(D|BSN_\nu\sigma[\ell-1]I) = (2\pi\sigma^2)^{-\frac{N}{2}} \exp\left\{-\frac{1}{2\sigma^2}\sum_{j=1}^{N}[d_j - \sum_{l=0}^{N_\nu} B_l G(S_l - r_j)]^2\right\}. \qquad (80)$$

Last, the likelihood for the data given the no-target hypothesis is given by

$$P(D|\ell B_0\sigma I) = (2\pi\sigma^2)^{-\frac{N}{2}} \exp\left\{-\frac{1}{2\sigma^2}\sum_{j=1}^{N}[d_j - B_0]^2\right\}. \qquad (81)$$

With the assignment of these likelihoods, all of the probabilities have now been assigned. The next task is to perform the indicated integrals and sums.

3.5 Evaluate The Integrals And Sums

All that remains to formally complete the problem is to apply the sum rule by evaluating the indicated integrals and sums. There are three types of hypotheses, so evaluating these integrals and sums must proceed in three steps. In these calculations only the multivariate Gaussian integrals may be evaluated in closed form. The remaining integrals must be evaluated numerically. Evaluating the multivariate Gaussian integrals for each of the three types of hypotheses is essentially identical. Consequently, the procedures needed will be demonstrated for the known targets and then the results will simply be given for the unknown and no-target hypotheses.

Evaluating The Integrals For The Known Targets

The posterior probability for the known target hypothesis is given by Eq. (67). The prior probability for the target hypothesis, $P(k|I)$, was assigned a uniform prior and because this term appears in all of the posterior probabilities its prior range cancels when these distributions are normalized. The prior probabilities for the range to the target, $P(R_c|I)$, and the orientation angle of the target, $P(\omega|I)$ are given by Eqs. (44,45). The prior probability for the dc offset was assigned a wide uniform prior and because this term also appears in every posterior probability exactly one time its prior range also cancels. The prior probabilities for the amplitudes, $P(B_l|I)$. are all given by Eq. (72). Last the likelihood function is given by Eq. (79). Gathering up these terms, the posterior probability for the known targets hypothesis is given by

$$\begin{aligned} P(k|DI) \propto & \int dBd\omega dR_c \\ & \times (2\pi\sigma_R^2)^{-\frac{1}{2}} \exp\left\{-\frac{[R_0 - R_c]^2}{2\sigma_R^2}\right\} \\ & \times (2\pi\sigma_\Omega^2)^{-\frac{1}{2}} \exp\left\{-\frac{[\Omega - \omega]^2}{2\sigma_\Omega^2}\right\} \qquad (1 \leq k \leq \ell - 2) \qquad (82) \\ & \times \left(\frac{1}{B_{max}}\right)^{N_k} \\ & \times (2\pi\sigma^2)^{-\frac{N}{2}} \exp\left\{-\frac{1}{2\sigma^2}\sum_{j=1}^{N}[d_j - \sum_{l=0}^{N_k} B_l G(S_{kl}\cos(\phi_{kl} - \omega) - r_j + R_c)]^2\right\} \end{aligned}$$

where each of the terms has been intentionally separated so they may be more readily identified. Additionally, the notation should have been modified to indicate that σ_R^2, σ_Ω^2, σ, R_0, ω and B_{max} are known quantities. However, to compress the notation, these quantities have been incorporated into the general background information I.

There are $N_k + 3$ integrals that must be evaluated. Of these only the $N_k + 1$ amplitude integrals may be evaluated in closed form. These integrals are multivariate Gaussian integrals and any integral of this form may be evaluated in closed form. Designating the amplitude integrals as $p_B(R_c, \omega)$, the integrals that must be evaluated are given by

$$p_B(R_c,\omega) = \int dB \exp\left\{-\frac{1}{2\sigma^2}\left[d\cdot d - 2\sum_{l=0}^{N_k} B_l T_l + \sum_{l=0}^{N_k}\sum_{\eta=0}^{N_k} B_l B_\eta g_{l\eta}\right]\right\} \tag{83}$$

where

$$d\cdot d \equiv \sum_{j=1}^{N} d_j^2, \tag{84}$$

$$T_l \equiv \sum_{j=1}^{N} d_j G(S_{kl}\cos(\phi_{kl} - \omega) - r_j + R_c), \tag{85}$$

and

$$g_{l\eta} = \sum_{j=1}^{N} G(S_{kl}\cos(\phi_{kl} - \omega) - r_j + R_c) G(S_{k\eta}\cos(\phi_{k\eta} - \omega) - r_j + R_c). \tag{86}$$

There are a number of different ways to evaluate these integrals; one of the easiest to understand is to introduce a change of variables that makes the $g_{l\eta}$ matrix diagonal, then all integrals uncouple and each may be done separately. The new variables, $\{A_0 \cdots A_{N_k}\}$, are defined as

$$A_l = \sqrt{\lambda_l}\sum_{\eta=0}^{N_k} B_\eta e_{l\eta} \quad \text{and} \quad B_l = \sum_{\eta=0}^{N_k} \frac{A_\eta e_{\eta l}}{\sqrt{\lambda_\eta}} \tag{87}$$

where λ_η is the ηth eigenvalue of the $g_{l\eta}$ matrix and $e_{\eta l}$ is the lth component of its ηth eigenvector. The eigenvalues and eigenvectors have the property that

$$\sum_{\eta=0}^{N_k} g_{l\eta} e_{\ell\eta} = \lambda_\ell e_{\ell l} \tag{88}$$

from which the $p_B(R_c, \omega)$ integral may be rewritten as

$$p_B(R_c,\omega) = \int dA \lambda_0^{-\frac{1}{2}} \cdots \lambda_{N_k}^{-\frac{1}{2}} \exp\left\{-\frac{1}{2\sigma^2}\left[d\cdot d - 2\sum_{l=0}^{N_k} A_l h_l + \sum_{l=0}^{N_k} A_l^2\right]\right\}, \tag{89}$$

where

$$h_l = \sum_{j=1}^{N} d_j H_l(r_j) \tag{90}$$

and

$$H_l(r_j) = \frac{1}{\sqrt{\lambda_l}} \sum_{\eta=0}^{N_k} e_{l\eta} G(S_{k\eta}\cos(\phi_{kl} - \omega) - r_j + R_c). \tag{91}$$

The model functions $H_l(r_j)$ are called orthonormal because they have the property

$$\sum_{j=1}^{N} H_l(r_j) H_\eta(r_j) = \delta_{l\eta} \tag{92}$$

where $\delta_{l\eta}$ is the Kronecker delta function.

This change of variables reduces the $p_B(R_c, \omega)$ integral to a series of independent Gaussian integrals. The integration limits are from zero to an upper bound. These limits are assumed so wide that the amount of probability contributed to the integral near the upper and lower bound is so small that the limits may be extended to plus and minus infinity and this extension will make only a negligible change in the results of the integral. Using this approximation one obtains

$$p_B(R_c, \omega) = (2\pi\sigma^2)^{\frac{N_k+1}{2}} \lambda_0^{-\frac{1}{2}} \cdots \lambda_{N_k}^{-\frac{1}{2}} \exp\left\{-\frac{d\cdot d - h\cdot h}{2\sigma^2}\right\}, \qquad (1 \le k \le \ell - 2) \tag{93}$$

where $h \cdot h$ is given by

$$h \cdot h = \sum_{l=0}^{N_k} h_l^2. \tag{94}$$

The quantity $h \cdot h$ plays the role of a sufficient statistic and summarizes all of the information in the data relevant to estimating the position and orientation angle of the target. Note that the sufficient statistic is a function of both R_c and ω even though this dependency has not been explicitly shown. Substituting $p_B(R_c, \omega)$ into the posterior probability for the known targets one obtains

$$\begin{aligned} P(k|DI) \;\propto\; & \frac{(2\pi\sigma^2)^{-\frac{N-N_k-1}{2}}}{2\pi\sigma_R\sigma_\Omega} \left(\frac{1}{B_{max}}\right)^{N_k} \\ \times\; & \int d\omega dR_c \exp\left\{-\frac{[R_0 - R_c]^2}{2\sigma_R^2} - \frac{[\Omega - \omega]^2}{2\sigma_\Omega^2}\right\} \qquad (1 \le k \le \ell - 2) \\ \times\; & \lambda_0^{-\frac{1}{2}} \cdots \lambda_{N_k}^{-\frac{1}{2}} \exp\left\{-\frac{d\cdot d - h\cdot h}{2\sigma^2}\right\}. \end{aligned} \tag{95}$$

The remaining two integrals must be evaluated numerically. In the numerical simulations, these integrals are approximated in a particularly simple way. Each integral is taken to be approximately the width of the integrand times its height. In this particular case this approximation is good enough because the data are extremely spiky. This results in an extraordinarily sharply peaked probability distribution. The widths are analogous to a prior penalty, and almost any values used for them will work (provided they are reasonable). Here reasonable means the widths must be within one or two orders of magnitude of the true values. Parameter estimates using probability theory as logic typically scale like one over root N, so the widths are easily set to the right order of magnitude.

Evaluating The Integrals For The Unknown Target

The process of evaluating the integrals for the unknown target hypothesis is essentially identical to what was done for the known target hypotheses. Consequently, only the results of these integrals are given. The posterior probability for the unknown target is given by

$$\begin{aligned} P(\ell-1|DI) &\propto \sum_{N_\nu=1}^{N} \int dBdS(N)^{-1}(N)^{-N_\nu}\left(\frac{1}{B_{max}}\right)^{N_\nu} \\ &\times (2\pi\sigma^2)^{-\frac{N-N_\nu-1}{2}}\lambda_0^{-\frac{1}{2}}\cdots\lambda_{N_\nu}^{-\frac{1}{2}}\exp\left\{-\frac{d\cdot d-h\cdot h}{2\sigma^2}\right\}, \end{aligned} \tag{96}$$

where the definitions of these quantities are analogous to those given in the preceding calculation. For example the sufficient statistic $h \cdot h$ is defined

$$h\cdot h=\sum_{l=0}^{N_\nu} h_l^2 \tag{97}$$

with

$$H_l(r_j)=\frac{1}{\sqrt{\lambda_l}}\sum_{\eta=0}^{N_\nu} e_{l\eta}G(S_{k\eta}-r_j) \tag{98}$$

and the eigenvalues and eigenvectors that appear in this calculation are formed from the interaction matrix associated with the unknown model function:

$$g_{l\eta}=\sum_{j=1}^{N} G(S_{kl}-r_j)G(S_{k\eta}-r_j). \tag{99}$$

For the known targets there was one sufficient statistic for each model, while here there is one for each value of the summation index N_ν. In principle, this is a long and tedious calculation. However, because of the spiked nature of the model function it is possible to implement this calculation using relatively simple approximations. In the numerical example two approximations were used: the sum was approximated by it largest term; while the integral was approximated by its height times its width. How these approximations worked in practice is the subject of the next Section.

Evaluating The Integrals For The No-target Model

The no-target model is particularly simple because the model contains only a single nuisance parameter. The posterior probability for the no-target model is given by

$$P(\ell|DI)\propto (N)^{-\frac{1}{2}}\left(\frac{1}{B_{max}}\right)(2\pi\sigma^2)^{-\frac{N-1}{2}}\exp\left\{-\frac{d\cdot d-N(\overline{d})^2}{2\sigma^2}\right\}, \tag{100}$$

where $\overline{d}$ is given by

$$\overline{d}=\frac{1}{N}\sum_{j=1}^{N} d_j. \tag{101}$$

With the completion of this integral the problem is formally completed. There are a number of integrals that must be evaluated numerically. In the case of the unknown, the entire calculation is so complicated that there is virtually no hope of implementing the exact calculation, and approximations must be used. However, unless one knows what one should aim for, it is hard to know how to make approximations, and this is one of the places where probability theory helps most. By telling one what to aim for, the problem is reduced to approximating the posterior probability to the best of one's ability. While making numerical approximations is difficult, it is less difficult than trying to guess the answer by intuition.

4 Numerical Methods

To demonstrate model selection, how to handle incomplete sets of hypotheses (the unknown hypothesis), and the feasibility of radar target identification, the identification calculations presented in this tutorial have been implemented in a numerical simulation. In this simulation there are three major routines: a data generation routine, an identification routine, and an output routine. In general terms the simulation is a loop. Each time through the loop a data set is generated, passed to the identification routines and the results of the simulation are written to an output file.

In this simulation there were 20 different hypotheses or targets; 18 known targets, one unknown, and one no-target hypothesis. The known target models were generated by a separate program and then used throughout the simulation. To do this the program used a random numbers generator to determine the angle and position of each scatter. The angular location of a scatter, ϕ_{kl}, was chosen to be between 0 and 2π, while radial location of a scatter, S_{kl}, was chosen to be between $-400 \leq R_c \leq 400$. There are 1024 data values, so the scatterers were chosen so that they always fit within the range window of the radar.

The data generation routine chooses one of the 20 targets at random. When it chooses the unknown target, the unknown is generated in a manner analogous to the known hypotheses. A uniform random number generator was used to randomly position between 3 and 10 scatterers within a range window of $-400 \leq R_c \leq 400$. Similarly, when the no-target model was chosen no scatterers were generated – only noise was placed in the simulated data. The amplitudes of the scatterers were set randomly. However, their amplitudes were scaled so that the *mean amplitude* to root-mean-square noise standard deviation was 20. The twenty target library is shown in Fig. 4 for one setting of the amplitudes, orientation angles, and positions of the targets.

In this simulation the real target identification problem was mimicked as closely as possible. To do this the data were generated and processed in a way that mimicked the effects encountered on a real radar. On a real radar, the radar will establish a track on a target, and only after the track has been established will the identification be attempted. In the process of tracking the target the radar will infer the vector position and orientation angle of the target. This information is available to the identification routines in the form of prior probabilities. Additionally, the amplitudes of the scatterers are extremely sensitive functions of the orientation angle of the target, changing by more than an order of magnitude for a change of only 0.1 degrees. For all practical purposes, this means the amplitude of a scatterer is completely unpredictable from one look to the next. To illustrate these effects, the same library targets (with a new unknown, and no-target model) were generated a second time. These targets are displayed in Fig. 5. While the positions of the

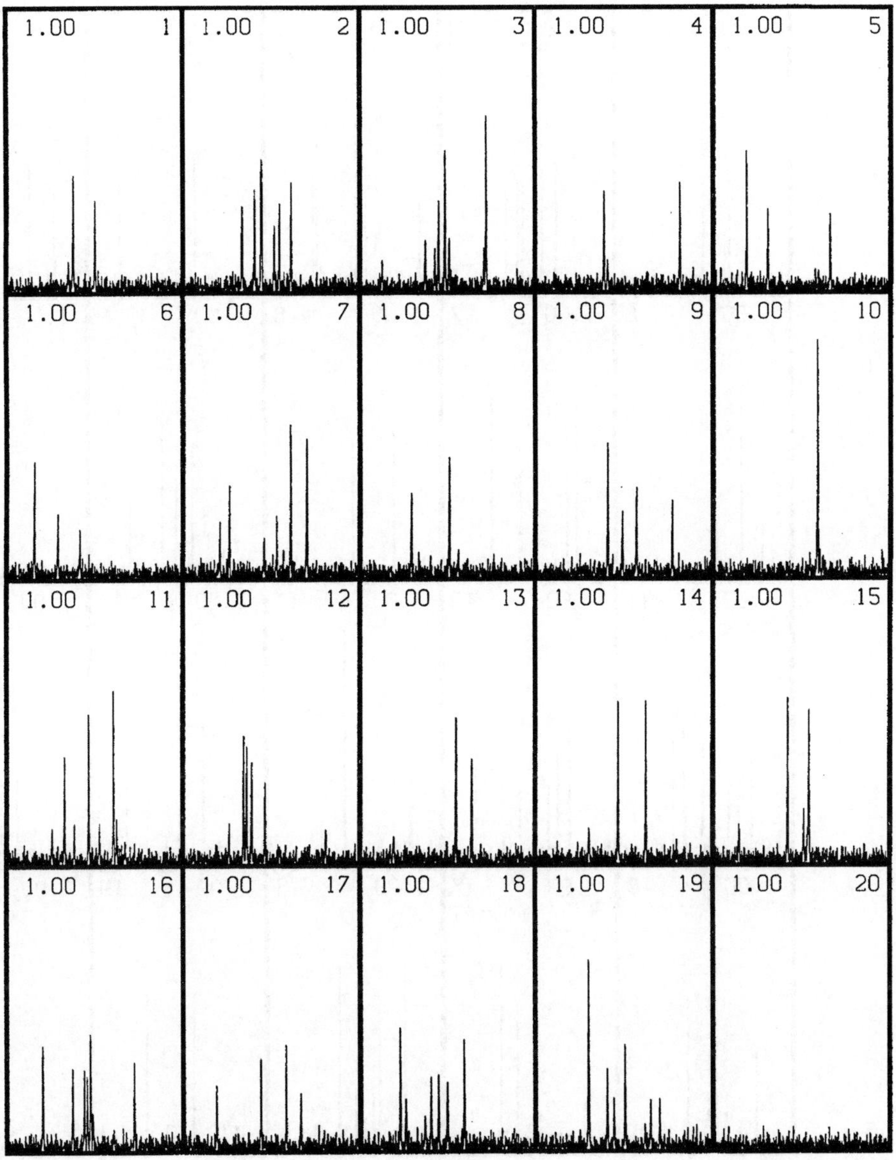

Fig. 4. The library used in the simulation consisted of 20 different targets. The first 18 of these are the known targets, corresponding to various types of aircraft. The amplitudes, location, and orientation angle of each known target is chosen randomly. Target 19 is the unknown. The locations, amplitudes, and number of scatterers for the this target are chosen randomly. Target 20 is the no-target model and contains no scatterers.

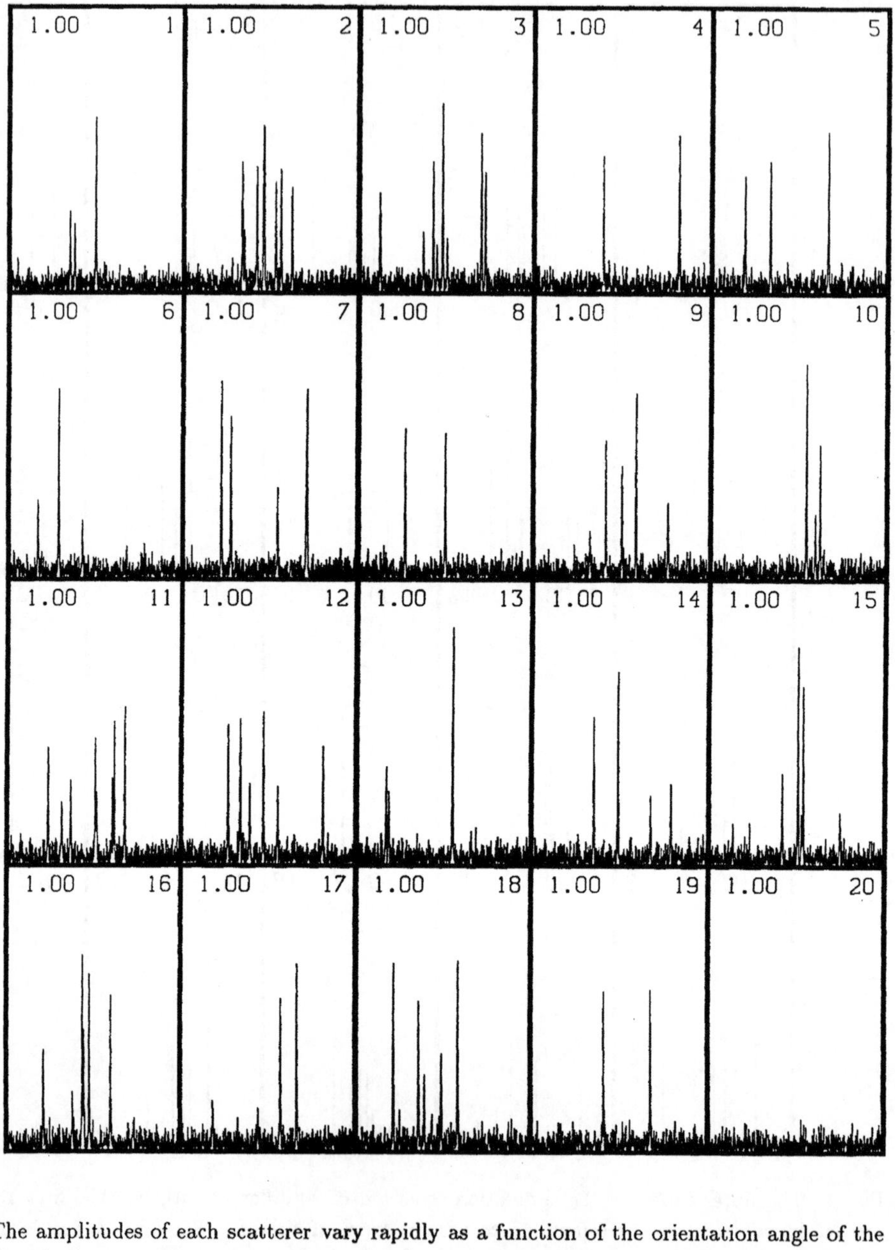

Fig. 5. The amplitudes of each scatterer vary rapidly as a function of the orientation angle of the target. Here is the same 20 targets as they might look at slightly different orientations angle. Try comparing these targets to those shown in Fig. 4 to see if you can tell that they are the same.

scatterers in Fig. 4 and Fig. 5 nearly overlap, the amplitudes are completely different. This effect is so striking that the author has difficulty telling that these are the same targets. The probability assigned to the *true* target is shown to three decimal places in the upper left hand corner of these figures. Note that in both Fig. 4 and Fig. 5 the correct target was identified to a probability of one to three decimal places in every case.

The simulation was run on 1000 simulated data sets, taking about 3 seconds for each simulation on an SGI Indigo. The first 20 simulated data sets are shown in Fig. 6. The full output from the simulation is shown in Table 1. This output consists of both a summary and detailed outputs. The summary output tells one the simulation number, i.e., 1, 2, 3 etc., the true target number, its probability, and the signal-to-noise ratio. The detailed output contains the unnormalized base 10 logarithm of the probability for each target. In the 1000 simulations the correct target was identified 999 times; there was only a single mis-identification. When the mis-identification was investigated, it was found that the generated target had most of its scatterers buried in the noise while two of them had exceptionally high signal-to-noise ratio. Under these conditions the unknown target *is* a better representation of the data than the true model. Thus the unknown target was understandably identified.

Table 1 illustrates very strongly why the unknown target hypothesis works. To understand it, look at the first simulation. The true target is number 5. The base 10 logarithm of its probability is 1991.3. Now look at the log probabilities for the other targets for this first simulation. The target with the second highest probability was the unknown, having a log probability of 1983.7, roughly seven orders of magnitude down from the true target. The target with the third highest probability is target 17, it has a log probability of 1456.0, more than 400 orders of magnitude down. Next, look at a second simulation, say simulation number 8. The true target is number 20, the no-target model, it's log probability is 145.49. The second highest log probability is again the unknown coming in at 139.83. Now examine all of the simulations in the table except simulation number 3. The unknown hypothesis is the second highest probability in every case! To understand this, note that the unknown target essentially fits all of the systematic detail in the data; its likelihood function is essentially identical to the likelihood of the true target (assuming the true target hypotheses is in the library) . But the unknown has many more parameters. In probability theory as logic these extra parameters carry a penalty in the form of the prior probabilities. The priors range for both the location and number of scatterers was $1/N$. If there were 3 scatterers on the target the unknown would have a prior penalty of $1/N^4$. The number of data values, N, was 1024 so the prior penalizes the unknown by a factor of approximately 10^{12}. This penalty is so large, that unless the true target is *not* present, the prior eliminates the unknown target from consideration. Now examine simulation number 3. The true target is the unknown. There is no known target present to prevent the unknown target from being identified.

5 Summary And Conclusions

To use probability theory as logic, one must relate the hypothesis of interest to the available evidence. This process is one of model building. While building the model one must state exactly what the hypotheses are and how they are related to the available evidence. In the case of radar target identification, this process has forced the radar target identification community to state exactly what is meant by the known, unknown and no-target

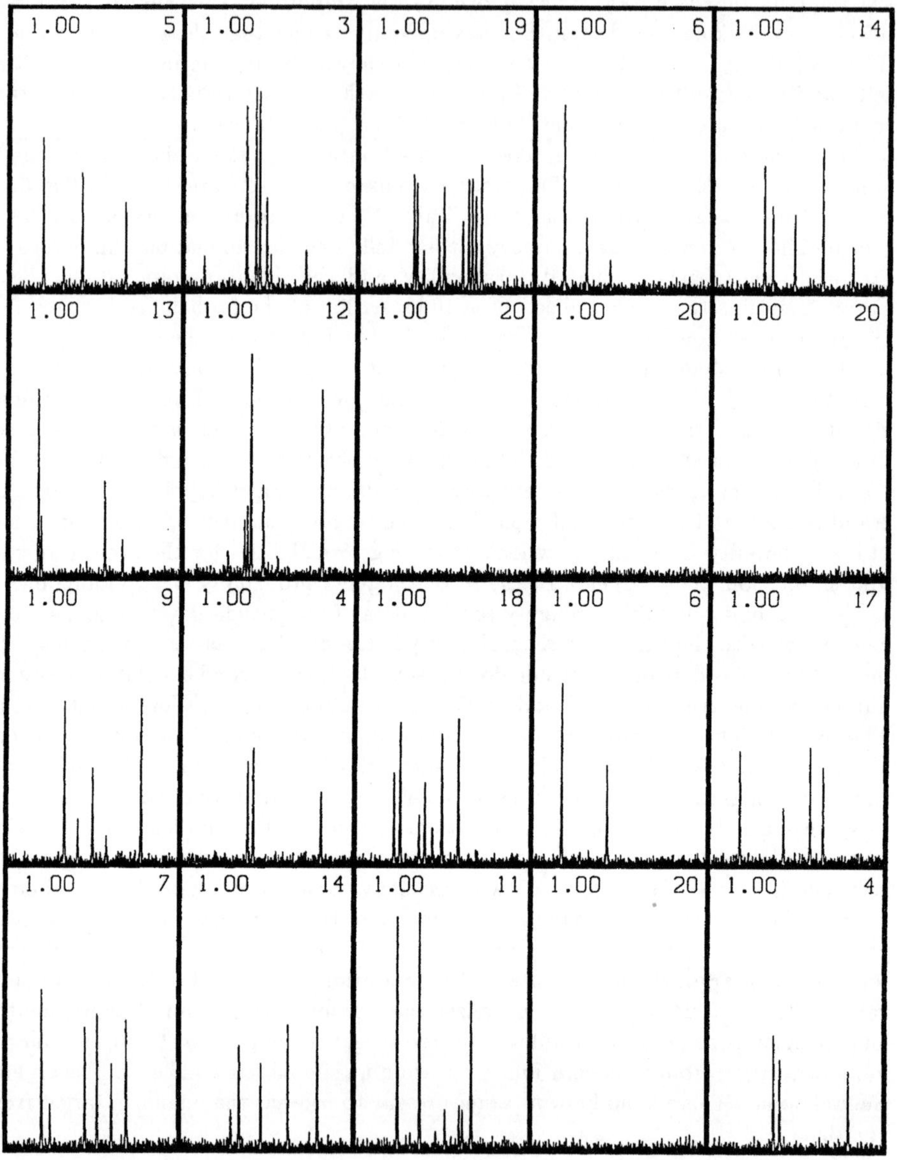

Fig. 6. This is the data generated in the first 20 simulations. The number in the upper right hand corner is the number of the true target. The number in the left hand corner is the probability assigned to this target. Note that the identification was perfect on these 20 targets.

Table 1: When Is The Unknown Target Identified?

1 True Target: 5, Its probability is:1.00; S/N=20.0}

1	342.21	2	823.30	3	819.98	4	828.76	5	1991.3
6	820.01	7	338.35	8	341.15	9	824.52	10	340.43
11	337.96	12	338.88	13	1194.3	14	353.80	15	339.31
16	336.93	17	1456.0	18	338.85	19	1983.7	20	346.57

2 True Target: 3, Its probability is:1.00; S/N=20.0}

1	1758.0	2	1981.4	3	5028.1	4	1982.5	5	2026.9
6	1983.0	7	655.49	8	627.86	9	2000.7	10	615.61
11	889.01	12	1753.9	13	615.23	14	1754.5	15	1990.0
16	2308.5	17	2031.3	18	3189.9	19	5015.5	20	615.86

}

3 True Target: Unknown(19), Its probability is:1.00; S/N=20.0}

1	1475.6	2	1631.9	3	1780.4	4	752.65	5	1475.5
6	751.85	7	1443.1	8	749.91	9	1331.7	10	1340.6
11	1605.3	12	1298.8	13	1197.2	14	1881.0	15	752.00
16	1244.8	17	1039.6	18	900.19	19	4168.9	20	750.25

4 True Target: 6, its probability is:1.00; S/N=20.0

1	289.23	2	284.32	3	310.25	4	315.44	5	314.46
6	1805.6	7	285.79	8	463.71	9	312.38	10	287.84
11	285.03	12	286.36	13	1584.2	14	465.75	15	312.17
16	285.62	17	309.19	18	286.62	19	1799.6	20	295.45

5 True Target: 14, its probability is:1.00; S/N=20.0

1	615.56	2	596.31	3	599.18	4	391.18	5	967.78
6	390.95	7	387.89	8	390.63	9	965.99	10	1158.6
11	1141.0	12	822.83	13	1155.8	14	2287.1	15	389.47
16	600.15	17	391.23	18	389.08	19	2277.0	20	395.64

6 True Target: 13, its probability is:1.00; S/N=20.0

1	331.36	2	662.09	3	327.42	4	331.51	5	436.72
6	1627.0	7	329.55	8	331.98	9	329.65	10	667.83
11	327.76	12	663.45	13	2141.4	14	372.26	15	665.66
16	327.18	17	2010.7	18	328.61	19	2133.4	20	337.16

7 True Target: 12, its probability is:1.00; S/N=20.0

1	633.44	2	630.05	3	633.77	4	2359.5	5	539.65
6	538.65	7	807.69	8	539.81	9	628.92	10	539.12
11	2632.3	12	4219.2	13	540.15	14	644.69	15	539.97
16	902.78	17	540.12	18	2448.6	19	4203.8	20	542.83

8 True Target: No Target(20), its probability is:1.00; S/N=20.0

1	138.97	2	132.39	3	133.21	4	139.01	5	137.88
6	139.23	7	134.28	8	138.86	9	136.85	10	138.45
11	133.77	12	134.32	13	137.98	14	136.93	15	135.92
16	133.91	17	138.39	18	134.35	19	139.83	20	145.49

9 True Target: No Target(20), its probability is:0.99999; S/N=20.0

1	138.26	2	131.78	3	132.10	4	137.90	5	137.16
6	137.82	7	133.13	8	137.71	9	136.05	10	136.61
11	132.07	12	133.46	13	136.55	14	135.67	15	134.67
16	131.81	17	137.37	18	132.92	19	138.88	20	144.26

10 True Target: No Target(20), its probability is:0.99999; S/N=20.0

1	126.72	2	120.33	3	122.27	4	126.99	5	126.08
6	127.59	7	122.20	8	126.86	9	124.99	10	126.00
11	121.17	12	122.16	13	126.27	14	124.92	15	123.54
16	122.42	17	125.93	18	122.54	19	127.77	20	133.32

11 True Target: 9, its probability is:1.00; S/N=20.0

1	1358.1	2	1409.0	3	734.61	4	436.37	5	1355.3
6	436.20	7	383.71	8	688.60	9	2703.1	10	394.53
11	1349.7	12	1348.7	13	385.26	14	677.26	15	383.00
16	677.75	17	397.55	18	682.65	19	2693.2	20	389.72

12 True Target: 4, its probability is:1.00; S/N=20.0

1	272.47	2	756.10	3	761.39	4	1321.6	5	765.64
6	767.68	7	267.81	8	272.65	9	764.52	10	270.85
11	626.02	12	782.11	13	271.55	14	270.32	15	762.95
16	625.45	17	757.66	18	627.50	19	1315.6	20	277.74

13 True Target: 18, its probability is:1.00; S/N=20.0

1	1153.0	2	1156.6	3	613.99	4	769.26	5	561.42
6	530.28	7	801.18	8	1766.5	9	1103.9	10	1261.6
11	1240.7	12	1462.8	13	530.03	14	1828.7	15	528.62
16	1188.0	17	1257.4	18	3248.9	19	3235.4	20	531.07

14 True Target: 6, its probability is:1.00; S/N=20.0

1	267.83	2	261.75	3	592.98	4	589.68	5	593.95
6	1789.7	7	265.15	8	268.20	9	595.83	10	268.10
11	262.97	12	263.90	13	1441.5	14	266.55	15	594.71
16	262.69	17	1774.2	18	265.63	19	1783.7	20	273.84

15 True Target: 17, its probability is:1.00; S/N=20.0

1	331.89	2	327.58	3	427.78	4	332.51	5	1118.6
6	430.92	7	329.53	8	799.51	9	801.72	10	799.39
11	328.05	12	328.45	13	1130.8	14	674.66	15	430.44
16	429.09	17	1730.6	18	853.47	19	1722.6	20	337.69

16 True Target: 7, its probability is:1.00; S/N=20.0

1	557.89	2	1289.4	3	1138.5	4	560.31	5	559.63
6	558.83	7	3525.4	8	598.40	9	560.97	10	758.86
11	1816.4	12	1134.8	13	557.64	14	557.65	15	1191.6
16	1952.9	17	557.83	18	1466.7	19	3511.5	20	561.05

17 True Target: 14, its probability is:1.00; S/N=20.0

1	813.47	2	657.30	3	656.92	4	420.90	5	463.77
6	420.31	7	1186.4	8	420.59	9	420.18	10	967.40
11	1187.8	12	1350.8	13	962.76	14	2240.8	15	419.36
16	658.86	17	420.70	18	418.35	19	2231.2	20	425.04

18 True Target: 11, its probability is:1.00; S/N=20.0

1	639.06	2	675.18	3	641.23	4	2718.8	5	1425.5
6	641.82	7	666.94	8	641.69	9	642.31	10	671.72
11	5732.9	12	4731.0	13	1432.4	14	1430.0	15	663.53
16	2759.3	17	669.44	18	2620.3	19	5704.7	20	642.51

19 True Target: No Target(20), its probability is:1.00; S/N=20.0

1	130.83	2	123.28	3	124.80	4	130.47	5	129.77
6	130.43	7	125.82	8	130.41	9	127.92	10	129.65
11	125.02	12	125.97	13	129.78	14	128.61	15	127.31
16	125.08	17	129.73	18	125.74	19	131.19	20	136.93

20 True Target: 4, its probability is:0.99997; S/N=20.0

1	272.19	2	585.49	3	588.66	4	1329.5	5	588.29
6	594.21	7	267.63	8	272.56	9	590.34	10	271.73
11	775.71	12	993.07	13	270.58	14	270.30	15	589.80
16	267.73	17	587.64	18	268.50	19	1325.0	20	277.63

Table 1 illustrates how the unknown target is identified. This is the detailed output from the first 20 simulations. The first line of each entry identifies the true target and its probability. Lines 2 thru 5 for each entry are the base 10 logarithm of the probability for each target. Target 19 is the unknown target. To see how, when and why the unknown is correctly identified, browse through this table and compare the log probability for the unknown to that of the true target. The log probability for the unknown is always less than the true target. However, it is always greater than any of the other targets. So the unknown is identified whenever the true target is not in the list of library targets.

hypotheses. In probability theory as logic there is *no such thing* as nonparametric statistics. Typically when this term is used, it is used to mean that the number of hypotheses grows very large. That is to say, the models are so general they can fit virtually any data. But this is not nonparametric statistics, indeed it is exactly the opposite: there are many more parameters than data. However, there are a few people who use the term to mean literally there are no models. Typically, the statistics advocated by these people cannot be derived from a correct application of the rules of probability theory and, at best, their results are intuitive and ad hoc.

Probability theory computes the probabilities for hypotheses. It computes the probability for parameters only in the sense that the parameter indexes a well defined hypothesis. Similarly, it test models only in the sense that models are statements of hypotheses. Thus there is no essential difference between model selection and parameter estimation. The differences are conceptual, not theoretical. These conceptual differences manifest themselves primarily in the prior probabilities. In parameter estimation it is often convenient and harmless to use improper priors (an improper prior is a function that is used as a prior probability that is not normalizable). It is convenient because improper priors often simplify the mathematics considerably, and harmless because the infinities so introduced cancel when the probabilities are normalized. Strictly speaking improper priors are not probabilities at all; rather they are the limit of a sequence of proper priors in the limit of infinite uncertainty in a hypothesis. As a limit, it must always be approached from well-defined finite mathematics to ensure one obtains a well behaved result. Use of an improper prior directly can and will result in disaster in model selection problems because the infinities don't generally cancel. For more on this point see Jaynes [11]. In parameter estimation, when using a uniform prior, the prior ranges cancel when the distribution is normalized. However, in model selection these prior ranges may or may not cancel. In the numerical simulation described in this tutorial, the prior range for the constant dc offset and which target was present canceled. The remaining prior ranges did not cancel, and so affect the results. These prior ranges essentially set the scale against which different models with differing parameterizations are compared. So it is vitally important that one think carefully about these quantities and set them based on the information one actually has.

The probability for a hypothesis C is computed conditional on the evidence $E_1 \cdots E_n$. This probability is given by $P(C|E_1 \cdots E_n)$. Every person who consistently follows the rules of probability theory will be lead to assign exactly the same probabilities conditional on that evidence. These probabilities are all of the form of prior probabilities. The distinction between data, strong prior information, weak prior information, and no prior information (which strictly speaking cannot exist in real problems) is purely artificial. Evidence is evidence and it is all used to assign prior probabilities! The principle of maximum entropy was used here to assign these priors because it assigns priors that are consistent with that evidence while remaining maximally uninformative. That is to say, the probabilities do not depend on things one does not know. This is particularly important when assigning the prior probability for the noise because it allows one to assign probabilities that depend only on what one actually knows about the true errors in the data and it renders the underlying sampling distribution of the noise completely irrelevant.

The calculations indicated by probability theory are often much too complicated to implement exactly. However, knowing what should be done enables one to reduce the problem from one of guessing the answer to one of numerical approximation. This is a tremendous simplification that often leads to simple numerical algorithms which, although not exact, capture the essence of the probability theory calculation and enable one to solve problems that would otherwise prove impossible.

ACKNOWLEDGMENTS. The author would like to thank Dr. C. R. Smith, and Dr. Jeffrey J. Neil for their valuable comments on preliminary versions of this paper. In particular I would like to extend my deepest thanks to Dr. Tom Loredo for his comments on Section 2. Without his assistance Section 2 would have been a mere shadow of the final version. The encouragement of Professor J. J. H. Ackerman is greatly appreciated. Additionally, the author would like to acknowledge a vary large debt to Professor E. T. Jaynes for his help and guidance over the years. Last, this work was supported by the U. S. Army through the Scientific Services Program.

References

[1] William of Ockham, *ca* 1340.

[2] Jeffreys, H., *Theory of Probability,* Oxford University Press, London, 1939; Later editions, 1948, 1961.

[3] Jaynes, E. T., *JASA*, Sept. 1979, p. 740, review of "Inference, Methods, and Decision: Towards a Bayesian Philosophy of Science." by R. D. Rosenkrantz, D. Reidel Publishing Co., Boston.

[4] Gull, S. F., "Bayesian Inductive Inference and Maximum Entropy," in *Maximum Entropy and Bayesian Methods in Science and Engineering"* **1**, pp. 53-75, G. J. Erickson and C. R. Smith *Eds.*, Kluwer Academic Publishers, Dordrecht the Netherlands, 1988.

[5] Bretthorst, G. Larry, "Bayesian Spectrum Analysis and Parameter Estimation," in *Lecture Notes in Statistics* **48**, Springer-Verlag, New York, New York, 1988.

[6] Bretthorst, G. Larry, "Bayesian Analysis I: Parameter Estimation Using Quadrature NMR Models," *J. Magn. Reson.,* **88**, pp. 533-551 (1990).

[7] Bretthorst, G. Larry, "Bayesian Analysis II: Model Selection," *J. Magn. Reson.*, **88**, pp. 552-570 (1990).

[8] Bretthorst, G. Larry, "Bayesian Analysis III: Spectral Analysis," *J. Magn. Reson.,* **88**, pp. 571-595 (1990).

[9] Tribus, M., *Rational Descriptions, Decisions and Designs,* Pergamon Press, Oxford, 1969.

[10] Zellner, A., *An Introduction to Bayesian Inference in Econometrics,* John Wiley and Sons, New York, 1971. Second edition (1987); R. E. Krieger Pub. Co., Malabar, Florida.

[11] Jaynes, E. T., "Probability Theory – The Logic of Science," in preparation. Copies of this TeX manuscript are available by anonymous FTP from "bayes.wustl.edu"

[12] Jaynes, E. T., "How Does the Brain do Plausible Reasoning?" unpublished Stanford University Microwave Laboratory Report No. 421 (1957); reprinted in *Maximum-Entropy and Bayesian Methods in Science and Engineering* **1**, pp. 1-24, G. J. Erickson and C. R. Smith *Eds.,* 1988.

[13] Bretthorst, G. Larry, "An Introduction to Parameter Estimation Using Bayesian Probability Theory," in *Maximum Entropy and Bayesian Methods,* Dartmouth College 1989, P. Fougère *ed.,* Kluwer Academic Publishers, Dordrecht the Netherlands, 1990.

[14] Bayes, Rev. T., "An Essay Toward Solving a Problem in the Doctrine of Chances," *Philos. Trans. R. Soc. London* **53**, pp. 370-418 (1763); reprinted in *Biometrika* **45**, pp. 293-315 (1958), and *Facsimiles of Two Papers by Bayes,* with commentary by W. Edwards Deming, New York, Hafner, 1963.

[15] Laplace, P. S., *A Philosophical Essay on Probabilities,* unabridged and unaltered reprint of Truscott and Emory translation, Dover Publications, Inc., New York, 1951, original publication date 1814.

[16] Jaynes, E. T., "Prior Probabilities," *IEEE Transactions on Systems Science and Cybernetics,* SSC-4, pp. 227-241 (1968); reprinted in [20].

[17] Shore J. E., R. W. Johnson, *IEEE Trans. on Information Theory,* **IT-26,** No. 1, pp. 26-37, 1981.

[18] Shore J. E., R. W. Johnson, *IEEE Trans. on Information Theory,* **IT-27,** No. 4, pp. 472-482, 1980.

[19] Jaynes, E. T., "Where Do We Stand On Maximum Entropy?" in *The Maximum Entropy Formalism,* R. D. Levine and M. Tribus *Eds.,* pp. 15-118, Cambridge: MIT Press, 1978; Reprinted in [20].

[20] Jaynes, E. T., *Papers on Probability, Statistics and Statistical Physics,* a reprint collection, D. Reidel, Dordrecht the Netherlands, 1983; second edition Kluwer Academic Publishers, Dordrecht the Netherlands, 1989.

[21] Jaynes, E. T., "Marginalization and Prior Probabilities," in *Bayesian Analysis in Econometrics and Statistics,* A. Zellner, *ed.,* North-Holland Publishing Company, Amsterdam, 1980; reprinted in [20].

[22] Shannon, C. E., "A Mathematical Theory of Communication," *Bell Syst. Tech. J.* **27,** pp. 379-423 (1948).

[23] Jaynes, E. T., 1989, "The Theory of Radar Target Discrimination," in MICOM Technical Report RD-AS-91-6, Feb. 1991.

[24] Bretthorst, G. Larry, "Radar Target Identification The Information Processing Aspects," Contract number DAAL03-92-c-0034, TCN 92060.

HYPERPARAMETERS: OPTIMIZE, OR INTEGRATE OUT?

David J.C. MacKay
Cavendish Laboratory,
Cambridge, CB3 0HE. United Kingdom.
mackay@mrao.cam.ac.uk

ABSTRACT. I examine two approximate methods for computational implementation of Bayesian hierarchical models, that is, models which include unknown hyperparameters such as regularization constants. In the 'evidence framework' the model parameters are *integrated* over, and the resulting evidence is *maximized* over the hyperparameters. The optimized hyperparameters are used to define a Gaussian approximation to the posterior distribution. In the alternative 'MAP' method, the true posterior probability is found by *integrating* over the hyperparameters. The true posterior is then *maximized* over the model parameters, and a Gaussian approximation is made. The similarities of the two approaches, and their relative merits, are discussed, and comparisons are made with the ideal hierarchical Bayesian solution.

In moderately ill-posed problems, integration over hyperparameters yields a probability distribution with a skew peak which causes significant biases to arise in the MAP method. In contrast, the evidence framework is shown to introduce negligible predictive error, under straightforward conditions.

General lessons are drawn concerning the distinctive properties of inference in many dimensions.

"Integrating over a nuisance parameter is very much like estimating the parameter from the data, and then using that estimate in our equations." *G.L. Bretthorst*

"This integration would be counter-productive as far as practical manipulation is concerned." *S.F. Gull*

1 Outline

In ill-posed problems, a Bayesian model $\mathcal{H}$ commonly takes the form:

$$P(D, \mathbf{w}, \alpha, \beta|\mathcal{H}) = P(D|\mathbf{w}, \beta, \mathcal{H})P(\mathbf{w}|\alpha, \mathcal{H})P(\alpha, \beta|\mathcal{H}), \tag{1}$$

where D is the data, $\mathbf{w}$ is the parameter vector, β defines a noise variance $\sigma_\nu^2 = 1/\beta$, and α is a regularization constant. In a regression problem, for example, D might be a set of data points, $\{\mathbf{x}, \mathbf{t}\}$, and the vector $\mathbf{w}$ might parameterize a function $f(\mathbf{x}; \mathbf{w})$. The model $\mathcal{H}$ states that for some $\mathbf{w}$, the dependent variables $\{\mathbf{t}\}$ are given by adding noise to $\{f(\mathbf{x}; \mathbf{w})\}$; the likelihood function $P(D|\mathbf{w}, \beta, \mathcal{H})$ describes the assumed noise process, parameterized by a noise level $1/\beta$; the prior $P(\mathbf{w}|\alpha, \mathcal{H})$ embodies assumptions about the spatial correlations and smoothness that the true function is expected to have, parameterized by a regularization constant α. The variables α and β are known as hyperparameters. Problems for which models can be written in the form (1) include linear interpolation with a fixed basis set

G. R. Heidbreder (ed.), Maximum Entropy and Bayesian Methods, 43–59.

(Gull 1988; MacKay 1992a), non-linear regression with a neural network (MacKay 1992b). non-linear classification (MacKay 1992c), and image deconvolution (Gull 1989).

In the simplest case (linear models, Gaussian noise), the first factor in (1), the likelihood, can be written in terms of a quadratic function of $\mathbf{w}$, $E_D(\mathbf{w})$:

$$P(D|\mathbf{w},\beta,\mathcal{H}) = \frac{1}{Z_D(\beta)}\exp(-\beta E_D(\mathbf{w})). \tag{2}$$

What makes the problem 'ill-posed' is that the hessian $\nabla\nabla E_D$ is ill-conditioned — some of its eigenvalues are very small, so that the maximum likelihood parameters depend undesirably on the noise in the data. The model is 'regularized' by the second factor in (1), the prior, which in the simplest case is a spherical Gaussian:

$$P(\mathbf{w}|\alpha,\mathcal{H}) = \frac{1}{Z_W(\alpha)}\exp(-\alpha\tfrac{1}{2}\mathbf{w}^{\mathrm{T}}\mathbf{w}). \tag{3}$$

The regularization constant α defines the variance $\sigma_w^2 = 1/\alpha$ of the prior for the components w_i of $\mathbf{w}$.

Much interest has centred on the question of how the constants α and β — or the ratio α/β — should be set, and Gull (1989) has derived an appealing Bayesian prescription for these constants (see also MacKay (1992a) for a review). This 'evidence framework' *integrates* over the *parameters* $\mathbf{w}$ to give the 'evidence' $P(D|\alpha,\beta,\mathcal{H})$. The evidence is then *maximized* over the *regularization constant* α *and noise level* β. A Gaussian approximation is then made with the hyperparameters fixed to their optimized values. This relates closely to the 'generalized maximum likelihood' method in statistics (Wahba 1975). This method can be applied to non-linear models by making appropriate local linearizations, and has been used successfully in image reconstruction (Gull 1989; Weir 1991) and in neural networks (MacKay 1992b; Thodberg 1993; MacKay 1994).

Recently an alternative procedure for computing inferences under the same Bayesian model has been suggested by Buntine and Weigend (1991), Strauss *et al.* (1993) and Wolpert (1993). In this approach, one *integrates* over the *regularization constant* α first to obtain the 'true prior', and over the *noise level* β to obtain the 'true likelihood'; then *maximizes* the 'true posterior' over the *parameters* $\mathbf{w}$. A Gaussian approximation is then made around this true probability density maximum. I will call this the 'MAP' method (for *maximum a posteriori*); this use of the term 'MAP' may not coincide precisely with its general usage.

The purpose of this paper is to examine the choice between these two Gaussian approximations, both of which might be used to approximate predictive inference. It is assumed that it is predictive *distributions* that are of interest, rather than point *estimates*. Estimation will only appear as a computational stepping stone in the process of approximating a predictive distribution. I concentrate on the simplest case of the linear model with Gaussian noise, but the insights obtained are expected to apply to more general non-linear models and to models with multiple hyperparameters. When a non-linear model has multiple local optima, one can approximate the posterior by a sum of Gaussians, one fitted at each optimum. There is then an analogous choice between either (a) optimizing α separately at each local optimum in $\mathbf{w}$, and using a Gaussian approximation conditioned on α (MacKay 1992b); or (b) fitting Gaussians to local maxima of the true posterior with the hyperparameter α integrated out.

2 The Alternative Methods

Given the Bayesian model defined in (1), we might be interested in the following inferences.
Problem A: Infer the parameters, *i.e.*, obtain a compact representation of $P(\mathbf{w}|D,\mathcal{H})$ and the marginal distributions $P(w_i|D,\mathcal{H})$.
Problem B: Infer the relative model plausibility, which requires the 'evidence' $P(D|\mathcal{H})$.
Problem C: Make predictions, *i.e.* obtain some representation of $P(D_2|D,\mathcal{H})$, where D_2, in the simplest case, is a single new datum.

Let us assume for simplicity that the noise level β is known precisely, so that only the regularization constant α is respectively optimized or integrated over. Comments about α can apply equally well to β.

The Ideal Approach

Ideally, if we were able to do all the necessary integrals, we would just generate the probability distributions $P(\mathbf{w}|D,\mathcal{H})$, $P(D|\mathcal{H})$, and $P(D_2|D,\mathcal{H})$ by direct integration over everything that we are not concerned with. The pioneering work of Box and Tiao (1973) used this approach to develop Bayesian robust statistics.

For real problems of interest, however, such exact integration methods are seldom available. A partial solution can still be obtained by using Monte Carlo methods to simulate the full probability distribution (see Neal (1993b) for an excellent review). Thus one can obtain (problem A) a set of samples $\{\mathbf{w}\}$ which represent the posterior $P(\mathbf{w}|D,\mathcal{H})$, and (problem C) a set of samples $\{D_2\}$ which represent the predictive distribution $P(D_2|D,\mathcal{H})$. Unfortunately, the evaluation of the evidence $P(D|\mathcal{H})$ with Monte Carlo methods (problem B) is a difficult undertaking. Recent developments (Neal 1993a; Skilling 1993) now make it possible to use gradient and curvature information so as to sample high dimensional spaces more effectively, even for highly non-Gaussian distributions. Let us come down from these clouds however, and turn attention to the two deterministic approximations under study.

The Evidence Framework

The evidence framework divides our inferences into distinct 'levels of inference':
Level 1: Infer the parameters $\mathbf{w}$ for a given value of α:

$$P(\mathbf{w}|D,\alpha,\mathcal{H}) = \frac{P(D|\mathbf{w},\alpha,\mathcal{H})P(\mathbf{w}|\alpha,\mathcal{H})}{P(D|\alpha,\mathcal{H})}. \tag{4}$$

Level 2: Infer α:

$$P(\alpha|D,\mathcal{H}) = \frac{P(D|\alpha,\mathcal{H})P(\alpha|\mathcal{H})}{P(D|\mathcal{H})}. \tag{5}$$

Level 3: Compare models:

$$P(\mathcal{H}|D) \propto P(D|\mathcal{H})P(\mathcal{H}). \tag{6}$$

There is a pattern in these three applications of Bayes' rule: at each of higher levels 2 and 3, the data-dependent factor (*e.g.* in level 2, $P(D|\alpha,\mathcal{H})$) is the normalizing constant (the 'evidence') from the preceding level of inference.

The inference problems listed at the beginning of this section are solved approximately using the following procedure.

- The level 1 inference is approximated by making a quadratic expansion, around a maximum of $P(\mathbf{w}|D,\alpha,\mathcal{H})$, of $\log P(D|\mathbf{w},\alpha,\mathcal{H})P(\mathbf{w}|\alpha,\mathcal{H})$; this expansion defines a Gaussian approximation to the posterior. The evidence $P(D|\alpha,\mathcal{H})$ is estimated by evaluating the appropriate determinant. For linear models the Gaussian approximation is exact.
- By maximizing the evidence $P(D|\alpha,\mathcal{H})$ at level 2, we find the most probable value of the regularization constant, $\alpha_{\rm MP}$, and error bars on it, $\sigma_{\log\alpha|D}$. (Because α is a positive scale variable, it is natural to represent its uncertainty on a log scale.)
- The value of $\alpha_{\rm MP}$ is substituted at level 1. This defines a probability distribution $P(\mathbf{w}|D,\alpha_{\rm MP},\mathcal{H})$ which is intended as a 'good approximation' to the posterior $P(\mathbf{w}|D,\mathcal{H})$. The solution offered for problem A is a Gaussian distribution around the maximum of this distribution, $\mathbf{w}_{\rm MP|\alpha_{MP}}$, with covariance matrix Σ defined by $\Sigma^{-1} = -\nabla\nabla\log P(\mathbf{w}|D,\alpha_{\rm MP},\mathcal{H})$. Marginals for the components of $\mathbf{w}$ are easily obtained from this distribution.
- The evidence for model $\mathcal{H}$ (problem B) is estimated using:

$$P(D|\mathcal{H}) \simeq P(D|\alpha_{\rm MP},H)P(\log\alpha_{\rm MP}|\mathcal{H})\sqrt{2\pi}\sigma_{\log\alpha|D}. \tag{7}$$

- Problem C: The predictive distribution $P(D_2|D,\mathcal{H})$ is approximated by using the posterior distribution with $\alpha = \alpha_{\rm MP}$:

$$P(D_2|D,\alpha_{\rm MP},\mathcal{H}) = \int d^k\mathbf{w}\; P(D_2|\mathbf{w},\mathcal{H})P(\mathbf{w}|D,\alpha_{\rm MP},\mathcal{H}). \tag{8}$$

  For a locally linear model with Gaussian noise, both the distributions inside the integral are Gaussian, and this integral is straightforward to perform.

As reviewed in MacKay (1992a), the most probable value of α satisfies a simple and intuitive implicit equation,

$$\frac{1}{\alpha_{\rm MP}} = \frac{\sum_1^k w_i^2}{\gamma} \tag{9}$$

where w_i are the components of the vector $\mathbf{w}_{\rm MP|\alpha_{MP}}$ and γ is the *number of well-determined parameters*, which can be expressed in terms of the eigenvalues λ_a of the matrix $\beta\nabla\nabla E_D(\mathbf{w})$:

$$\gamma = k - \alpha{\rm Trace}\Sigma = \sum_1^k \frac{\lambda_a}{\lambda_a+\alpha}. \tag{10}$$

This quantity is a number between 0 and k. Recalling that α can be interpreted as the variance σ_w^2 of the distribution from which the parameters w_i come, we see that equation (9) corresponds to an intuitive prescription for a variance estimator. The idea is that we are estimating the variance of the distribution of w_i from only γ well-determined parameters, the other $(k-\gamma)$ having been set roughly to zero by the regularizer and therefore not contributing to the sum in the numerator.

In principle, there may be multiple optima in α, but this is not the typical case for a model well matched to the data. Under general conditions, the error bars on $\log\alpha$ are $\sigma_{\log\alpha|D} \simeq \sqrt{2/\gamma}$ (MacKay 1992a) (see section 5). Thus $\log\alpha$ is well-determined by the data if $\gamma \gg 1$.

The central computation can be summarised thus:

Evidence approximation: find the self-consistent solution $\{\mathbf{w}_{\text{MP}|\alpha_{\text{MP}}}, \alpha_{\text{MP}}\}$ such that $\mathbf{w}_{\text{MP}|\alpha_{\text{MP}}}$ maximizes $P(\mathbf{w}|D, \alpha_{\text{MP}}, \mathcal{H})$ and α_{MP} satisfies equation (9).

*Justification for the Evidence Approximation*The central approximation in this scheme can be stated as follows: when we integrate out a parameter, the effect for most purposes is to estimate the parameter from the data, and then constrain the parameter to that value (Box and Tiao 1973; Bretthorst 1988). When we predict an observable D_2, the predictive distribution is dominated by the value $\alpha = \alpha_{\text{MP}}$. In symbols,

$$P(D_2|D,\mathcal{H}) = \int P(D_2|D,\alpha,\mathcal{H})P(\log\alpha|D,\mathcal{H})\,d\log\alpha \simeq P(D_2|D,\alpha_{\text{MP}},\mathcal{H}).$$

This approximation is accurate as long as $P(D_2|D,\alpha,\mathcal{H})$ is insensitive to changes in $\log\alpha$ on a scale of $\sigma_{\log\alpha|D}$, so that the distribution $P(\log\alpha|D,\mathcal{H})$ is effectively a delta function. This is a well-established idea.

A similar equivalence of two probability distributions arises in statistical thermodynamics. The 'canonical ensemble' over all states r of a system,

$$P(r) = \exp(-\beta E_r)/Z, \tag{11}$$

describes equilibrium with a heat bath at temperature $1/\beta$. Although the energy of the system is not fixed, the probability distribution of the energy is usually sharply peaked about the mean energy $\bar{E}$. The corresponding 'microcanonical ensemble' describes the system when it is isolated and has fixed energy:

$$P(r) = \begin{cases} 1/\Omega & E_r \in [\bar{E} \pm \delta E/2] \\ 0 & \text{otherwise} \end{cases}. \tag{12}$$

Under these two distributions, a particular microstate r may have numerical probabilities that are completely different. For example, the most probable microstate under the canonical ensemble is always the ground state, for any temperature $1/\beta \geq 0$; whereas its probability under the microcanonical ensemble is zero. But it is well known (Reif 1965) that for most macroscopic purposes, if the system has a large number of degrees of freedom, the two distributions are indistinguishable, because most of the probability *mass* of the canonical ensemble is concentrated in the states in a small interval around $\bar{E}$.

The same reasoning justifies the evidence approximation for ill-posed problems, with particular values of $\mathbf{w}$ corresponding to microstates. If the number of well-determined parameters is large, then α, like the energy above, is well-determined. This does not imply that the two densities $P(\mathbf{w}|D,\mathcal{H})$ and $P(\mathbf{w}|D,\alpha_{\text{MP}},\mathcal{H})$ are numerically close in value, but we have no interest in the probability of the high dimensional vector $\mathbf{w}$. For practical purposes, we only care about distributions of low-dimensional quantities (*e.g.*, an individual parameter w_i or a new datum); what matters, and what is asserted here, is that when we project the distributions down in order to predict low-dimensional quantities, the approximating distribution $P(\mathbf{w}|D,\alpha_{\text{MP}},\mathcal{H})$ puts most of its probability mass in the right place. A more precise discussion of this approximation is given in section 5.

The MAP Method

The alternative procedure is first to integrate out α to obtain the true prior:

$$P(\mathbf{w}|\mathcal{H}) = \int d\alpha\, P(\mathbf{w}|\alpha,\mathcal{H})P(\alpha|\mathcal{H}). \tag{13}$$

We can then write down the true posterior directly (except for its normalizing constant):

$$P(\mathbf{w}|D,\mathcal{H}) \propto P(D|\mathbf{w},\mathcal{H})P(\mathbf{w}|\mathcal{H}). \tag{14}$$

This posterior can be maximized to find the MAP parameters, $\mathbf{w}_{\mathrm{MP}}$. How does this relate to the desired inferences listed at the head of this section? Not all authors describe how they intend the true posterior to be used in practical problems (*e.g.*, Wolpert (1993)); here I describe a method based on the suggestions of Buntine and Weigend (1991).

Problem A: The posterior distribution $P(\mathbf{w}|D,\mathcal{H})$ is approximated by a Gaussian distribution, fitted around the most probable parameters, $\mathbf{w}_{\mathrm{MP}}$; to find the Hessian of the posterior, one needs the Hessian of the prior, derived below. A simple evaluation of the factors on the right hand side of (14) is not a satisfactory solution of problem A, since (a) the normalizing constant is missing; (b) even if the r.h.s. of (14) were normalized, the ability to evaluate the local value of this density would be of little use as a summary of the distribution in the high-dimensional space; how, for example, is one to obtain marginal distributions over w_i from (14)?

Problem B: An estimate of the evidence is obtained from the determinant of the covariance matrix of this Gaussian distribution.

Problem C: The parameters $\mathbf{w}_{\mathrm{MP}}$ with error bars are used to generate predictions as in (8).

A simple example will illustrate that this approach actually gives results qualitatively very similar to the evidence framework. If we apply the improper prior $P_{\mathrm{Imp}}(\log\alpha) = \mathrm{const}$ and evaluate the true prior, we obtain:[1]

$$P_{\mathrm{Imp}}(\mathbf{w}|\mathcal{H}) = \int_{\alpha=0}^{\infty} \frac{e^{-\alpha\sum_{i=1}^{k} w_i^2/2}}{Z_W(\alpha)}\, d\log\alpha \;\propto\; \frac{1}{(\sum_i w_i^2)^{k/2}}. \tag{15}$$

The derivative of the true log prior with respect to $\mathbf{w}$ is $-(k/\sum_i w_i^2)\mathbf{w}$. This 'weight decay' term can be directly viewed in terms of an 'effective α',

$$\frac{1}{\alpha_{\mathrm{eff}}(\mathbf{w})} = \frac{\sum_i w_i^2}{k}. \tag{16}$$

Any maximum of the true posterior $P(\mathbf{w}|D,\mathcal{H})$ is therefore also a maximum of the conditional posterior $P(\mathbf{w}|D,\alpha,\mathcal{H})$, with α set to α_{eff}. The similarity of equation (16) to equation (9) of the evidence framework is clear. We can therefore describe the MAP method thus:

> **MAP method** (improper prior): find the self-consistent solution $\{\mathbf{w}_{\mathrm{MP}},\alpha_{\mathrm{eff}}\}$ such that $\mathbf{w}_{\mathrm{MP}}$ maximizes $P(\mathbf{w}|D,\alpha_{\mathrm{eff}},\mathcal{H})$ and α_{eff} satisfies equation (16).

This procedure is suggested in (MacKay 1992b) as a 'quick and dirty' approximation to the evidence framework.

---

[1] If a uniform prior over α from 0 to ∞ is used (instead of a prior over $\log\alpha$) then the resulting exponent changes from $k/2$ to $(k/2+1)$.

THE EFFECTIVE α AND THE CURVATURE OF A GENERAL PRIOR

We have just established that, when the improper prior (15) is used, the MAP solution lies on the 'alpha trajectory' — the graph of $\mathbf{w}_{\text{MP}|\alpha}$ — for a particular value of $\alpha = \alpha_{\text{eff}}$. This result still holds when a proper prior over α is used to define the true prior (13). The effective $\alpha(\mathbf{w})$, found by differentiation of $\log P(\mathbf{w}|\mathcal{H})$, is:

$$\alpha_{\text{eff}}(\mathbf{w}) = \int d\alpha \, \alpha \, P(\alpha|\mathbf{w}, \mathcal{H}). \tag{17}$$

In general there may be multiple local probability maxima, all of which lie on the alpha trajectory. In summary, optima $\mathbf{w}_{\text{MP}}$ found by the MAP method can be described thus:

MAP method (proper prior): find the self-consistent solution $\{\mathbf{w}_{\text{MP}}, \alpha_{\text{eff}}\}$ such that $\mathbf{w}_{\text{MP}}$ maximizes $P(\mathbf{w}|D, \alpha_{\text{eff}}, \mathcal{H})$ and α_{eff} satisfies equation (17).

The curvature of the true prior is needed for evaluation of the error bars on $\mathbf{w}$ in the MAP method. By direct differentiation of the true prior (13), we find:

$$-\nabla\nabla \log P(\mathbf{w}|\mathcal{H}) = \alpha_{\text{eff}}\mathbf{I} - \sigma_{\alpha}^2(\mathbf{w})\mathbf{w}\mathbf{w}^{\text{T}}, \tag{18}$$

where $\alpha_{\text{eff}}(\mathbf{w})$ is defined in (16), and the effective variance of α is:

$$\sigma_{\alpha}^2(\mathbf{w}) = \overline{\alpha^2}(\mathbf{w}) - \alpha_{\text{eff}}(\mathbf{w})^2 = \int d\alpha \, \alpha^2 \, P(\alpha|\mathbf{w}, \mathcal{H}) - \left(\int d\alpha \, \alpha \, P(\alpha|\mathbf{w}, \mathcal{H})\right)^2. \tag{19}$$

This is an intuitive result: if α were fixed to α_{eff}, then the curvature would just be the first term in (18), $\alpha_{\text{eff}}\mathbf{I}$. The fact that α is uncertain depletes the curvature in the radial direction $\hat{\mathbf{w}} = \mathbf{w}/|\mathbf{w}|$.

3 Pros and Cons

The algorithms for finding the evidence framework's $\mathbf{w}_{\text{MP}|\alpha_{\text{MP}}}$ and the MAP method's $\mathbf{w}_{\text{MP}}$ have been seen to be very similar. Is there any real distinction to be drawn between these two aproaches?

The MAP method has the advantage that it involves no approximations until after we have found the MAP parameters $\mathbf{w}_{\text{MP}}$; in contrast, the evidence framework approximates an integral over α.

In the MAP method the integrals over α and β need only be performed once and can then be used repeatedly for different data sets; in the evidence framework, each new data set has to receive individual attention, with a sequence of (Gaussian) integrations being performed each time α and β are optimized.

So why not always integrate out hyperparameters whenever possible? Let us answer this question by magnifying the systematic differences between the two approaches. With sufficient magnification it will become evident to the intuition that the approximation of the evidence framework is superior to the MAP approximation. The distinction between $\mathbf{w}_{\text{MP}}$ and $\mathbf{w}_{\text{MP}|\alpha_{\text{MP}}}$ is similar to that between the two estimators of standard deviation on a calculator, σ_N and σ_{N-1}, the former being the (biased) maximum likelihood estimator, whereas the latter is unbiased. The true posterior distribution has a skew peak, so that the MAP parameters are not representative of the whole posterior distribution. This is best illustrated by an example.

The Widget Example

A collection of widgets $i = 1..k$ have a property called 'wibble', w_i, which we measure, widget by widget, in noisy experiments with a known noise level $\sigma_\nu = 1.0$. Our model for these quantities is that they come from a Gaussian prior $P(w_i|\alpha, \mathcal{H})$, where $\alpha = 1/\sigma_w^2$ is not known. Our prior for this variance is flat over $\log \sigma_w$ from $\sigma_w = 0.1$ to $\sigma_w = 10$.

Scenario 1. Suppose four widgets have been measured and give the following data: $\{d_1, d_2, d_3, d_4\} = \{3.2. \text{-}3.2, 2.8, \text{-}2.8\}$. The task is (problem A) to infer the wibbles of these four widgets, *i.e.* to produce a representative $\mathbf{w}$ with error bars. On the back of an envelope, or in a computer algebra system, we find the following answers using equations (9) and (16/17):

Evidence framework: $\alpha_{\text{MP}} = 0.124$, $\mathbf{w}_{\text{MP}|\alpha_{\text{MP}}} = \{2.8, -2.8, 2.5, -2.5\}$, each with error bars ± 0.9.

MAP method: $\alpha_{\text{eff}} = 0.145$, $\mathbf{w}_{\text{MP}} = \{2.8, -2.8, 2.4, -2.4\}$, each with error bars ± 0.9. These answers are insensitive to the details of the prior over σ_w.

So far so good: $\mathbf{w}_{\text{MP}|\alpha_{\text{MP}}}$ is slightly less regularized than $\mathbf{w}_{\text{MP}}$, but there is not much disagreement when all the parameters are well-determined.

Scenario 2. Suppose in addition to the four measurements above we are now informed that there are an additional four unmeasured widgets in a box next door. Thus we now have both well-determined and ill-determined parameters, as in an ill-posed problem. Intuitively, we would like our inferences about the well-measured widgets to be negligibly affected by this vacuous information about the unmeasured widgets, just as the true Bayesian predictive distributions are unaffected. But clearly with $k = 8$, the difference between k and γ in equations (9) and (16) is going to become significant. The value of α_{eff} will be substantially greater than that of α_{MP}.

In the evidence framework the value of γ is exactly the same, since each of the ill-determined parameters has $\lambda = 0$ and adds nothing to the sum in (10). So the value of α_{MP} and the predictive distributions are unchanged.

In contrast, the MAP parameter vector $\mathbf{w}_{\text{MP}}$ is squashed close to zero. The precise value of $\mathbf{w}_{\text{MP}}$ is sensitive to the prior over α. Solving equation (17) in a computer algebra system, we find: $\alpha_{\text{eff}} = 79.2$, $\mathbf{w}_{\text{MP}} = \{0.040, -0.040, 0.035, -0.035, 0, 0, 0, 0\}$, with marginal error bars on all eight parameters $\sigma_{w|D} = 0.11$.

Thus the MAP Gaussian approximation is terribly biased towards zero. The final disaster of this approach is that the error bars on the parameters are also correspondingly small.

This is not a contrived example. It contains the basic feature of ill-posed problems: that there are both well-determined and poorly-determined parameters. To aid comprehension, the two sets of parameters are separated. This example can be transformed into a typical ill-posed problem simply by rotating the basis to mix the parameters together. In neural networks, a pair of scenarios identical to those discussed above can arise if there are a large number of poorly determined parameters which have been set to zero by the regularizer, and we consider 'pruning' these parameters. In scenario 1, the network is pruned, removing the ill-determined parameters. In scenario 2, the parameters are retained, and assume their most probable value, zero. In each case, what is the optimal setting of the weight decay rate α (assuming the traditional regularizer $\mathbf{w}^{\text{T}}\mathbf{w}/2$)? We would expect the answer to be unchanged. Yet the MAP method effectively sets α to a much larger value in the second scenario.

The MAP method may locate the true posterior maximum, but it fails to capture most of the true probability mass.

4 Inference in Many Dimensions

In many dimensions, therefore, new intuitions are needed.

Nearly all of the volume of a k-dimensional hypersphere is in a thin shell near its surface. For example, in 1000 dimensions, 90% of a hypersphere of radius 1.0 is within a depth of 0.0023 of its surface. A central core of the hypersphere, with radius 0.5, contains less than $1/10^{300}$ of the volume.

This has an important effect on high-dimensional probability distributions. Consider a Gaussian distribution $P(\mathbf{w}) = (1/\sqrt{2\pi}\sigma_w)^k \exp(-\sum_1^k w_i^2/2\sigma_w^2)$. Nearly all of the probability mass of a Gaussian is in a thin shell of radius $r = \sqrt{k}\sigma_w$ and of thickness $\propto r/\sqrt{k}$. For example, in 1000 dimensions, 90% of the mass of a Gaussian with $\sigma_w = 1$ is in a shell of radius 31.6 and thickness 2.8. However, the probability *density* at the origin is $e^{k/2} \simeq 10^{217}$ times bigger than the density at this shell where most of the probability mass is.

Consider two Gaussian densities in 1000 dimensions which differ in σ_w by 1%, and which contain equal total probability mass. In each case 90% of the mass is located in a shell which differs in radius by only 1% between the two distributions. The maximum probability density, however, is greater at the centre of the Gaussian with smaller σ_w, by a factor of $\sim \exp(0.01k) \simeq 20{,}000$.

In summary, probability density maxima often have very little associated probability mass, even though the value of the probability density there may be immense, because they have so little associated volume. If a distribution is composed of a mixture of Gaussians with different σ_w, the probability density maxima are strongly dominated by smaller values of σ_w. This is why the MAP method finds a silly solution in the widget example.

Thus the locations of probability density maxima in many dimensions are generally misleading and irrelevant. Probability densities should only be maximized if there is good reason to believe that the location of the maximum conveys useful information about the whole distribution, *e.g.*, if the distribution is approximately Gaussian.

Condition Satisfied by Typical Samples

The conditions (9) and (16), satisfied by the optima $(\alpha_{\text{MP}}, \mathbf{w}_{\text{MP}|\alpha_{\text{MP}}})$ and $(\alpha_{\text{eff}}, \mathbf{w}_{\text{MP}})$ respectively, are complemented by an additional result concerning typical samples from posterior distributions conditioned on α. The maximum $\mathbf{w}_{\text{MP}|\alpha}$ of a Gaussian distribution is not typical of the posterior: the maximum has an atypically small value of $\mathbf{w}^{\text{T}}\mathbf{w}$, because, as discussed above, nearly all of the mass of a Gaussian is in a shell at some distance surrounding the maximum.

Consider samples $\mathbf{w}$ from the Gaussian posterior distribution with α fixed to α_{MP}, $P(\mathbf{w}|D, \alpha_{\text{MP}}, \mathcal{H})$. The average value of $\mathbf{w}^{\text{T}}\mathbf{w} = \sum_i w_i^2$ for these samples satisfies:

$$\alpha_{\text{MP}} = \frac{k}{\langle \sum_i w_i^2 \rangle_{|D,\alpha_{\text{MP}}}}. \tag{20}$$

Proof: The deviation $\Delta\mathbf{w} = \mathbf{w} - \mathbf{w}_{\mathrm{MP}|\alpha_{\mathrm{MP}}}$ is Gaussian distributed with $\Delta\mathbf{w}\Delta\mathbf{w}^{\mathrm{T}} = \mathbf{\Sigma}$. So $\alpha_{\mathrm{MP}}\langle\sum_i w_i^2\rangle_{|D,\alpha_{\mathrm{MP}}} = \alpha_{\mathrm{MP}}(\mathbf{w}_{\mathrm{MP}|\alpha_{\mathrm{MP}}} + \Delta\mathbf{w})^{\mathrm{T}}(\mathbf{w}_{\mathrm{MP}|\alpha_{\mathrm{MP}}} + \Delta\mathbf{w}) = \alpha_{\mathrm{MP}}\mathbf{w}^2_{\mathrm{MP}|\alpha_{\mathrm{MP}}} + \alpha_{\mathrm{MP}}\mathrm{Trace}\mathbf{\Sigma} = k$, using equations (9) and (10).

Thus a typical sample from the evidence approximation prefers just the same value of α as does the evidence.

5 Conditions for the Evidence Approximation

We have observed that the MAP method can lead to absurdly biased answers if there are many ill-determined parameters. In contrast, I now discuss conditions under which the evidence approximation works. I discuss the case of linear models with Gaussian probability distributions. This includes the case of image reconstruction problems that have separable Gaussian distributions in the Fourier domain.

What do we care about when we approximate a complex probability distribution by a simple one? My definition of a good approximation is a practical one, concerned with (A) estimating parameters; (B) estimating the evidence accurately; and (C) getting the predictive mass in the right place. Estimation of individual parameters (A) is a special case of prediction (C), so in the following I will address only (C) and (B).

For convenience let us work in the eigenvector basis where the prior (given α) and the likelihood are both diagonal Gaussian functions. The curvature of the log likelihood is represented by eigenvalues $\{\lambda_a\}$. For a typical ill-posed problem these eigenvalues cover several orders of magnitude in value. Without loss of generality let us assume k data measurements $\{d_a\}$, such that $d_a = \sqrt{\lambda_a}w_a + \nu$, where the noise standard deviation is $\sigma_\nu = 1$. We define the probability distribution of everything by the product of the distributions:

$$P(\log\alpha|\mathcal{H}) = \frac{1}{\log(\alpha_{\max}/\alpha_{\min})}, \quad P(\mathbf{w}|\alpha,\mathcal{H}) = \left(\frac{\alpha}{2\pi}\right)^{k/2}\exp\left(-\frac{1}{2}\alpha\sum_1^k w_a^2\right), \text{ and}$$

$$P(D|\mathbf{w},\mathcal{H}) = (2\pi)^{-k/2}\exp\left\{-\frac{1}{2}\sum_1^k\left(\sqrt{\lambda_a}w_a - d_a\right)^2\right\}.$$

In the case of a deconvolution problem the eigenvectors are the Fourier set and the point spread function in Fourier space is given by $\sqrt{\lambda_a}$.

The discussion proceeds in two steps. First, the posterior distribution over α must have a single sharp peak at α_{MP}. No general guarantee can be given for this to be the case, but various pointers are given. Second, given a sharp Gaussian posterior over $\log\alpha$, it is proved that the evidence approximation introduces negligible error.

Concentration of $P(\log\alpha|D,\mathcal{H})$ in a Single Maximum

Condition 1 *In the posterior distribution over* $\log\alpha$*, all the probability mass should be contained in a single sharp maximum.*

For this to hold, several sub-conditions are needed. If there is any doubt whether these conditions are sufficient, it is straightforward to iterate all the way down the α trajectory, explicitly evaluating $P(\log\alpha|D,\mathcal{H})$.

The prior over α must be such that the posterior has negligible mass at $\log\alpha \to \pm\infty$. In cases where the signal to noise ratio of the data is very low, there may be a significant tail in the evidence for large α. There may even be no maximum in the evidence, in which case the evidence framework gives singular behaviour, with α going to infinity. But often the tails of the evidence are small, and contain negligible mass if our prior over $\log\alpha$ has cutoffs at some $\alpha_{\min}$ and $\alpha_{\max}$ (surrounding α_{MP}). For each data analysis problem, one may evaluate the critical $\alpha_{\max}$ above which the posterior is measurably affected by the large α tail of the evidence (Gull 1989). Often, as Gull points out, this critical value of $\alpha_{\max}$ has bizarre magnitude.

Even if a flat prior between appropriate $\alpha_{\min}$ and $\alpha_{\max}$ is used, it is possible in principle for the posterior $P(\log\alpha|D,\mathcal{H})$ to be multi-modal. However this is not expected when the model space is well matched to the data. Examples of multi-modality only arise if the data are grossly at variance with the likelihood and the prior. For example, if some large eigenvalue measurements give small $d_{a(l)}$, and some measurements with small eigenvalue give large $d_{a(s)}$, then the posterior over α can have two peaks, one at large α which nicely explains $d_{a(l)}$, but must attribute $d_{a(s)}$ to unusually large amounts of noise, and one at small α which nicely explains $d_{a(s)}$, but must attribute $d_{a(l)}$ to $w_{a(l)}$ being unexpectedly close to zero. I now suggest a way of formalizing this concept into a quantitative test.

If we accept the model, then we believe that there is a true value of $\alpha = \alpha_{\text{T}}$, and that given α_{T}, the data measurements d_a are the sum of two independent Gaussian variables $\sqrt{\lambda_a}w_a$ and ν_a, so that $P(d_a|\alpha_{\text{T}},\mathcal{H}) = \text{Normal}(0,\sigma^2_{a|\alpha_T})$, where $\sigma^2_{a|\alpha_T} = \frac{\lambda_a}{\alpha_T}+1$. The expectation of d_a^2 is $\langle d_a^2\rangle = \frac{\lambda_a}{\alpha_T}+1$. We therefore expect that there is an α_{T} such that the quantities $\{d_a^2/\sigma^2_{a|\alpha_T}\}$ are independently distributed like χ^2 with one degree of freedom.

Definition 1 *A data set* $\{d_a\}$ *is* grossly at variance *with the model for a given value of* α, *if any of the quantities* $j_a = d_a^2/(\frac{\lambda_a}{\alpha}+1)$ *is not in the interval* $[e^{-\tau}, 1+\tau]$; *where* τ *is the significance level of this test.*

It is conjectured that if we find a value of $\alpha = \alpha_{\text{MP}}$ which locally maximizes the evidence, and with which the data are not grossly at variance, then there are no other maxima over α.

Conversely, if the data are grossly at variance with a local maximum α_{MP}, then there may be multiple maxima in α, and the evidence approximation may be inaccurate. In these circumstances one might also suspect that the entire model is inadequate in some way.

Assuming that $P(\log\alpha|D,\mathcal{H})$ has a single maximum over $\log\alpha$, how sharp is it expected to be? I now establish conditions under which the $P(\log\alpha|D,\mathcal{H})$ is locally Gaussian and sharp.

Definition 2 *The symbol* n_e *is defined by:*

$$n_e \equiv \sum_a \frac{4\lambda_a\alpha_{\text{MP}}}{(\lambda_a+\alpha_{\text{MP}})^2}. \tag{21}$$

This is a measure of the number of eigenvalues λ_a *within approximately e-fold of* α_{MP}.

In the following, I will assume that $n_e \ll \gamma$, but this condition is not essential for the evidence approximation to be valid. If $n_e \ll \gamma$, and the data are not grossly at variance with α_{MP}, then we find on Taylor-expanding $\log P(\alpha|D,\mathcal{H})$ about $\alpha = \alpha_{\text{MP}}$, that the second derivative is large, and that the third derivative is relatively small:

$$\begin{aligned}
\left.\frac{\partial \log P(D|\alpha,\mathcal{H})}{\partial \log \alpha}\right|_{\alpha_{\text{MP}}} &= \frac{1}{2}\left(\gamma - \alpha \mathbf{w}^2_{\text{MP}|\alpha_{\text{MP}}}\right) = 0 \\
\left.\frac{\partial^2 \log P(D|\alpha,\mathcal{H})}{\partial (\log \alpha)^2}\right|_{\alpha_{\text{MP}}} &\simeq -\alpha \mathbf{w}^2_{\text{MP}|\alpha_{\text{MP}}} = -\frac{\gamma}{2} \\
\left.\frac{\partial^3 \log P(D|\alpha,\mathcal{H})}{\partial (\log \alpha)^3}\right|_{\alpha_{\text{MP}}} &\simeq -\alpha \mathbf{w}^2_{\text{MP}|\alpha_{\text{MP}}} = -\frac{\gamma}{2}.
\end{aligned}$$

The first derivative is exact, assuming that the eigenvalues λ_a are independent of α, which is true in the case of a Gaussian prior on $\mathbf{w}$ (Bryan 1990). The second and third derivatives are approximate, with terms proportional to n_e being omitted. The third derivative *is* relatively small (even though it is equal to the second derivative), since in the expansion $P(l) = \exp(-\frac{c}{2}l^2 + \frac{d}{6}l^3 + \ldots)$, the second term gives a negligible perturbation for $l \sim c^{-1/2}$ if $d \ll c^{3/2}$. In this case, since $d = c = \gamma \gg 1$, the perturbation introduced by the higher order terms is $O(\gamma^{-1/2})$. Thus the posterior distribution over $\log \alpha$ has a maximum that is both locally Gaussian and sharp if $\gamma \gg 1$ and $n_e \ll \gamma$. The expression for the evidence (7) follows.

Error of Low-dimensional Predictive Distributions

I will now assume that the posterior distribution $P(\log \alpha|D,\mathcal{H})$ is Gaussian with standard deviation $\sigma_{\log \alpha|D} = 1/\sqrt{\kappa\gamma}$, with $\kappa\gamma \gg 1$, and $\kappa = O(1)$.

Theorem 1 *Consider a scalar which depends linearly on* $\mathbf{w}$, $y = \mathbf{g} \cdot \mathbf{w}$. *The evidence approximation's predictive distribution for* y *is close to the exact predictive distribution, for nearly all projections* $\mathbf{g}$. *In the case* $\mathbf{g} = \mathbf{w}$, *the error (measured by a cross-entropy) is of order* $\sqrt{n_e/\kappa\gamma}$. *For all* $\mathbf{g}$ *perpendicular to this direction, the error is of order* $\sqrt{1/\kappa\gamma}$.

A similar result is expected still to hold when the dimensionality of y is greater than one, provided that it is much less than $\sqrt{\gamma}$.

Proof: At 'level 1', we infer $\mathbf{w}$ for a fixed value of α:

$$P(\mathbf{w}|D,\alpha,\mathcal{H}) \propto \exp\left\{-\frac{1}{2}\sum_a (\lambda_a + \alpha)\left(w_a - \frac{\sqrt{\lambda_a} d_a}{\lambda_a + \alpha}\right)^2\right\}. \tag{22}$$

The most probable $\mathbf{w}$ given this value of α is: $w_a^{\text{MP}|\alpha} = \sqrt{\lambda_a} d_a/(\lambda_a + \alpha)$. The posterior distribution is Gaussian about this most probable $\mathbf{w}$. We introduce a *typical* $\mathbf{w}$, that is, a sample from the posterior for a particular value of α:

$$w_a^{\text{TYP}|\alpha} = \frac{\sqrt{\lambda_a} d_a}{\lambda_a + \alpha} + \frac{r_a}{\sqrt{\lambda_a + \alpha}}, \tag{23}$$

where r_a is a sample from Normal(0,1).

Now, assuming that $\log\alpha$ has a Gaussian posterior distribution with standard deviation $1/\sqrt{\kappa\gamma}$, a typical α, *i.e.*, a sample from this posterior, is given to leading order by

$$\alpha^{\text{TYP}} = \alpha_{\text{MP}}\left(1 + \frac{s}{\sqrt{\kappa\gamma}}\right), \tag{24}$$

where s is a sample from Normal(0,1). We now substitute this α^{TYP} into (23) and obtain a typical $\mathbf{w}$ from the true posterior distribution, which depends on $k+1$ random variables $\{r_a\}, s$. We expand each component of this vector $\mathbf{w}^{\text{TYP}}$ in powers of $1/\gamma$:

$$\begin{aligned} w_a^{\text{TYP}} &= \frac{\sqrt{\lambda_a}d_a}{\lambda_a+\alpha_{\text{MP}}}\left(1 - \frac{s}{\sqrt{\kappa\gamma}}\frac{\alpha_{\text{MP}}}{\lambda_a+\alpha_{\text{MP}}} + \frac{s^2}{\kappa\gamma}\frac{\alpha_{\text{MP}}^2}{(\lambda_a+\alpha_{\text{MP}})^2} + \ldots\right) + \\ & \frac{r_a}{\sqrt{\lambda_a+\alpha_{\text{MP}}}}\left(1 - \frac{1}{2}\frac{s}{\sqrt{\kappa\gamma}}\frac{\alpha_{\text{MP}}}{\lambda_a+\alpha_{\text{MP}}} + \frac{3}{8}\frac{s^2}{\kappa\gamma}\frac{\alpha_{\text{MP}}^2}{(\lambda_a+\alpha_{\text{MP}})^2}\cdots\right) \end{aligned} \tag{25}$$

We now examine the mean and variance of $y^{\text{TYP}} = \sum_a g_a w_a^{\text{TYP}}$. Setting $\langle r_a^2\rangle = \langle s^2\rangle = 1$ and dropping terms of higher order than $1/\gamma$, we find that whereas the evidence approximation gives a Gaussian predictive distribution for y which has mean and variance:

$$\mu_0 = \sum_a g_a w_a^{\text{MP}|\alpha_{\text{MP}}}, \quad \sigma_0^2 = \sum_a \frac{g_a^2}{\lambda_a+\alpha_{\text{MP}}},$$

the true predictive distribution is, to order $1/\gamma$, Gaussian with mean and variance:

$$\mu_1 = \mu_0 + \frac{1}{\kappa\gamma}\sum_a g_a w_a^{\text{MP}|\alpha_{\text{MP}}}\frac{\alpha_{\text{MP}}^2}{(\lambda_a+\alpha_{\text{MP}})^2},$$

$$\sigma_1^2 = \sigma_0^2 + \frac{1}{\kappa\gamma}\left\{\left(\sum_a g_a w_a^{\text{MP}|\alpha_{\text{MP}}}\frac{\alpha_{\text{MP}}}{(\lambda_a+\alpha_{\text{MP}})}\right)^2 + \sum_a \frac{g_a^2}{\lambda_a+\alpha_{\text{MP}}}\frac{\alpha_{\text{MP}}^2}{(\lambda_a+\alpha_{\text{MP}})^2}\right\}.$$

How wrong can the evidence approximation be? Since both distributions are Gaussian, it is simple to evaluate various distances between them. The cross entropy between $p_0 = \text{Normal}(\mu_0, \sigma_0^2)$ and $p_1 = \text{Normal}(\mu_1, \sigma_1^2)$ is

$$H(p_0, p_1) \equiv \int p_1 \log\frac{p_1}{p_0} = \frac{1}{2}\frac{(\mu_1-\mu_0)^2}{\sigma_0^2} + \frac{1}{4}\left(\frac{\sigma_1^2-\sigma_0^2}{\sigma_0^2}\right)^2 + O\left\{\left(\frac{\sigma_1^2-\sigma_0^2}{\sigma_0^2}\right)^3\right\}.$$

We consider the two dominant terms separately.

$$\frac{(\mu_1-\mu_0)^2}{\sigma_0^2} = \frac{1}{\kappa^2\gamma^2}\left(\sum_a h_a w_a^{\text{MP}|\alpha_{\text{MP}}}\frac{\alpha_{\text{MP}}^2}{(\lambda_a+\alpha_{\text{MP}})^{3/2}}\right)^2 \Big/ \sum h_a^2, \tag{26}$$

where $h_a = g_a/\sqrt{\lambda_a+\alpha_{\text{MP}}}$. The worst case is given by the direction $\mathbf{g}$ such that $h_a = w_a^{\text{MP}|\alpha_{\text{MP}}}\frac{\alpha_{\text{MP}}^2}{(\lambda_a+\alpha_{\text{MP}})^{3/2}}$. This worst case gives an upper bound to the contribution to the cross entropy:

$$\begin{aligned} \frac{(\mu_1-\mu_0)^2}{\sigma_0^2} &\le \frac{1}{\kappa^2\gamma^2}\sum_a \frac{{w_a^{\text{MP}|\alpha_{\text{MP}}}}^2\alpha_{\text{MP}}^4}{(\lambda_a+\alpha_{\text{MP}})^3} & (27) \\ &< \frac{\alpha_{\text{MP}}}{\kappa^2\gamma^2}\sum {w_a^{\text{MP}|\alpha_{\text{MP}}}}^2 = \frac{1}{\kappa^2\gamma} \ll 1 & (28) \end{aligned}$$

So the change in μ *never* has a significant effect.

The variance term can be split into two terms:

$$\left(\frac{\sigma_1^2-\sigma_0^2}{\sigma_0^2}\right)^2 = \frac{1}{\kappa\gamma}\left\{\left(\sum_a \frac{h_a w_a^{\text{MP}|\alpha_{\text{MP}}}\alpha_{\text{MP}}}{\sqrt{\lambda_a+\alpha_{\text{MP}}}}\right)^2 + \sum_a h_a^2 \frac{\alpha_{\text{MP}}^2}{(\lambda_a+\alpha_{\text{MP}})^2}\right\} \Big/ \sum_a h_a^2,$$

where, as above, $h_a = g_a/\sqrt{\lambda_a+\alpha_{\text{MP}}}$.

For the first term, the worst case is the direction $h_a = w_a^{\text{MP}|\alpha_{\text{MP}}} \frac{\alpha_{\text{MP}}}{\sqrt{\lambda_a+\alpha_{\text{MP}}}}$, *i.e.*, the radial direction $\mathbf{g} = \alpha_{\text{MP}}\mathbf{w}_{\text{MP}|\alpha_{\text{MP}}}$. Substituting in this direction, we find:

$$\text{First term} \le \frac{1}{\kappa\gamma}\sum_a w_a^{\text{MP}|\alpha_{\text{MP}}\,2} \frac{\alpha_{\text{MP}}^2}{\lambda_a+\alpha_{\text{MP}}} \tag{29}$$

$$< \frac{\alpha_{\text{MP}}}{\kappa\gamma}\sum_a w_a^{\text{MP}|\alpha_{\text{MP}}\,2} = \frac{1}{\kappa} = O(1) \tag{30}$$

We can improve this bound by substituting for $w_a^{\text{MP}|\alpha_{\text{MP}}}$ in terms of d_a and making use of the definition of n_e. Only n_e of the terms in the sum in equation (29) are significant. Thus

$$\text{First term} \lesssim \frac{n_e}{\kappa\gamma}. \tag{31}$$

So this term can give a significant effect, but only in one direction; for any direction orthogonal (in $\mathbf{h}$) to this radial direction, this term is zero.

Finally, we examine the second term:

$$\frac{1}{\kappa\gamma}\sum_a h_a^2 \frac{\alpha_{\text{MP}}^2}{(\lambda_a+\alpha_{\text{MP}})^2} \Big/ \sum_a h_a^2 < \frac{1}{\kappa\gamma} \ll 1. \tag{32}$$

So this term never has a significant effect.

*Conclusion*The evidence approximation affects the mean and variance of properties y of $\mathbf{w}$, but only to within $O(\gamma^{-1/2})$ of the property's standard deviation; this error is insignificant, for large γ. The sole exception is the direction $\mathbf{g} = \mathbf{w}_{\text{MP}|\alpha_{\text{MP}}}$, along which the variance is erroneously small, with a cross-entropy error of order $O(n_e/\gamma)$.

A Correction Term

This result motivates a straightforward term which could be added to the inverse Hessian of the evidence approximation, to correct the predictive variance in this direction. The predictive variance for a general $y = \mathbf{g}^{\text{T}}\mathbf{w}$ could be estimated by

$$\sigma_y^2 = \mathbf{g}^{\text{T}}\left(\Sigma + \sigma_{\log\alpha|D}^2 \mathbf{w}'_{\text{MP}|\alpha}\mathbf{w}'^{\text{T}}_{\text{MP}|\alpha}\right)\mathbf{g}, \tag{33}$$

where $\mathbf{w}'_{\text{MP}|\alpha} \equiv \partial\mathbf{w}_{\text{MP}|\alpha}/\partial(\log\alpha) = \alpha\Sigma\mathbf{w}_{\text{MP}|\alpha}$, and $\sigma_{\log\alpha|D}^2 = \frac{2}{\gamma}$. With this correction, the predictive distribution for any direction would be in error only by order $O(1/\gamma)$. If the noise variance $\sigma_\nu^2 = \beta^{-1}$ is also uncertain, then the factor $\sigma_{\log\alpha|D}^2$ is incremented by $\sigma_{\log\beta|D}^2 = \frac{2}{N-\gamma}$.

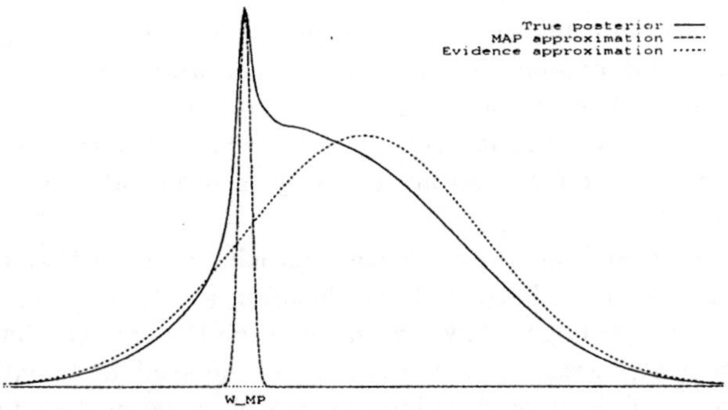

Figure 1: Approximating complicated distributions with a Gaussian
This is a schematic illustration of the properties of a multi-dimensional distribution. A typical posterior distribution for an ill-posed problem has a skew peak. A Gaussian fitted at the MAP parameters is a bad approximation to the distribution: it is in the wrong place, and its error bars are far too small. Additional features of the true posterior distribution not illustrated here are that it typically has spikes of high probability density at the origin $\mathbf{w}=0$ and at the maximum likelihood parameters $\mathbf{w} = \mathbf{w}_{\mathrm{ML}}$. The evidence approximation gives a Gaussian distribution which captures most of the probability mass of the true posterior.

6 Discussion

The MAP method, though exact, is capable of giving MAP parameters which are unrepresentative of the true posterior. In high dimensional spaces, maxima are misleading. MAP estimates play no fundamental role in Bayesian inference, and they can change arbitrarily with arbitrary re-parameterizations. The problem with MAP estimates is that they maximize the probability *density*, without taking account of the complementary *volume* information. Figure 1 attempts, in one dimension, to convey this difference between the two Gaussian approximations.

When there are many ill-determined parameters, the MAP method's integration over α yields a $\mathbf{w}_{\mathrm{MP}}$ which is over-regularized.[2]

There are two general take-home messages.

(1) When one has a choice of which variables to integrate over and which to maximize over, one should integrate over as many variables as possible, in order to capture the relevant volume information. There are typically far fewer regularization constants and other hyperparameters than there are 'level 1' parameters.

(2) If practical Bayesian methods involve approximations such as fitting a Gaussian to a posterior distribution, then one should think twice before integrating out hyperparameters (Gull 1988). The probability density which results from such an integration typically has a skew peak; a Gaussian fitted at the peak may not approximate the distribution well. In

[2]Integration over the noise level $1/\beta$ to give the true likelihood leads to a bias in the other direction. These two biases may cancel: the evidence framework's $\mathbf{w}_{\mathrm{MP}|\alpha_{\mathrm{MP}},\beta_{\mathrm{MP}}}$ coincides with $\mathbf{w}_{\mathrm{MP}}$ if the number of well-determined parameters happens to obey the condition $\gamma/k = N/(N+k)$.

contrast, optimization of the hyperparameters can give a Gaussian approximation which, for predictive purposes, puts most of the probability mass in the right place.

The evidence approximation, which sets hyperparameters so as to maximize the evidence, is not intended to produce an accurate numerical approximation to the true posterior distribution over **w**; and it does not. But what matters is whether low-dimensional properties of **w** (*i.e.*, predictions) are seriously mis-calculated as a result of the evidence approximation.

The main conditions for the evidence approximation are that the data should not be grossly at variance with the likelihood and the prior, and that the number of well-determined parameters γ should be large. How large depends on the problem, but often a value as small as $\gamma \simeq 3$ is sufficient, because this means that α is determined to within a factor of e (recall $\sigma_{\log\alpha|D} \simeq \sqrt{2/\gamma}$); predictive distributions are often insensitive to changes of α of this magnitude. Thus the approximation is usually good if we have enough data to determine a few parameters.

If satisfactory conditions do not hold for the evidence approximation (*e.g.*, if γ is too small), then it should be emphasized that this would not then motivate integrating out α first. The MAP approximation is systematically inferior to the evidence approximation. It would probably be most convenient numerically to retain α as an explicit variable, and integrate it out *last* (Bryan 1990).

A final point in favour of the evidence framework is that it can be naturally extended (at least approximately) to more elaborate priors such as mixture models; it would be difficult to integrate over the mixture hyperparameters in order to evaluate the 'true prior' in these cases.

Acknowledgments

I thank Radford Neal, David R.T. Robinson, Steve Gull, and Martin Oldfield for helpful discussions, and John Skilling for invaluable contributions to the proof in section 5. I am grateful to Anton Garrett for comments on the manuscript.

References

Box, G. E. P., and Tiao, G. C. (1973) *Bayesian inference in statistical analysis.* Addison–Wesley.

Bretthorst, G. (1988) *Bayesian spectrum analysis and parameter estimation.* Springer.

Bryan, R. (1990) Solving oversampled data problems by Maximum Entropy. In *Maximum Entropy and Bayesian Methods, Dartmouth, U.S.A., 1989*, ed. by P. Fougere, pp. 221–232. Kluwer.

Buntine, W., and Weigend, A. (1991) Bayesian back–propagation. *Complex Systems* 5: 603–643.

Gull, S. F. (1988) Bayesian inductive inference and maximum entropy. In *Maximum Entropy and Bayesian Methods in Science and Engineering, vol. 1: Foundations*, ed. by G. Erickson and C. Smith, pp. 53–74, Dordrecht. Kluwer.

GULL, S. F. (1989) Developments in maximum entropy data analysis. In *Maximum Entropy and Bayesian Methods, Cambridge 1988*, ed. by J. Skilling, pp. 53–71, Dordrecht. Kluwer.

MACKAY. D. J. C. (1992a) Bayesian interpolation. *Neural Computation* **4** (3): 415–447.

MACKAY. D. J. C. (1992b) A practical Bayesian framework for backpropagation networks. *Neural Computation* **4** (3): 448–472.

MACKAY. D. J. C. (1992c) The evidence framework applied to classification networks. *Neural Computation* **4** (5): 698–714.

MACKAY. D. J. C. (1994) Bayesian non-linear modelling for the 1993 energy prediction competition. In *Maximum Entropy and Bayesian Methods, Santa Barbara 1993*, ed. by G. Heidbreder, Dordrecht. Kluwer.

NEAL, R. M. (1993a) Bayesian learning via stochastic dynamics. In *Advances in Neural Information Processing Systems 5*, ed. by C. L. Giles, S. J. Hanson, and J. D. Cowan, pp. 475–482, San Mateo, California. Morgan Kaufmann.

NEAL, R. M. (1993b) Probabilistic inference using Markov chain Monte Carlo methods. Technical Report CRG–TR–93–1, Dept. of Computer Science, University of Toronto.

REIF, F. (1965) *Fundamentals of statistical and thermal physics*. McGraw–Hill.

SKILLING. J. (1993) Bayesian numerical analysis. In *Physics and Probability*, ed. by W. T. Grandy, Jr. and P. Milonni, Cambridge. C.U.P.

STRAUSS. C. E. M., WOLPERT, D. H., and WOLF, D. R. (1993) Alpha, evidence, and the entropic prior. In *Maximum Entropy and Bayesian Methods, Paris 1992*, ed. by A. Mohammed-Djafari, Dordrecht. Kluwer.

THODBERG, H. H. (1993) Ace of Bayes: application of neural networks with pruning. Technical Report 1132 E, Danish meat research institute.

WAHBA, G. (1975) A comparison of GCV and GML for choosing the smoothing parameter in the generalized spline smoothing problem. *Numer. Math.* **24**: 383–393.

WEIR, N. (1991) Applications of maxmimum entropy techniques to HST data. In *Proceedings of the ESO/ST–ECF Data Analysis Workshop, April 1991*.

WOLPERT. D. H. (1993) On the use of evidence in neural networks. In *Advances in Neural Information Processing Systems 5*, ed. by C. L. Giles, S. J. Hanson, and J. D. Cowan, pp. 539–546, San Mateo, California. Morgan Kaufmann.

WHAT BAYES HAS TO SAY ABOUT THE EVIDENCE PROCEDURE

David H. Wolpert
Santa Fe Institute
1399 Hyde Park Road
Santa Fe, NM 87501 USA (dhw@santafe.edu)

Charlie E. M. Strauss
Los Alamos National Laboratory
Los Alamos, NM 87545 USA (cems@lanl.gov)

ABSTRACT. The "evidence" procedure for setting hyperparameters is essentially the same as the techniques of ML-II and generalized maximum likelihood. Unlike those older techniques however, the evidence procedure has been justified (and used) as an approximation to the hierarchical Bayesian calculation. We use several examples to explore the validity of this justification. Then we derive upper and (often large) lower bounds on the difference between the evidence procedure's answer and the hierarchical Bayesian answer, for many different quantities. We prescribe a simple, easy to compute, test that can check the validity of the approximation after the fact. We also touch on subjects like the close relationship between the evidence procedure and maximum likelihood, and the self-consistency of deriving priors by "first-principles" arguments that don't set the values of hyperparameters.

"... any inference must be based on strict adherence to the laws of probability theory, because any deviation automatically leads to inconsistency." - S. Gull, in [5]

"(Some have) estimated alpha from the data and then proceeded as if alpha is known. It is better to use the standard methods of Bayesian statistics and integrate out alpha." B. D. Ripley, in [13]

1. Introduction

In many statistics problems one has one or more "hyperparameters" (sometimes called "nuisance parameters") which occur in the distributions of interest but may not be of direct interest themselves. Examples are a choice of model, a noise level, a regularization constant in a regression problem, and "α" in maxent image reconstruction.

How to deal with a hyperparameter? A full Bayesian approach is to marginalize out the hyperparameter. (This is "hierarchical Bayes" - see [1, 3].) A non-Bayesian approach might set the hyperparameter to a single value, and use that value throughout the subsequent analysis. For example, one might choose the hyperparameter via maximum likelihood - choose the hyperparameter γ such that the conditional probability $P(D \mid \gamma)$ (or alternatively $P(\gamma \mid D)$) is maximized, where D is one's data. Recently it has been claimed that this kind of non-Bayesian approach is a good approximation to the full Bayesian approach whenever $P(\gamma \mid D)$ is peaked as a function of γ [9, 11]. In the context of this claim, setting γ to the value maximizing $P(\gamma \mid D)$ is known as "the evidence procedure" [9, 11, 12, 14].

G. R. Heidbreder (ed.), Maximum Entropy and Bayesian Methods, 61–78.

Even though the evidence procedure has become popular amongst some Bayesians, the validity of its claim to approximate the Bayesian approach has never been thoroughly discussed. Consequently the accuracy of the procedure as such an approximation is rarely checked or reported. Perhaps even more remarkably, for some applications the full Bayesian answer is easier to calculate and apply [16, 20, 3]. Yet many researchers jump straight to the approximation of the evidence procedure, without checking if the exact answer is tractable, or if not, if perhaps some approximation other than the evidence procedure is preferable.

In the first part of this paper we state the evidence procedure, giving both an intuitive argument that it is a good approximation and an intuitive argument that it is not. We then explore the validity of the procedure in a simple Gaussians example. In this example the procedure fails miserably for certain objects of interest, but works for others. We end with a formal discussion giving lower and upper bounds on the approximation error incurred with the evidence procedure. The bounds concern error in evaluating the posterior at a point, in evaluating the full posterior (both supremum norm and L^n norm error), in estimating the predictive distribution, and in estimating expectation values. This discussion demonstrates explicitly that the informal justifications for the evidence procedure found in the literature are inadequate. It also has implications for the self-consistency of any "first-principles" argument for a prior that does not fix all hyperparameters in that prior.

A recurring theme throughout the paper is that for many quantities of interest, the evidence procedure becomes more accurate as the object of interest becomes more dominated by the likelihood distribution. In other words, for those quantities the procedure is most accurate when the prior is irrelevant, so that there is no need for Bayesian analysis.

We emphasize that here we only analyze how well the evidence procedure approximates the full Bayesian answer. We are not concerned with whether the procedure meets non-Bayesian desiderata. (E.g., desiderata like requiring that one's answer doesn't change when additional irrelevant information is introduced, or like the desiderata in Section 6.5 of [11] that actually argue for the use of maximum likelihood in *all* contexts, not just those related to hyperparameters.) Nor do we make any claims concerning how one should use the posterior (e.g., take its mean vs. take its mode), an issue properly addressed by decision theory. Moreover, we make no claims about how well the procedure works in practice. (A procedure's being non-Bayesian does not mean it works poorly in practice.) Studies empirically comparing the evidence procedure to other methods for setting hyperparameters have given mixed results [?]he procedure in a simple Gaussians exademoment,fortier,ripley,sibisi,strauss,thompson1,thompson2,wahba,me93. However in evidence's defense we note that MacKay has recently won a prediction competition [12] by using the evidence procedure, albeit in conjunction with some new techniques like stacking [2] and the use of different regularization hyperparameters for different parts of the space.

2. What is the evidence procedure?

To illustrate the evidence procedure, consider the case where the hyperparameter parameterizes the prior distribution over the hypothesis space of vectors f. (To distinguish it from the generic hyperparameter γ, this kind of hyperparameter is indicated by α.) Some examples are the MaxEnt and Gaussian distributions: $P(f \mid \alpha) = \exp(\alpha S(f))/Z_s(\alpha)$, and $P(f \mid \alpha) \propto \alpha^{N/2}\, e^{-\alpha|\vec{f}\,|^2}$, respectively.

Write the posterior distribution as

$$P(f \mid D) = \frac{1}{P(D)} \int P(\alpha, f, D)\, d\alpha. \tag{1}$$

Multiply and divide the integrand in (1) by $P(\alpha \mid D)$:

$$P(f \mid D) \propto \int \frac{P(\alpha, f, D)}{P(\alpha \mid D)}\, P(\alpha \mid D)\, d\alpha \propto \int P(f \mid \alpha, D)\, P(\alpha \mid D)\, d\alpha. \tag{2}$$

When $P(\alpha \mid D)$ is sharply peaked about α_{ev} it's natural to treat it as a delta function about α_{ev} and collapse the last integral in (2). The idea of collapsing Bayesian integrals this way is old, going back at least as far as [6]. It forms the conventional justification for the view that the evidence procedure is an approximation to the full Bayesian approach; the evidence procedure says that

$$P(f \mid D) \approx P(f \mid \alpha_{ev}, D) \propto P(f \mid \alpha_{ev})\, P(D \mid f). \tag{3}$$

Under many circumstances (e.g., relatively flat $P(\alpha)$) this kind of reasoning also appears to support the idea of setting $P(f \mid D)$ to $P(f \mid D, \text{argmax}_\alpha P(D \mid \alpha))$, so long as $P(D \mid \alpha)$ is a peaked function of α. (In fact, this kind of reasoning appears to support setting α to the maximum of almost any distribution over α and D that is a peaked function of α.) So there is ambiguity in what peak we should set α to, i.e., in how to define α_{ev} (ambiguity that is reflected in the literature). Accordingly, when it's helpful for illustrative purposes, we will consider $P(D \mid \alpha)$ rather than $P(\alpha \mid D)$ and will take the term "evidence" to mean $P(D \mid \alpha)$ rather than (our default meaning) $P(\alpha \mid D)$.

Stripped of the context of equation (3), the idea of setting the hyperparameter to the value α_{ev} is essentially identical to the techniques of ML-II and generalized maximum likelihood [4, 1, 19]. The primary difference between the evidence procedure and those older techniques is that those older techniques do not attempt to justify themselves with the approximation in equation (3), but rather view setting $\alpha = \alpha_{ev}$ as *a priori* reasonable.

As it turns out, there are reasons to doubt the validity of equation (3). One such reason is that in general the change of variables $\alpha = \eta(\alpha')$ results in the evidence procedure returning $P(f \mid \alpha, D)$ for an α different from α_{ev}. That is, the Jacobian of the variable transformation can change the distribution's mode while still leaving it peaked. In general there will be functions η for which $P(\alpha' \mid D)$ is highly peaked about an α' which does not equal $\eta^{-1}(\alpha_{ev})$. For such an η the evidence procedure used with the hyperparameter variable α' returns a posterior distribution for f given by $P(f \mid \alpha'_{ev}, D)$ where $\alpha'_{ev} \neq \alpha_{ev}$. [22] So the answer of the evidence procedure can change under a variable transformation of the hyperparameter, whereas the true posterior can not (cf. equation (1)). This suggests that the reasoning embodied in equations (1) through (3) must be flawed. More is needed than simply having a distribution over α and D that is a sharply peaked function of α.

Another reason to doubt the accuracy of the approximation in (3) arises from considering the evidence procedure from a graphical perspective. The contour plots in figure 1 show two hypothetical $P(\alpha, f \mid D)$'s, for one-dimensional f. The projections of these distributions onto the α and f axes are $P(\alpha \mid D)$ and $P(f \mid D)$, respectively. In both plots $P(\alpha \mid D)$ is peaked, about $\alpha = \alpha_{ev}$. The evidence procedure's posterior distribution is given by the

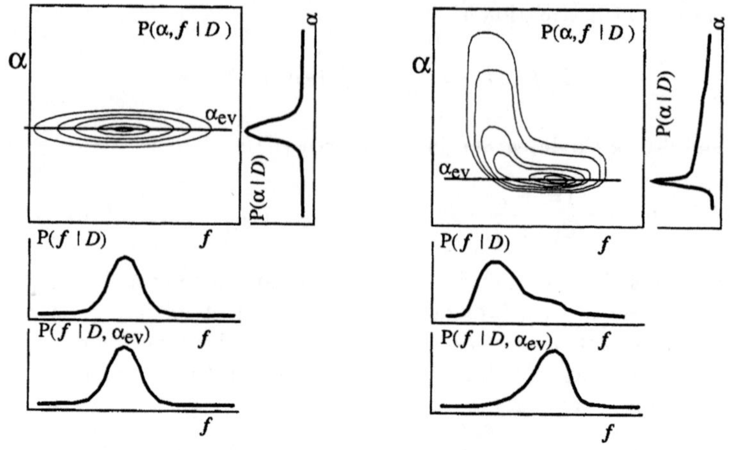

Figure 1: *Contour sketches of hypothetical $P(\alpha, f \mid D)$'s along with their projections onto the α and f axes. The bottom plots are (proportional to) slices of the distributions through $\alpha = \alpha_{ev}$. The left sketch is a success of the evidence procedure, and the right a failure. The right sketch is similar to what one would get for the Gaussian scenario discussed below.*

slice of the original distribution through $\alpha = \alpha_{ev}$. In the left plot that slice resembles the true posterior projection. But in the right plot it does not. Again we see that $P(\alpha \mid D)$'s being peaked cannot be the sole criterion for the validity of the evidence approximation.

These problems are partially due to the fact that $P(\alpha \mid D)$ appeared in the integrand in (2) only after we multiplied and divided by it. So no matter how peaked the numerator $P(\alpha \mid D)$, it is exactly canceled by the denominator $P(\alpha \mid D)$. This suggests that the function $P(f \mid \alpha, D)$ appearing in equation (2) is just as rapidly varying a function of α as $P(\alpha \mid D)$, in which case collapsing the integral at α_{ev} is unjustified.

Note though that if the α-peak of $P(\alpha, f, D)$ is close to α_{ev}, there might be a fortuitous cancellation of peaks that renders $P(f \mid \alpha, D)$ a slowly varying function of α. (See equation (2).) While it is usually difficult to check whether precise cancellation occurs, at a minimum the peaks must overlap substantially for such cancellation to be possible. (This is proven formally in Section 5.) When there is such overlap it's possible that the evidence procedure closely approximates the Bayesian answer. Ironically, whereas the intuition behind equation (3) suggests that the procedure works better for more highly peaked $P(\alpha \mid D)$, the need for that narrow peak to overlap with the peak of $P(\alpha, f, D)$ suggests that the opposite is true. (Theorem 4 below proves that that "opposite" is indeed true; the evidence procedure fails for almost all f in the regime of sufficiently peaked $P(\alpha \mid D)$.)

The previous observation offers a simple test that can be applied to one's result to check the evidence procedure. If the α-peak of $P(f, \alpha, D)$ does not overlap with the α-peak of the evidence then collapsing the integral in equation 2 to the sharp peak of the evidence is unjustified. To illustrate this we consider Gull's famous Susie reconstruction [9]. Figure 2 plots $P(\alpha \mid D)$ and $P(f, \alpha, D)$ as functions of α for the f (i.e., the image) at the peak of the evidence procedure's posterior in Gull's Susie reconstruction. The two peaks clearly do not cancel, which means the argument leading to the evidence procedure does not hold.

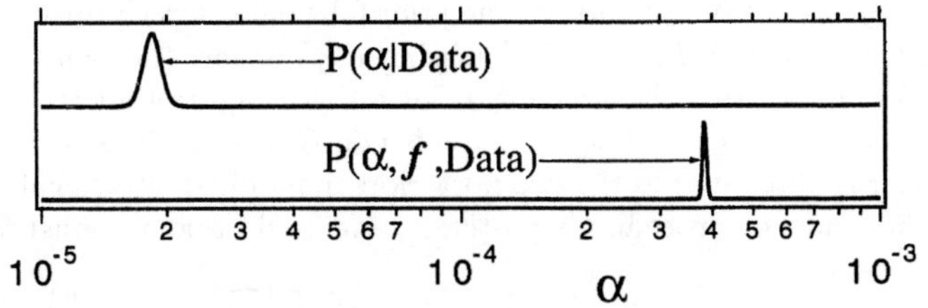

Figure 2: *A comparison of $P(\alpha \mid D)$ and $P(\alpha, f, D)$ as functions of α shows they do not overlap. The data is taken from Gull's Susie reconstruction: f here is the MAP of the evidence procedure posterior f presented in Gull's article (see text).*

It turns out that even when peaks cancel and $P(\alpha \mid D)$ is highly peaked, we still can't conclude that equation (3) is necessarily a good approximation. This is because $P(f \mid \alpha, D)$ need not be normalized over α, so the contribution to the integral from the (often very long) tails of the integrand in equation (2) can be as sizable as the contribution from around α_{ev}.

As a final example of the subtleties involved in equation (3) note that with enough hyperparameters the evidence procedure can produce a posterior that is highly peaked about the maximum likelihood f. (Nothing in the intuition behind equation (3) presumes α is low-dimensional. Indeed, some researchers have used the evidence procedure with high-dimensional α.) This follows from the equality $P(D \mid \vec{\alpha}) = \int df P(D \mid f)P(f \mid \vec{\alpha})$. This equality shows that for a sufficiently high-dimensional $\vec{\alpha}$ (i.e., sufficiently flexible $P(f \mid \alpha)$), to find the α maximizing $P(D \mid \vec{\alpha})$ one simply finds the α for which $P(f \mid \vec{\alpha})$ is highly peaked about the maximum likelihood f (i.e., about the mode of $P(D \mid f)$). Consequently, for that α, $P(f \mid D, \alpha)$ is also highly peaked about the maximum likelihood f.

3. The Gaussian distributions case

In this section we will focus on a particular example in which both the likelihood and the conditional prior distribution are Gaussians. For simplicity the likelihood does not involve convolutions. The prior is centered on the origin and the likelihood is centered at a point D all of whose components have equal magnitude d. (These restrictions entail no loss of generality due to the translational and rotational invariance of Gaussians). Accordingly, with N the dimension of f, the likelihood and (conditional) prior are given by

$$P(D \mid f) \propto \beta^{N/2} e^{-\beta|\vec{f} - D|^2}, \text{ and } P(f \mid \alpha) \propto \alpha^{N/2} e^{-\alpha|\vec{f}|^2} \tag{4}$$

To agree with common usage, we will take the prior over α to equal $1/\alpha$ from α_{min} to α_{max} and zero elsewhere. We will be interested in the common case where α_{min} is very close to zero. Since our analysis won't depend on the exact value of α_{min} (the primary effect of that value is to set the overall normalization), here we will set it equal to 0. Also, for this section, we will treat α_{ev} as though it equaled $\text{argmax}_\alpha P(D \mid \alpha)$. It is straightforward to redo the analysis under different restrictions.

Evaluating $\int d\alpha P(f, \alpha)$ gives $P(f)$ in terms of the incomplete gamma function:

$$\begin{aligned} P(f) &\propto \frac{1}{|f|^N} \Gamma\left((N/2), \alpha_{\max}|f|^2\right) \\ &\approx \frac{1}{|f|^N} \text{ when } \alpha_{\max}|f|^2 \gg N/2. \end{aligned} \tag{5}$$

Note that for f away from the origin, the prior falls off as a reciprocal power of distance from the origin; even though $P(f \mid \alpha)$ is Gaussian $P(f)$ is not. (See Theorem 1 below for a proof of the generality of this phenomenon.) Since the true posterior is proportional to the product of the prior with the likelihood, it too is non-Gaussian. However the evidence procedure's posterior is Gaussian, so the two posteriors must differ. To calculate the difference we must find the evidence procedure's posterior, and to do that we must first evaluate

$$P(D \mid \alpha) = \int df P(f, D \mid \alpha) \propto \left[\sqrt{\frac{\alpha\beta}{\alpha+\beta}} e^{-\frac{\alpha\beta}{\alpha+\beta}d^2}\right]^N. \quad (6)$$

We can solve for the peak of this distribution, α_{ev}:

$$\alpha_{ev} = \frac{\beta}{2\beta d^2 - 1}. \quad (7)$$

So the evidence procedure's posterior is a Gaussian centered between the peaks of the prior and likelihood (i.e., between $f = 0$ and $f = d$):

$$P(f \mid D, \alpha_{ev}) \propto (\alpha_{ev}\beta)^{N/2} e^{-\beta|f - D|^2 - \alpha_{ev}|f|^2} \propto e^{-(\beta+\alpha_{ev})\left|f - \frac{\beta}{\alpha_{ev}+\beta}D\right|^2}. \quad (8)$$

Note that d is the distance along any coordinate separating the peaks of the prior and the likelihood. Therefore $2\beta d^2$ is the separation between the peaks measured in units of the likelihood's width. But equation (7) only has a meaningful solution if $2\beta d^2 > 1$; unless the peaks are separated by more than the width of likelihood, there isn't a peak in the evidence. In this sense the evidence procedure is not even well-defined unless the data are unexpected. (We use the term "unexpected" a bit loosely here; more formally - and laboriously - one could analyze how "unexpected" the data are by considering the width of the prior predictive distribution rather than the width of the likelihood.) Moreover, as the separation increases beyond two widths, so that $2\beta d^2 > 2$, the value α_{ev} becomes smaller than β. Yet as α_{ev} shrinks below β the evidence procedure's approximation to the posterior approaches the likelihood distribution. So as we pass the condition allowing the evidence procedure to be well-defined, the data become more unexpected, and the evidence procedure produces a posterior which increasingly approximates the likelihood.

We can apply the test from Section 2. The α-width of $P(f, \alpha, D)$ can be estimated from its curvature at the peak as $\Delta\alpha_{joint} \sim \frac{4\alpha_{joint}}{\sqrt{N}}$, where α_{joint} is the peak position. Appyling the test at the peak value of f from the evidence procedure's posterior we discover, surprisingly, that the peaks, α_{joint} and α_{ev}, only lie within the half width, $\frac{\Delta\alpha_{joint}}{2}$, of each other when $\alpha_{ev} \lesssim \frac{2\beta}{\sqrt{N}}$. That is, in general, the α-peaks of the joint probability distribution and the evidence procedure won't overlap as required for the evidence procedure's approximation to be justified. Moreover, when they do overlap, the posterior is solidly in the likelihood dominated regime (for large N). Section 5 discusses the overlap criteria formally.

These and related effects are illustrated in figure 3. Since the evidence approximated posterior is a symmetric Gaussian it is fully characterized by any single one-dimensional slice through its peak. This is not the case with the true posterior unfortunately, since that posterior is not symmetric about its peak. Nonetheless, we can learn a lot about the true posterior by looking at a slice through it going from the origin out along the D direction in

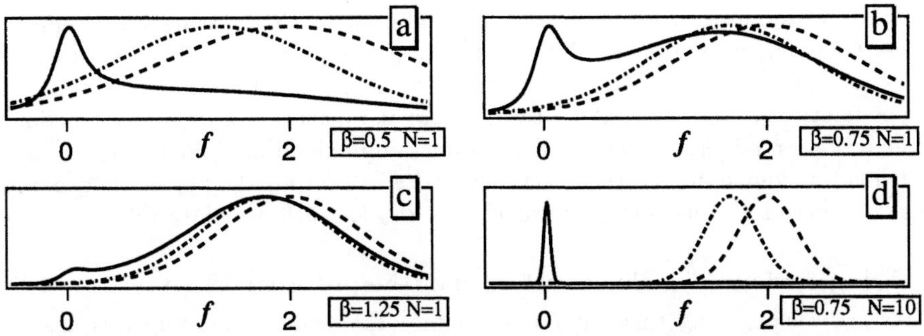

Figure 3: **Solid line**: *True posterior,* $P(f \mid D)$*;* **Dot-Dash**: *Evidence procedure's posterior* $P(f \mid \alpha_{ev}, D)$*;* **Dashed**: *Likelihood* $P(D \mid f)$*. Going from figure (a) through (c), there is increasing distance (i.e., increasing* $2\beta D^2$*) between the peaks of the prior and the likelihood. For* $2\beta d^2 < 1, \alpha_{ev}$ *is undefined. Figure (d) increases the dimension from N = 1 to N = 10; the mismatch between the distributions becomes worse.* $(\alpha_{max} = 100, d = 2)$

f space. Figure 3 shows this slice and the corresponding slice of the evidence procedure's posterior for various separations, i.e., various values of $2\beta d^2$. The likelihood is also shown. The plots for other slice directions exhibit similar behavior.

These plots show that the evidence and true posteriors have different symmetries, peak positions and widths. Moreover the true posterior can have two peaks whereas the evidence procedure's posterior only has one, and the true posterior tends to have (sometimes much) more of its probability "mass" near the origin. Also note that the neither the peak position nor peak widths of the two distributions approach one another until the distributions start to converge on the likelihood - at which point the true posterior is about as well approximated by the likelihood as it is by the evidence procedure's posterior.

For large enough α_{max} and α_{min} close to 0, as N increases the peaks of the true posterior and of the evidence procedure's posterior don't move, nor does the position of the peak of the evidence move. But all those distributions—and in particular the plots in figure 3—become sharper (cf. equations (4, 5, and 8), and compare figures 3b and 3d). (Due to this sharpening of peaks the plots for high N values aren't very informative; this is why the plots are for low N values even though the evidence isn't very peaked for low N values.) So as N increase, the evidence becomes more peaked. But at the same time the discrepancy between the true posterior and the evidence procedure's posterior gets *worse*, not better.

Given all this, it seems fair to say that the evidence procedure's posterior is a poor representation of the true posterior—except for in the case when the prior doesn't matter (i.e., when things are likelihood dominated). Nonetheless, in some circumstances, the evidence procedure's posterior could provide a good approximation for calculating low-dimensional expectation values. This will occur if erroneous behavior in the tails of the distribution "compensates" for erroneous behavior in the central regions. (See Section 4 below.)

Finally, we point out that it is a simple matter to calculate the true prior (and therefore the posterior) not only when the conditioned prior is Gaussian, but also when it is entropic (see equation (5) and [16]). Moreover, for both scenarios one can often directly approximate the exact posterior with a convenient form. Equation (5) presents an example of this for the Gaussian prior case, and for the entropic prior such a direct approximation is $P(f) \sim 1/S(f)^{N/2}$, where S is the entropy (see [16]). Nonetheless, one can not rule out the possibility that there might be cases where the evidence procedure's functional form for the posterior is more convenient than "direct approximations" for the posterior. On the other hand of course, unlike the exact calculation's form for the posterior, generating the evidence procedure's form entails recalculating α_{ev} for each new data set.

4. Using evidence for things other than the posterior

Interestingly enough, all this doesn't mean that the evidence procedure is useless. This is because even though it gets the posterior wrong, *when certain conditions are met* the evidence procedure's approximation for low-dimensional expectation values can be excellent.

As an example, consider the posterior expected value of a function $g(f)$: $\langle g \rangle \equiv \int d\alpha \int df\; g(f)\; P(f, \alpha \mid D)$. Suppose that g is a simple function of a single coordinate f_j, and that $P(f, D \mid \alpha)$ factors as $\Pi_{k=1}^{N} P(f_k, D_k \mid \alpha)$ (as it does in our Gaussians example). Then by equation (2),

$$\langle g \rangle = \int_{\alpha_{\min}}^{\alpha_{\max}} d\alpha \int df g(f_j) \frac{P(f, \alpha \mid D)}{P(\alpha \mid D)} P(\alpha \mid D).$$

Cancelling terms between the numerator $P(f, \alpha \mid D)$ and the denominator $P(\alpha \mid D) = \int df P(f, \alpha \mid D)$ (recall the assumption that $P(f, D \mid \alpha)$ factors), we see that

$$\langle g \rangle = \int_{\alpha_{\min}}^{\alpha_{\max}} d\alpha P(\alpha \mid D) R(\alpha) \tag{9}$$

where $R(\alpha) \equiv \frac{\int df_j g(f_j) P(f_j, D_j \mid \alpha)}{\int df_j P(f_j, D_j \mid \alpha)} = \int df_j g(f_j) P(f_j \mid D_j, \alpha)$.

Equations (9) and (2) have the same form, except that in equation (9) the ratio occurring in the integrand ($R(\alpha)$) only involves one-dimensional quantities. As a result, often equation (9) does not give us the same difficulty that equation (2) did; since in equation (9) the denominator of the ratio is a one-dimensional integral, it is often not strongly peaked, so to have the ratio be smooth on the scale of the peak of the evidence does not require that the numerator of that ratio be strongly peaked, as it did in equation (2). So as long as: α_{max} is not too large (so that the tails don't contribute much); $R(\alpha)$ is not a rapidly varying function (a condition often met for simple expectation values like the mean); and $P(\alpha \mid D)$ is a highly peaked function of α (cf. equation (6)); then calculating the expected g by collapsing the integral over α down to the peak of $P(\alpha \mid D)$ might be justified.

$R(\alpha)$ and $P(\alpha \mid D)$ for the Gaussians case are sketched in figure 4 for $g(f) = f$ (so $\langle g \rangle$ is the posterior average f). To highlight the important aspects of the plot, $P(\alpha)$ is flat between 0 and α_{max} rather than Jeffreys. These plots show that slowly-varying $R(\alpha)$ and peaked $P(\alpha \mid D)$ are not uncommon, provided one has appropriate choices of α_{max} and the like. (Note that this is not the behavior of all the plots however.) So in some circumstances the evidence procedure can accurately estimate low-dimensional expectation values even if it poorly approximates the (high-dimensional) posterior distribution. To help understand this in light of the preceding discussion, note that $P(\alpha \mid D)$ is usually only

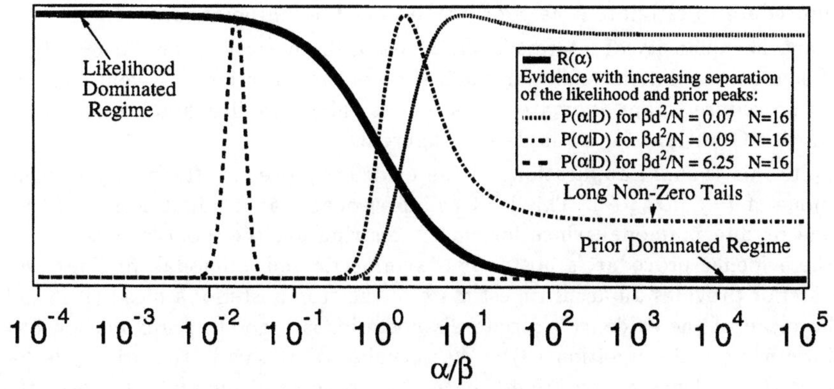

Figure 4: *$R(\alpha)$ makes a smooth transition from the prior-dominated to the likelihood-dominated regime. It is weighted by $P(\alpha \mid D)$ in the integral giving $\langle g \rangle$. The long tails of $P(\alpha \mid D)$ can outweigh the peak of $P(\alpha \mid D)$ in the integral, particularly when that peak lies beyond the crossover point from the likelihood-dominated regime.*

highly peaked on the likelihood-dominated side of the midpoint in $R(\alpha)$. And of course in the likelihood-dominated regime we are free to introduce some error into the prior.

Of course, all of this depends on the tails in figure 4 being relatively unimportant, which usually holds only if α_{max} is not too large. For example, in the Gaussians case, for large enough α_{max} the tails of $P(\alpha \mid D)$ will provide more weight in the integral over α than the peak does. (Note the logarithmic scale of the α/β-axis that "compresses" the tails.) In such a situation, we are not justified in "collapsing the integral down to the peak", and the evidence's procedure's approximation for the expectation value is poor.

Unfortunately though, there is a lot of confusion about how to choose α_{max}. In particular, while a large α_{max} does indeed result in a less informative $P(\alpha)$, it results in a *more* informative $P(f)$. This is because the larger α_{max} is, the narrower $P(f)$ becomes. (Similar "conjugate" behavior in a different context has been discussed by Jaynes [10].) This is a special example of the following more general rule: if one knows the physical meaning of a hyperparameter, then one can set the prior over it directly, without concern for how that prior affects $P(f)$. However if the hyperparameter has no physical meaning, and if one sets the prior over it without taking into account how that prior affects $P(f)$, then one is introducing (usually fictitious) prior "knowledge" concerning the ultimate object of interest, f. This problem is particularly pronounced if $P(f \mid \alpha)$ is somewhat ad hoc, like in the case of neural nets, where f is an input-output mapping, and $P(\alpha)$ only sets $P(f)$ indirectly, by means of an intermediate distribution over "weight vectors" [21].

There are many other quantities of interest in addition to the posterior and its low-dimensional marginalizations. Two such quantities are the posterior over a single coordinate (i.e., $P(f_i \mid D)$) and the predictive distribution for new data given old data (i.e., $P(\text{new data set} = D' \mid D)$). Since the posterior over a single coordinate is a low-dimensional marginalization of the full posterior, we expect the evidence procedure to estimate it accurately when it estimates other low-dimensional marginalizations well. On the other hand, the predictive distribution is a high-dimensional object, and therefore we expect the evidence procedure to estimate it as poorly as it does the full posterior.

Yet another quantity of interest is the mode of the posterior, the "MAP" f. Since the MAP f is not a low-dimensional marginalization of the posterior, one would not expect the evidence procedure to approximate it well unless things are likelihood dominated. This is the case with Gaussians for example - see figure 3.

Despite this though, applications of the evidence procedure frequently concentrate on the f-mode of $P(f \mid \alpha_{ev}, D)$. This isn't as unreasonable as it might seem if $P(f \mid \alpha_{ev}, D)$ is symmetric and unimodal, since for such a distribution the mode equals the mean. So when the evidence procedure's posterior is symmetric and unimodal, finding the mode of that posterior provides an accurate estimate of the true posterior's mean (if it so happens that the mean of the evidence procedure's posterior is a good approximation of the true posterior's mean - cf. equation (9)). We speculate that this is the origin of the cryptic claim that the evidence procedure estimates "where most of the mass is" correctly.

So in these symmetric and unimodal circumstances it is indeed sensible to concentrate on the mode of the evidence procedure's posterior. However when the evidence procedure's posterior is either asymmetric or multimodal, the peak of the procedure's posterior does not equal its mean. For such cases the mode of the procedure's posterior has no special significance, and there is no reason to concentrate on that mode. In particular, this problem affects use of the evidence procedure with the entropic prior, and with (highly multi-modal) neural nets. Ironically, these are two situations in which it happens to be particularly common for researchers to concentrate on modes of the evidence procedure's posterior.

As a final example of a quantity of interest, note that in many applications one is more concerned with unusual events than with likely events. (For example, a battleship's captain might not be interested in a "typical" reconstruction of a radar-image, but rather in the probability that that image was created by an approaching periscope.) In such a case we are interested in the behavior in the tails of the probability distribution. However in general there is no reason to believe that the evidence procedure the ratio of the true posterior to the evidence procedure's posterior goes to infinity in the tails of f (cf. equations (5, 8)). In the final analysis, whether or not a particular use of the evidence procedure is sound depends on what one wants to know (which in turn is determined by one's loss function).

5. Formal bounds on evidence's error

This section presents a formal analysis of upper and lower bounds on the error incurred by using the evidence procedure. (Some of these results correct deficiencies in the results reported in [20].) In most of this analysis we will not restrict attention to hyperparameters which occur in the conditional prior, so we denote hyperparameters by γ rather than α. Also, although most of this analysis goes through essentially unchanged when γ is multi-dimensional, for simplicity only the one-dimensional γ case is presented here.

This section is organized as follows. First it is proven that $P(f)$ can not be of the form $P(f \mid \gamma = \kappa)$ for some constant κ (i.e., marginalizing out a hyperparameter can never be equivalent to setting it to some particular value). It is argued that this means that "first-principles" arguments for a prior which don't set the value of the hyperparameter are not self-consistent. It also means that the evidence procedure will *always* have some error.

Next the reasoning of Section 2 is formalized to derive an upper bound on the error of the evidence procedure. Like many of the other results presented in this section, this upper bound applies to a wide variety of possible uses of the evidence procedure.

Then it is shown that the separation between the γ-peaks of $P(f, \gamma \mid D)$ and $P(\gamma \mid D)$ must be small or the evidence procedure's error will be large (cf. the discussion of "fortuitous cancellation of peaks" near the end of Section 2). This is done by both showing that the upper bound on the error increases with that separation, and then by deriving a lower bound on the error which increases with that separation. So by measuring the separation one can test the evidence procedure. In addition, the lower bound can be used to show that when $P(\gamma \mid D)$ is highly peaked—exactly the situation which traditionally was thought to justify the evidence procedure—the evidence procedure can give an accurate estimate of the entire posterior $P(f \mid D)$ only if that posterior is likelihood-dominated. Finally, we discuss how well the evidence procedure performs when one uses error measures like the L^n difference between the correct posterior and the evidence procedure's guess for that posterior.

5.1. The evidence procedure never gets the posterior right

We start with a proof that for a broad class of $P(f \mid \gamma)$'s, there is no non-pathological scenario for which the evidence procedure's approximation to $P(f)$ is correct:

Theorem 1: Assume that for those γ for which it does not equal zero, $P(f \mid \gamma) \propto e^{-\gamma U(f)}$ for some function $U(.)$. Then the only way that one can have $P(f) \propto e^{-\kappa U(f)}$ for some constant κ is if $P(\gamma) = 0$ for all $\gamma \neq \kappa$.

Proof: *Our proposed equality is* $e^{-\kappa U} = \int d\gamma T(\gamma) \times e^{-\gamma U}$, *where the integration limits are implicitly restricted to the region where* $P(f \mid \gamma) \neq 0$, *and where* $T(\gamma) \equiv P(\gamma) \times \int df e^{-\kappa U(f)} / \int df e^{-\gamma U(f)}$. *(Note that for both* $P(f)$ *and* $P(f \mid \gamma)$ *to be properly defined, both integrals in the definition of* $T(.)$ *must be greater than zero and finite.) We must find a* κ *and* $T(\gamma)$ *such that this equality holds for all realizable values of* U. *Let* u *be such a realizable value of* U. *Take the derivative with respect to* U *of both sides of the proposed equality* t *times, and evaluate for* $U = u$. *The result is* $\kappa^t = \int d\gamma (\gamma)^t \times R(\gamma)$ *for any integer* $t \geq 0$, *where* $R(\gamma) = T(\gamma) \times e^{u(\kappa - \gamma)}$. *Therefore* $\int d\gamma (\gamma - \kappa)^2 \times R(\gamma)) = 0$. *Since both* $R(\gamma)$ *and* $(\gamma - \kappa)^2$ *are nowhere negative, this means that for all* γ *for which* $(\gamma - \kappa)^2 \neq 0$, $R(\gamma)$ *must equal zero. Therefore* $P(\gamma)$ *must equal zero for all* $\gamma \neq \kappa$. QED.

Theorem 1 has two important consequences. First, consider any "first principles" argument which says that the prior over f is proportional to $K(f)e^{-\gamma U(f)}$ for some $U(.)$ and $K(.)$ but does not fix γ. Our ignorance concerning γ implies a non-delta function distribution $P(\gamma)$. By Theorem 1, such a distribution ensures that $P(f)$ is not proportional to $K(f)e^{-\kappa U(f)}$ for some κ. So in a certain sense, such a "first-principles" argument for a prior is not self-consistent. In particular, the first principles arguments which have been offered in favor of the so-called "entropic prior" but which do not fix γ (e.g., (Skilling 1989)) suffer from this problem. As another example, with $U(f) = -\log[V(f)]$, Theorem 1 implies that a Dirichlet prior with an unspecified exponent (i.e., a non-delta function $P(\gamma)$) is not a Dirichlet prior. (A similar point is made in [10].)

Second, if the likelihood is nowhere-zero, Theorem 1 says that there is a non-zero lower bound on the error of using evidence to set the posterior. The only question is how low the bound is. To address this make the definition $P(f \mid D) = P(D \mid f)[P_E(f) + Er(f)] \,/\, P(D)$, where "$P_E(f)$" means the evidence procedure's approximation to $P(f)$. So if $P(D) \simeq P_E(D)$, the error in the evidence procedure's estimate for the posterior equals $P(D \mid f) \times Er(f)/P(D)$. Therefore we can have arbitrarily large $Er(f)$ for a particular f and not introduce sizable error into the posterior of that f, but only if the likelihood is small for that f. As D varies, the set of those f whose likelihood is not small varies. And as such a set of f varies, the γ (if there is one) such that for those f $P(f \mid \gamma)$ is a good approximation to $P(f)$ varies. When it works, the $\gamma(D)$ returned by the evidence procedure reflects this changing of γ with D.

5.2. Upper bounds on evidence's error

In general though, one needn't use the evidence procedure to estimate a posterior, but might instead use it for other purposes (see Section 4). To circumvent the issue of how the posterior gets used, we will examine the evidence procedure's error as an estimator of an expectation value $\int df' A(f') \times P(f' \mid D)$, where f' is a dummy f variable, and $A(.)$ is determined by the use we have in mind for the posterior. For example, $A(f') = f'$ if we're interested in the posterior average f. If we're interested in the posterior directly, then $A(f') = A(f, f') = \delta(f - f')$, and expected A is a function of f as well as f'. As a final example, if we're interested in the predictive distribution, then $A(f') = P(\text{new data set} = D' \mid f')$, and A is a function of D' as well as f'.

To analyze such expectation values, let expressions of the form "$E_f(A \ldots \text{stuff})$" mean $\int df' A(f') \times P(f' \ldots \text{stuff})$, where "stuff" can involve f', conditional bars, or whatever; E_f expectation values are over f alone. So for example $E_f(A \mid D) \equiv \int df' A(f') \times P(f' \mid D)$, and $E_f(A, \gamma \mid D) \equiv \int df' A(f') \times P(f', \gamma \mid D)$. (This is slightly non-standard use of the "$E(.)$" notation.) Also, take expressions like "$P(\gamma^* + \delta, \ldots)$" to be shorthand for "$P(\gamma = \gamma^* + \delta, \ldots)$".

The intuition for when the evidence procedure works for expectation values is analogous to the intuition for posteriors; the posteriors intuition is based on equation (2), and the expectation values intuition is based on the very similar equation

$$E_f(A \mid D) = \int d\gamma \frac{E_f(A, \gamma \mid D)}{P(\gamma \mid D)} P(\gamma \mid D) \propto \int d\gamma E_f(A \mid \gamma, D) P(\gamma \mid D). \tag{10}$$

Just like equation (3), equation (10) suggests (!) that if $P(\gamma \mid D)$ is sharply peaked about γ^* and $E_f(A \mid \gamma, D)$ is slowly varying, then $E_f(A \mid D) \simeq E_f(A \mid \gamma^*, D)$.

We now present several theorems which formalize this intuitive reasoning. These theorems give upper and lower bounds on the error induced by using the evidence procedure. In these theorems we never need to specify $A(.)$. In addition, we don't need to assume anything special about the probability distributions, e.g., that they're linear Gaussian models.

We will consider three properties:

1) How sharp the γ-peak of $P(\gamma \mid D)$ is.
2) How much $E_f(A \mid \gamma, D) = E_f(A, \gamma \mid D)/P(\gamma \mid D)$ varies around that peak of $P(\gamma \mid D)$. (This provides the scale for measuring the peakedness of $P(\gamma \mid D)$.)
3) How $E_f(A, \gamma \mid D)$ behaves for γ significantly far from that peak of $P(\gamma \mid D)$.

(This - not peakedness of $P(\gamma \mid D)$ - determines if we are justified in ignoring the tails in our integrals.)

Formally, first choose a γ^* and a $\delta > 0$. In practice these will usually serve as the peak position and peak width of $P(\gamma \mid D)$ respectively, and we will loosely refer to them as such. (Note though that we make no such stipulations in their definitions, and the theorems presented below don't rely on their serving those functions.)

Our first two definitions characterize the "peakedness" of $P(\gamma \mid D)$; the smaller λ and/or ρ, the more "peaked" the distribution.

$\lambda \equiv \max\,[\frac{P(\gamma^*+\delta)|D)}{P(\gamma^*|D)}, \frac{P(\gamma^*-\delta|D)}{P(\gamma^*|D)}]$;

We will say "condition (i) holds" if λ is small. It is usually assumed that $\lambda < 1$.

$\rho \equiv 1 - \int_{\gamma^*-\delta}^{\gamma^*+\delta} d\gamma P(\gamma \mid D)$;

We will say "condition (i') holds" if ρ is small.

Our next definition characterizes how slowly varying $E_f(A \mid \gamma, D)$ is across the peak; the smaller τ, the more slowly varying $E_f(A \mid \gamma, D)$ is across $[\gamma^* - \delta, \gamma^* + \delta]$.

$\tau \equiv \max |E_f(A \mid \gamma, D) - E_f(A \mid \gamma^*, D)|$ across $\gamma \in [\gamma^* - \delta, \gamma^* + \delta]$;

We will say "condition (ii) holds" if τ is small.

Our next two definitions characterize how much tails over γ matter; the smaller ϵ and/or B, the less those tails matter.

$\epsilon \equiv |E_f(A \mid D) - \int_{\gamma^*-\delta}^{\gamma^*+\delta} d\gamma E_f(A, \gamma \mid D)|$;

ϵ is the contribution to $E_f(A \mid D)$ arising from $E_f(A, \gamma \mid D)$ lying outside $[\gamma^* - \delta, \gamma^* + \delta]$.

We will say "condition (iii) holds" if ϵ is small.

$B \equiv \max |E_f(A \mid \gamma, D)|$ across $\gamma \notin [\gamma^* - \delta, \gamma^* + \delta]$;

B measures how big $E_f(A \mid \gamma, D)$ can get when γ is outside of $[\gamma^* - \delta, \gamma^* + \delta]$;

We will say "condition (iv) holds" if B is not too large.

"Evidence's error" is the magnitude of the difference between the full Bayesian answer and the evidence procedure's answer: $|E_f(A \mid D) - E_f(A \mid \gamma^*, D)|$. We will say that "evidence works" if evidence's error is small.

We can now formalize the intuition for when evidence works by writing down an upper bound on evidence's error:

Theorem 2: Evidence's error $\leq \epsilon + \tau(1 - \rho) + E_f(A \mid \gamma^*, D) \times |\rho|$.

Proof: $|E_f(A \mid D) - \int_{\gamma^*-\delta}^{\gamma^*+\delta} d\gamma[E_f(A \mid \gamma, D) \times P(\gamma \mid D)]| = \epsilon$, *by definition of* ϵ. *By the definition of* τ, $|\int_{\gamma^*-\delta}^{\gamma^*+\delta} d\gamma[E_f(A \mid \gamma, D)P(\gamma \mid D)] - E_f(A \mid \gamma^*, D)\int_{\gamma^*-\delta}^{\gamma^*+\delta} d\gamma P(\gamma \mid D)| \leq \tau \int_{\gamma^*-\delta}^{\gamma^*+\delta} d\gamma P(\gamma \mid D)$. *Combining*, $|E_f(A \mid D) - E_f(A \mid \gamma^*, D)\int_{\gamma^*-\delta}^{\gamma^*+\delta} d\gamma P(\gamma \mid D)| \leq \epsilon + \tau \int_{\gamma^*-\delta}^{\gamma^*+\delta} d\gamma P(\gamma \mid D)$. *Therefore* $E_f(A \mid \gamma^*, D) - E_f(A \mid D) \leq \epsilon + \tau(1-\rho) + E_f(A \mid \gamma^*, D) \times \rho$. QED.

One can find some sufficiency conditions for evidence to work in the literature. These are specific to certain kinds of distributions, and are derived by evaluating the evidence procedure's answer and the exact answer and seeing if the two differ. Of course, if you can evaluate the exact answer, there's no need for an approximation like the evidence procedure in the first place. In contrast, Theorem 2 provides us with some sets of sufficiency conditions which don't rely on evaluating the exact answer.

For example, if conditions (i'), (ii) and (iii) hold, and $E_f(A \mid \gamma^*, D)$ is not too large, then Theorem 2 tells us that evidence's error is small. (We have no guarantees that it's easy to evaluate whether those conditions hold, of course.) Intuitively, condition (iii) is what lets us restrict attention to the region immediately surrounding the peak of $P(\gamma \mid D)$. Condition (ii) then tells us that $E_f(A \mid \gamma, D)$ doesn't vary across that region, and can therefore be evaluated at $\gamma = \gamma^*$ and pulled out of the integral. The overall error introduced by the value of that remaining integral is reflected in the $E_f(A \mid \gamma^*, D) \times |\rho|$ term.

Note that this remaining error can be minimized either by having a sharp peak (ρ small) or by having $E_f(A \mid \gamma^*, D)$ - the guess of the evidence procedure - be close to zero. So we don't need to have condition (i') hold (i.e., have $P(\gamma \mid D)$ peaked) for evidence to work. (There are a number of other situations in which the evidence procedure can be justified even though $P(\gamma \mid D)$ is not peaked; see [22].) On the other hand, in Section 1 we saw that peaked $P(\gamma \mid D)$ does not guarantee the accuracy of the evidence procedure. Summarizing, the evidence procedure sometimes works even when $P(\gamma \mid D)$ isn't peaked, and there are also circumstances for which it doesn't work despite $P(\gamma \mid D)$'s being peaked.

All of this notwithstanding, when evidence works in practice usually condition (iii) is met by having ρ small, with $E_f(A \mid \gamma, D)$ staying reasonably bounded for γ outside of $[\gamma^* - \delta, \gamma^* + \delta]$. Formally, $\epsilon \leq B \times \rho$, so that conditions (i') and (iv) give condition (iii). In such scenarios, peakedness of $P(\gamma \mid D)$ does go hand in hand with evidence working.

5.3. Lower bounds on evidence's error

We now turn to the issue of lower bounds on the error of the evidence procedure. Intuitively, one might think that since γ^* is the "dominant contributing γ", the evidence procedure should work for peaked $P(\gamma \mid D)$ in general. The problem is that one can just as easily argue that the "dominant contributing γ" *for what we are interested in* (namely $E_f(A \mid D)$) is given by $\text{argmax}_\gamma E_f(A, \gamma \mid D)$, not $\text{argmax}_\gamma P(\gamma \mid D)$. After all, $E_f(A \mid D)$ is the γ-integral of $E_f(A, \gamma \mid D)$, not of $P(\gamma \mid D)$. This suggests that for evidence to work, γ^* must (nearly) maximize $E_f(A, \gamma \mid D)$.

Indeed, recall that the intuitive justification of the evidence procedure outlined in equation (10) required that the peaks of $E_f(A, \gamma \mid D)$ and $P(\gamma \mid D)$ nearly coincide, lest τ be too large. This reasoning is formalized in the following theorem, which provides a lower bound on τ based on the peak separation, and which uses the λ measure of peakedness.

Theorem 3: If $E_f(A, \gamma, D)$ does not have a γ-peak somewhere within δ of γ^*, then $\tau \geq E_f(A \mid \gamma^*, D)(1 - \lambda) / \lambda$.

Proof: *By hypothesis $E_f(A, \gamma^*, D)$ has no local maximum in $(\gamma^* - \delta, \gamma^* + \delta)$. Therefore we can't have both $E_f(A, \gamma^* - \delta, D)$ and $E_f(A, \gamma^* + \delta, D)$ less than $E_f(A, \gamma^*, D)$. Without loss of generality, assume $E_f(A, \gamma^*, D) \leq E_f(A, \gamma^* + \delta, D)$. Now examine the ratio of expectation values $E_f(A \mid \gamma^* + \delta, D)/E_f(A \mid \gamma^*, D)$, which we can write as the product of ratios $[P(\gamma^* \mid D)/P(\gamma^* + \delta \mid D)] \times [E_f(A, \gamma^* + \delta, D)/E_f(A, \gamma^*, D)]$. By our assumption, the second term in square brackets ≥ 1. However by definition of λ, the first term in square brackets $\geq 1/\lambda$. Therefore $E_f(A \mid \gamma^* + \delta, D) \geq E_f(A \mid \gamma^*, D)/\lambda$, and the difference $E_f(A \mid \gamma^* + \delta, D) - E_f(A \mid \gamma^*, D) \geq E_f(A \mid \gamma^*, D) \times (\lambda^{-1} - 1)$. Using the definition of τ, this means that $E_f(A \mid \gamma^*, D) \times (\lambda^{-1} - 1) \leq \tau$.* QED.

In terms of equation (1), large τ means that around $\gamma = \gamma^*$, $E_f(A \mid \gamma, D)$ is *not* slowly varying on the scale of the width of the peak of $P(\gamma \mid D)$. Recall though that if τ is large, then the intuition behind the evidence procedure—that $P(\gamma \mid D)$ "picks out" $E_f(A \mid \gamma, D)$ evaluated at $\gamma = \gamma^*$—is faulty. Formally, if τ is large Theorem 2 gives a weak upper bound. And by Theorem 3 τ is always large if we have a wide separation between our peaks.

In fact, we can use distance between the peaks to give a *lower* bound on evidence's error, to go with the upper bound of Theorem 2. To do this, define Γ as the magnitude of the distance between γ^* and that γ-maximum of $E_f(A, \gamma, D)$ which lies closest to γ^*.

Theorem 4: If $E_f(A, \gamma \mid D)$ is non-negative for all γ, it follows that evidence's error $\geq E_f(A \mid \gamma^*, D) \times [\Gamma P(\gamma^* \mid D) - 1]$. Equivalently, it follows that evidence's error $\geq E_f(A \mid D) \times [1 - (1 / \Gamma P(\gamma^* \mid D))]$.

Proof: *Since evidence's error is non-negative, if $\Gamma = 0$, the theorem trivially holds. If $\Gamma > 0$, γ^* isn't a maximum of $E_f(A, \gamma, D)$. Accordingly, $E_f(A, \gamma, D)$ must either grow as γ increases past γ^* or as it decreases below γ^*. ("Grow" here is taken to mean "stays level or rises".) Without loss of generality assume it grows as γ increases past γ^*. Then the soonest it could stop growing is at $\gamma = \gamma^* + \Gamma$. Therefore $\int_{\gamma^*}^{\gamma^*+\Gamma} d\gamma E_f(A, \gamma, D) \geq \Gamma E_f(A, \gamma^*, D)$, which implies that $\int_{\gamma^*}^{\gamma^*+\Gamma} d\gamma E_f(A, \gamma \mid D) \geq \Gamma E_f(A, \gamma^* \mid D)$. Recall our hypothesis that $E_f(A, \gamma \mid D)$ is non-negative, which implies that $E_f(A \mid D) = \int d\gamma E_f(A, \gamma \mid D) \geq \int_{\gamma^*}^{\gamma^*+\Gamma} d\gamma E_f(A, \gamma \mid D)$; $E_f(A \mid D) \geq \Gamma E_f(A, \gamma^* \mid D)$. So $E_f(A \mid D) - E_f(A \mid \gamma^*, D) \geq E_f(A \mid \gamma^*, D) \times [\Gamma P(\gamma^* \mid D) - 1]$, which proves the first bound. Now define Δ as the evidence's error and use the fact that $E_f(A \mid \gamma^*, D) \geq E_f(A \mid D) - \Delta$ to convert our lower bound on $E_f(A \mid D)$ to $E_f(A \mid D) \geq \Gamma P(\gamma^* \mid D) \times [E_f(A \mid D) - \Delta]$. Rearranging gives the second bound.* QED.

Theorem 4 shows why having the γ-peaks far apart is bad for the evidence procedure. However, Theorem 4 does not mean that a small separation between the peaks implies that evidence works. Note that it is even possible for the magnitude of evidence's error to be small when the peaks are well separated; the overall multaplicative factor might be tiny. However, even then, the peak separation must be small if one wants the proportional error of the evidence procedure to be small.

Note that our two peaks are the maximizers over γ of two very similar integrals: $\int df' A(f') P(f', \gamma, D)$ and $\int df' P(f', \gamma, D)$. Accordingly, often if one can evaluate the peak of the evidence, one can also evaluate the peak of $E_f(A, \gamma, D)$, and therefore one can evaluate Γ. So if one can use the evidence procedure, usually one can test its validity.

Example: *Consider the case where the hyperparameter, β, sets the noise level in an N-dimensional Gaussian likelihood (see section 3). The joint probability distribution is given by $P(f, \beta, D) \propto P(f) P(\beta) \beta^{\frac{N}{2}} e^{-\beta_{ev} \chi^2}$, where $\chi^2 \equiv |f - D|^2$ is the usual squared error term. Solving for the β-peak we obtain a relation which holds at the peak: $2\beta\chi^2 = N + 2\beta \frac{\partial \log P(\beta)}{\partial \beta}$.*

For the usual case, where $P(\beta)$ is either flat or the Jeffries prior, the last term in the relation is small, either 0 or 2, respectively. Now assume that N is fairly large and that $P(\beta_{ev} \mid D)$ and $P(f \mid \beta_{ev}, D)$ are as well. Theorem 4 then tells us that the only f's for *which the evidence procedure's approximation might be valid are those corresponding to* $2\beta_{ev}\chi^2 \approx N$. *For instance, if one reconstructs an image using the evidence procedure to set the noise level and then finds that* $2\beta_{ev}\chi^2$ *for the image of interest is not* $\sim N$ *then the reconstruction is unjustified. A corollary to this is that it makes sense to skip the evidence procedure entirely and use the* β*-peak of* $P(f, \beta, D)$ *instead of the peak of the evidence; after all, if they aren't close, then the evidence procedure is unjustified anyhow.*

Naturally, this test also applies to the case where one has used the evidence procedure to set α, *a hyperparameter in a Gaussian conditional prior (c.f. Section 3). The analogous relation that must be satisfied is,* $2\alpha_{ev}\chi^2 \approx N$, *where one now takes* $\chi^2 \equiv |f - \hat{f}|^2$, *the squared error between f and the peak of the conditional prior,* $\hat{f}$. *An example (drawn from the literature) of the evidence procedure failing to meet this criteria is shown in figure 2.*

In some cases in fact, it's easier to evaluate the peak of $E_f(A, \gamma, D)$ than it is to evaluate the evidence peak (e.g., for the entropic prior - see [16]). In such circumstances, if one has reason to believe that the evidence procedure is valid (so that Γ must be small), it is easier to evaluate α_{ev} by finding the mode of $E_f(A, \gamma, D)$ than by finding the mode of $P(\gamma \mid D)$.

The need for the peaks to coincide can set strong restrictions on the use of the evidence procedure. For example, take $A(f') = \delta(f - f')$, so that expectation values of A are probabilities of f. Assume $P(\gamma \mid D)$ is quite peaked. Say we want to use the evidence procedure to estimate $E_f(A \mid D) = P(f \mid D)$ for some particular f, $\hat{f}$. Then Theorem 4 tells us that for evidence to work, if $P(\hat{f} \mid D)$ is non-negligible (or equivalently the evidence procedure's prediction $P(\hat{f} \mid \gamma^*, D)$ is non-negligible), then Γ must be quite small for $\hat{f}$, i.e., the peak of $P(\hat{f}, \gamma, D) = P(D \mid \hat{f}, \gamma)P(\hat{f} \mid \gamma)P(\gamma)$ must lie close to γ^* (as measured on the scale of $1/P(\gamma^* \mid D)$). Setting the peaks exactly equal gives us an equation for $\hat{f}$ in terms of D (γ^* being a function of D). In general this equation will have a highly restricted solution for $\hat{f}$, $F(D)$ (i.e., $F(D)$ is a low-dimensional manifold in f-space). For example, in the case of the entropic prior, $F(D)$ is a set of f all sharing the same entropy (that entropy value being set by D). In our Gaussians case, $F(D)$ is a set of points all sharing the same $|f|^2$ (where again the precise value is set by D - see Theorem 4 of [20]).

So for sufficiently peaked evidence, unless those f with non-negligible posterior all lie in a highly restricted region ($F(D)$), the evidence procedure is guaranteed to have sizable error for some f. Therefore for sufficiently peaked evidence, if the evidence procedure is to correctly estimate the full posterior, that posterior must be highly peaked (i.e., its support must be confined to a highly restricted region). This in turn usually implies that we're in a likelihood dominated regime - in which case there's little reason to apply Bayesian analysis.

These effects can be envisioned with the help of figure 3. Recall that as N rises, the only effect is that all (!) distributions (over both α and f) become more peaked; the shapes of the distributions and in particular the positions of their peaks do not change. This means that the curves in figure 3 get more peaked—but otherwise do not change—as the evidence gets more peaked (cf. parts b and d of figure 3). Accordingly, as the evidence gets more peaked, the set of f which both have non-zero posterior and which have their posterior well approximated by the evidence procedure becomes tightly restricted. Indeed, that set is empty in part d of figure 3. In fact, of the three β's in figure 3, it is only for the β of part c that the "tightly restricted set of f" doesn't quickly vanish with rising N. Yet it is precisely that value of β in part c that is the largest of those depicted in the figure. This illustrates the fact that when the evidence procedure correctly estimates the full posterior we have high β, and that this effect becomes more pronounced as the evidence becomes more peaked (i.e., as N rises). Rephrasing, things must be likelihood-dominated for the evidence procedure to work, especially when the evidence is peaked.

5.4. Other kinds of error

Finally, it is worth briefly discussing those scenarios where one isn't directly concerned with "evidence's error" as defined heretofore. Most such scenarios have $A(.)$ be a function of f as well as f', so our expectation values are functions of f. (Recall that this is the case when posterior expected $A(.)$ is equivalent to the posterior probability of f, for example.). To avoid confusion, in addressing these scenarios we will write expressions like $E_{f'}(A_f \mid D) \equiv \int df' A_f(f')P(f' \mid D)$; since $A(.)$ is a function of two arguments, the subscript on the "E" is modified to indicate exactly which argument is being marginalized, and a subscript is introduced onto the $A(.)$ to indicate the remaining free variable.

For this kind of $A(.)$ one might wish to measure the accuracy of the evidence procedure over all f, rather than just at one particular f. One way to do this is to evaluate a functional of the two functions $E_{f'}(A_f \mid D)$ and $E_{f'}(A_f \mid \gamma^*, D)$. So for example we might be interested in the least upper bound (over all f) of $|E_{f'}(A_f \mid D) - E_{f'}(A_f \mid \gamma^*, D)|$. Since Theorem 2 holds for any individual f, this least upper bound is bounded above by the quantity $\max_f(\, \epsilon(f) + \tau(f)(1-\rho) + E_{f'}(A_f \mid \gamma^*, D)\, |\rho|\,)$ (ϵ and τ have dependence on f through their dependence on $A(.)$). This gives the largest possible gap (across f) between the evidence approximation to the posterior and the correct posterior.

Arguments similar to this least upper bound (lub) one can be used to directly bound $\int df |E_{f'}(A_f \mid D) - E_{f'}(A_f \mid \gamma^*, D)|$. More generally, we can use a bound (however arrived at) on $\text{lub}_f(\, |E_{f'}(A_f \mid D) - E_{f'}(A_f \mid \gamma^*, D)|\,)$ to get bounds on the L^n difference between $E_{f'}(A_f \mid D)$ and $E_{f'}(A_f \mid \gamma^*, D)$ for any n. To illustrate this, consider the case where $A(f, f') = \delta(f - f')$, so that the expectation value we're examining is the posterior distribution of f. Define "$L^n(x(f) - y(f))$" to mean the L^n difference between $x(f)$ and $y(f)$. Let μ be an upper bound on $\text{lub}_f(\, |P(f \mid D) - P(f \mid \gamma^*, D)|\,)$. Then $L^n[P(f \mid D) - P(f \mid \gamma^*, D)] \leq \mu \times [2/\mu]^{1/n}$ [22].

ACKNOWLEDGMENTS: DHW was supported in part by NLM grant F37 LM00011. We thank David Wolf, Tim Wallstrom and Richard Silver for helpful discussions.

References

[1] J. Berger, *Statistical Decision Theory and Bayesian Analysis*, Springer-Verlag, 1985.

[2] L. Breiman, "Stacked regression," *University of California, Berkeley, Dept. of Statistics*, TR-92-367, 1992.

[3] W. Buntine, A. Weigend, "Bayesian back-propagation," *Complex Systems*, Vol. 5, p. 603, 1991.

[4] A.R. Davies, R.S. Anderssen, R.S., "Optimization in the regularization of ill-posed problems," *J. Australian Math. Soc. Ser. B*, Vol. 28, p. 114, 1986.

[5] D. L. Donoho et al., "Maximum entropy and the nearly black object," *J. R. Stat. Soc. B*, Vol. 54, pp: 41-81, 1992.

[6] R. Duda, P. Hart, *Pattern Classification and Scene Analysis*, Wiley and Sons, 1973.

[7] G. Demoment, "Image reconstruction and restoration: overview of common estimation structures and problems," *IEEE Transactions on Acoustics, Speech, and Signal Processing*, Vol. 37, pp: 2024-2036, 1989.

[8] N. Fortier et al., "GCV and ML methods of determining parameters in image restoration by regularization: fast computation in the spatial domain and experimental comparison," *Journal of visual communication and image representation*, Vol. 4, pp: 157-170, 1993.

[9] S. Gull, "Developments in maximum entropy data analysis," In *Maximum Entropy and Bayesian Methods*, J. Skilling (Ed.). Kluwer Academics Publishers, 1989.

[10] E. Jaynes, "Monkeys, kangaroos and alpha," In *Maximum Entropy and Bayesian Methods*, Kluwer Academic Publishers, 1988.

[11] D.J.C. MacKay, "Bayesian Interpolation," "A Practical Framework for Backpropagation Networks," *Neural Computation*, Vol. 4, pp: 415 and 448, 1992.

[12] D.J.C. MacKay, "Bayesian non-linear modeling for the energy prediction competition," In these proceedings.

[13] B.D. Ripley, "Statistical Aspects of Neural Networks," In *Networks and Chaos—Statistical and Probabilistic Aspects*, O.E. Barndorff-Nielsen et al. (Eds.) Chapman and Hall, 1993.

[14] S. Sibisi, "Regularization and inverse problems," In *Maximum Entropy and Bayesian Methods*, J. Skilling (Ed.). Kluwer Academics publishers. 1989.

[15] J. Skilling, "Classic maximum entropy," In *Maximum Entropy and Bayesian Methods*, J. Skilling (Ed.). Kluwer Academic publishers. 1989.

[16] C.E.M. Strauss, D.H. Wolpert, and D.R. Wolf, "Alpha, Evidence, and the Entropic Prior," In *Maximum Entropy and Bayesian Methods*, A. Mohammed-Djafari (Ed.). Kluwer Academics publishers, 1993.

[17] A. M. Thompson et al., "A study of methods of choosing the smoothing parameter in image restoration by regularization," *IEEE Transactions on Pattern Analysis and Machine Intelligence*, Vol. 13, pp: 326-339, 1991.

[18] A.M. Thompson, J. Kay, "On some Bayesian choices of regularization parameter in image restoration," Technical Report from The University of Edinburgh, no number.

[19] G. Wahba, "A comparison of GCV and GML for choosing the smoothing parameter in the generalized spline smoothing problem," *The Annals of Statistics*, Vol. 13, pp: 1378-1402, 1985.

[20] D.H. Wolpert, "On the use of evidence in neural networks," In *Advances in Neural Information Processing Systems 5*, Giles et al. (Eds.), Morgan Kauffman Publishers, 1993.

[21] D.H. Wolpert, "Bayesian backpropagation over I-O functions rather than weights," In *Advances in Neural Information Processing Systems 6*, Cowan et al. (Eds.), Morgan Kauffman Publishers, 1994.

[22] D.H. Wolpert, C.E.M. Strauss "What Bayes has to say about the evidence procedure", SFI TR 94-07-001. (This TR is an extended version of this paper.)

RECONCILING BAYESIAN AND NON-BAYESIAN ANALYSIS

David H. Wolpert
The Santa Fe Institute, 1660 Old Pecos Trail
Santa Fe, NM, 87501, USA (dhw@santafe.edu)

ABSTRACT. This paper is an attempt to reconcile Bayesian and non-Bayesian approaches to statistical inference, by casting both in terms of a broader formalism. In particular, this paper is an attempt to show that when one extends conventional Bayesian analysis to distinguish the truth from one's guess for the truth, one gains a broader perspective which allows the inclusion of non-Bayesian formalisms. This perspective shows how it is possible for non-Bayesian techniques to perform well, despite their handicaps. It also highlights some difficulties with the "degree of belief" interpretation of probability.

1. Introduction

Why should one want to reconcile Bayesian and non-Bayesian analysis? Why be bothered with non-Bayesian techniques? Bayesian analysis forces one to makes one's assumptions explicit; it ensures self-consistency; it provides a single unified approach to all inference problems; if one is very sure of the prior (e.g., as an extreme, you constructed the data-generating mechanism yourself) it is essentially impossible to beat; and in some ways most important of all (sociologically speaking), Bayesian analysis is in some senses more elegant than non-Bayesian analysis.

For these very reasons I have used Bayesian techniques in the past and will do so again in the future. As compelling as these reasons are though, none of them constitute a proof that Bayesian techniques perform better than non-Bayesian techniques in the real world. Indeed, there are many examples — some constructed by self-professed Bayesians — which cast doubt on such a guarantee. For example, as is discussed separately at this conference [1], although the "evidence" procedure sometimes works well in practice [2, 3], and although it is championed by fervent Bayesians, careful scrutiny reveals that it is a non-Bayesian technique. In particular, in [3], in a section entitled "Why Bayes can't systematically reject the truth", MacKay presents a theoretical argument for the evidence procedure's setting hyperparameters by maximum likelihood. However this argument can be extended to "justify" setting *any* parameter by maximum likelihood, not just a hyperparameter. It is hard to imagine a more non-Bayesian line of reasoning. As another example, despite being a self-professed fervent Bayesian, MacKay recently won a prediction competition using an extension of the non-Bayesian technique of cross-validation [4].

Another reason not to dismiss non-Bayesian techniques arises from the problem of setting the probability distribution $P(\text{truth}= t, \text{data}= d)$.[1] If—as is often the case in the real world—we already know the likelihood $P(d \mid t)$, in what ways can we fix the remaining degrees of freedom in the joint distribution while ensuring consistency with the laws of probability theory? One way to do this is to provide the prior distribution $P(t)$—this is the basis for conventional Bayesian analysis. But there are other ways as well. For example,

G. R. Heidbreder (ed.), Maximum Entropy and Bayesian Methods, 79–86.

consider the case where there is a data set $d = d'$ such that $P(d' \mid t)$ does not exactly equal zero for any t. If we now provide the values of $P(t \mid d')$ for all possible t, we will have fixed the entire joint distribution, for all possible data sets. (This follows from the equality $P(t_i, d_j) = \frac{P(t_i|d')P(d_j|t_i)/P(d'|t_i)}{\sum_{k,m} P(t_k|d')P(d_m|t_k)/P(d'|t_k)}$.) In particular, by setting $P(t \mid d')$, we will have fixed $P(t \mid d)$ for any $d \neq d'$.

With this alternative scheme we would be assured of self-consistency, our assumptions would be explicit, etc.; this scheme possesses all the formal strengths of conventional Bayesian analysis. However rather than use pseudo-intuitive arguments to set $P(t)$, as in the conventional prior-based Bayesian approach, with this alternative scheme we use such arguments to set $P(t \mid d)$ for one particular d. For example, one could set $P(t \mid d)$ for one particular d using "pseudo-intuitive" cross-validation type arguments. One might even be able to use "desiderata" to set $P(t \mid d)$ for one specific d, rather than to set $P(t)$. There is no reason prior knowledge has to concern a prior probability; one can have prior knowledge that is expressed directly as a posterior. For example, my "prior knowledge" might consist of knowing that cross-validation works well for a certain class of problems.

In fact many practicing statisticians do implicitly exploit "prior knowledge" directly concerning the posterior. However they do so in conjunction with a approximation; they have their prior knowledge set all of $P(t \mid d)$ at once, and therefore they (usually) violate strict consistency with the laws of probability. For historical reasons, such approximations are usually called "non-Bayesian". However they are closely analogous to using a conventional (i.e., prior-based) Bayesian analysis which involves calculational approximations, and which therefore also violates strict consistency with the laws of probability. So the question arises of how accurate the approximations in a Bayesian technique must be to "beat" a particular non-Bayesian technique. (From here on the term "Bayesian" will mean conventional, prior-based Bayesian.) To address this and related issues we need to use a new formalism.

2. A Formalism for Reconciling Bayes and Non-Bayes

In most inference problems there are four quantities of interest: the data d, the truth t (which might be a probability distribution), one's "guess for the truth" g, and a real world "cost" or "loss" or "utility" accompanying a particular use of one's inference technique. (For many scenarios cost only depends on t and g, and g is formally called a "decision".) Accordingly, the inference process is governed by $P(t, g, d, c)$.

Now conventional Bayesian analysis doesn't distinguish t from g—it does not analyze joint distributions over those two variables. Therefore one must be careful in relating $P(t, g, d, c)$ to the distributions used in Bayesian analysis. In particular, note that the "posterior" of Bayesian analysis is $P(t \mid d)$, not $P(g \mid d)$. This follows from how a Bayesian uses Bayes' theorem to set the "posterior" in terms of the likelihood. Since the likelihood is $P(d \mid \text{truth} = t)$, not $P(d \mid \text{guess} = g)$, the "posterior" must be $P(t \mid d)$.

$P(g \mid d)$ is a different kind of object which has no analogue in Bayesian analysis. It is the probability of making a guess g given data d. In other words, it is one's algorithm for performing statistical inference. A priori, it need have nothing to do with Bayesian techniques, and need not even be expressible in "Bayesian" terms. (For example, the evidence procedure's $P(g \mid d)$ can not be expressed this way, since there is always necessarily some difference between it and full hierarchical Bayesian analysis—see [1].) As such, $P(g \mid d)$ is the object which allows one to expand the discussion to consider non-Bayesian techniques.

One nice feature of this "extended" Bayesian framework is that in it, the difference between conventional Bayesian analysis and (most forms of) non-Bayesian analysis is no longer some quasi-philosophical preference for different statistical dogmas. Instead that difference reduces to simply what conditional distribution the two formalisms choose to evaluate. Bayesian analysis is concerning with finding the $P(g \mid d)$ that optimizes $P(c \mid d)$, and sampling theory statistics with evaluating $P(c \mid t, m)$ (m being the data set size). It is only with the extended Bayesian framework that one can consider both at once, and thereby investigate the subtle connections between the two [7].

The implicit view in this extended framework is that inference is a 2-person game pitting you, the statistician, against the data-generating mechanism, aka the universe. Your opponent draws truths t at random, according to $P(t)$, and then randomly produces a data set from t, according to $P(d \mid t)$. This d is shown to you. Based on d, you guess a g according to $P(g \mid d)$. We then use some cost function to determine how well g matches t. Note that if you know $P(t)$ and $P(d \mid t)$, then you can use that information to perform optimally. But if you don't know $P(t)$ exactly (!) and therefore have to guess it—as in the real world—you have no such assurance.

In fact, extended Bayesian analysis can be used to prove the following (see [5, 6])):

Theorem 1: $P(c \mid d) = \sum_{g,t} P(g \mid d)\ P(t \mid d)\ M_{c,d}(g,t)$, for some matrix M parameterized by c and d.

(A similar result holds if g and t are not countable.)

In many situations M is symmetric, in which case theorem (1) means that $P(c \mid d)$ is given by an inner product between the posterior and one's inference algorithm. In other words, how well your algorithm performs is determined by how "aligned" it is with the true posterior. In particular, theorem (1) allows that a Bayesian's $P(g \mid d)$ might not be predicated on the actual $P(t \mid d)$, and therefore might perform poorly—perhaps even worse than a non-Bayesian $P(g \mid d)$. Such mismatch between the Bayesian's $P(g \mid d)$ and $P(t \mid d)$ can occur even if the Bayesian somehow knows $P(t)$ and $P(d \mid t)$, if the Bayesian's $P(g \mid d)$ uses those distributions in conjunction with calculational approximations. So in general there are two issues confronting both the Bayesian and the non-Bayesian: i) how accurately $P(g \mid d)$—based as it is on assumptions and approximations—aligns with $P(t \mid d)$, and ii) how probability of cost varies with changes in that accuracy.

In fact, if the inference problem is to build a classifier, so that both g and t are mappings from features vectors to classification labels, and if one's cost is determined by how well g matches t for feature vectors outside of the data set, one has the following theorem [6]:

Theorem 2: Let $E(\cdot)$ indicate an expectation value, and m the size of the "training set" d. For any two inference algorithms $P_1(g \mid d)$ and $P_2(g \mid d)$, independent of the noise,

i) if there exists a t such that $E(c \mid t, m)$ is lower for $P_1(g \mid d)$, then there exists a different t such that $E(c \mid t, m)$ is lower for $P_2(g \mid d)$;

ii) if there exists a t and a d such that $E(c \mid t, d)$ is lower for $P_1(g \mid d)$, then there exists a different t and d such that $E(c \mid t, d)$ is lower for $P_2(g \mid d)$;

iii) if there exists a $P(t)$ and a d such that $E(c \mid d)$ is lower for $P_1(g \mid d)$, then there exists a different $P(t)$ such that $E(c \mid d)$ is lower for $P_2(g \mid d)$;

iv) if there exists a $P(t)$ such that $E(c \mid m)$ is lower for $P_1(g \mid d)$, then there exists a different $P(t)$ such that $E(c \mid m)$ is lower for $P_2(g \mid d)$.

All of this holds whether or not the inference algorithms in question are constructed in a Bayesian manner. Moreover these (and associated) results don't just say that a non-Bayesian algorithm might beat a Bayesian in one particular trial, by luck. Rather a non-Bayesian algorithm might win on average. In fact, not only does theorem (2) not rely on pathological trials; it also doesn't rely on pathological processes generating the trials. For example 2(iv) can be recast as "averaged over all $P(t)$, $E(c \mid m)$ is the same for all learning algorithms". So for any two inference algorithms, there are "just as many" $P(t)$'s (loosely speaking) for which algorithm one has a lower expected cost as there are for which algorithm two's expected cost is lower. Unless you somehow know $P(t)$ rather than just guess it, your being a Bayesian provides no guarantees.

From this perspective, the Bayesian approach is the approach of choice only if there is no alternative (non-Bayesian) approach which is sufficiently compelling in comparison. (The comparison being between how compelling is a $P(g \mid d)$ based on a guess for $P(t)$ vs. a $P(g \mid d)$ based on other considerations.) Those (not at all uncommon) scenarios in which the Bayesian approach works well compared to non-Bayesian techniques do not reflect some inherent "*a priori* superiority" of Bayesian techniques. Rather they reflect the fact that at least some aspects of the non-Bayesian techniques considered in those scenarios are not sufficiently powerful in comparison to the corresponding Bayesian techniques. Indeed, the utility of using "sufficiently powerful" non-Bayesian approaches when possible is explicitly acknowledged in several variations of Bayesian analysis, like empirical Bayes and ML-II [9].

A particularly important implication of this is that there is nothing inherently bad about using a non-Bayesian algorithm to choose between Bayesian and/or non-Bayesian techniques. For example, if we have little information concerning $P(t)$ (and especially in the limiting case of no knowledge—a case sometimes dealt with via an "uninformative prior"), then it makes sense to be suspicious of any guess for $P(t)$ (even a guess that $P(t)$ is "uninformative"). Therefore it is reasonable to be suspicious of any $P(g \mid d)$ constructed under that guess. In such a scenario, one need not be shy about using something like cross-validation to choose amongst the techniques, or even about using stacked generalization to combine them [8].

All this provides suggestions of what some non-Bayesian formalisms are "getting at". For example, if one knows $P(t)$ exactly, then Bayesian techniques incorporating that knowledge into $P(g \mid d)$ always win, on average (assuming we also know the likelihood). However imagine we have limited information concerning $P(t)$. In this case we will inevitably be off a bit in the guess which we make for $P(t)$ and then incorporate into our $P(g \mid d)$. According to theorem (1), this means we will perform sub-optimally. So there is a correlation between how much we know about $P(t)$ and how assured we are that a Bayesian technique using our guess for $P(t)$ is superior to a particular non-Bayesian technique. This can be viewed as introducing a distinction between an assumption for a probability and one's "confidence" in that assumption.[2] It's conceivable that this is what advocates of Dempster-Schaffer theory, fuzzy logic, and the like are getting at with notions like "plausibility vs. probability".

Another example of what non-Bayesian formalisms might be "getting at" arises if we take $P(t, g, c, d)$ to mean $P(t, g, c, d \mid$ prior information $I)$, so we must define the space of possible I. We *could* say that I fixes the precise statistical problem p that we are considering. As an example, that problem may be predicting the change in the value of the Dow Jones average across some precise date, given the current values of *all* physical variables within the light cone of the space-time coordinate {Wall Street, the date in question}, and all of that information is in I.

However if we ignore quantum mechanical issues for the moment, then physics tells us that for such a "precise problem" the outcome is fixed rather than random, regardless of whether that outcome's already occurred or not. Now as usually defined probability distributions must equal 1 for true events and 0 for false events. (Note that such definitions pay no attention whatsoever to whether we happen to know what's true and false). Accordingly, for a "precise problem" probability distributions are delta functions, and statistics becomes vacuous. (This difficulty is similar to the common complaint of non-Bayesians that Bayesians treat parameters as random variables even though they aren't.)

However, tautologically, we're only interested in that information we have concerning the precise problem p that affects how we would guess for p. Accordingly, one could require that I is only that information we have concerning the problem p such that the g and/or d dependence of $P(g \mid d, I)$ would differ if that information were left out of I. (To agree with common usage, I'm taking d to not be part of the "prior information".) Such a choice of the prior information I fixes $P(g \mid d)$ but not necessarily vice-versa.

Now in practice the extra bits fixing the "precise problem" don't affect how we guess. (E.g., this is true for the bits concerning the vast majority of the physical variables within the light cone of the space-time coordinate {Wall Street, the date in question}). Accordingly, those bits aren't in I. This means that $P(t \mid I)$ is not a delta function, and we don't have the vacuous-statistics problem. (I doesn't even include whether we will actually make a guess, since that information doesn't affect $P(g \mid d)$.)

Under this restriction on I, the posterior "$P(t \mid d, I)$" is a distribution defined for the set of all possible problems with the same guess-affecting information as p. It is not defined solely for the precise problem p. So this restriction suggests a set of multiple problems, just as a frequentist might. In fact, this kind of multiple problem $P(t \mid d, I)$ is exactly the starting point for the conventional frequentist view of statistical physics.

On the other hand, if due to his/her beliefs the guess of statistician A depends on the value of variable Q, whereas that of statistician B does not, they have different I's. (From the frequentist perspective, they are concerned with different sets of problems, in only one of which is the value of Q held constant.) So although it suggests frequentism, the concern of some Bayesians for "beliefs" is also reflected in this definition of "prior information" I.

3. The "degree of belief" interpretation of probability

Some researchers interpret "probability" as synonymous with a subjective "degree of belief" (the precise meaning of this expression — to the degree there is one — isn't relevant for current purposes). Bayesians have often used this interpretation to argue for the superiority of their techniques. The reasoning is that under this interpretation, $P(t)$ is your belief in proposition t, i.e., you automatically know $P(t)$ exactly. Therefore — under this interpretation — if you also know the likelihood you know $P(t \mid d)$, and you can use this to set $P(g \mid d)$ in such a way that you have minimal expected cost (up to calculational approximations). (Sometimes the vague caveat is added to this argument that one's beliefs must be "rational".)

This seems to imply that Bayesian and non-Bayesian analysis are not reconcilable, that Bayesian approaches to statistics are definitionally superior to non-Bayesian ones. However the degree of belief (dob) interpretation justifies non-Bayesian techniques just as readily as Bayesian ones: interpret a non-Bayesian's $P(t \mid d)$ as his/her "degree of belief" in t given d, so (s)he "automatically knows $P(t \mid d)$ exactly", and can use that knowledge to guess with "minimal expected cost", again up to various approximations. In this, the dob interpretation does not play favorites between Bayesian and non-Bayesian approaches.

However there is another more major flaw in this supposed irreconcilability implication: there are foundational problems with the dob interpretation of probability itself. This flaw is the subject of the rest of this section. Fortunately, we don't have to adopt any particular alternative (invariably contentious) interpretation of probability to address it.

The first such foundational problem is that the analysis of the previous section is exactly correct if you're playing a real two-person game. So if in the honored tradition of probability theory one is investigating gambling, then the analysis of the previous section and all of its implications are tautologically correct. In particular, in gambling your "degree of belief" involves $P(g)$ and *a priori* need have nothing to do with $P(t)$, which is instead determined by the other player (the house). Even if you arrive at your beliefs through sophisticated, almost "indisputable" group/information - theoretic arguments, if it turns out that your priors disagree with those of the house, well, then you lose. Your beliefs might be a good *approximation* to $P(t)$, if arrived at rationally and based on extensive prior knowledge (presumably this is the case in those scenarios where Bayesian analysis works well). But that doesn't mean the two quantities are definitionally equal. It doesn't somehow mean that you rather than your opponent fix the probability that your opponent is bluffing.[3]

So the question arises of whether there is a fundamental distinction, with concrete ramifications, between gambling and all real world statistical problems. If there isn't—and it's hard to imagine how there could be—then Thm.'s 1 and 2 imply that there are no guarantees of optimality for the dob Bayesian.

More generally, a "truth" t and a guess g are different objects. Therefore their distributions need not be related *a priori*. (This is reflected in the theorems of the previous section.) Accordingly, a formal statement connecting $P(t \mid d)$ and $P(g \mid d)$ corresponds to an extra assumption concerning $P(t, g, c, d)$, an assumption not demanded by the mathematics. In particular, the dob interpretation is such an extra unjustified assumption.

Another difficulty with the dob interpretation is that if we had sufficient knowledge of the laws of physics (in particular, of the boundary conditions of the universe) and of the (resultant) laws of human psychology, and if we were sufficiently competent to perform the appropriate quantum mechanical calculations, then we might say that we could calculate $P(t)$ exactly. In other words, one possible interpretation is that $P(t)$ is the "real" $P(t)$, determined by the laws of quantum mechanics applied to the universe as a whole. *A priori*, such a $P(t)$ need have nothing to do with one's (pre-calculation) degrees of belief.

Indeed, anyone can *imagine* that quantum mechanics is correct, even if they don't *believe* that to be the case. So we can self-consistently imagine that the universe evolves in accord with equations governing "absolute, objective" probabilities, since those are the building blocks of quantum mechanics. This simple fact that we can self-consistently (!) imagine quantum mechanics shows that there is no formal problem with quantum mechanics' implicit notion of absolute, objective probabilities, which exist independently of any particular person's degree of belief. So there is nothing mathematically necessary about the dob interpretation of probability.

In this regard, note that nothing in Cox's axioms forces a particular interpretation of probability. Those axioms only say that any (reasonable) calculus of uncertainty must obey the laws of probability theory. They do not tell us how to assign the probability values in the first place. One *could* interpret probability as degree of belief. In such a case, Bayesian analysis becomes a set of rules for telling you what structure your beliefs must have to be self-consistent. But the math does not force us to that interpretation.

All of this agrees with Bayesianism as practiced; the actions of a practicing dob Bayesian are indistinguishable from those of someone who thinks $P(t)$ is independent of $P(g \mid d)$, and therefore is not "automatically known" but rather has to be discovered. It's just that for a dob Bayesian, the to-be-discovered $P(t)$ is rather disingenuously considered to be the distribution which "best reflects prior knowledge". To the dob Bayesian, as our understanding of statistics improves, as we get a better understanding of what "uninformative" means, etc., we get a more accurate idea of that $P(t)$. To an outsider, the dob Bayesian is simply changing his/her guess for $P(t)$.

As an example, some of the more prominent attendees at this conference have spent much of their careers looking for arguments to establish what priors to use for certain scenarios. Moreover, they've changed their views on this several times. Each time they act as though they were assuming an "incorrect" $P(t)$ before, despite the fact that that old assumption for $P(t)$ properly reflected their old degrees-of-belief. And each time they tend to look askance at any laggards still using the old guess for $P(t)$, despite the fact that said laggards are directly following along with their beliefs. This behavior is consistent with the idea that degree-of-belief Bayesians do not, deep down, view probabilities as just degrees of personal belief, but rather view them as possessing some degree of objective reality.

Indeed, for a century Bayesianism was in disrepute, and the current consensus is that is was in disrepute because it was used with "bad choices of priors". Just translate "bad choice of" to "incorrect assumption for", and you have the theorems of the previous section, with their implication that Bayesianism can be sub-optimal.

Alternatively, note that transferring from an "incorrect" $P(t)$ to a better one is really nothing more than the process accompanying the (in)famous "opportunity to learn" which one encounters when one's Bayesian analysis leads to poor results. Or to put it another way, having $P(t \mid d)$ poorly reflected in $P(g \mid d)$ is an opportunity to learn. If you assume these distributions are always "automatically" connected, you're assuming you never have an opportunity to learn. (As an aside, note that a mismatch in the distributions is an "opportunity to learn" whether or not $P(g \mid d)$ is based on Bayesian analysis—Bayesians have no monopoly on the use of the concept "opportunity to learn" as a cover for poor performance of their statistical algorithms.)

Finally, note that there might well be a way to embed the reasonableness/desiderata arguments often used by dob Bayesians to set priors inside a complete mathematical framework (e.g., there might be a framework which maps any (!) I to a unique prior distribution).

If we had such a framework, *then* one might claim that such reasonableness arguments are a well-principled way to assign probabilities. Without that framework in hand though, we have no assurance that any particular reasonableness argument assigns the same values to probabilities as that framework would. In particular we have no assurance that there isn't some lurking reasonableness argument which contradicts our current arguments. In short, at present "degrees of belief" set by desiderata arguments do not constitute mathematics. They constitute philosophy.

Endnotes

1. For the purposes of this paper there is no reason to specify whether the notation "$P(.)$" refers to a probability, a probability density function, or some other similar object.

2. I'm speaking loosely here, and have not defined "confidence" formally. In particular, I have not defined confidence in a probability with probability of a probability.

3. Note that assigning a "degree of belief" to a proposition and making an assumption for the probability of that proposition (as one might do in gambling against the house) are very similar things. Both are subjective declarations concerning how reasonable the researcher thinks the proposition is. This might be why people have confused them so easily. There is an important distinction between the two concepts however: in declaring one's degree of belief in a proposition one is tautologically correct, whereas there is no such notion of tautological correctness to making an assumption. It is the claim of this paper that $P(t)$ is something one can assume as opposed to something one can simply declare.

ACKNOWLEDGMENTS: This work was supported in part by NLM grant F37 LM00011, by the Santa Fe Institute, and by TXN Inc.

References

[1] D.H. Wolpert, C.E.M. Strauss, C.E., and D.R. Wolf, "What Bayes has to say about the evidence procedure," These proceedings, 1994.

[2] S. Gull, "Developments in maximum entropy data analysis," in "Maximum-entropy and Bayesian methods," J. Skilling (Ed.). Kluwer Academics publishers, 1989.

[3] D.J.C. MacKay, "Bayesian Interpolation," "A Practical Framework for Backpropagation Networks," *Neural Computation*, Vol. 4, pp: 415 and 448, 1992.

[4] D.J.C. MacKay, "Bayesian non-linear modeling for the energy prediction competition," These proceedings.

[5] D.H. Wolpert, "On the connection between in-sample testing and generalization error," *Complex Systems*, Vol. 6, pp: 47-94, 1992.

[6] D.H. Wolpert, "On overfitting avoidance as bias," SFI TR 93-03-016.

[7] D.H. Wolpert, "The Relationship Between PAC, the Statistical Physics framework, the Bayesian framework, and the VC framework", in *The Mathematics of Generalization*, D.H. Wolpert (Ed.), Addison Wesley, 1994.

[8] L. Breiman, "Stacked regression," *University of California, Berkeley, Dept. of Statistics*, TR-92-367, 1992.

[9] J. Berger, "Statistical Decision Theory and Bayesian Analysis," Springer-Verlag, 1985.

BAYESIAN ROBUSTNESS: A NEW LOOK FROM GEOMETRY

Carlos C. Rodríguez
Department of Mathematics and Statistics
State University of New York at Albany
Albany NY 12222, USA
Email: carlos@math.albany.edu

ABSTRACT. The geometric concept of the Lie derivative is introduced as the natural way of quantifying the intrinsic robustness of a hypothesis space. Prior and posterior probability measures are interpreted as differential forms defined invariantly on the hypothesis space. Rates of change with respect to local deformations of the model are computed by means of Lie derivatives of tensors defined on the model (like the information metric, prior, posterior, etc.). In this way a field theory of inference is obtained. The class of deformations preserving the state of total ignorance is introduced and characterized by a partial differential equation. For location models this equation is the familiar $\nabla \cdot \xi = 0$. A simple condition for the robustness of prior (or posterior) distributions is found: There is robustness when the deformation is along level surfaces of the prior (or posterior) density. These results are then applied to the class of entropic priors. It is shown that the hyper parameter controls the sensitivity with respect to local deformations. It is also shown that entropic priors are only sensitive to deformations that change the intrinsic form of the model around the initial guess.

1. Introduction

The robustness, of a statistical procedure, is commonly defined as the stability with respect to small changes in the assumptions. This notion has immediate intuitive appeal and it has even been equated to the Holy Grail of Statistics (see [5]).

There is general agreement about the desirability of a consistent theory of statistical robustness (Bayesian and non-Bayesian) and the large number of articles and books dedicated to the subject testify it. The technical definition of robustness is still controversial, however. For a serious criticism to the definitions of Hampel [3, p. 1980] and Huber [4, p. 10] see [6, p. 17].

In this paper, the geometric concept of the Lie derivative is introduced as a technical tool for quantifying robustness. The great arsenal of modern geometry tools provide a flexible, rigorous, and powerful framework for developing statistical inference in general and Bayesian robustness in particular.

The geometrization of statistics is possible in part due to the fact that statistical models have a natural manifold structure. Fisher information endows the models with a Riemannian metric and the Kullback number (entropy) generates this and many other natural geodesic metrics on the model (see [1] and [7]).

The main idea is to exploit the rich geometric structure available in the hypothesis space for the quantification of robustness. Once differential geometry is permitted to be the operational framework, a number of consequences for robustness are straight forward and inevitable. This paper concentrates on the quantification of the robustness of probability

G. R. Heidbreder (ed.), Maximum Entropy and Bayesian Methods, 87–96.

distributions defined on the model. The same technique can be used for quantifying the stability of any tensorial quantity defined on the space.

In this approach, there are important differences with traditional methods. First, everything is *intrinsic* to the model. There is no need for postulating super models, nonparametric neighborhoods, or anything outside the given hypothesis space. The model is an enclosed universe that is assumed to include *all* the relevant probability measures for the observed data. The possibility of encoding deformations of the model without reference to an outside, is a remarkable achievement of modern geometry. Second, n does not have to go to infinity for the methods to make sense. In fact they even make sense in the absence of all data. We can quantify the sensitivity of a prior distribution with respect to deformations of the model independently of the observations.

The paper is divided into four sections. In Section 1 we introduce the notation and provide a summary of the main definitions and results from geometry that will be needed later. In Section 2 we introduce probability distributions over the parameter space as differential forms defined on the model and their Lie derivatives are computed by using the methods of Section 1. We also compute explicitly the sensitivity of the class of entropic priors with respect to deformations of the model. Finally, in Section 3, we workout the example of inference in the one dimensional Gaussian model with entropic priors. We conclude in Section 4 with some comments on the possible future developments of these methods.

2. Local Deformations, Lie Derivatives of Tensors and Volume Elements

We collect here some classical results from the geometry of vector fields on manifolds. The material in this section can be found in most books on modern geometry. We follow the presentation and notation of [2, chap. 23].

Regular (finite dimensional) parametric statistical models will be denoted by $\mathcal{P} = \{P_\theta : \theta \in \Theta\}$. They are Riemannian manifolds. The parameterization $\Theta \subset \Re^k$ plays the role of a coordinate system. The tangent space at $P \in \mathcal{P}$ is modeled by the linear space generated by the partial derivatives (w.r. to θ) of the log-likelihoods. In this way the tangent space at P is a subspace of $L^2(P)$ and it inherits the inner product from it. It turns out that the Riemannian metric on the tangent space at P_θ, $g_{ij}(\theta)$, coincides with the Fisher information matrix at θ, see [1] and [7] for detailed definitions.

A vector field ξ on the manifold $\mathcal{P}$ is a mapping that assigns to each $P \in \mathcal{P}$ a tangent vector at P. For a pictorial representation think of the model as a k-dimensional (curved) surface and the vector field as the velocity field of a fluid moving on the surface. If the field ξ is smooth (as a map between manifolds) the theory of ordinary differential equations warranties the existence and uniqueness of the following associated autonomous system of differential equations:

$$\begin{aligned} \frac{d\theta^i}{dt} &= \xi^i\left(\theta^1(t),\ldots,\theta^k(t)\right), \quad i = 1,\ldots,k \\ \theta^i|_{t=t_0} &= \theta^i_0 \end{aligned} \tag{1}$$

where θ^i and ξ^i denote the components of P_θ and ξ in the coordinate system Θ and θ_0 is the initial condition. The solution to this system is known as the integral curve of ξ passing through P_{θ_0}. We denote it by $F_t(\theta_0) = \theta(t)$. For a given t the map, $F_t : \theta_0 \mapsto \theta(t)$, defined in a neighborhood of the point θ_0, represents the new position after time t of a particle of fluid which is initially at θ_0. The theory of ordinary differential equations assures that for t small enough the maps F_t are diffeomorphisms (i.e. one to one and with continuous differential both ways). More precisely they form a local one parameter group of diffeomorphisms with group operation $F_t \circ F_s = F_{t+s}$, inverse $F_t^{-1} = F_{-t}$ and identity F_0. Each transformation F_t defines (at least locally) a change of coordinates from θ_0 to $\theta(t)$ i.e. a local re-labeling of the elements of $\mathcal{P}$. In this way if T is a quantity defined in terms of the labels $\theta(t)$ (but providing intrinsic information about the points $P \in \mathcal{P}$) it will have an expression $(F_t T)$ in terms of the labels θ_0, satisfying the rules of transformation for tensors. We have the following:

Definition 1 *The Lie derivative of a tensor T along a vector field ξ is the tensor $L_\xi T$ given by*

$$L_\xi T = \left[\frac{d}{dt}(F_t T)\right]_{t=0} \tag{2}$$

Again, in visual terms the Lie derivative of T along ξ gives the rate of change of T as it is seen when moving with the fluid. Or equivalently, standing at θ_0 we see the components of T change due to the (time dependent) deformation of the space given by F_t and the Lie derivative is just the rate of change (with respect to time) of what we see. By applying the rules of transformations for tensors and using the smoothness of the F_t's (by Taylor's theorem $F_t(\theta_0) = \theta_0 + t\xi(\theta_0) + o(t)$), we can find the components of the Lie derivative of the tensor $T^{i_1 \ldots i_p}_{j_1 \ldots j_q}$. They are given by:

$$\begin{aligned} L_\xi T^{i_1 \ldots i_p}_{j_1 \ldots j_q} &= \xi^s \frac{\partial T^{i_1 \ldots i_p}_{j_1 \ldots j_q}}{\partial \theta^s} + T^{i_1 \ldots i_p}_{k j_2 \ldots j_q} \frac{\partial \xi^k}{\partial \theta^{j_1}} + \ldots + T^{i_1 \ldots i_p}_{j_1 \ldots j_{q-1} k} \frac{\partial \xi^k}{\partial \theta^{j_q}} \\ & -T^{l i_2 \ldots i_p}_{j_1 \ldots j_q} \frac{\partial \xi^{i_1}}{\partial \theta^l} - \ldots - T^{i_2 \ldots i_{p-1} l}_{j_1 \ldots j_q} \frac{\partial \xi^{i_p}}{\partial \theta^l}. \end{aligned} \tag{3}$$

where here, as in the rest of the paper, the standard implicit summation over repeated indices is assumed. As special cases of this formula we have:

I). **Scalar field**. $T = f$

$$L_\xi f = \xi \cdot \nabla f = \xi^i \frac{\partial f}{\partial \theta^i} \tag{4}$$

II). **Vector field**. $T = \eta^i$

$$L_\xi \eta^i = [\xi, \eta]^i = \xi^j \frac{\partial \eta^i}{\partial \theta^j} - \eta^j \frac{\partial \xi^i}{\partial \theta^j} \tag{5}$$

where $[\xi, \eta]$ denotes the commutator between the vector fields.

III). **Covector field**. $T = T_j = \frac{\partial f}{\partial \theta^j}$

$$\begin{aligned} (L_\xi T)_j &= \xi^k \frac{\partial T_j}{\partial \theta^k} + T_k \frac{\partial \xi^k}{\partial \theta^j} \\ &= d(L_\xi f)_j = L_\xi (df)_j \end{aligned} \tag{6}$$

i.e. Lie derivatives commute with differentials.

IV). **Bilinear form.** $T = g_{ij}$

$$L_\xi g_{ij} = \xi^s \frac{\partial g_{ij}}{\partial \theta^s} + g_{kj} \frac{\partial \xi^k}{\partial \theta^i} + g_{ik} \frac{\partial \xi^k}{\partial \theta^j} = u_{ij} \tag{7}$$

This is known as the strain tensor.

V). **Volume element.**

$$\begin{aligned} T = T_{i_1 \dots i_k} &= \sqrt{|g|}\, \epsilon_{i_1 \dots i_k} \\ &= \sqrt{|g|}\, d\theta^{i_1} \wedge \dots \wedge d\theta^{i_k} \\ &= \pm\sqrt{|g|}\, d\theta^1 \wedge \dots \wedge d\theta^k \quad \text{if no two indices are equal.} \end{aligned} \tag{8}$$

Where $\epsilon_{i_1 \dots i_k}$ is the *Levi-Civita* tensor defined as $+1, -1, 0$ depending on the indices forming an even, odd or no permutation of the first k integers. We denote by $|g|$ the absolute value of the determinant of the metric tensor $g_{ij}(\theta)$.

The volume element plays a central role in Bayesian inference. Geometrically, it gives the *surface area* of a small patch on the k-dimensional surface. For this reason is the analogous of the Lebesgue (uniform) measure on flat space. It behaves like a totally anti symmetric (skew-symmetric) tensor under coordinate transformations preserving a given orientation of the space. Volume elements can then be interpreted as differential forms of order k and as *total ignorance priors* in statistics. After some simplifications formula (3) gives,

$$L_\xi \left(\sqrt{|g|}\, d\theta^1 \wedge \dots \wedge d\theta^k \right) = \frac{1}{2} g^{im} \left(L_\xi g_{im} \right) \sqrt{|g|}\, d\theta^1 \wedge \dots \wedge d\theta^k \tag{9}$$

Where g^{im} denotes the inverse of the (Fisher information) matrix, g_{im}. Expressions involving g are always functions of θ but this will be kept implicit to simplify the notation. Notice that the effect of taking the Lie derivative of the volume element is to multiply it by one half the *trace* of the strain tensor defined in (7).

VI). **Leibniz' rule.** If T and R are arbitrary tensors and $T \otimes R$ denotes the tensor product between them, then,

$$L_\xi (T \otimes R) = (L_\xi T) \otimes R + T \otimes (L_\xi R) \tag{10}$$

3. Robustness of Probability Distributions Defined on the Parameter Space

Probability measures defined on Θ (e.g. priors and posteriors) can be seen as providing alternative ways of measuring the surface area of patches on the manifold. The parameterization Θ is only a convenient artifact to be able to write the formulas explicitly without having to perform the integration directly over the functional space of probability measures. But the parameterization is arbitrary and therefore it must be immaterial. The formulas should show this invariance under reparameterization up front.

From this point of view, it is necessary to leave tradition and move from the usual interpretation of probability densities as functions with transformation rules governed by the so called *change of variables theorem* to scalar fields with no transformation rules whatsoever. Traditionally, density functions are integrated with *integrals of the second kind* which are just multiple integrals to be handled independently of any metric which may be defined on the space. But if we move to densities as scalar fields, then they have to be integrated with *integrals of the first kind* with respect to the volume element in the desired parameterization. A probability measure on Θ will then be written as a differential form,

$$\pi(\theta)\sqrt{|g|}\, d\theta^1 \wedge \ldots \wedge d\theta^k \tag{11}$$

where $\pi(\theta)$ is a scalar field. Notice that $\pi(\theta)$ is just the Radon-Nikodym derivative of the probability measure defined by 11 with respect to the volume element measure. In other words the density (as a scalar field) is given relative to *total ignorance.* Notice also that (11) is wonderfully invariant. If what was called θ we now call θ' all we need to do to (11) is to prime the θ's and we get the formula in the new coordinate system.

An example may help to fix the ideas. For example, $\frac{1}{2\pi}e^{-\frac{1}{2}r^2}$ is the probability density scalar field of the standard bivariate Gaussian on the Euclidean plane parameterized with polar coordinates r, θ. The same density in cartesian coordinates is just $\frac{1}{2\pi}e^{-\frac{1}{2}(x^2+y^2)}$. i.e. the point on the Euclidean plane with the two labels (x, y) and $[r, \theta]$ has exactly the same density relative to (euclidean) ignorance since $r^2 = x^2 + y^2$.

Noteworthy, this almost trivial change in point of view, helps to clarify an old puzzle of inference: *How come that complete ignorance about a value $x \in [0,1]$ is not complete ignorance about $y = x^2 \in [0,1]$?*. In other words, the *change of variable theorem* transforms the uniform density of x into the non uniform density $\frac{1}{2}y^{-1/2}$ for y. This is regarded as paradoxical, for, it is claimed, indifference about the number x should produce indifference about the number $y = x^2$. When considering *densities as scalar fields* there is no puzzle. The puzzle arises from the insistence, of the change of variables theorem, to keep the underlying measure to be the same (Lebesgue measure on $[0,1]$ in this case) for x and for y. But, x and $y = x^2$ are just two different numerical labels for events (perhaps measurements of the same thing but in two systems of units) so whatever it was labeled $\frac{1}{2}$, say, with x is relabeled as $\frac{1}{4}$ by y. Therefore, the labels $x = 0.5$ and $y = 0.25$ must have the same chance of occurrence. In fact, they do. But the change of variables theorem hides it by shifting the Jacobian from the volume element, where it belongs, to the density, where it does not belong. Our formula (11), assigns constant density to the numbers in $[0,1]$ in **all coordinate systems**, linear or non linear transformations of x.

Formula (11) is composed as the product of two invariants. The scalar field density and the volume element. Remember that the volume element is invariant under all reparameterizations preserving orientation. When changing coordinate systems, the two parts remain the same.

3.1. The Robustness of Total Ignorance

The rate of change of the total ignorance prior along a deformation of the model given by a vector field ξ is given by (9). Replacing (7) into (9) and simplifying, we can write

$$L_\xi\left(\sqrt{|g|}\, d\theta^1 \wedge \ldots \wedge d\theta^k\right) = \left(\frac{1}{2}g^{ij}\nabla g_{ij} + \nabla\right)\cdot\xi\left(\sqrt{|g|}\, d\theta^1 \wedge \ldots \wedge d\theta^k\right) \tag{12}$$

Thus, a local deformation of the hypothesis space, does not change the state of total ignorance if it is along a vector field ξ solving the partial differential equation:

$$\left(\frac{1}{2} g^{ij} \nabla g_{ij} + \nabla\right) \cdot \xi = 0 \tag{13}$$

Therefore, if the metric tensor is independent of θ (e.g. for location models), equation 13 reduces to the familiar:

$$\nabla \cdot \xi = 0 \tag{14}$$

Equation (14), and the more general equation, (13), encode the idea of *conservation of ignorance.* Deformations of the model that satisfy them, are precisely those that do not create nor destroy information.

3.2. Robustness of Priors and Posteriors

To compute the Lie derivative of an arbitrary distribution over the parameter space we apply Leibniz' rule (10), to the differential form (11),

$$\begin{aligned} L_\xi \left(\pi \sqrt{|g|}\, d\theta^1 \wedge \ldots \wedge d\theta^k\right) &= \pi L_\xi \left(\sqrt{|g|}\, d\theta^1 \wedge \ldots \wedge d\theta^k\right) + \left(\sqrt{|g|}\, d\theta^1 \wedge \ldots \wedge d\theta^k\right) L_\xi \pi \\ &= \left(\frac{1}{2} g^{im} L_\xi g_{im} + L_\xi \log \pi\right) \pi \sqrt{|g|}\, d\theta^1 \wedge \ldots \wedge d\theta^k \end{aligned} \tag{15}$$

where we have used the fact that tensor multiplication by a scalar field is just regular multiplication and equations (5), and (9). Thus, robustness is obtained when π, ξ and the metric g_{im} are connected through the partial differential equation:

$$L_\xi \log \pi = -\frac{1}{2} g^{im} L_\xi g_{im} \tag{16}$$

This is again an equation expressing conservation of information. There is invariance along ξ when the gradient of the log-likelihood (of the prior or posterior π) projected onto ξ exactly eliminates the sources of information created by the deformation. But, if the deformation ξ does not artificially create information, i.e. if it preserves the state of complete ignorance then, by (9) and (4), the general equation (16) simplifies to,

$$\xi \cdot \nabla \pi = 0 \tag{17}$$

This is not surprising. Since the gradient, $\nabla \pi$, is always orthogonal to the level surfaces $\{\theta : \pi(\theta) = c\}$ we can rewrite (17) as,

Theorem 1 *Let ξ be an ignorance preserving vector field. Then, a probability distribution on Θ with scalar field density π, is robust with respect to deformations along ξ iff π puts constant probability mass on the integral curves of ξ.*

In other words, prior (or posterior) probabilities do not change, only when the deformations remain inside the level surfaces of the density.

3.3. Robustness of Entropic Priors

The name and the derivation of entropic priors for the manifold of discrete distributions are due to Skilling (see [11]). The generalization to arbitrary regular parametric models appears in the same volume in [7], see also [9], [8], [10].

Entropic priors are defined by their scalar field density. In the coordinate system of the θ's they are given by

$$\pi(\theta) = \frac{1}{c} e^{-\alpha I(\theta : \theta_0)} \tag{18}$$

where, $I(\theta : \theta_0)$ is the Kullback number between the distributions labeled by θ and a given initial value θ_0. The parameter $\alpha \geq 0$ has to be large enough, so that the constant of integration, c, is finite. Equation (18) has an easy interpretation: The chance of θ decreases exponentially fast with the Kullback distance from θ_0 and the parameter α controls the sensitivity to changes in the distance. Since the density is given as a scalar field, this interpretation holds in **all coordinate systems** i.e. for all parameterizations of the model.

To compute the sensitivity of entropic priors, with respect to deformations of the model, we need only to replace (18) into (15). If we denote by Π the entropic prior probability measure, we have:

$$\frac{dL_\xi \Pi}{d\Pi} = \frac{1}{2} g^{im} L_\xi g_{im} - \alpha \xi \cdot \nabla I(\theta : \theta_0) \tag{19}$$

where the left hand side denotes the Radon-Nikodym derivative of the (signed) measure $L_\xi \Pi$ with respect to Π. If ξ preserves ignorance, the first term of the sum in (19) is zero and

$$\frac{dL_\xi \Pi}{d\Pi} = -\alpha \xi \cdot \nabla I(\theta : \theta_0). \tag{20}$$

Equation (20) contains a lot of information about the nature of entropic priors. Firstly, notice that the parameter α controls the size of the derivative. In other words, the smaller α is, the more robust the inferences are. Jeffreys priors appear as tautological winners: *Ignorance priors are robust with respect to deformations preserving ignorance.* Besides this tautological robustness, obtained when $\alpha = 0$, we have robustness when ξ is orthogonal to ∇I. In other words, when the integral curves of ξ are located on the surface of *entropy spheres* centered at θ_0 i.e. $\{\theta : I(\theta : \theta_0) = \text{const.}\}$. This justifies the following:

Definition 2 *A vector field defined on the statistical model is said to be* information preserving at θ_0 *if it does not change ignorance and has integral curves contained in the level surfaces of* $I(\theta : \theta_0)$.

This definition makes true the following:

Theorem 2 *Entropic priors are robust with respect to deformations preserving information at the initial guess,* θ_0

It is well known that the Kullback number generates the Riemannian metric (see [7] or [9]). In fact, a simple Taylor expansion of the Kullback number produces:

$$\begin{aligned} \xi \cdot \nabla I(\theta : \theta_0) &= \langle \xi, \theta - \theta_0 \rangle_\theta + o(|\theta - \theta_0|) \\ &= \xi^i v^j g_{ij}(\theta) + o(|\theta - \theta_0|) \end{aligned} \tag{21}$$

where $v = \theta - \theta_0$ is in fact a tangent vector at θ when θ_0 approaches θ. From here, we obtain the following

Theorem 3 *Entropic priors are robust with respect to local isometries at θ_0*

By local isometries at θ_0 we mean deformations that close to θ_0 do not change the metric. These deformations just send points in spherical orbits around θ_0. The previous theorem shows that entropic priors, as opposed to other classes of priors, are very compatible with the intrinsic Riemannian geometry of the hypothesis space. Entropic priors are sensitive only to deformations that change the *intrinsic form* of the model around θ_0.

4. Example: The Gaussians

The main purpose of this section is to illustrate some of the formulas introduced in this paper with a concrete example. An in depth analysis of the robustness of Gaussians, however, is beyond the scope of the present article.

The gaussian distributions form a two dimensional Riemannian space. The metric tensor (Fisher information matrix) in the coordinate system $\theta = (\theta^1, \theta^2) = (\mu, \sigma)$ is diagonal with $g_{11} = 1/\sigma^2$, $g_{22} = 2/\sigma^2$. Thus,

$$g^{ij} \nabla g_{ij} = \sigma^2 \left(0, \frac{-2}{\sigma^3}\right) + \frac{\sigma^2}{2} \left(0, \frac{-4}{\sigma^3}\right) \tag{22}$$

The vector fields $\xi = (\xi^1, \xi^2)$ that preserve ignorance are given, from (13) and (22), by

$$\frac{\partial \xi^1}{\partial \mu} + \frac{\partial \xi^2}{\partial \sigma} - \frac{2}{\sigma} \xi^2 = 0 \tag{23}$$

This can be easily shown to have the general solution:

$$\xi = \left(h(\sigma) + \frac{2}{\sigma}\psi - \frac{\partial \psi}{\partial \sigma}, \frac{\partial \psi}{\partial \mu}\right) \tag{24}$$

where h is an arbitrary differentiable function of σ, and ψ is an arbitrary differentiable function of μ and σ, with continuous second order partial derivatives.

4.1. Entropic Prior on the Gaussians

Consider the entropic prior model on the manifold of Gaussians with initial guess $\theta_0 = (0, 1)$ i.e. the standard normal distribution. Straight forward computations show (18) to be,

$$\pi(\mu, \sigma) = \frac{1}{c(\alpha)} \sigma^\alpha e^{-\frac{\sigma^2}{2\alpha}} e^{-\frac{\mu^2}{2\alpha}} \tag{25}$$

Equation (25) is the scalar field density relative to the volume element:

$$\sigma^{-2}\, d\mu \wedge d\sigma \tag{26}$$

The symbolic manipulator MAPLE shows that, for $\alpha > 1$

$$c(\alpha) = 2^{(\alpha/2-1)}\sqrt{\pi}\alpha^{-\alpha/2}\Gamma\left(\frac{1}{2}(\alpha - 1)\right) \tag{27}$$

The integral of (25) with respect to (26) diverges for $\alpha \leq 1$. I believe this to be the reason of why Jeffreys, and many others after him, thinks that the volume element (26) (i.e. $\alpha = 0$) is too uninformative. The prior distribution obtained at the first divergent value of c, when $\alpha = 1$, produces posterior inferences remarkably similar to the popular conjugate prior for this case. Even for two observations. This suggests to extend the definition of uninformative prior to include all the entropic priors with divergent c. The boundary value of α (in this case $\alpha = 1$) can be used to approximate the frequentist methods.

The level curves of (25), are closed curves on the upper half plane (μ, σ) with equations,

$$\frac{\mu^2}{2\alpha} + \frac{\sigma^2}{2\alpha} - \alpha \log \sigma = k \tag{28}$$

where k depends on α. Computer experiments show these curves to be similar to ellipses centered about $(0, 1 - \epsilon(\alpha))$ with ϵ tending to zero as α increases. Therefore, there is robustness when the velocity vectors (24) are tangent to the level curves, (28). This happens when

$$\left(\sigma^2 - \alpha^2\right) \frac{1}{\mu} \frac{\partial \Psi}{\partial \mu} - \sigma \frac{\partial \Psi}{\partial \sigma} - 2\Psi + \sigma h(\sigma) = 0. \tag{29}$$

Preliminary analysis indicates that equation (29) imposes a heavy restriction on the deformations for which there is robustness. The group of isometries of the Gaussians, together with theorem 3, could be used to find the desired deformations around $\theta_0 = (0, 1)$. It is possible to show, that the group of direct isometries of the Gaussians, is that of the Lobachevskian plane. This group, is known to be isomorphic to the orthochronous connected component of the identity of the Lorentz group for three dimensional space-time (i.e. a space with metric: $x^2 + y^2 - t^2$, see [9]).

5. Conclusions

In retrospect, this paper should be considered a first attempt to demonstrate that it makes sense to use Lie derivatives for quantifying Bayesian robustness. No doubt, the geometrization of inference provides a powerful language for asking questions about statistical procedures. As usual, geometry brings the paraphernalia of visual imagery that embodies the objects of study and allows to *see* the theorems. Looking ahead, to the (immediate) future, we can anticipate that many of the successful applications of modern geometry to physics might be reproduced, for the theory of Inference. Stokes theorem will begin to play a central role in Bayesian robustness, for the very simple reason that the Lie derivatives of priors and posteriors are again differential forms ready to be integrated over patches. There is also room for connections and gauges, square roots of Laplacians, Lie algebras and index theorems. We need to find the people with the *guts* to do it.

References

[1] Shun-ichi Amari. *Differential-Geometrical Methods in Statistics*, volume 28 of *Lecture Notes in Statistics*. Springer-Verlag, 1985.

[2] B.A. Dubrovin, A.T. Fomenko, and S.P. Novikov. *Modern Geometry–Methods and Applications, Part-I*, volume GTM 93 of *Graduate Texts in Mathematics*. Springer-Verlag, 1984.

[3] F. R. Hampel. A general qualitative definition of robustness. *Ann. Math. Statist.*, 42:1887–1896, 1971.

[4] P. J. Huber. *Robust Statistics*. John Wiley and Sons, 1981.

[5] J. B. Kadane, editor. *Robustness of Bayesian Analyses*. North-Holland, 1981.

[6] J. Pfanzagl. *Contributions to a general asymptotic statistical theory*. Lecture Notes in Statistics. Springer-Verlag, New York, 1982.

[7] Carlos C. Rodríguez. The metrics induced by the Kullback number. In John Skilling, editor, *Maximum Entropy and Bayesian Methods*. Kluwer Academic Publishers, 1989.

[8] Carlos C. Rodríguez. Objective Bayesianism and geometry. In Paul F. Fougère, editor, *Maximum Entropy and Bayesian Methods*. Kluwer Academic Publishers, 1990.

[9] Carlos C. Rodríguez. Entropic priors. Available in electronic form on the Internet "gopher", Oct. 1991. `gopher cscgoph2.albany.edu 2-4-12-1-1`.

[10] Carlos C. Rodríguez. From Euclid to entropy. In W. T. Grandy, Jr., editor, *Maximum Entropy and Bayesian Methods*. Kluwer Academic Publishers, 1991.

[11] John Skilling. Classical Max Ent data analysis. In John Skilling, editor, *Maximum Entropy and Bayesian Methods*. Kluwer Academic Publishers, 1989.

LOCAL POSTERIOR ROBUSTNESS WITH PARAMETRIC PRIORS : MAXIMUM AND AVERAGE SENSITIVITY

Sanjib Basu
Department of Mathematical Sciences
University of Arkansas, Fayetteville, AR 72701

Sreenivasa Rao Jammalamadaka *
Department of Statistics and Applied Probability
University of California, Santa Barbara, CA 93106

and Wei Liu
Ciba Geigy, Summit, NJ 07901

ABSTRACT. The local sensitivity of a posterior quantity $\rho(P)$ to the choice of the prior P is considered. When the prior P_λ is indexed by parameter λ, a natural measure is the total derivative of $\rho(P_\lambda)$ w.r.t. λ. Total derivative, however, is direction specific. To measure the local sensitivity of $\rho(P_\lambda)$ to specification of λ, one may either use the norm (maximum over all directions) of the total derivative or alternatively, the average sensitivity which evaluates the average of this total derivative over all directions. Simple expressions are given for the maximum and average sensitivity which make their evaluations very easy. Discussion and several examples illustrate implications of these ideas.

1. Introduction

Bayesian paradigm requires one to specify two parametric models; the sampling density $f(X|\theta)$ and the prior $P(\theta)$. However, in practice, knowledge about these models are never accurate, and such specifications are only approximations or guesses at best. Hence, sensitivity of the final action to deviations of these various inputs from their idealized models is of much concern. As Tukey (1960) writes, "A tacit hope in ignoring deviations from ideal models was that they would not matter; that statistical procedures which are optimal under the strict model would still be approximately optimal under the approximate model. Unfortunately, it turned out that this hope was often drastically wrong; even mild deviations often have much larger effects than were anticipated by most statisticians".

Robustness studies, in both Classical and Bayesian statistics, can broadly be divided into two subgroups; global sensitivity analysis and local or infinitesimal approach. The former examines the effect of misspecification, when the true model may or may not be close to the idealized one. In the Bayesian context, global sensitivity to misspecification of the prior has been expounded by many, see Berger (1993), Wasserman (1992), Basu and DasGupta (1992), Rivier et al. (1990), and the references therein. In contrast, local sensitivity studies explore the effect of infinitesimal perturbations from the idealized model. Recent advances in this area include Rodríguez (1994), Ruggeri and Wasserman (1993), and

*Research supported in part by ONR Grant number N00014-93-1-0174.

G. R. Heidbreder (ed.), Maximum Entropy and Bayesian Methods, 97–106.

Skilling (1990). Our efforts in this article will be directed towards studying local sensitivity of Bayesian analysis to the choice of the prior.

Formally, we observe data X from the sampling density $f(x|\theta)$. The observed likelihood function $f(X|\theta)$ will be denoted by $\ell(\theta)$ (with conditioning X understood), and $P(\cdot)$ will denote the prior distribution on θ. Let $m(P) = \int_\Theta \ell(\theta)\, dP(\theta)$ denote the marginal w.r.t. prior P. Given the likelihood $\ell(\cdot)$ and the prior $P(\cdot)$, the posterior probability distribution, defined as $P(A|X) = \frac{1}{m(P)} \int_A \ell(\theta)\, dP(\theta)$ for any set A, will be denoted by $P(\cdot|X)$ (with dependence on $\ell(\theta)$ understood). Similarly, $\pi(\cdot)$ and $\pi(\cdot|X)$ will respectively denote the prior and the posterior densities (whenever appropriate). We will use $\rho(P)$ or ρ_P to denote a posterior quantity (such as the posterior mean) corresponding to the prior P.

As we mentioned before, prior specification is typically imprecise. Thus, in reality, we have a multiplicity of P as possible choices of the prior, from which we choose a single P_0 as our idealized prior. We will use $\mathcal{P}$ to denote the class of all plausible priors. Sometimes, the prior class $\mathcal{P}$ is indexed by a parameter. For example, we may decide to use $P = N(\mu, \tau^2)$, but are not sure about any specific values of μ and τ^2, thus leading to the class $\{ N(\mu,\tau^2) : (\mu,\tau^2)^T \in (-\infty,\infty) \otimes (0,\infty) \}$. Such parametric classes will be denoted by $\mathcal{P}_\Lambda = \{P_\lambda : \lambda \in \Lambda\}$. We will often assume that the indexing set $\Lambda \subseteq \Re^k$. In other situations, when any particular parametric form for the prior is not apparent, one uses a nonparametric class, such as an ε-contamination class $\mathcal{P}^\varepsilon$. An ε-contamination class arises when one is $100(1-\varepsilon)\%$ certain about the idealized P_0 as the choice of the prior, and $100\,\varepsilon\%$ uncertain $(0 \le \varepsilon < 1)$, thus resulting in the class $\mathcal{P}^\varepsilon = \{P : P = (1-\varepsilon)P_0 + \varepsilon Q\}$ where Q is any arbitrary prior distribution.

When we have a class $\mathcal{P}$ of plausible priors, and an idealized prior P_0, the first question that comes to mind is : "if the true prior Q in $\mathcal{P}$ is close to the idealized P_0, is it guaranteed that $\rho(Q)$ will be close to $\rho(P_0)$?" In a limiting sense, this amounts to continuity of $\rho(P)$ (as a function of P) at $P = P_0$, and in the terminology of classical robustness literature, this corresponds to Hampel's (1971) definition of *qualitative robustness.* Note that we are posing the question in terms of $\rho(P)$, however, an exactly similar question can be posed in terms of the posterior distribution $P(\cdot|X)$. If *qualitative robustness* is achieved, a second natural question to ask would be : "is the change in $\rho(P)$ bounded by the change in P ?". To formalize this question, suppose $d(\cdot,\cdot)$ is a metric on the space of priors, and $\nu(\cdot,\cdot)$ is a metric on the space of the posterior quantities $\rho(P)$. Then, we can pose our question as follows: "does $\exists$ an $\alpha > 0$ such that $\nu(\rho(P),\rho(P_0)) \le M\,[d(P(\cdot),P_0(\cdot)]^\alpha$ for some $M > 0$?". Mathematically, this is a Lipschitz condition of order α. Basu, Jammalamadaka and Liu (1993) termed this second notion as *stability*, and studied the *qualitative robustness* and *stability* of $\rho(P)$ and $P(\cdot|X)$.

Qualitative robustness and *stability* are very necessary but rather weak characterizations of robustness. A local sensitivity study should also explore the rate of change of $\rho(P)$ as P deviates infinitesimally from the idealized P_0. If the prior class $\mathcal{P} = \mathcal{P}_\Lambda$ is parametric and $\Lambda \subseteq \Re$, this is easy. For $P_0 = P_{\lambda_0}$ and $\rho(P_\lambda) = \rho(\lambda)$, one simply computes the derivative $\rho'(\lambda) = \frac{d}{d\lambda}\rho(\lambda)$ at $\lambda = \lambda_0$. If $\rho'(\lambda_0)$ is small, it suggests that $\rho(\lambda)$ is not sensitive to mild perturbations of P_λ around $\lambda = \lambda_0$. The situation gets complicated when $\Lambda \subseteq \Re^k$. We consider a more complex setup when $\boldsymbol{\rho}$ is also multidimensional, i.e., $\boldsymbol{\rho} = [\rho_1,\ldots,\rho_n]^T$. A proper concept of derivative in such multivariate situations is the total derivative. In section

2.1., we establish sufficient conditions for total differentiability of a posterior quantity $\rho(\boldsymbol{\lambda})$. However, total derivative is direction specific, its value depends on the direction of deviation $\boldsymbol{\lambda}$ from the idealized value $\boldsymbol{\lambda}_0$. We thus evaluate the norm of the total derivative, or its maximum value over all directions. Theorem 2 supplies an easy formula for evaluation of this norm. An alternative viewpoint would suggest computing the average of the total derivative over all directions. This leads us to average sensitivity. Section 2.3. discusses this issue and again supplies simple expressions for ease of computation. Several univariate and multivariate applications are explored in section 3.. Finally, section 4. briefly discusses the issue of quantification of local sensitivity over nonparametric prior classes.

Use of derivatives to quantify the sensitivity of a posterior quantity is not new. Diaconis and Freedman (1986), and Ruggeri and Wasserman (1993) evaluated norm of Fréchet derivatives over the class of all signed measures and/or its appropriate nonparametric subclasses. Rodríguez (1994) used the concept of Lie derivative to quantify the intrinsic robustness of a hypothesis space. To our knowledge, such explorations over parametric classes have not been explicitly considered before.

2. Parametric prior classes

2.1. Total derivative

Mathematical and numerical convenience often attracts one to use a prior of a special parametric form (this is more true in multivariate situations). For example, in a linear model setup : $\underset{\sim}{Y} \sim N(X\underset{\sim}{\beta}, \sigma^2 I)$ with $\sigma^2 > 0$ known, it is common to use a $N(\underset{\sim}{\mu}, \Gamma)$ prior for $\underset{\sim}{\beta}$. Even if such a formulation is justified, specification of the prior hyperparameters poses a secondary problem, which is often handled through Empirical Bayes and/or Hierarchical Bayes methods, or the hyperparameters are specified as inputs by the user. Again, these inputs are never exactly accurate, so that local sensitivity to a particular choice of the hyperparameters is of concern.

Let $\boldsymbol{\lambda} = [\lambda_1, \ldots, \lambda_k]^T$ denote a generic element of Λ, and let $\mathcal{P}_\Lambda = \{P_{\boldsymbol{\lambda}} : \boldsymbol{\lambda} \in \Lambda\}$ be the class of all plausible parametric priors from which we choose $P_{\boldsymbol{\lambda}_0}$ as an idealized prior. We will assume that Λ is an open subset in $\Re^k$ so that for each $\boldsymbol{\lambda}_0 \in \Lambda$, $\exists$ a neighborhood N_0 of $\boldsymbol{\lambda}_0$ such that $\boldsymbol{\lambda}_0 \in N_0 \subseteq \Lambda$. Let $\rho(P_{\boldsymbol{\lambda}}) = \rho(\boldsymbol{\lambda})$ be the posterior quantity of interest. $\rho(\boldsymbol{\lambda})$ may be univariate (a single posterior quantity), or multivariate (a vector of such quantities); in general, we will assume that $\boldsymbol{\rho}$ is n-dimensional and $\boldsymbol{\lambda}$ is k-dimensional, i.e., $\boldsymbol{\rho} = [\rho_1, \ldots, \rho_n]^T : \Lambda \subseteq \Re^k \mapsto \Re^n$. Often, we will focus on ratio-linear posterior quantities, i.e., $\boldsymbol{\rho}(\boldsymbol{\lambda}) = [\rho_1(\boldsymbol{\lambda}), \ldots, \rho_n(\boldsymbol{\lambda})]^T = [\frac{1}{m(P_{\boldsymbol{\lambda}})} \int h_i(\theta)\, \ell(\theta)\, dP_{\boldsymbol{\lambda}}(\theta)\,]_{i=1}^n$. Such quantities will be denoted by $\boldsymbol{\rho}^{\boldsymbol{h}}(\boldsymbol{\lambda})$.

Our concern is the local sensitivity of the posterior quantity $\boldsymbol{\rho}(\boldsymbol{\lambda})$ to the particular choice of the parameter $\boldsymbol{\lambda} = \boldsymbol{\lambda}_0$. The weaker local sensitivity properties of $\boldsymbol{\rho}(\boldsymbol{\lambda})$, namely, *qualitative robustness* and *stability*, are explored in Basu, Jammalamadaka and Liu (1993). Here, we focus on measuring the rate of change of $\boldsymbol{\rho}(\boldsymbol{\lambda})$ to small perturbations in $\boldsymbol{\lambda}$, in other words, the derivative of $\boldsymbol{\rho}(\boldsymbol{\lambda})$ w.r.t. $\boldsymbol{\lambda}$ at $\boldsymbol{\lambda} = \boldsymbol{\lambda}_0$. Let $\nabla\boldsymbol{\rho}(\boldsymbol{\lambda}_*) = [[\frac{\delta \rho_i(\boldsymbol{\lambda}_*)}{\delta \lambda_j}]_{j=1}^k]_{i=1}^n$ denote the matrix of partial derivatives of $\boldsymbol{\rho}$ w.r.t. $\boldsymbol{\lambda}$ at $\boldsymbol{\lambda} = \boldsymbol{\lambda}_*$. However, the appropriate derivative in multivariate calculus is not the partial derivative, but rather, the *total derivative* $T\boldsymbol{\rho}_{\boldsymbol{\lambda}_*}$. The function $\boldsymbol{\rho} : \Lambda \subseteq \Re^k \mapsto \Re^n$ is called (totally) differentiable at $\boldsymbol{\lambda}_* \in \Lambda$ if $\exists$ a linear function $T\boldsymbol{\rho}_{\boldsymbol{\lambda}_*} : \Re^k \mapsto \Re^n$ such that $\frac{\|\boldsymbol{\rho}(\boldsymbol{\lambda}_*+\boldsymbol{v}) - \boldsymbol{\rho}(\boldsymbol{\lambda}_*) - T\boldsymbol{\rho}_{\boldsymbol{\lambda}_*}(\boldsymbol{v})\|_n}{\|\boldsymbol{v}\|_k} \to 0$ as $\|\boldsymbol{v}\|_k \to 0$ ($\|\boldsymbol{v}\|_k = \sqrt{v_1^2 + \ldots + v_k^2}$ denotes the standard Euclidean norm on the k-dimensional space $\Re^k$). Note that each $\boldsymbol{\lambda}_* \in \Lambda$ gives rise to a distinct linear transformation $T\boldsymbol{\rho}_{\boldsymbol{\lambda}_*}$.

The existence of the total derivative $T\boldsymbol{\rho}_{\boldsymbol{\lambda}_*}$, however, is easier to prove through the existence and continuity of the partial derivatives $\frac{\delta \rho_i(\boldsymbol{\lambda}_*)}{\delta \lambda_j}$. A well known result in differential calculus states that the total derivative $T\boldsymbol{\rho}$ exists over a neighborhood N_0 of $\boldsymbol{\lambda}_0$ and is continuous on the space $\mathcal{L}(\Re^k, \Re^n)$ of linear transformations from $\Re^k \mapsto \Re^n$ iff the partial derivatives $\frac{\delta \rho_i}{\delta \lambda_j}$ exist and are continuous on N_0 $\forall 1 \le i \le n$, $1 \le j \le k$ (Rudin (1976), p 219). We use this result to investigate differentiability of the ratio-linear posterior quantity $\boldsymbol{\rho}^{\boldsymbol{h}}(\boldsymbol{\lambda})$ in Theorem 1. It is easier to state the result in terms of densities, thus we will assume that each $P_{\boldsymbol{\lambda}} \in \mathcal{P}_\Lambda$ has a density $\pi_{\boldsymbol{\lambda}}(\theta) = \pi(\theta, \boldsymbol{\lambda})$.

Theorem 1 *Let N_0 be a neighborhood of $\boldsymbol{\lambda} \in \Lambda$. Assume $|\ell(\theta)| \le M_0$, and for all $1 \le i \le n$, $|h_i(\theta)\,\ell(\theta)| \le M_i$ $\forall \theta \in \Theta$. We further assume the following :*

(i) For each $1 \le j \le k$, the partial derivative $\frac{\delta}{\delta \lambda_j}\pi(\theta, \boldsymbol{\lambda})$ exists $\forall$ $(\theta, \boldsymbol{\lambda}) \in \Theta \otimes N_0$, and is continuous as a function of $\boldsymbol{\lambda}$ for every $\theta \in \Theta$.

(ii) For every $1 \le j \le k$, $\exists$ a function $g_j(\theta)$ on Θ such that (a) $g_j(\theta) \ge 0$ $\forall \theta \in \Theta$, (b) $\int_\Theta g_j(\theta\, d\mu(\theta) \le L_j < \infty$, and (c) $|\frac{\delta}{\delta \lambda_j}\pi(\theta, \underset{\sim}{\lambda})| \le g_j(\theta)$ $\forall\, (\theta, \underset{\sim}{\lambda}) \in \Theta \otimes N_0$.

Then the total derivative $T\boldsymbol{\rho}^{\boldsymbol{h}}_{\boldsymbol{\lambda}}$ of the posterior quantity $\boldsymbol{\rho}^{\boldsymbol{h}}(\boldsymbol{\lambda})$ exists for $\boldsymbol{\lambda} \in N_0$ and $T\boldsymbol{\rho}^{\boldsymbol{h}}$ is continuous on $\mathcal{L}(\Re^k, \Re^n)$.

Proof : Let $N_i(\boldsymbol{\lambda}) = \int h_i(\theta)\,\ell(\theta)\,\pi(d\theta, \boldsymbol{\lambda})$, thus $\rho_i(\boldsymbol{\lambda}) = \frac{N_i(\boldsymbol{\lambda})}{m(\pi_{\boldsymbol{\lambda}})}$, $1 \le i \le n$. The conditions of the theorem ensure that for $1 \le i \le n$, $1 \le j \le k$, and $\forall \boldsymbol{\lambda} \in N_0$, the partial derivative $\frac{\delta}{\delta \lambda_j} N_i(\boldsymbol{\lambda})$ exists and $= \int h_i(\theta)\,\ell(\theta)\,\frac{\delta}{\delta \lambda_j}\pi(\theta, \boldsymbol{\lambda})$ by the Dominated Convergence theorem. Continuity of $\xi_j(\boldsymbol{\lambda}) = \frac{\delta}{\delta \lambda_j}\pi(\theta, \boldsymbol{\lambda})$ and another application of D.C.T. prove that $\frac{\delta}{\delta \lambda_j} N_i(\boldsymbol{\lambda})$ is continuous in $\boldsymbol{\lambda} \in N_0$. Similarly, $\frac{\delta}{\delta \lambda_j} m(\pi_{\boldsymbol{\lambda}})$, and hence $\frac{\delta}{\delta \lambda_j}\rho_i^{\boldsymbol{h}}(\boldsymbol{\lambda})$ exist and are continuous in $\boldsymbol{\lambda}$ for every $1 \le i \le n$, $1 \le j \le k$. The proof of the theorem follows ∎

2.2. Maximum sensitivity

Our interest lies in measuring the rate of change of $\boldsymbol{\rho}(\boldsymbol{\lambda})$ as $\boldsymbol{\lambda}$ deviates from $\boldsymbol{\lambda}_0$. In particular, since we are not sure about any specific direction of deviation, we would like to find the maximum rate of change of $\boldsymbol{\rho}(\boldsymbol{\lambda})$ over all directions. However, note that the total derivative $T\boldsymbol{\rho}_{\boldsymbol{\lambda}_0}$ is a linear function of $\boldsymbol{v} \in \Re^k$, i.e., even if we fix a direction $\boldsymbol{v}$, $T\boldsymbol{\rho}_{\boldsymbol{\lambda}_0}(c\cdot\boldsymbol{v}) = c\cdot T\boldsymbol{\rho}_{\boldsymbol{\lambda}_0}(\boldsymbol{v})$ for any $c > 0$. Hence, $\sup_{\text{all } \boldsymbol{v} \neq \boldsymbol{0}} \|T\boldsymbol{\rho}_{\boldsymbol{\lambda}_0}(\boldsymbol{v})\|_n$ is clearly infinite. What we need is the concept of the *norm* of a linear functional, defined by $\|T\boldsymbol{\rho}_{\boldsymbol{\lambda}_0}\| = \sup_{\boldsymbol{v} \neq \boldsymbol{0}} \frac{1}{\|\boldsymbol{v}\|_k} \|T\boldsymbol{\rho}_{\boldsymbol{\lambda}_0}(\boldsymbol{v})\|_n = \sup_{\|\boldsymbol{v}\|_k = c} \frac{1}{c} \|T\boldsymbol{\rho}_{\boldsymbol{\lambda}_0}(\boldsymbol{v})\|_n$. Here $c > 0$ can be chosen arbitrarily small to make sure that $\boldsymbol{\rho}(\boldsymbol{\lambda}_0 + \boldsymbol{v})$ is well defined for all $\{\boldsymbol{v} : \|\boldsymbol{v}\|_k = c\}$ (see definition of $T\boldsymbol{\rho}_{\boldsymbol{\lambda}_0}$). Also, note that if $\boldsymbol{\lambda}$ and $\boldsymbol{\rho}$ are univariate, i.e., $k = n = 1$, then $\frac{1}{||v||_1} ||T\rho_{\lambda_0}(v)||_1 = |\frac{d\rho_\lambda(\lambda_0)}{d\lambda}|$. Thus, $\frac{1}{||\boldsymbol{v}||_k} ||T\boldsymbol{\rho}_{\boldsymbol{\lambda}_0}(\boldsymbol{v})||_n$ has an intuitive interpretation as the rate of change of $\boldsymbol{\rho}(\boldsymbol{\lambda})$ at $\boldsymbol{\lambda}_0$ in the direction of $\boldsymbol{\lambda}_0 + \boldsymbol{v}$, and we are trying to find the maximum rate over all such directions $\boldsymbol{v}$.

Direct evaluations of the total derivative $T\boldsymbol{\rho}_{\boldsymbol{\lambda}}$ and its norm, however, are hard. The next theorem expresses $||T\boldsymbol{\rho}_{\boldsymbol{\lambda}_0}||$ as a function of the partial derivatives $\frac{\delta\rho_i(\boldsymbol{\lambda})}{\delta\lambda_j}$, which are much easier to calculate.

Theorem 2 *Let Λ, $\boldsymbol{\rho}(\boldsymbol{\lambda})$, and $\bigtriangledown\boldsymbol{\rho}(\boldsymbol{\lambda})$ be as defined before. Assume $\boldsymbol{\rho}(\cdot) : \Lambda \mapsto \Re^n$ is totally differentiable at an interior point $\boldsymbol{\lambda}_0$ of Λ. Then $||T\boldsymbol{\rho}_{\boldsymbol{\lambda}_0}||^2$ = maximum eigenvalue of the $k \times k$ nonnegative definite matrix $\bigtriangledown\boldsymbol{\rho}(\boldsymbol{\lambda}_0)^T \bigtriangledown\boldsymbol{\rho}(\boldsymbol{\lambda}_0)$.*

Proof : It is well known that the total derivative $T\boldsymbol{\rho}_{\boldsymbol{\lambda}_0}$ is a linear combination of the partial derivatives, i.e., $T\boldsymbol{\rho}_{\boldsymbol{\lambda}_0}(\boldsymbol{v}) = \bigtriangledown\boldsymbol{\rho}(\boldsymbol{\lambda}_0)\,\boldsymbol{v}$ (Rudin (1976), pp. 215). Hence, $||T\boldsymbol{\rho}_{\boldsymbol{\lambda}_0}||^2 = \sup_{\boldsymbol{v}^T\boldsymbol{v}\neq 0} \frac{1}{\boldsymbol{v}^T\boldsymbol{v}} \boldsymbol{v}^T \bigtriangledown\boldsymbol{\rho}(\boldsymbol{\lambda}_0)^T \bigtriangledown\boldsymbol{\rho}(\boldsymbol{\lambda}_0)\,\boldsymbol{v}$ = maximum eigenvalue of $\bigtriangledown\boldsymbol{\rho}(\boldsymbol{\lambda}_0)^T \bigtriangledown\boldsymbol{\rho}(\boldsymbol{\lambda}_0)$ (see, for instance, Rao, C.R. (1973), p 62) ■

Corollary 1 *Suppose we consider a single posterior quantity, i.e., $\rho(\cdot) : \Lambda \subseteq \Re^k \mapsto \Re$. Then $||T\rho_{\boldsymbol{\lambda}_0}|| = \sqrt{\sum_{i=1}^{k} [\frac{\delta}{\delta\lambda_i}\rho(\boldsymbol{\lambda}_0)]^2}$.*

Proof : Immediate from Theorem 2 ■

2.3. Average sensitivity

It should be mentioned that the norm of the total derivative, or the maximum sensitivity, is a very conservative estimate in the sense that it tries to guard against large changes in $\boldsymbol{\rho}(\boldsymbol{\lambda})$ by computing the fastest rate of change over all possible directions. Another less conservative concept would be to average the rate of change over all directions. Mathematically, this amounts to evaluating $\int_{\{||\boldsymbol{v}||_k=1\}} ||T\boldsymbol{\rho}_{\boldsymbol{\lambda}_0}(\boldsymbol{v})||_n\, d\boldsymbol{v}$. But since this integral is hard to compute, we square the integrand and evaluate $\int_{\{||\boldsymbol{v}||_k=1\}} ||T\boldsymbol{\rho}_{\boldsymbol{\lambda}_0}(\boldsymbol{v})||_n^2\, d\boldsymbol{v}$ instead. The choice of "1" as the radius of the hypersphere is completely arbitrary here; any other radius leads to an equivalent definition (through the linear structure of $T\boldsymbol{\rho}_{\boldsymbol{\lambda}_0}(\boldsymbol{v})$).

Definition 1 *Assume $\boldsymbol{\rho}(\cdot) : \Lambda \subseteq \Re^k \mapsto \Re^n$ is totally differentiable at an interior point $\boldsymbol{\lambda}_0$ of Λ. Then the* average sensitivity *of the posterior quantity $\boldsymbol{\rho}(\boldsymbol{\lambda}) = \boldsymbol{\rho}(P_{\boldsymbol{\lambda}})$ w.r.t. the choice of the prior parameter $\boldsymbol{\lambda} = \boldsymbol{\lambda}_0$ is defined to be $\overline{T\boldsymbol{\rho}_{\boldsymbol{\lambda}_0}} = \frac{1}{r^3} \int_{\{||\boldsymbol{v}||_k=r\}} ||T\boldsymbol{\rho}_{\boldsymbol{\lambda}_0}(\boldsymbol{v})||_n^2\, d\boldsymbol{v}$. Here $r > 0$ is arbitrary (the definition is independent of the choice of r).*

The next theorem shows how to evaluate $\overline{T\boldsymbol{\rho}_{\boldsymbol{\lambda}_0}}$ for a totally differentiable posterior quantity $\boldsymbol{\rho}(\boldsymbol{\lambda})$.

Theorem 3 *Assume the setup of Theorem 2 with $\Lambda \subseteq \Re^k$. Then $\overline{T\boldsymbol{\rho}_{\boldsymbol{\lambda}_0}} = \frac{w_k}{k} \times$ {sum of eigenvalues of the $k \times k$ matrix $\bigtriangledown\boldsymbol{\rho}(\boldsymbol{\lambda}_0)^T \bigtriangledown\boldsymbol{\rho}(\boldsymbol{\lambda}_0)$}, where w_k = surface area of the hypersphere $\{\boldsymbol{v} : ||\boldsymbol{v}||_k = 1\} = \frac{2\pi^{k/2}}{\Gamma(k/2)}$.*

Proof : Since $T\boldsymbol{\rho}_{\boldsymbol{\lambda}_0}(\boldsymbol{v}) = \nabla\boldsymbol{\rho}(\boldsymbol{\lambda}_0)\,\boldsymbol{v}$, $||T\boldsymbol{\rho}_{\boldsymbol{\lambda}_0}(\boldsymbol{v})||^2 = \boldsymbol{v}^T A\,\boldsymbol{v}$ with $A_{(k\times k)} = \nabla\boldsymbol{\rho}(\boldsymbol{\lambda}_0)^T\,\nabla\boldsymbol{\rho}(\boldsymbol{\lambda}_0)$. Let $\beta_1,\ldots,\beta_k$ be the eigenvalues of A, i.e., $A = P^T\,D\,P$ where $D_{(k\times k)} = \text{diag}\{\beta_1,\ldots,\beta_k\}$ and P is an orthogonal matrix. Let $\boldsymbol{u} = (u_1,\ldots,u_k)^T = P\,\boldsymbol{v}$. Then $\overline{T\boldsymbol{\rho}_{\boldsymbol{\lambda}_0}} = \int_{\{||\boldsymbol{v}||_k=1\}} (\boldsymbol{v}^T A\,\boldsymbol{v})\,d\boldsymbol{v} = \sum_{i=1}^{k}\{\beta_i \int_{\{||\boldsymbol{u}||_k=1\}} u_i^2\,d\boldsymbol{u}\}$. Clearly, $S = \int_{\{||\boldsymbol{u}||_k=1\}} u_i^2\,d\boldsymbol{u}$ is independent of "i", and $k\,S = \int_{\{||\boldsymbol{u}||_k=1\}} \{\sum_{i=1}^{k} u_i^2\}d\boldsymbol{u} = w_k$. This completes the proof of the theorem ■

Corollary 2 *Suppose $\rho(\lambda)$ is univariate, i.e, $\rho(\cdot) : \Lambda \subseteq \Re^k \mapsto \Re$. Then*
$$\frac{k}{w_k}\,\overline{T\boldsymbol{\rho}_{\boldsymbol{\lambda}_0}} = \sum_{i=1}^{k}[\tfrac{\delta}{\delta\,\lambda_i}\,\rho(\boldsymbol{\lambda}_0)]^2 = ||T\rho_{\boldsymbol{\lambda}_0}||^2.$$

Proof : Follows trivially from Theorem 3 ■
Remark : It is clear that for $n = 1$, these two concepts of maximum and average sensitivity are equivalent (see Corollaries 1 and 2) □

3. Examples

We look at several applications of Theorem 2 and Theorem 3 in this section. The first three examples evaluate the maximum sensitivity of posterior quantities, while Example 4 examines average sensitivity.

Example 1 : Suppose we observe X from $N(\theta,\sigma^2)$, where $\sigma^2 > 0$ is known, and decide to use a $N(\mu,\tau^2)$ prior for θ. Thus, $P_{\boldsymbol{\lambda}} = N(\mu,\tau^2)$ with $\boldsymbol{\lambda} = (\mu,\tau^2)^T \in \Re\otimes(0,\infty)$. Our interest is the Bayes estimate of θ under squared-error loss, i.e., $\rho(\mu,\tau) = E_{\boldsymbol{\lambda}}(\theta\,|\,X) = \frac{\tau^2\,X + \sigma^2\,\mu}{\tau^2+\sigma^2}$. To evaluate local sensitivity of $\rho(\mu,\tau)$ w.r.t. a particular choice of the prior location parameter μ and scale parameter τ, we evaluate the total derivative of ρ. Clearly, $\frac{\delta}{\delta\,\mu}\rho = \frac{\sigma^2}{\tau^2+\sigma^2}$ and $\frac{\delta}{\delta\,\tau}\rho = \frac{2\,\tau\,\sigma^2\,(X-\mu)}{(\tau^2+\sigma^2)^2}$, thus, by Corollary 1, $||T\rho_{(\mu,\tau)}|| = \frac{\sigma^2}{\tau^2+\sigma^2}\sqrt{1 + \frac{4\tau^2\,(X-\mu)^2}{(\tau^2+\sigma^2)^2}}$. Notice that the local sensitivity index $||T\rho_{(\mu,\tau)}||$ decreases as $|X-\mu|$ decreases and/or as τ increases (subject to $\tau \geq \sigma$). Thus, for this particular example, our evaluation of $||T\rho_{(\mu,\tau)}||$ mathematically justifies the popular belief that if the center of the prior matches with that of the likelihood and/or if the prior has a flat tail, then (generally) posterior robustness (w.r.t. the prior) is achieved □

Example 2 : Let X be observed from $N(\theta,1)$, and the user or a finite elicitation process specifies the prior median and quartiles of θ at 0 and ± 1 respectively. Several distributions satisfy these requirements (see Basu and DasGupta (1992)). For comparison, we only consider the sharp tailed $\pi^n(\mu,\tau^2) = N(\mu,\tau^2)$ with $\mu = 0, \tau = 1.48$, and the flat tailed $\pi^c(\mu,\tau^2) = \text{Cauchy}(\mu,\tau^2)$ with $\mu = 0, \tau = 1$. However, the specifications of median $= 0$ and quartiles $= \pm 1$ often can not be taken as exactly accurate. We thus consider the local sensitivity of the specification $\mu = 0, \tau = 1.48$ in the class of all $N(\mu,\tau^2)$ priors, and compare it with the sensitivity of the specification $\mu = 0, \tau = 1$ in the class of all Cauchy(μ,τ^2) priors. Let $\rho^n(\mu,\tau^2)$ and $\rho^c(\mu,\tau^2)$ denote the posterior means w.r.t. $\pi^n(\mu,\tau^2)$ and $\pi^c(\mu,\tau^2)$ respectively. The local sensitivity in the Normal class, i.e., $||T\rho^n_{(\mu=0,\tau=1.48)}||$, can be easily found from the calculations done in Example 1. Let $\rho^c(\mu,\tau^2) = \frac{N^c(\mu,\tau^2)}{D^c(\mu,\tau^2)}$, where $N^c(\mu,\tau^2) = \int \theta\ell(\theta)\pi^c(\theta|\mu,\tau^2)\,d\theta$, $D^c(\mu,\tau^2) = \int \ell(\theta)\pi^c(\theta|\mu,\tau^2)\,d\theta$, and $\ell(\theta)$ is the appropriate likelihood. $N^c(\mu,\tau^2)$ is difficult to compute analytically. However, it is easy to check that the condition for interchange of derivative and integral is satisfied, i.e., $\frac{\delta}{\delta\mu} N^c(\mu,\tau^2) = \int \theta\ell(\theta)\,[\frac{\delta}{\delta\mu}\pi^c(\theta|\mu,\tau^2)]\,d\theta$. Similar result holds for $D^c(\mu,\tau^2)$. Now, $\frac{\delta}{\delta\mu}\rho^c(\mu,\tau^2) = \frac{1}{D^c(\mu,\tau^2)^2}\{D^c(\mu,\tau^2)\frac{\delta}{\delta\mu}N^c(\mu,\tau^2) - N^c(\mu,\tau^2)\frac{\delta}{\delta\mu}D^c(\mu,\tau^2)\}$, and each term in the above expression involves a simple numerical integration. Same is true for $\frac{\delta}{\delta\tau}\rho^c(\mu,\tau^2)$. Thus, $||T\rho^c_{(\mu=0,\tau=1)}||$ can be obtained with little numerical work.

Table 1: $||T\rho||$ for $N(0, 2.19)$ and Cauchy$(0,1)$ priors

X	0.5	1.0	1.5	2.0	2.5	3.0	3.5	4.0
$\lVert T\rho^n\rVert$	0.346	0.428	0.537	0.661	0.792	0.927	1.065	1.205
$\lVert T\rho^c\rVert$	0.481	0.497	0.512	0.511	0.476	0.402	0.309	0.225

Table 1 shows the values of $||T\rho^n_{(\mu=1,\tau=1.48)}||$ and $||T\rho^c_{(\mu=0,\tau=1)}||$ for different values of X. As can be seen, the value of $||T\rho^n||$ increases with X, and is large for $X \geq 2.5$, whereas $||T\rho^c||$ fluctuates very little. Thus, misspecification of the prior parameters results in much less sensitivity for the heavy tailed Cauchy prior than for the sharp tailed Normal prior (especially when the the center of the prior and the likelihood do not match), which again agrees with prevalent beliefs □

<u>Example 3 :</u> Consider a standard linear model setup : $\underset{\sim}{Y} \sim N_m(X\underset{\sim}{\beta}, \Sigma)$. Here, $\underset{\sim}{Y}_{m\times 1}$ is an observed vector, $X_{m\times k}$ is a known design matrix, Σ is a known positive definite matrix, and $\underset{\sim}{\beta}_{k\times 1}$ is an unknown parameter vector. Under the Bayesian paradigm, we assume a $N_k(\underset{\sim}{\mu}, \Gamma)$ prior for $\underset{\sim}{\beta}$. It is well known that in this setup, the posterior mean for $\underset{\sim}{\beta}$ is $\underset{\sim}{\beta}^* = [\Gamma^{-1} + X^T\Sigma^{-1}X]^{-1}[\Gamma^{-1}\underset{\sim}{\mu} + X^T\Sigma^{-1}X\underset{\sim}{b}]$, where $\underset{\sim}{b} = [X^T\Sigma^{-1}X]^{-1}X^T\Sigma^{-1}\underset{\sim}{Y}$ is the generalized least square estimate (or *mle*) of $\underset{\sim}{\beta}$. For notational simplicity, we denote $X^T\Sigma^{-1}X$ by A from now on. However, specification of the prior parameters $\underset{\sim}{\mu}$ and Γ is again of concern. First, we assume Γ is exactly known, and find the local sensitivity of $\underset{\sim}{\beta}^*$ w.r.t. misspecifications of $\underset{\sim}{\mu}$. Clearly, $[\frac{\delta\underset{\sim}{\beta}^*}{\delta\underset{\sim}{\mu}}]_{k\times k} = [\Gamma^{-1} + A]^{-1}\Gamma^{-1}$, thus $||T\underset{\sim}{\beta}^*_{\underset{\sim}{\mu}}||^2 =$ maximum eigenvalue of $[\Gamma^{-1} + A]^{-1}\Gamma^{-1}\Gamma^{-1}[\Gamma^{-1} + A]^{-1}$. Surprisingly, this local sensitivity $||T\underset{\sim}{\beta}^*_{\underset{\sim}{\mu}}||$ does not depend on $\underset{\sim}{\mu}$ or on the observed value of $\underset{\sim}{Y}$.

We next assume that $\underset{\sim}{\mu}$ is correctly specified and examine the sensitivity of $\underset{\sim}{\beta}^*$ to misspecifications of Γ. In particular, we presume that Γ has a equicorrelated structure, i.e., $\Gamma = \sigma\{(1-r)I + r\,\underset{\sim}{1}\,\underset{\sim}{1}^T\}$, thus specification of Γ requires specifying the variance term σ and the correlation term r (The following calculations can also be done for a general positive definite Γ, but with increased complexity). For ease in calculations, we write $\Gamma = \tau\{I + \rho\,\underset{\sim}{1}\,\underset{\sim}{1}^T\}$, thus $\tau = \sigma(1-r)$, $\rho = \frac{r}{1-r}$, and $\Gamma^{-1} = \frac{1}{\tau}[I - \frac{\rho}{1+k\rho}\underset{\sim}{1}\,\underset{\sim}{1}^T]$. Calculation of $\frac{\delta\underset{\sim}{\beta}^*}{\delta\tau}$ and $\frac{\delta\underset{\sim}{\beta}^*}{\delta\rho}$, however, requires use of matrix derivatives. In particular, we need : (i) if V and W (both matrices) are functions of a matrix $U_{m\times n}$, then $\frac{d(VW)}{dU} = (\frac{dV}{dU})(W\otimes I_n) + (V\otimes I_m)(\frac{dW}{dU})$, and (ii) if V is invertible, then $\frac{d(V^{-1})}{dU} = -(V^{-1}\otimes I_m)(\frac{dV}{dU})(V^{-1}\otimes I_n)$. Here, $\otimes$ denotes a Kronecker product and, for $V_{p\times q}$, $U_{m\times n}$, $[\frac{dV}{dU}]_{mp\times nq} = V\otimes\frac{d}{dU}$ where $\frac{d}{dU}$ is a matrix of derivative operators $[\frac{\delta}{\delta u_{ij}}]_{m\times n}$ (see MacRae (1974), Polasek (1985) for more on matrix derivatives). Using these formulae, we find $\frac{\delta \underset{\sim}{\beta}^*}{\delta\tau} = \frac{1}{\tau}[\Gamma^{-1}+A]^{-1}\,\Gamma^{-1}\,\{[\Gamma^{-1}+A]^{-1}\,[\Gamma^{-1}\mu + A\,\underset{\sim}{b}] - \underset{\sim}{\mu}\}$ and $\frac{\delta \underset{\sim}{\beta}^*}{\delta\rho} = \frac{1}{(1+k\rho)^2\,\tau}\,[\Gamma^{-1}+A]^{-1}\,\underset{\sim}{1}\,\underset{\sim}{1}^T\,\{[\Gamma^{-1}+A]^{-1}\,[\Gamma^{-1}\mu + A\,\underset{\sim}{b}] - \underset{\sim}{\mu}\}$. Going back to our original parameters, we have $\nabla\underset{\sim}{\beta}^*(\sigma,r) = [\frac{\delta\underset{\sim}{\beta}^*}{\delta(\sigma,r)}]_{k\times 2} = [\frac{\delta\underset{\sim}{\beta}^*}{\delta(\tau,\rho)}]_{k\times 2}\,[\frac{\delta(\tau,\rho)}{\delta(\sigma,r)}]_{2\times 2} = [\frac{\delta\underset{\sim}{\beta}^*}{\delta\tau}, \frac{\delta\underset{\sim}{\beta}^*}{\delta\rho}]\begin{pmatrix}1-r & -\sigma\\ 0 & 1/(1-r)^2\end{pmatrix}$. Moreover, $||T\underset{\sim}{\beta}^*_{(\sigma,r)}||^2$ = maximum eigenvalue of $[\nabla\underset{\sim}{\beta}^*]^T\,[\nabla\underset{\sim}{\beta}^*]$. Notice that the matrix on the r.h.s is only 2×2, so that the maximum eigenvalue can be found easily.

For example, suppose we consider a simple linear regression model : $Y_i = \beta_0 + \beta_1\,x_i + \varepsilon_i$, $i = 1,\ldots,m$, where ε_i's are i.i.d. $N(0,1)$ and $|x_i| \le 1$. Using an optimal design strategy, we take 10 observations at $x_i = 1$ and 10 at $x_i = -1$. Thus, $m = 20$, $k = 2$, $\Sigma = I$, and $X^T X = 20\,I$. In this setup, $||T\underset{\sim}{\beta}^*_{\underset{\sim}{\mu}}|| = \frac{1}{1+20\sigma\,(1-|r|)}$. Notice, we did not require to specify either $\underset{\sim}{Y}$ or $\underset{\sim}{\mu}$ to evaluate $||T\underset{\sim}{\beta}^*_{\underset{\sim}{\mu}}||$. Moreover, $||T\underset{\sim}{\beta}^*_{\underset{\sim}{\mu}}||$ decreases, i.e., $\underset{\sim}{\beta}^*$ becomes less sensitive to specification of $\underset{\sim}{\mu}$ as the variance term σ increases and/or as the correlation r gets close to 0. Evaluation of $||T\underset{\sim}{\beta}^*_{(\sigma,r)}||$, however, requires us to know $\underset{\sim}{\mu}$, and $\underset{\sim}{Y}$, or equivalently, the least square estimate $\underset{\sim}{b}$. We specify $\underset{\sim}{\mu} = (0,0)^T$, and evaluate $||T\underset{\sim}{\beta}^*_{(\sigma,r)}||$ in Table 2 for three different values of $\underset{\sim}{b}$, namely, $\underset{\sim}{b} = (1,1)^T, (1,3)^T$, and $(3,3)^T$. From Table 2, we see that $\underset{\sim}{\beta}^*$ becomes less sensitive to specifications of (σ, r) as σ increases and/or r gets close to 0. However, positive and negative r values have different effects. Also, $\underset{\sim}{\beta}^*$ is less sensitive to (σ,r) for $\underset{\sim}{b} = (1,1)^T$ (which is close to the prior specification $\underset{\sim}{\mu} = (0,0)^T$) than for other values of $\underset{\sim}{b}$ □

Example 4 (Example 3 continued) : As before, consider a linear model setup : $\underset{\sim}{Y} \sim$

Table 2: $||T\underset{\sim}{\beta}^*_{(\sigma,r)}||$ for different values of $\underset{\sim}{b}$, σ and r

	$\sigma = 1$					$\sigma = 2$				
r	−.75	−.5	0	.5	.75	−.75	−.5	0	.5	.75
$\underset{\sim}{b} = (1,1)^T$	.22	.13	.09	.27	.30	.08	.04	.04	.14	.15
$\underset{\sim}{b} = (1,3)^T$	.44	.26	.18	.58	.99	.16	.09	.08	.30	.56
$\underset{\sim}{b} = (3,3)^T$	.66	.38	.27	.81	.90	.25	.13	.11	.41	.46

$N_m(X\underset{\sim}{\beta}, \Sigma)$ with $\underset{\sim}{\beta}_{k\times 1} \sim N_k(\underset{\sim}{\mu}, \Gamma)$. When Γ is known and we focus on the average sensitivity of $\underset{\sim}{\beta}^*$ (the posterior mean of $\underset{\sim}{\beta}$) to specification of $\underset{\sim}{\mu}$, we have : $\overline{T\underset{\sim}{\beta}^*_{\underset{\sim}{\mu}}} = \frac{w_k}{k} \times$ {sum of eigenvalues of $[\Gamma^{-1} + A]^{-1}\,\Gamma^{-1}\,\Gamma^{-1}[\Gamma^{-1} + A]^{-1}$ }, where $A = X^T\,\Sigma^{-1}\,X$. If $\underset{\sim}{\mu}$ is correctly specified, and we want to evaluate the average sensitivity of $\underset{\sim}{\beta}^*$ w.r.t. σ and r (where $\Gamma = \sigma\{(1-r)I + r\,\underset{\sim}{1}\,\underset{\sim}{1}^T\}$), then $\overline{T\underset{\sim}{\beta}^*_{(\sigma,r)}} = \pi\times$ {sum of eigenvalues of $[\nabla\underset{\sim}{\beta}^*(\sigma,r)]^T\,[\nabla\underset{\sim}{\beta}^*(\sigma,r)]$ } (see Example 3).

In particular, if we consider the specific example : $m = 20$, $k = 2$, $\Sigma = I$, and $X^T X = 20\,I$, then $\overline{T\underset{\sim}{\beta}^*_{\underset{\sim}{\mu}}} = \pi\times\{\frac{1}{[1+20\sigma\,(1+r)]^2} + \frac{1}{[1+20\sigma\,(1-r)]^2}\}$. Notice, $\overline{T\underset{\sim}{\beta}^*_{\underset{\sim}{\mu}}}$ increases as σ decreases. It also increases as $|r|$ increases. For $\underset{\sim}{\mu} = (0,0)^T$, we also evaluated $\overline{T\underset{\sim}{\beta}^*_{(\sigma,r)}}$, and plotted it against r for different values of $\underset{\sim}{b}$ (the least square estimate of $\underset{\sim}{\beta}$) and σ (plot not shown). These plots showed that the average sensitivity decreases with increase of σ. However, the effect of r was somewhat surprising, $\overline{T\underset{\sim}{\beta}^*_{(\sigma,r)}}$ (for fixed $\underset{\sim}{b}$ and σ) did not attain its minimum at $r = 0$ as was expected $\square$

4. Nonparametric classes

An important issue in prior elicitation is that a parametric functional form of the prior is generally hard to determine. Recent attention in robust Bayesian analysis is thus more focused towards nonparametric prior classes. Our technique of computing the total derivative to quantify the sensitivity of $\rho(P)$ fails here, since the relevant domain of $\rho(P)$ is no longer a Euclidean space, but a general polish space $\mathcal{M}$ of all probability measures on Θ. Thus, the notion of functional derivatives, in particular, Fréchet derivatives enters the picture. Diaconis and Freedman (1986), and Ruggeri and Wasserman (1990) quantified the local sensitivity of a posterior quantity $\rho(P)$ by computing the norm of its Fréchet derivative over the class of all signed measures or its appropriate subclasses. Srinivasan and Truszczynska (1990) used Fréchet derivatives to approximate ranges of posterior quantities.

Fréchet derivatives are defined on normed linear spaces, or more generally, on topological vector spaces. However, the posterior quantity $\rho(P)$ is defined on $\mathcal{M}$ which is convex, but not linear. Thus ρ has to be artificially extended to the linear space of all signed measures Δ before the notion of Fréchet differentiability could be applied to ρ.

A different line of attack was proposed by Huber (1981) and others who generalized the definition of Fréchet derivatives to encompass the case when ρ is defined only on $\mathcal{M}$. We find this approach more natural from a statistical viewpoint. This generalized definition, however, comes with a price since we can not use strong theorems which are available for Fréchet derivatives on vector spaces. In our current ongoing work, we have established (Huber's) Fréchet differentiability of ratio-linear posterior quantities. We have also argued that since $\mathcal{M}$ is only convex, a direct maximization of the Fréchet derivative is more intuitive rather than treating $\mathcal{M}$ as a subspace of the linear space Δ and computing the norm of the Fréchet derivative over $\mathcal{M}$. We are in the process of developing methods for computing this maximum over different subclasses of $\mathcal{M}$.

Acknowledgement : The authors thank Benny Cheng for suggesting an improvement in the proof of Theorem 3.

References

[1] Basu, S., Jammalamadaka, S.R., and Liu, W. (1993), "Qualitative robustness and stability of posterior distributions and posterior quantities", Technical Report, **238**, Department of Statistics and Applied Probability, University of California, Santa Barbara.

[2] Basu, S. and DasGupta, A. (1992), "Bayesian analysis with distribution bands : the role of the loss function", verbally accepted in *Statist. and Decisions*

[3] Berger, J. (1993), "An overview of robust Bayesian analysis", Technical Report, **93-53**, Department of Statistics, Purdue University.

[4] Diaconis, P. and Freedman, D. (1986), "On the consistency of Bayes estimates", *Ann. Statist.*, **14**, 1-67.

[5] Hampel, F.R. (1971), "A general qualitative definition of robustness", *Ann. Math. Statist.*, **42**, 1887-1896.

[6] Huber, P.J. (1981), *Robust Statistics*, John Wiley : New York.

[7] MacRae, E.C (1974), "Matrix derivatives with an application to an adaptive linear decision problem", *Ann. Statist.*, **2**, 337-346.

[8] Polasek, W. (1985), "A dual approach for matrix-derivatives", *Metrika*, **32**, 275-292.

[9] Rao, C.R. (1973), *Linear statistical inference and its applications*, Wiley.

[10] Rivier, N., Englman, R., and Levine, R. D. (1990), "Constructing priors in maximum entropy methods", In *Maximum Entropy and Bayesian Methods*, P.F. Fougére (Ed.), 233-242, Kluwer Academic Publishers.

[11] Rodríguez, C.C. (1994), "Bayesian robustness : a new look from geometry", to appear in *Maximum Entropy and Bayesian Statistics*, G. Heidbreder (Ed.), Kluwer Academic Publishers.

[12] Rudin, W. (1976), *Principles of mathematical analysis*, McGraw-Hill.

[13] Ruggeri, F. and Wasserman, L. (1993), "Infinitesimal sensitivity of posterior distributions", *Canad. J. Statist.*, **21**, 195-203.

[14] Skilling, J. (1990), "Quantified maximum entropy", In *Maximum Entropy and Bayesian Methods*, P.F. Fougére (Ed.), 341-350, Kluwer Academic Publishers.

[15] Srinivasan, C. and Truszczynska, H. (1990), "Approximation to the range of a ratio-linear posterior quantity based on Fréchet derivative", Technical Report, **289**, Department of Statistics, University of Kentucky.

[16] Tukey, J.W. (1960), "A survey of sampling from contaminated distributions", In *Contributions to Statistics and Probability*, I. Olkin et al (Ed.), 448-485, Stanford University Press, Stanford, California.

[17] Wasserman, L. (1992), "Recent Methodological advances in robust Bayesian inference", In *Bayesian Statistics 4*, J.M. Bernardo, et. al. (Eds.), Oxford University Press, Oxford.

TREE-STRUCTURED CLUSTERING VIA THE MINIMUM CROSS ENTROPY PRINCIPLE

David Miller and Kenneth Rose
Department of Electrical and Computer Engineering
University of California
Santa Barbara, CA 93106

ABSTRACT. We propose a new interdisciplinary approach to the tree-structured clustering problem, wherein structural constraints are imposed in order to reduce the classification search complexity of the resulting statistical classifier. Most known methods are greedy and optimize nodes of the tree one at a time to minimize a local cost. By constrast, we develop a joint optimization method, derived based on information-theoretic principles and closely related to known methods in statistical physics. The approach is inspired by the deterministic annealing method for unstructured clustering, which was based on maximum entropy inference. The new approach is based on the principle of minimum cross entropy, using informative priors to approximate the unstructured clustering solution while imposing the structural constraint. As in the original deterministic annealing method, the number of distinct representatives (and hence the tree) grows in a non-heuristic fashion by a sequence of phase transitions which occur so as to optimize the effective free energy cost. Examples demonstrate considerable improvement over known methods.

1 Introduction

The problem of clustering involves the partitioning of data into groups or clusters in order to maximize the homogeneity within each group, and to also maximize the discrimination between groups. For a review of the clustering problem, see [1] and [2]. The impact of the basic problem extends over a variety of disciplines, with important applications in pattern recognition, data compression, statistics, image analysis, as well as other fields. In pattern recognition, clustering is often posed directly as the problem of choosing a partition of the training set so as to minimize a cost function. The minimum cost partition is then used as a classifier for the feature space. The most widely used clustering objective is the sum of squared distances

$$D = \sum_j \sum_{x \in C_j} |x - y_j|^2, \tag{1}$$

where C_j is the jth cluster with representative (mean) y_j and x an element of the training set. The standard approaches to optimizing this cost function are the Isodata algorithm [3] and its sequential relative, the K-means algorithm [4]. Related approaches have also been derived for fuzzy clustering [5, 6]. Since most important cost functions are non-convex, clustering is a hard optimization problem and conventional descent methods (like Isodata) produce solutions that are highly dependent on the initialization.

While minimizing D is the primary clustering objective, an important concern, especially for problems involving numerous natural clusters and high dimensional spaces, is the classification complexity of the resulting classifier. In the pattern recognition field,

G. R. Heidbreder (ed.), Maximum Entropy and Bayesian Methods, 107–120.

classification complexity is an important concern which has led to methods for designing reduced-complexity trees for purposes of classification and regression, see e.g. [7, 8]. In the data compression community, the pervasiveness of the complexity problem is also evident from research devoted to designing structured quantizers, see e.g. [9, 10]. Unlike the unstructured problem, in the case of tree-structured clustering there are no known methods guaranteeing convergence to even a locally optimal solution. The standard approaches are greedy procedures which build a tree one node at a time by minimizing a local, heuristic cost. Even for relatively simple clustering problems these methods may fail. Thus, there is strong motivation to develop improved structured clustering methods and this is the focus of our paper. Our method is inspired by the deterministic annealing method (DA) [11, 12] for the unstructured clustering problem. Therefore, before addressing the structured problem we will review DA.

2 Deterministic Annealing

An important class of algorithms for solving hard optimization problems was inspired by annealing processes in chemistry and physics. The stochastic, simulated annealing algorithm has been applied to a variety of challenging problems with much success, though at high computational cost. In order to reduce the computational burden, deterministic approximations to simulated annealing have been suggested within several different contexts [13, 14, 15]. In [12], a deterministic annealing approach to clustering (DA) was derived within information theory, but with analogy to statistical physics. The method is independent of initialization and generates a sequence of solutions corresponding to distinct temperature scales. By direct analogy to the chemical process, an effective free energy is minimized, starting from high temperature for which the energy cost is convex. As the temperature is lowered, the solution is tracked to avoid local minima inherent in the cost function. In the limit of low temperature, the energy function converges to the desired nonconvex cost and the solution forms a "hard partitioning" of the data space. The method has been demonstrated to obtain significant improvement over conventional clustering methods, both within pattern recognition [16] and the data compression field [11]. Recently, the method has been given a more fundamental interpretation within rate-distortion theory, and an algorithm based on DA has been suggested as a practical alternative to the Blahut algorithm for rate-distortion function computation [17]. Moreover, the DA approach has also been generalized to attack a larger class of optimization problems by adding constraints on the cluster representatives [18]. Here, we briefly review the basic method and its derivation.

Even if one is interested in a "hard" (i.e. non-fuzzy) clustering solution, still it may be useful within the context of an optimization method to consider points associated *in probability* with clusters. In deterministic annealing, no underlying assumptions are made about the data distribution. In order to obtain a set of association probabilities relating representatives and data points given no prior knowledge, our best recourse is to invoke the principle of maximum entropy. In [19], Jaynes gives a strong argument that, in some sense, the maximum entropy distribution is most probable – i.e., it is the distribution which explains the data set and the system constraints in an overwhelmingly greater number of ways than any other distribution. If we can assume the representatives' set $Y = \{y_j\}$ is given, then we seek the set of association probabilities $P[x \in C_j]$ satisfying some average clustering distortion constraint. Here, $x \in C_j$ means that data point x belongs to representative j. Applying the principle of maximum entropy, for each data point we optimize

$$\max\{-\sum_j P[x \in C_j] \log P[x \in C_j]\} \tag{2}$$

subject to

$$< D_x >= \sum_j P[x \in C_j] d(x, y_j). \tag{3}$$

In (3), $d(\cdot, \cdot)$ is a specified distance measure. The optimization problem specified by (2) and (3) assumes that the association probabilities for distinct data points are independent. The solution is the Gibbs distribution

$$P[x \in C_j] = \frac{e^{-\beta d(x,y_j)}}{\sum_k e^{-\beta d(x,y_k)}}, \tag{4}$$

where the denominator is a partition function from statistical physics. The Lagrange multiplier β which determines $< D_x >$ can be interpreted as an inverse temperature, controlling the degree of fuzziness of the probability distribution, i.e., the amount of influence distant data points have on the representatives. At $\beta = 0$, all codevectors are equally associated with the data set, while for large β, the $P[x \in C_j]$ become "hard" 0-1 associations, specifying a nearest neighbor partition.

While we have obtained association probabilities given a fixed set of representatives, the more interesting problem involves optimization of *both* the representatives' set and associations. Therefore, we will consider the system composed of the representatives' set $Y = \{y_j\}$, the data set $X = \{x_i\}$, and the specification of a "hard" partition $V = \{v_{ij}\}$. The $\{v_{ij}\}$ are defined by

$$v_{ij} = \begin{cases} 1 & \text{if } x_i \in C_j \\ 0 & \text{otherwise.} \end{cases}$$

The clustering distortion associated with this partitioning is

$$D(Y, V) = \sum_i \sum_j v_{ij} d(x_i, y_j). \tag{5}$$

The pair (Y, V) determines a particular instance of a solution and the probability of this instance is represented by the joint distribution $P[Y, V]$. We seek the maximum entropy distribution over all deterministic "hard" clustering solutions subject to an expected distortion constraint, i.e.

$$\max\{-\sum_{Y,V} P[Y, V] \log P[Y, V]\} \tag{6}$$

subject to

$$< D >= \sum_{Y,V} P[Y, V] D(Y, V). \tag{7}$$

Applying the method of Lagrange multipliers, we again obtain a Gibbs distribution

$$P[Y, V] = \frac{e^{-\beta D(Y,V)}}{\sum_{Y'} \sum_{V'} e^{-\beta D(Y',V')}}. \tag{8}$$

Maximizing $P[Y, V]$ is equivalent to minimizing the argument $\beta D(Y, V)$ with respect to Y and V. For cardinalities $|Y|$ and $|X|$ reasonably large, this minimization is clearly not practical. Alternatively, since $|Y|$ is almost always much smaller than $|X|$, it may be a practical objective to find the most probable set of representatives by considering the corresponding marginal $P[Y]$, found by summing over the set V:

$$P[Y] = \frac{Z(Y)}{\sum_{Y'} Z(Y')}. \tag{9}$$

The numerator is the partition function associated with a particular Y,

$$Z(Y) = \sum_V e^{-\beta D(Y,V)} = \prod_x \sum_j e^{-\beta d(x,y_j)}, \tag{10}$$

and the denominator is the partition function comprising all solution instances. Equation (9) can be re-written as the Gibbs distribution

$$P[Y] = \frac{e^{-\beta F(Y)}}{\sum_{Y'} e^{-\beta F(Y')}}, \tag{11}$$

by defining

$$F(Y) \equiv -\frac{1}{\beta} \log Z(Y) = -\frac{1}{\beta} \sum_x \log \sum_j e^{-\beta d(x,y_j)}. \tag{12}$$

Therefore, maximizing $P[Y]$ is equivalent to minimizing the energy $\beta F(Y)$. This $F(Y)$ is the free energy in the physical analogy, minimized at isothermal equilibrium. The set Y which minimizes the free energy satisfies the centroid rule

$$\sum_x P[x \in C_j] \frac{\partial}{\partial y_j} d(x, y_j) = 0, \tag{13}$$

where $P[x \in C_j]$ was defined earlier in (4). Note that (13) really specifies a set of scalar equations, one for each component direction, since y_j is a vector. At $\beta = 0$, the associations $P[x \in C_j]$ are the same for all data points and all representatives, and there is a unique global minimum solution with all representatives lying at the global centroid of the data set. Effectively, at $\beta = 0$, there is one natural cluster – i.e., one distinct cluster representative. As β increases, the emphasis on minimizing distortion increases, prompting phase transitions and cluster splits. For the sum of squared distances measure, it is shown in [12] that the first phase transition is initiated at β satisfying

$$\det[I - 2\beta_c R_{xx}] = 0, \tag{14}$$

where R_{xx} is the covariance matrix of the data set. Thus, the critical temperature is determined by the variance along the largest principal axis of the distribution. Subsequent phase transitions are initiated in a similar way, dependent on the covariance matrix of the data "owned" by the natural cluster undergoing the split. The amount of cluster splitting is only limited by the number of representatives assumed by the system as the temperature is decreased to zero. At zero temperature, the free energy is the "hard" clustering distortion and the algorithm becomes equivalent to known descent methods.

3 A Structured Clustering Formulation

A tree- structured solution can be represented symbolically by a tree diagram, shown for a simple binary tree of depth two in Figure 1. The test vectors $\{s_j\}$ at the non-leaf nodes

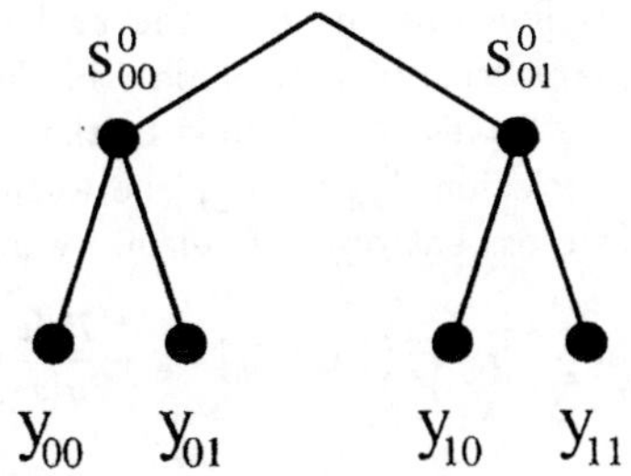

Figure 1: A tree diagram for a binary tree of depth two.

in the tree are used to specify the hierarchical partitioning of the data space and the $\{y_{jk}\}$ at the leaf nodes are the cluster representatives. Here, the index j refers to a node in the first layer and index k refers to a sibling at the leaf layer. In this example, the non-leaf nodes determine a nearest neighbor partition via the equation $d(x, s_0) = d(x, s_1)$, cutting the input space in half based on the dissimilarity ("distance") measure $d(\cdot, \cdot)$. The leaf layer, based on the preceding, higher level's decision boundary, further partitions the input space into four regions whose cluster representatives are the vectors $\{y_{jk}\}$.

To make the tree-structured clustering problem mathematically precise, the problem can be posed as the optimization of the hierarchy $\{s_j\}$ and the representatives at the leaf layer $\{y_{jk}\}$ to minimize a well-defined cost objective such as D in (1). One greedy approach for tree-structured clustering popular in both the data compression and pattern recognition literature is the splitting method, e.g. [20, 10]. While a number of variants of splitting exist, the basic approach involves the successive application of a clustering algorithm such as basic Isodata to hierarchically partitioned data subsets. In [21], a simple example was used to illustrate the potential shortcomings of splitting and a crude heuristic method was proposed for improving tree-structured solutions. The basic idea was to re-optimize the hierarchy to better agree with nearest neighbor (i.e., minimum risk) ownership at the leaf layer. Here, we will derive a more general approach inspired by the deterministic annealing algorithm for unstructured clustering. The structured clustering approach is developed in two parts, analogous to the two optimization steps in basic Isodata (i.e., the nearest neighbor and centroid rules). We first consider optimization of the leaf layer given a fixed hierarchy and then consider optimization of the hierarchy given fixed leaves. The structure (e.g., balanced) and size of the tree (i.e., number of leaves) are fixed in this derivation.

3.1 Optimization of the Leaf Layer

For the unstructured problem, no prior knowledge existed other than that embodied in the clustering distortion constraint. Accordingly, we invoked the principle of maximum entropy. By contrast, in the structured case there is a significant constraint imposed on the solution which does not appear to readily conform to the maximum entropy framework.

The principle of minimum cross entropy, though, is a generalization of the Maxent principle to include priors. In [22], it it shown that the principle of minimum cross entropy is *the* consistent principle of inference given new information. Here, we apply the principle to tree-structured clustering. Accordingly, we seek to express the structural constraint as an informative prior probability distribution which influences the leaf codevector/data association probabilities. Define this prior as the probability of association $P_H[x \in S_j]$ where S_j is the set of all points classified to node j in the layer of the tree directly above (i.e., the parent layer of) the leaves. Analogous to the unstructured DA formulation, we seek the probabilities of association $P_L[x \in C_{jk}]$ between leaf representatives y_{jk} and the data samples. By minimum cross entropy inference, we pose the problem

$$\min_{P_L} \sum_j \sum_k P_L[x \in C_{jk}] \log \frac{P_L[x \in C_{jk}]}{(P_H[x \in S_j]/K)} \tag{15}$$

subject to

$$< D_x >= \sum_j \sum_k P_L[x \in S_j] d(x, y_{jk}). \tag{16}$$

Here, the prior at the parent node is assumed equally split between all K of its descendants at the leaves (justifiable by the principle of maximum entropy). The solution is the so-called "tilted" distribution, i.e.

$$P_L[x \in C_{jk}] = \frac{P_H[x \in S_j] e^{-\beta d(x, y_{jk})}}{\sum_{lm} P_H[x \in S_l] e^{-\beta d(x, y_{lm})}}. \tag{17}$$

As in the unstructured case, the Lagrange multiplier β determines the average level of clustering distortion $< D_x >$. It also has its usual interpretation as an inverse "temperature" influencing the degree of fuzziness of the distribution. For $\{P_H[x \in S_j]\}$ uniform and $\beta = 0$, the association probabilities are uniform over all partition regions C_{jk}. At the other extreme, if $\{P_H[x \in S_j]\}$ specifies a "hard" classifier with a tree structure then, as $\beta \to \infty$, $\{P_L[\cdot]\}$ specifies a tree-structured partition associated with a "hard" clustering solution. Moreover, for $P_H[\cdot]$ uniform, the associations revert to the unstructured DA associations of (4), as we expect. We note in passing that the distribution $P_L[x \in C_{jk}]$ can also be given a Bayesian interpretation as the posterior $p[C_{jk}|x]$, where $\{P_H[\cdot]\}$ is the prior and $e^{-\beta d(x, y_{jk})}$ is proportional to the density $p[x|C_{jk}]$. The partition function associated with a single datum is the denominator of (17). Assuming datum independence, the total partition function is then the product

$$Z' = \prod_x Z'_x = \prod_x \sum_l P_H[x \in S_l] \sum_m e^{-\beta d(x, y_{lm})} \tag{18}$$

Correspondingly, the free energy is

$$F' \equiv -\frac{1}{\beta} \log Z' = -\frac{1}{\beta} \sum_x \log \sum_l P_H[x \in S_l] \sum_m e^{-\beta d(x, y_{lm})}. \tag{19}$$

Note that if $\{P_H[\cdot]\}$ specifies a "hard" tree-structured partition, then for $\beta \to \infty$ the free energy is equivalent to the tree-structured clustering distortion. Thus, as in the unstructured method, optimization of the representatives is realized by an annealing approach, minimizing F' starting from high temperature (small β) and tracking the solution while gradually lowering the temperature. The condition for optimizing the free energy at any temperature is

$$\frac{\partial F'}{\partial y_{jk}} = 0, \qquad \forall j, k, \tag{20}$$

or the centroid rule

$$\sum_x P_L[x \in C_{jk}] \frac{\partial}{\partial y_{jk}} d(x, y_{jk}) = 0, \qquad \forall j, k. \tag{21}$$

For the squared distances measure, we may write

$$y_{jk} = \frac{\sum_x x P_L[x \in C_{jk}]}{\sum_x P_L[x \in C_{jk}]}, \tag{22}$$

which must be iterated until a fixed point is reached.

3.2 Optimization of the Hierarchy

While the centroid rule is a descent step in the clustering distortion D in both the unstructured and structured clustering cases, there is no direct analogue of the nearest neighbor paritioning rule for structured clustering. In fact, the optimal *tree-structured* partitioning given fixed leaves is obtained by solving a challenging risk minimization problem, see e.g. [1], involving joint optimization over the tree. Rather than view the problem in this way, the conventional approaches such as the method in [10] optimize nodes of the hierarchy one at a time and to minimize local, heuristic cost functions which often poorly reflect the minimum risk objective. While directly finding an optimal partition is a formidable goal, there is a practical paradigm which is closely related and which leads to significant improvement over the conventional greedy methods.

We will consider the simple case of a two layer binary tree and note that the derivation is extended to more general tree structures in [23]. For the optimization over the leaves, the prior knowledge was in the form of a prior distribution over the fixed hierarchy. Now, likewise, we interpret fixing the leaf layer to mean specification of an ideal prior distribution over the leaves, $P_I[x \in C_{jk}]$. The goal of the hierarchy is to produce correct (i.e., minimum risk) classification to the leaf layer. Within the probabilistic framework. the hierarchy should be chosen so as to produce a distribution which approximates the ideal prior over the leaves as closely as possible within the imposed, structural constraints. We need to choose a form for the hierarchical probabilities which embodies the structural constraint and we need to specify the ideal prior. For the simple two layer binary tree, if squared distance is used in the first layer, then the "hard" structured non-leaf partition is simply a hyperplane. Thus, we seek a parametrization of $P_H[x \in S_j]$ consistent with a hyperplane decision boundary as the probabilities become "hard". A reasonable choice. justified by the principle of maximum entropy, is the Gibbs distribution

$$P_H[x \in S_j] = \frac{e^{-\gamma d(x,s_j)}}{e^{-\gamma d(x,s_j)} + e^{-\gamma d(x,s_j^c)}}, \tag{23}$$

which for $\gamma \to \infty$ determines a hyperplane partition. Here, s_j^c denotes the sibling node of s_j. The parameters s_j, s_j^c, and γ should be chosen in order to agree as closely as possible with $P_I[x \in C_{jk}]$. Now, we must specify the prior. Some inspiration is gained by noting that given fixed leaf representatives, the *optimal* partition is just the nearest neighbor partition induced by the leaf representatives. Within our probabilistic annealing framework, given the "temperature" β, the corresponding optimal (unstructured) prior distribution is the maximum entropy distribution over the leaves, i.e.

$$P_I[x \in C_{jk}] = \frac{e^{-\beta d(x, y_{jk})}}{\sum\limits_{l,m} e^{-\beta d(x, y_{lm})}}. \tag{24}$$

Summing over leaf siblings, the prior over the parent layer is

$$P_I[x \in S_j] = \sum_k P_I[x \in C_{jk}]. \tag{25}$$

To minimize the distance between distributions we again appeal to the principle of minimum cross entropy and pose the problem

$$\min J(\gamma, \{s\}) \equiv \min_{\gamma, \{s\}} \sum_x \sum_j P_I[x \in S_j] \log \frac{P_I[x \in S_j]}{P_H[x \in S_j]}. \tag{26}$$

Taking the gradient with respect to a test vector we obtain

$$\nabla_{s_j} J = \sum_x (x - s_j)\{P_I[x \in S_j](1 - P_H[x \in S_j]) - (1 - P_I[x \in S_j])P_H[x \in S_j]\}. \tag{27}$$

This simple rule suggests that *moving a test vector in the negative gradient direction implies "pulling" the test vector toward all points that "should be" in S_j (i.e., those points with large $P_I[\cdot]$) but which "are not" currently in S_j (i.e., those points with small $P_H[\cdot]$). Similarly, the test vector is "pushed away" from all points that "should not be" in S_j but which currently "are" in S_j.* The optimization over the scale parameter γ tries to match the degree of fuzziness in the hierarchical probabilities with that of the prior probabilities. Note that in the limit as both sets of probabilities become "hard", the gradient realizes a batch version of the Perceptron weight update rule. Although the Perceptron does not converge for non-separable classes, a convergent method results so long as the probabilities retain some diminishing degree of fuzziness. Thus, we note that the invocation of a principle of inference from information theory designed to incorporate prior knowledge leads to a probabilistic generalization of a well-known supervised learning rule. Accordingly, we can view the prior distribution $P_I[\cdot]$ as a "supervising" distribution.

When generalized to trees of any depth and node branching factor, a similar, intuitive interpretation results for optimization of the structured hierarchy. In the general case, any non-leaf test vector is moved towards all points its descendants at the leaves "own" in a nearest neighbor sense (via the $P_I[\cdot]$) but which it currently does not "own" via its structured decision boundary. Likewise, the test vector is moved away from all points that its descendants at the leaves do not "own" but which it currently "owns" with its structured boundary.

We can now summarize our annealing approach for tree-structured clustering. It is listed in pseudo-code in Table I.

Initialize the test vectors and leaf vectors to the global centroid of the data set.
Initialize β and the scale parameters to small values.
do {
  do {
    Minimize the free energy $F'(Y)$ of (19) to obtain new leaves.
    Minimize the cross entropy of (26) to obtain a new hierarchy.
  } while not converged
  Increase β.
} while $\beta < \beta_{max}$

Table 1: A summary of the basic tree-structured DA method.

3.3 Phase Transitions and Tree Growth

A limitation of the method discussed heretofore is the assumption that the *structure* of the clustering solution is fixed. Even if (as we have assumed thus far) the representatives' set is of fixed size, the best tree-structured clustering solution of given size and given maximal depth will have an a priori unknown structure – in particular, it is unlikely to be a balanced tree. Optimal unbalanced trees may approach the performance of *unstructured* clustering, and often with a negligible increase in classification search. The standard methods for tree growing use the splitting algorithm in conjunction with heuristic decision rules which determine the order of node splitting, see e.g. [24]. These trees are then typically pruned in an optimal fashion by the method in [25, 7] to achieve the best cost/complexity tradeoff *given the initial tree.* While these methods do obtain performance gains over balanced tree design, their greedy nature is a potential source of sub-optimality. Thus, we are motivated to seek an extension of our basic approach for designing unbalanced trees of a prior unknown structure.

Let us briefly consider the phase transitions in the process. It was shown in [12] for the basic unstructured DA method that the number of representatives grows by a sequence of phase transitions in the process, with the first transition along the principal data axis. The critical temperature for this transition, assuming the squared distance measure, is $\beta_c = \frac{1}{2\lambda_{max}}$ where λ_{max} is the associated maximum eigenvalue. It is easily seen (by noting that the $P_H[\cdot]$ are initially uniform and hence $F' = F$ before the first phase transition) that this condition for the first phase transition is the same for the tree-structured clustering case. Moreover, a general condition for all growing phase transitions can also be derived; see [23, 17]. For the tree-structured case, this condition can be written as

$$(2\beta R_{xx}^{jk} - I)y_{jk} = 0, \tag{28}$$

where R_{xx}^{jk} is the sample covariance matrix for codevector y_{jk} based on the distribution $P_L[x \in C_{jk}]$. This is just an eigenvalue equation with solution first occurring for

$$\beta_{crit} = \frac{1}{2\lambda_{max}}. \tag{29}$$

Thus, our condition for all phase transitions in the process is a probabilistic generalization of the rule for the first phase transition. We note that in [26], a heuristic for splitting along the principle axis was suggested but *only* in the usual context of greedy growing where the splits determine both leaf and non-leaf nodes. By contrast, our condition is only for the leaf representatives. Our method for growing unbalanced trees determines structure directly based on the phase transitions in the process. So long as the size of the tree is not limited a priori, the natural tree structure (i.e., the natural number of distinct branches) emerges at each β. Another possibility is to start from a zero layer tree with one representative and grow new intermediate nodes and leaves as needed, based on satisfaction of the condition for phase transitions.

It is important to note here that unlike the conventional methods, tree growth in our method is not obtained in a greedy fashion. Rather, splits occur as a *direct consequence* of optimizing the free energy F' at critical β and may be interpreted as phase transitions in the annealing process. Thus, we suggest that an optimal or a near optimal tree-structured clustering solution of fixed number of leaves and given maximal depth is obtained directly from tree growing via our method, without considering subsequent pruning.

4 Results

Here we present a few examples of our method, demonstrating the performance in comparison with the conventional splitting algorithm [10], and demonstrating the tree growth via phase transitions. The data for Figure 2 is generated by randomly sampling from a normal mixture distribution with eight components. An X in the figures denotes an actual mixture center and an O denotes a cluster representative. The cost D is the sum of squared distances. Figure 2a shows the performance of the splitting algorithm, which cuts through one natural cluster with its highest level decision boundary and fails to distinguish three natural clusters towards the left of the figure. In the splitting method, the non-leaf test vectors are suboptimally chosen to minimize *their* distortion on the data set, rather than being chosen to minimize the risk associated with classification of data to the leaf layer. Our structured DA solution is shown in Figure 2b and obtains considerable improvement, distinguishing all natural clusters and incurring a much smaller sum of squared distances cost.

In Figure 3 we show an "evolution" of solutions corresponding to tree growing in our method. The curved partitions are equiprobable contours, denoting membership probability above a threshold p within a given cluster. In Figure 3a and 3g, $p = 0.5$ (i.e, the "hard" partition is shown) while $p = 0.33$ in all other figures to demonstrate the probabilistic associations obtained by our method. The number of distinct cluster representatives grows in Figure 3 by a sequence of phase transitions, which occur in a non-heuristic fashion so as to directly optimize the free energy cost and determine the structure of the tree. The method can be used to search for the optimal "hard" tree-structured solution of given size by fixing the number of representatives and lowering the temperature toward zero. Alternatively the hierarchy of probabilistic solutions at intermediate temperatures is potentially useful for exploratory data analysis and for exploring issues of clustering validity.

We note that a more comprehensive description of the method and results can be found in [23]. There, we found that our method consistently outperformed conventional tree-structured clustering methods. Moreover, despite its structural handicap, our method also outperformed conventional *unstructured* clustering methods such as basic Isodata as well. For complex data distributions, there are numerous local minima to trap conventional

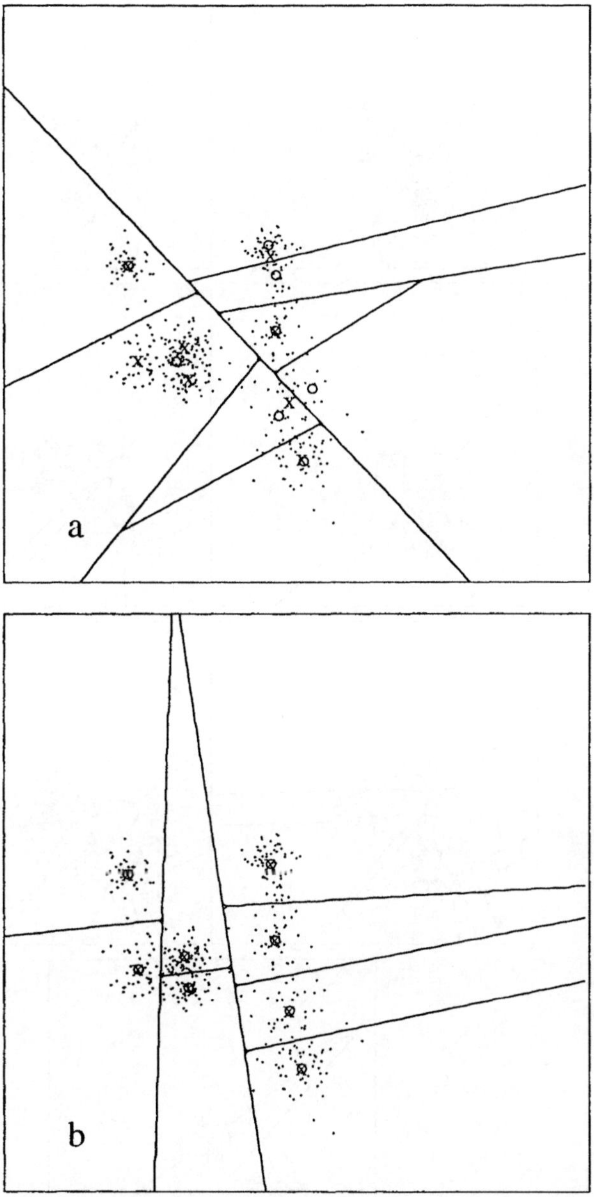

Figure 2: An example involving a mixture of eight normal distributions. a) The splitting solution with D=0.73. b) The tree-structured DA solution with D=0.50.

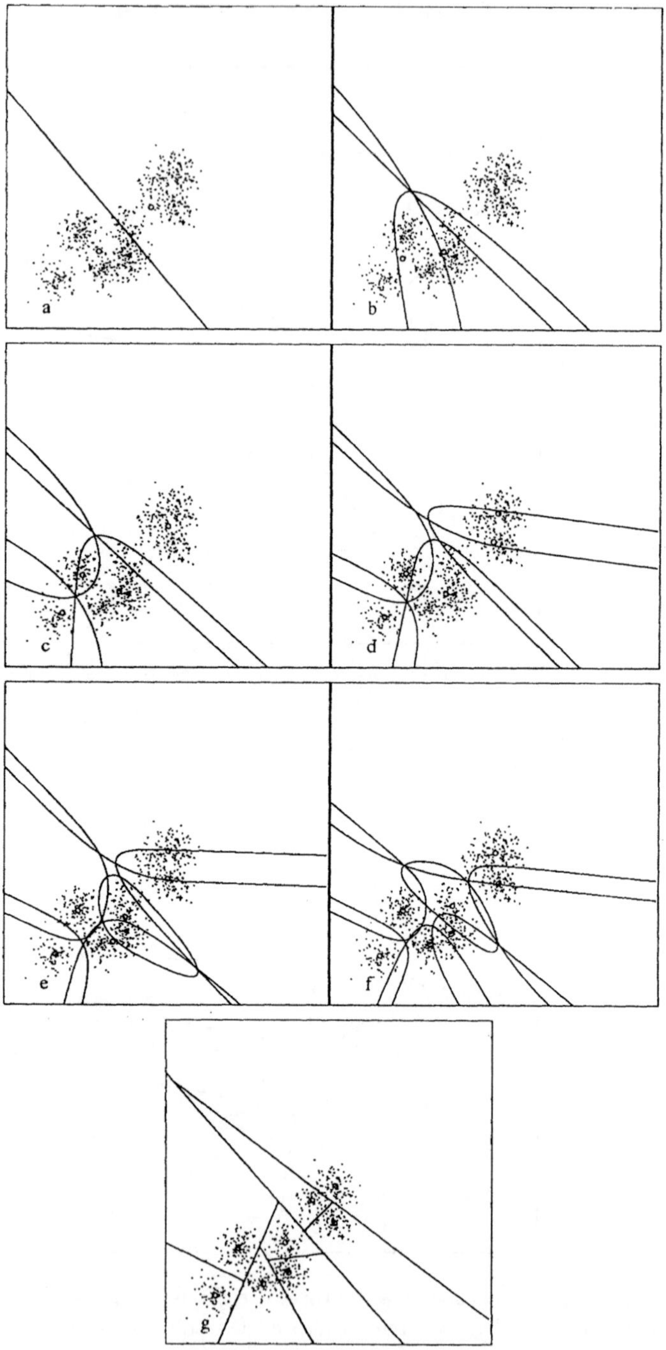

Figure 3: A hierarchy of tree-structured solutions generated by unbalanced tree growing.

descent-based methods and even the best of numerous solutions based on random initialization within the data may be suboptimal relative to our method.

5 Conclusions

In this paper,we have extended the deterministic annealing approach to address the problem of structurally constrained clustering. Whereas the original approach was developed using the principle of maximum entropy, the new method is based on minimum cross entropy inference, which is a convenient "language" for expressing the joint objectives of enforcing a tree-structured solution and approximating the optimal (unstructured) solution. Our method improves upon the conventional tree-structured methods in several important respects. We consider a joint optimization over the hierarchy to minimize a cost objective closely related to risk minimization. Moreover, we "imbed" the problem within a probabilistic annealing framework to avoid local optima of the cost. One outgrowth of our method is a probabilistic generalization of the Perceptron algorithm and its connection with minimum cross entropy inference. Our basic approach is shown to obtain significant improvement over the conventional splitting method. We then discussed a more general unbalanced tree growing method, for which tree growth is *non-heuristic* and occurs via phase transitions in the annealing process which occur so as to directly optimize the effective free energy cost. This method was proposed to search for the optimal tree-structured solution of given cluster size and maximal depth.

References

[1] R. O. Duda and P. E. Hart, *Pattern classification and scene analysis.* New York, NY: Wiley-Interscience, 1974.

[2] A. K. Jain and R. C. Dubes, *Algorithms for clustering data.* Englewood Cliffs, NJ: Prentice Hall, 1988.

[3] G. Ball and D. Hall, "A clustering technique for summarizing multivariate data," *Behavioral Science*, vol. 12, pp. 153–155, 1967.

[4] J. MacQueen, "Some methods for classification and analysis of multivariate observations," in *Proc. Fifth Berkeley Symposium on Math. Stat. and Prob.*, vol. I, pp. 281–297, 1967.

[5] J. C. Dunn, "A fuzzy relative of the ISODATA process and its use in detecting compact well-separated clusters," *J. Cybern.*, vol. 3, pp. 32–57, 1974.

[6] J. C. Bezdek, "A convergence theorem for the fuzzy ISODATA clustering algorithms," *IEEE Trans. on Patt. Anal. and Mach. Intell.*, vol. PAMI-2, pp. 1–8, 1980.

[7] L. Breiman, J. Friedman, R. Olshen, and C. Stone, *Classification and regression trees.* Belmont,CA.: The Wadsworth Statistics/Probability Series, 1980.

[8] P. A. Chou, "Optimal partitioning for classification and regression trees," *IEEE Trans. on Patt. Anal. and Mach. Intell.*, vol. 13, pp. 340–354, 1991.

[9] B. Juang and A. Gray, "Multiple stage vector quantization for speech coding," in *Proc. of the Intl. Conf. on Acous., Speech, and Sig. Proc.*, vol. 1, pp. 597–600, 1982.

[10] A. Buzo, A. Gray, Jr., R. Gray, and J. Markel, "Speech coding based on vector quantization," *IEEE Trans. on Acous., Speech, and Sig. Proc.*, vol. 28, pp. 562–574, 1980.

[11] K. Rose, E. Gurewitz, and G. C. Fox, "Vector quantization by deterministic annealing," *IEEE Trans. on Inform. Theory*, vol. 38, pp. 1249–1258, 1992.

[12] K. Rose, E. Gurewitz, and G. C. Fox, "Statistical mechanics and phase transitions in clustering," *Phys. Rev. Lett.*, vol. 65, pp. 945–948, 1990.

[13] A. L. Yuille, "Generalized deformable models, statistical physics, and matching problems," *Neural Computation*, vol. 2, pp. 1–24, 1990.

[14] D. Geiger and F. Girosi, "Parallel and deterministic algorithms from MRFs: Surface reconstruction," *IEEE Trans. on Patt. Anal. and Mach. Intell.*, vol. 13, pp. 401–412, 1991.

[15] P. D. Simic, "Statistical mechanics as the underlying theory of elastic and neural optimization," *Network*, vol. 1, pp. 89–103, 1990.

[16] K. Rose, E. Gurewitz, and G. C. Fox, "A deterministic annealing approach to clustering," *Pattern Rec. Lett.*, vol. 11, pp. 589–594, 1990.

[17] K. Rose, "A mapping approach to rate-distortion computation and analysis." (To appear in IEEE Trans. on Inform. Theory, Nov. 1994.).

[18] K. Rose, E. Gurewitz, and G. C. Fox, "Constrained clustering as an optimization method," *IEEE Trans. on Patt. Anal. and Mach. Intell.*, vol. 15, pp. 785–794, 1993.

[19] E. T. Jaynes, "Information theory and statistical mechanics," in *Papers on probability, statistics and statistical physics* (R. D. Rosenkrantz, ed.), Dordrecht, The Netherlands: Kluwer Academic Publishers, 1989. (Reprint of the original 1957 papers in *Physical Review*).

[20] J. A. Hartigan, *Clustering Algorithms.* New York: John Wiley and Sons, 1975.

[21] K. Rose and D. Miller, "Hierarchical clustering using deterministic annealing," in *Proc. of the Intl. Joint Conf. on Neural Networks*, pp. IV–85–90, 1992.

[22] J. E. Shore and R. W. Johnson, "Axiomatic derivation of the principle of maximum entropy and the principle of minimum cross-entropy," *IEEE Transactions on Information Theory*, vol. IT-26, pp. 26–37, 1980.

[23] D. Miller and K. Rose, "Hierarchical, unsupervised learning with growing via phase transitions." (Submitted to Neural Computation), 1994.

[24] E. Riskin and R. Gray, "A greedy growing algorithm for the design of variable rate vector quantizers," *IEEE Trans. on Sig. Proc.*, vol. 39, pp. 2500–2507, 1991.

[25] P. Chou, T. Lookabaugh, and R. Gray, "Optimal pruning with applications to tree-structured source coding and modeling," *IEEE Trans. on Inform. Theory*, vol. 35, pp. 299–315, 1989.

[26] X. Wu and K. Zhang, "A better tree-structured vector quantizer," in *Proc. of the Data Comp. Conf.*, pp. 392–401, 1991.

A SCALE INVARIANT BAYESIAN METHOD TO SOLVE LINEAR INVERSE PROBLEMS

Ali Mohammad-Djafari and Jérôme Idier
Laboratoire des Signaux et Systèmes (CNRS-ESE-UPS)
École Supérieure d'Électricité,
Plateau de Moulon, 91192 Gif-sur-Yvette Cédex, France

ABSTRACT. In this paper we propose a new Bayesian estimation method to solve linear inverse problems in signal and image restoration and reconstruction problems which has the property to be scale invariant. In general, Bayesian estimators are *nonlinear* functions of the observed data. The only exception is the Gaussian case. When dealing with linear inverse problems the linearity is sometimes a too strong property, while *scale invariance* often remains a desirable property. As everybody knows one of the main difficulties with using the Bayesian approach in real applications is the assignment of the direct (prior) probability laws before applying the Bayes' rule. We discuss here how to choose prior laws to obtain scale invariant Bayesian estimators. In this paper we discuss and propose a family of generalized exponential probability distributions functions for the direct probabilities (the prior $p(\boldsymbol{x})$ and the likelihood $p(\boldsymbol{y}|\boldsymbol{x})$), for which the posterior $p(\boldsymbol{x}|\boldsymbol{y})$, and, consequently, the main posterior estimators are scale invariant. Among many properties, generalized exponentials can be considered as the maximum entropy probability distributions subject to the knowledge of a finite set of expectation values of some known functions.

1. Introduction

We address a class of linear inverse problems arising in signal and image reconstruction and restoration problems which is to solve integral equations of the form:

$$g_{ij} = \int_D f(\boldsymbol{r}')\, h_{ij}(\boldsymbol{r}')\, \mathrm{d}\boldsymbol{r}' + b_{ij}, \quad i,j = 1,\cdots,M, \tag{1}$$

where $\boldsymbol{r}' \in \mathbb{R}^2$, $f(\boldsymbol{r}')$ is the object (image reconstruction problems) or the original image (image restoration problems), g_{ij} are the measured data (the projections in image reconstruction or the degraded image in image restoration problems), b_{ij} are the measurement noise samples and $h_{ij}(\boldsymbol{r}')$ are known functions which depend only on the measurement system. To show the generality of this relation, we give in the following some applications we are interested in:

- Image restoration:

$$g(x_i, y_j) = \iint_D f(x', y') h(x_i - x', y_j - y')\, \mathrm{d}x'\, \mathrm{d}y' + b(x_i, y_j) \quad , \quad \begin{array}{l} i = 1,\cdots,N \\ j = 1,\cdots,M \end{array},$$

  where $g(x_i, y_j)$ are the observed degraded image pixels and $h(x, y)$ is the point spread function (PSF) of the measurement system.

G. R. Heidbreder (ed.), Maximum Entropy and Bayesian Methods, 121–134.

- X-ray computed tomography (CT):

$$g(r_i,\phi_j) = \iint_D f(x,y)\delta(r_i - x\cos\phi_i - y\sin\phi_i)\,\mathrm{d}x\,\mathrm{d}y + b(r_i,\phi_j) \quad , \quad \begin{array}{l} i=1,\cdots,N \\ j=1,\cdots,M \end{array} ,$$

  where $g(r_i,\phi_j)$ are the projections along the axis $r_i = x\cos\phi_i - y\sin\phi_i$, having the angle ϕ_j, and which can be considered as the samples of the Radon transform (RT) of the object function $f(x,y)$.

- Fourier Synthesis in radio astronomy, in SAR imaging and in diffracted wave tomographic imaging systems:

$$g(u_j,v_j) = \iint_D f(x,y)\exp\left[-j(u_j x + v_j y)\right]\,\mathrm{d}x\,\mathrm{d}y + b(u_j,v_j), \quad j=1,\cdots,M,$$

  where $\boldsymbol{u}_j = (u_j,v_j)$ is a radial direction and $g(u_j,v_j)$ are the samples of the complex valued visibility function of the sky in radio astronomy or the Fourier transform of the measured signal in SAR imaging.

Other examples can be found in [6, 7, 5, 8, 9].

In all these applications we have to solve the following ill-posed problem: how to estimate the function $f(x,y)$ from some finite set of measured data which may also be noisy, because there is no experimental measurement device, even the most elaborate, which could be entirely free from uncertainty, the simplest example being the finite precision of the measurements.

The numerical solution of these equations needs a discretization procedure which can be done by a quadrature method. The linear system of equations resulting from the discretization of an ill-posed problem is, in general, very ill-conditioned if not singular. So the problem is to find a unique and stable solution for this linear system. The general methods which permit us to find a unique and stable solution to an ill-posed problem by introducing an *a priori* information on the solution are called regularization . The *a priori* information can be either in a deterministic form (positivity) or in a stochastic form (some constraints on the probability density functions).

When discretized, these problems can be described by the following:

> "Estimate a vector of the parameters $\boldsymbol{x} \in \mathbb{R}^n$ (pixel intensities in an image for example) given a vector of measurements $\boldsymbol{y} \in \mathbb{R}^m$ (representing, for example, either degraded image pixel values in restoration problems or the projections values in reconstruction problems) and a linear transformation $\boldsymbol{A}$ relating them by:
>
> $$\boldsymbol{y} = \boldsymbol{A}\boldsymbol{x} + \boldsymbol{b}, \tag{2}$$
>
> where $\boldsymbol{b}$ represents the discretization errors and the measurement noise which is supposed to be zero-mean and additive."

In this paper we propose to use the Bayesian approach to find a regularized solution to this problem. Note that the Bayesian theory only gives us a framework for the formulation of the inverse problem, not a solution of it. The main difficulty is, in general, before the application of the Bayes' formula, *i.e.*; how to formulate appropriately the problem and how to assign the direct probabilities. Keeping this fact in mind, we propose the following organization to this paper: In section 2. we give a brief description of the Bayesian approach with detailed calculations of the solution in the special case of Gaussian laws. In section 3. we discuss the *scale invariance* property and propose a family of prior probability density functions (*pdf*) which insure this property for the solution. Finally, in section 4., we present some special cases and give detailed calculations for the solution.

2. General Bayesian approach

A general Bayesian approach involves the following steps:

- Assign a prior probability law $p(\boldsymbol{x})$ to the unknown parameter to express our incomplete *a priori* information (prior beliefs) about these parameters;
- Assign a direct probability law pyx to the measured data to express the lack of total precision and the inevitable existence of the measurement noise;
- Use the Bayes' rule to calculate the posterior law $p(\boldsymbol{x}|\boldsymbol{y})$ of the unknown parameters;
- Define a decision rule to give values $\widehat{\boldsymbol{x}}$ to these parameters.

To illustrate the whole procedure, let us to consider an example; the Gaussian case. If we suppose that what we know about the unknown input $\boldsymbol{x}$ is its mean $\mathrm{E}\{\boldsymbol{x}\} = \boldsymbol{x}_0$ and its covariance matrix $\mathrm{E}\left\{(\boldsymbol{x}-\boldsymbol{x}_0)(\boldsymbol{x}-\boldsymbol{x}_0)^t\right\} = \boldsymbol{R}_x = \sigma_x^2\boldsymbol{P}$, and what we know about the measurement noise $\boldsymbol{b}$ is also its covariance matrix $\mathrm{E}\left\{\boldsymbol{b}\boldsymbol{b}^t\right\} = \boldsymbol{R}_b = \sigma_b^2\boldsymbol{I}$, then we can use the maximum entropy principle to assign:

$$p(\boldsymbol{x}) \propto \exp\left[-\frac{1}{2}(\boldsymbol{x}-\boldsymbol{x}_0)^t\boldsymbol{R}_x{}^{-1}(\boldsymbol{x}-\boldsymbol{x}_0)\right], \tag{3}$$

and

$$p(\boldsymbol{y}|\boldsymbol{x}) \propto \exp\left[-\frac{1}{2}(\boldsymbol{y}-\boldsymbol{A}\boldsymbol{x})^t\boldsymbol{R}_b{}^{-1}(\boldsymbol{y}-\boldsymbol{A}\boldsymbol{x})\right]. \tag{4}$$

Now we can use the Bayes' rule to find:

$$p(\boldsymbol{x}|\boldsymbol{y}) \propto p(\boldsymbol{y}|\boldsymbol{x})\, p(\boldsymbol{x}), \tag{5}$$

and use, for example, the maximum a posteriori (MAP) estimation rule to give a solution to the problem, *i.e.*;

$$\widehat{\boldsymbol{x}} = \arg\max_{\boldsymbol{x}}\left\{p(\boldsymbol{x}|\boldsymbol{y})\right\}, \tag{6}$$

Other estimators are possible. In fact, all we want to know is started in the posterior law. In general, one can construct a Bayesian estimator by defining a cost (or utility) function $C(\widehat{\boldsymbol{x}},\boldsymbol{x})$ and by minimizing its mean value

$$\widehat{\boldsymbol{x}} = \arg\min_{\boldsymbol{z}}\left\{\mathrm{E}_{X|Y}\left\{C(\boldsymbol{z},\boldsymbol{x})\right\}\right\} = \arg\min_{\boldsymbol{z}}\left\{\int C(\boldsymbol{z},\boldsymbol{x})p(\boldsymbol{x}|\boldsymbol{y})\,\mathrm{d}\boldsymbol{x}\right\}.$$

The two classical estimators:

- Posterior mean (PM): $\widehat{\boldsymbol{x}} = \mathrm{E}_{X|Y}\{\boldsymbol{x}\} = \int \boldsymbol{x}\, p(\boldsymbol{x}|\boldsymbol{y})\,\mathrm{d}\boldsymbol{x}$,
  is obtained when defining $C(\widehat{\boldsymbol{x}},\boldsymbol{x}) = (\widehat{\boldsymbol{x}}-\boldsymbol{x})^t(\widehat{\boldsymbol{x}}-\boldsymbol{x})$, and
- Maximum *a posteriori* (MAP): $\widehat{\boldsymbol{x}} = \arg\max_{\boldsymbol{x}}\{p(\boldsymbol{x}|\boldsymbol{y})\}$,
  is obtained when defining $C(\widehat{\boldsymbol{x}},\boldsymbol{x}) = 1-\delta(\widehat{\boldsymbol{x}}-\boldsymbol{x})$.

Now, let us go a little further inside the calculations. Replacing (3), and (4) in (5), we calculate the posterior law:

$$p(\boldsymbol{x}|\boldsymbol{y}) \propto \exp\left[-\frac{1}{2\sigma_b^2}J(\boldsymbol{x})\right], \quad \text{with } J(\boldsymbol{x}) = (\boldsymbol{y}-\boldsymbol{A}\boldsymbol{x})^t(\boldsymbol{y}-\boldsymbol{A}\boldsymbol{x}) + \lambda(\boldsymbol{x}-\boldsymbol{x}_0)^t\boldsymbol{P}^{-1}(\boldsymbol{y}-\boldsymbol{x}_0),$$

where $\lambda = \sigma_b^2/\sigma_x^2$. The posterior is then also a Gaussian. We can now use any decision rule to obtain a solution. For example the maximum a posteriori (MAP) solution is obtained by:

$$\widehat{\boldsymbol{x}} = \arg\max_{\boldsymbol{x}}\{p(\boldsymbol{x}|\boldsymbol{y})\} = \arg\min_{\boldsymbol{x}}\{J(\boldsymbol{x})\}. \tag{7}$$

Note that in this special Gaussian case both estimators, *i.e.*; the posterior mean (PM) and the MAP estimators are the same:

$$\widehat{\boldsymbol{x}} = \mathrm{E}_{X|Y}\{\boldsymbol{x}\} = \arg\max_{\boldsymbol{x}}\{p(\boldsymbol{x}|\boldsymbol{y})\} \tag{8}$$

and the minimization of the criterion $J(\boldsymbol{x})$, which can also be written in the form

$$J(\boldsymbol{x}) = ||\boldsymbol{y}-\boldsymbol{A}\boldsymbol{x}||^2 + \lambda||\boldsymbol{x}-\boldsymbol{x}_0||^2_{\boldsymbol{P}}, \tag{9}$$

can be considered as a regularization procedure to the inverse problem (2). Indeed, the Bayesian approach yields a new interpretation of the regularization parameter in terms of the signal to noise ratio, *i.e.*; $\lambda = \sigma_b^2/\sigma_x^2$.

$J(\boldsymbol{x})$ is a quadratic function of $\boldsymbol{x}$. The solution $\widehat{\boldsymbol{x}}$ is then a linear function of the data $\boldsymbol{y}$. This is due to the fact that the problem is linear and all the probability laws are Gaussian. In general, the Bayesian estimators are not linear functions of the observations $\boldsymbol{y}$. However, we may not need that the solution be a linear function of the data $\boldsymbol{y}$, but the *scale invariance* is the minimum property which is often needed.

3. Scale invariant Bayesian estimators

What we are proposing in this paper is to study under what conditions we can obtain estimators which are scale invariant. Note that *linearity* is the combination of

$$\textit{additivity:} \quad \left\{\begin{array}{l}\boldsymbol{y}_1 \mapsto \widehat{\boldsymbol{x}}_1,\\ \boldsymbol{y}_2 \mapsto \widehat{\boldsymbol{x}}_2\end{array}\right. \Longrightarrow \boldsymbol{y}_1+\boldsymbol{y}_2 \mapsto \widehat{\boldsymbol{x}}_1+\widehat{\boldsymbol{x}}_2,$$

and

$$\textit{scale invariance:} \quad \boldsymbol{y}_1 \mapsto \widehat{\boldsymbol{x}}_1 \Longrightarrow \forall k>0,\ k\boldsymbol{y}_1 \mapsto k\widehat{\boldsymbol{x}}_1.$$

In a linear inverse problem what is often necessary is that the solution be scale invariant. As we have seen in the last section, when all the probability laws are Gaussian, the Bayesian estimators are linear functions of the data, so that the methods based on this assumption do not have to take care of the scale of the measured data. The Gaussian assumption is very restrictive. On the other hand, more general priors yield Bayesian estimators which are nonlinear functions of data, so the results of the inversion method depend on the absolute values of the measured data. In other words, two users of the method using two different scale factors would not get the same results, even rescaled:

$$\boldsymbol{y} \longrightarrow \boxed{k_1} \longrightarrow \boxed{\text{Estimation}} \longrightarrow \widehat{\boldsymbol{x}}_1$$

$$\frac{\widehat{\boldsymbol{x}}_2}{k_2} \neq \frac{\widehat{\boldsymbol{x}}_1}{k_1}$$

$$\boldsymbol{y} \longrightarrow \boxed{k_2} \longrightarrow \boxed{\text{Estimation}} \longrightarrow \widehat{\boldsymbol{x}}_2$$

A general nonlinear (scale variant) estimation method

What we want to specify in this paper is a family of probability laws for which these estimators are scale invariant. Then the user of the inversion method can process the data without worrying about rescaling them to an arbitrary level and two users of the method at two different scales will obtain the proportional results:

$$\boldsymbol{y} \longrightarrow \boxed{k_1} \longrightarrow \boxed{\text{Estimation}} \longrightarrow \widehat{\boldsymbol{x}}_1$$

$$\frac{\widehat{\boldsymbol{x}}_2}{k_2} = \frac{\widehat{\boldsymbol{x}}_1}{k_1}$$

$$\boldsymbol{y} \longrightarrow \boxed{k_2} \longrightarrow \boxed{\text{Estimation}} \longrightarrow \widehat{\boldsymbol{x}}_2$$

A scale invariant estimation method

To do this let us note

- $\boldsymbol{\theta}$, all the unknown parameters defining our measuring system (noise variance σ^2 and the prior law parameters for example),
- $p_1(\boldsymbol{x}_1|\boldsymbol{y}_1;\boldsymbol{\theta}_1)$ and $p_k(\boldsymbol{x}_k|\boldsymbol{y}_k;\boldsymbol{\theta}_k)$, the two expressions of the posterior law for scale 1 and for scale k with

$$\boldsymbol{x}_k = k\boldsymbol{x}_1, \quad \boldsymbol{y}_k = k\boldsymbol{y}_1.$$

Then, what we need is the following:

$$\exists \boldsymbol{\theta}_k = f(\boldsymbol{\theta}_1, k) \mid \forall k > 0, \forall \boldsymbol{x}_1, \boldsymbol{y}_1, \quad p_k(\boldsymbol{x}_k|\boldsymbol{y}_k;\boldsymbol{\theta}_k) = \frac{1}{k^n} p_1(\boldsymbol{x}_1|\boldsymbol{y}_1;\boldsymbol{\theta}_1), \tag{10}$$

which means that the functional form of the posterior law remains unchanged when the measurement's scale is changed. Only we have to modify the parameters $\boldsymbol{\theta}_k = f(\boldsymbol{\theta}_1, k)$ which is only a function of $\boldsymbol{\theta}_1$ and the scale factor k.

However, not all estimators based on this posterior will be scale invariant. The cost function must also have some property to obtain a scale invariant estimator. So, the main result of this paper can be stated in the following theorem:

Theorem: If $\exists \boldsymbol{\theta}_k = f(\boldsymbol{\theta}_1, k) \mid \forall k > 0, \forall \boldsymbol{x}_1, \boldsymbol{y}_1$,

$$p_k(\boldsymbol{x}_k|\boldsymbol{y}_k;\boldsymbol{\theta}_k) = \frac{1}{k^n} p_1(\boldsymbol{x}_1|\boldsymbol{y}_1;\boldsymbol{\theta}_1),$$

then any Bayesian estimator with a cost function $C(\widehat{\boldsymbol{x}}, \boldsymbol{x})$ satisfying:

$$C(\widehat{\boldsymbol{x}}_k, \boldsymbol{x}_k) = a_k + b_k C(\widehat{\boldsymbol{x}}_1, \boldsymbol{x}_1),$$

is a scale invariant estimator, *i.e.*;

$$\widehat{\boldsymbol{x}}_k(\boldsymbol{y}_k; \boldsymbol{\theta}_k) = k\widehat{\boldsymbol{x}}_1(\boldsymbol{y}_1; \boldsymbol{\theta}_1).$$

Proof: In fact, it is easy to see the following:

$$\begin{aligned}
\widehat{\boldsymbol{x}}_k(\boldsymbol{y}_k; \boldsymbol{\theta}_k) &= \arg\min_{\boldsymbol{z}_k} \left\{ \int C(\boldsymbol{z}_k, \boldsymbol{x}_k) p_k(\boldsymbol{x}_k | \boldsymbol{y}_k; \boldsymbol{\theta}_k) \, \mathrm{d}\boldsymbol{x}_k \right\} \\
&= k \arg\min_{\boldsymbol{z}_1} \left\{ \int [b_k C(\boldsymbol{z}_1, \boldsymbol{x}_1) + a_k] \frac{1}{k^n} p_1(\boldsymbol{x}_1 | \boldsymbol{y}_1; \boldsymbol{\theta}_1) k^n \, \mathrm{d}\boldsymbol{x}_1 \right\} \\
&= k \arg\min_{\boldsymbol{z}_1} \left\{ b_k \int C(\boldsymbol{z}_1, \boldsymbol{x}_1) p_1(\boldsymbol{x}_1 | \boldsymbol{y}_1; \boldsymbol{\theta}_1) \, \mathrm{d}\boldsymbol{x}_1 + a_k \right\} \\
&= k \arg\min_{\boldsymbol{z}_1} \left\{ \int C(\boldsymbol{z}_1, \boldsymbol{x}_1) p_1(\boldsymbol{x}_1 | \boldsymbol{y}_1; \boldsymbol{\theta}_1) \, \mathrm{d}\boldsymbol{x}_1 \right\} \\
&= k \, \widehat{\boldsymbol{x}}_1(\boldsymbol{y}_1; \boldsymbol{\theta}_1)
\end{aligned}$$

Note the great significance of this result. Even if the estimator $\widehat{\boldsymbol{x}}(\boldsymbol{y}; \boldsymbol{\theta})$ is a nonlinear function of the observations $\boldsymbol{y}$ it stays scale invariant.

Now, the task is to search for a large family of probability laws $p(\boldsymbol{x})$ and $p(\boldsymbol{y}|\boldsymbol{x})$ in a manner that the posterior law $p(\boldsymbol{x}|\boldsymbol{y})$ remains scale invariant. We propose to do this search in the generalized exponential family for two reasons:

- First the generalized exponential probability density functions form a very rich class, and
- Second, they can be considered as the maximum entropy prior laws subject to a finite number of constraints (linear or nonlinear).

Note also that if $p(\boldsymbol{x})$ and $p(\boldsymbol{y}|\boldsymbol{x})$ are scale invariant then the posterior $p(\boldsymbol{x}|\boldsymbol{y})$ is also scale invariant and there is a symmetry for $p(\boldsymbol{x})$ and $p(\boldsymbol{y}|\boldsymbol{x})$ so that it is only necessary to find the scale invariance conditions for one of them. In the following, without loss of generality, we consider the case where $p(\boldsymbol{y}|\boldsymbol{x})$ is Gaussian,

$$p(\boldsymbol{y}|\boldsymbol{x}; \sigma^2) \propto \exp\left[-\chi^2(\boldsymbol{x}, \boldsymbol{y}; \sigma^2)\right], \quad \text{with } \chi^2(\boldsymbol{x}, \boldsymbol{y}; \sigma^2) = \frac{1}{2\sigma^2}[\boldsymbol{y} - \boldsymbol{H}\boldsymbol{x}]^t[\boldsymbol{y} - \boldsymbol{H}\boldsymbol{x}], \tag{11}$$

and find the conditions for $p(\boldsymbol{x})$ to be scale invariant. We choose the generalized exponential *pdf*'s for $p(\boldsymbol{x})$, *i.e.*;

$$p(\boldsymbol{x}; \boldsymbol{\lambda}) \propto \exp\left[-\sum_{i=1}^{r} \lambda_i \phi_i(\boldsymbol{x})\right], \tag{12}$$

and find the conditions on the functions $\phi_i(\boldsymbol{x})$ for which $p(\boldsymbol{x})$ is scale invariant.

Note that these laws can be considered as the maximum entropy prior laws if our prior knowledge is as follows:

- What we know about $\boldsymbol{x}$ is:

$$\mathrm{E}\{\phi_i(\boldsymbol{x})\} = d_i, \quad i = 1, \cdots, r,$$

- and what we know about the noise $\boldsymbol{b}$ is:

$$\begin{cases} \mathrm{E}\{\boldsymbol{b}\} = 0, \\ \mathrm{E}\{\boldsymbol{b}\boldsymbol{b}^t\} = \boldsymbol{R}_b = \sigma^2 \boldsymbol{I}, \end{cases}$$

where $\boldsymbol{R}_b$ is the covariance matrix of $\boldsymbol{b}$.

Now, using the equations (11) and (12) and noting that $\boldsymbol{\theta} = (\sigma^2, \lambda_1, \cdots, \lambda_r)$, that $\boldsymbol{\lambda} = (\lambda_1, \cdots, \lambda_r)$, and that $\boldsymbol{\phi}(\boldsymbol{x}) = (\phi_1(\boldsymbol{x}), \cdots, \phi_r(\boldsymbol{x}))$, we have

$$p(\boldsymbol{x}|\boldsymbol{y};\boldsymbol{\theta}) \propto \exp\left[-\chi^2(\boldsymbol{x},\boldsymbol{y};\sigma^2) - \boldsymbol{\lambda}^t\boldsymbol{\phi}(\boldsymbol{x})\right], \tag{13}$$

and the scale invariance condition becomes:

$$\forall k > 0, \forall \boldsymbol{x}_1, \boldsymbol{y}_1, \quad \chi_k^2(\boldsymbol{x}_k, \boldsymbol{y}_k; \sigma_k^2) + \boldsymbol{\lambda}_k^t \boldsymbol{\phi}(\boldsymbol{x}_k) = \chi_1^2(\boldsymbol{x}_1, \boldsymbol{y}_1; \sigma_1^2) + \boldsymbol{\lambda}_1^t \boldsymbol{\phi}(\boldsymbol{x}_1) + cte.$$

But with the Gaussian choice for the noise *pdf* we have

$$\forall k > 0, \forall \boldsymbol{x}_1, \boldsymbol{y}_1,\ \chi_k^2(\boldsymbol{x}_k, \boldsymbol{y}_k; \sigma_k^2) = \frac{1}{2\sigma_k^2}\,||\boldsymbol{y}_k - \boldsymbol{H}\boldsymbol{x}_k||^2 = \frac{1}{2k^2\sigma_1^2}\,k^2\,||\boldsymbol{y}_1 - \boldsymbol{H}\boldsymbol{x}_1||^2 = \chi_1^2(\boldsymbol{x}_1, \boldsymbol{y}_1; \sigma_1^2),$$

and so the condition becomes

$$\forall k > 0, \forall \boldsymbol{x}, \quad \boldsymbol{\lambda}_k^t \boldsymbol{\phi}(\boldsymbol{x}_k) = \boldsymbol{\lambda}_1^t \boldsymbol{\phi}(\boldsymbol{x}_1) + cte, \tag{14}$$

or equivalently,

$$p_k(\boldsymbol{x}_k; \boldsymbol{\lambda}_k) = \frac{1}{k^n}\, p_1(\boldsymbol{x}_1; \boldsymbol{\lambda}_1) \quad \text{with} \quad \boldsymbol{\lambda}_k = f(\boldsymbol{\lambda}_1, k).$$

Thus, in the case of a centered Gaussian *pdf* for the noise, to have a scale invariant posterior law it is sufficient to have a scale invariant prior law.

Now, assuming interchangeable (independent) pixels, *i.e.*,

$$p(\boldsymbol{x}; \boldsymbol{\lambda}) = \exp\left[\lambda_0 + \sum_{i=1}^{r} \boldsymbol{\lambda}_i \phi_i(\boldsymbol{x})\right] = \prod_{j=1}^{N} p(x_j; \boldsymbol{\lambda}), \tag{15}$$

or equivalently,

$$\phi_i(\boldsymbol{x}) = \sum_{j=1}^{N} \phi_i(x_j), \tag{16}$$

we have to find the conditions on the scalar functions $\phi_i(x)$ of scalar variables x which satisfy the equation (14) or equivalently

$$\forall k > 0, \forall x, \quad \sum_{i=1}^{r} \lambda_i(k)\, \phi_i(kx) = \sum_{i=1}^{r} \lambda_i(1)\, \phi_i(x) + cte. \tag{17}$$

We have shown (see appendix) that, the functions $\phi_i(x)$ which satisfy these conditions are all either the powers of x or the powers of $\ln x$ or a multiplication of them. The general expressions for these functions are:

$$\phi(x) = \sum_{m=1}^{M} \left(\sum_{n=0}^{N_m - 1} c_{mn} (\ln x)^n\right) x^{\alpha_m} + \sum_{n=0}^{N_0} c_{0n} (\ln x)^n, \quad \text{with } M \le r \text{ and } \sum_{m=0}^{M} N_m = r, \tag{18}$$

where M and N_m are integer numbers, and c_{mn}, c_{0n} and α_m are real numbers. For a geometrical interpretation and more details see the appendix. The following examples show some special and interesting cases.

One parameter laws: Consider the case of $r = 1$. In this case we have

$$p(x;\lambda) \propto \exp\left[-\lambda\phi(x)\right]. \tag{19}$$

Applying the general rule with

$$r = 1 \longrightarrow \begin{cases} M = 0, N_0 = 1, & \longrightarrow c_{00} + c_{01}\ln x \\ M = 1, N_0 = 0, N_1 = 1, & \longrightarrow c_{00} + c_{10}x^{\alpha_1} \end{cases}$$

we find that the only functions who satisfy these conditions are:

$$\left\{\phi(x)\right\} = \left\{x^{\alpha}, \ln x\right\} \tag{20}$$

where α is a real number. There are two interesting special cases:

- $\phi(x) = x^{\alpha}$, resulting in $p(x) \propto \exp\left[-\lambda x^{\alpha}\right]$, $\alpha > 0, \lambda > 0$, which is a generalized Gaussian *pdf*, and
- $\phi(x) = \ln x$, resulting in $p(x) \propto \exp\left[-\lambda \ln x\right]$, which is a special case of the Beta *pdf*.

Note that the famous *entropic* prior law $p(x) \propto \exp\left[-\lambda x \ln x\right]$ of Gull and Skilling [11, 4] does not verify the scale invariance property. But, if we add one more parameter,

$$p(x) \propto \exp\left[-\lambda x \ln x + \mu x\right],$$

it will then satisfy this condition as we can see in the next section.

Two parameters laws: This is the case where $r = 2$ and we have

$$p(x;\lambda) \propto \exp\left[-\lambda\phi_1(x) - \mu\phi_2(x)\right]. \tag{21}$$

Applying the general rule:

$$r = 2 \longrightarrow \begin{cases} M = 2, N_0 = 0, N_1 = 1, N_2 = 1, & \longrightarrow c_{00} + c_{10}x^{\alpha_1} + c_{20}x^{\alpha_2} \\ M = 1, N_0 = 0, N_1 = 2, & \longrightarrow c_{00} + c_{10}x^{\alpha_1} + c_{11}x^{\alpha_1}\ln x \\ M = 1, N_0 = 1, N_1 = 1, & \longrightarrow c_{00} + c_{10}x^{\alpha_1} + c_{01}\ln x \\ M = 0, N_0 = 2, & \longrightarrow c_{00} + c_{01}\ln x + c_{02}\ln^2 x, \end{cases}$$

we see that in this case the only functions (ϕ_1, ϕ_2) which satisfy these conditions are

$$\left\{(\phi_1(x), \phi_2(x))\right\} = \left\{(x^{\alpha_1}, x^{\alpha_2}), (x^{\alpha_1}, x^{\alpha_1}\ln x), (x^{\alpha_1}, \ln x), (\ln x, \ln^2 x)\right\} \tag{22}$$

where α_1 and α_2 are two real numbers. Special cases are obtained when we choose $\phi_2(x) = x$. The only possible functions for $\phi_1(x)$ are then

$$\left\{x^{\alpha}, \ln x, x\ln x\right\}, \tag{23}$$

and we have the following interesting cases:

- $\phi_1(x) = x^2$, resulting in $p(x) \propto \exp\left[-\lambda x^2 - \mu x\right] \propto \exp\left[-\lambda\left(x + \frac{\mu}{2\lambda}\right)^2\right]$, which is a Gaussian *pdf* $\mathcal{N}\left(m = \frac{-\mu}{\lambda}, \sigma^2 = \frac{1}{2\lambda}\right)$,
- $\phi_1(x) = \ln x$, resulting in $p(x) \propto \exp\left[-\lambda \ln x - \mu x\right] = x^{-\lambda}\exp\left[-\mu x\right]$, which is the Gamma *pdf*, and finally,
- $\phi_1(x) = x \ln x$, resulting in $p(x) \propto \exp\left[-\lambda x \ln x - \mu x\right]$. which is known as the *entropic pdf*.

Three parameters laws: This is the case where $r = 3$. Once more applying the general rule we find

$$r = 3 \rightarrow \begin{cases} M = 3, N_0 = 0, N_1 = 1, N_2 = 1, N_3 = 1, & \rightarrow c_{00} + c_{10}x^{\alpha_1} + c_{20}x^{\alpha_2} + c_{30}x^{\alpha_3} \\ M = 2, N_0 = 0, N_1 = 1, N_2 = 2, & \rightarrow c_{00} + c_{10}x^{\alpha_1} + c_{20}x^{\alpha_2} + c_{21}x^{\alpha_2}\ln x \\ M = 2, N_0 = 1, N_1 = 1, N_2 = 1, & \rightarrow c_{00} + c_{01}\ln x + c_{10}x^{\alpha_1} + c_{20}x^{\alpha_2} \\ M = 1, N_0 = 0, N_1 = 3, & \rightarrow c_{00} + c_{10}x^{\alpha_1} + c_{11}x^{\alpha_1}\ln x + c_{12}x^{\alpha_1}\ln^2 x \\ M = 1, N_0 = 1, N_1 = 2, & \rightarrow c_{00} + c_{01}\ln x + c_{10}x^{\alpha_1} + c_{11}x^{\alpha_1}\ln x \\ M = 1, N_0 = 2, N_1 = 1, & \rightarrow c_{00} + c_{01}\ln x + c_{02}\ln^2 x + c_{10}x^{\alpha_1} \\ M = 0, N_0 = 3, & \rightarrow c_{00} + c_{01}\ln x + c_{02}\ln^2 x + c_{03}\ln^3 x \end{cases},$$

which means

$$\left\{(\phi_1(x), \phi_2(x), \phi_3(x))\right\} = \left\{ \begin{array}{l} (x^{\alpha_1}, x^{\alpha_2}, x^{\alpha_3}),\ (x^{\alpha_1}, x^{\alpha_2}, \ln x),\ (x^{\alpha_1}, x^{\alpha_1}\ln x, x^{\alpha_1}\ln^2 x), \\ (x^{\alpha_1}, x^{\alpha_1}\ln x, \ln x),\ (x^{\alpha_1}, x^{\alpha_2}, x^{\alpha_2}\ln x),\ (x^{\alpha_1}, \ln x, \ln^2 x), \\ \hfill (\ln x, \ln^2 x, \ln^3 x) \end{array}\right\}, \tag{24}$$

where α_1, α_2 and α_3 are three real numbers.

4. Proposed method

The general procedure of the inversion method we propose can be summarized as follows:

- Choose a set of functions $\phi_i(x)$ from the possible ones described in the last section and assign the prior $p(\boldsymbol{x})$. In many imaging applications we proposed and used successfully the following one with two parameters:

$$p(\boldsymbol{x}; \boldsymbol{\lambda}) \propto \exp\left[-\lambda_1 H(\boldsymbol{x}) - \lambda_2 S(\boldsymbol{x})\right], \quad \text{with } H(\boldsymbol{x}) = \sum_{j=1}^{N}\phi_1(x_j), \text{ and } S(\boldsymbol{x}) = \sum_{j=1}^{N}\phi_2(x_j),$$

  where $\phi_1(x)$ and $\phi_2(x)$ were chosen from the possible ones in (22) or (23).

- When what we know about the noise $\boldsymbol{b}$ is only its covariance matrix $\mathrm{E}\left\{\boldsymbol{b}\boldsymbol{b}^t\right\} = \boldsymbol{R}_b = \sigma_b^2\boldsymbol{I}$, then using the maximum entropy principle we have:

$$p(\boldsymbol{y}|\boldsymbol{x}) \propto \exp\left[-\frac{1}{2}Q(\boldsymbol{x})\right], \quad \text{with } Q(\boldsymbol{x}) = (\boldsymbol{y} - \boldsymbol{A}\boldsymbol{x})^t\boldsymbol{R}_b^{-1}(\boldsymbol{y} - \boldsymbol{A}\boldsymbol{x}).$$

  We may note that $p(\boldsymbol{y}|\boldsymbol{x})$ is also a scale invariant probability law.

- Using the Bayes' rule and MAP estimator the solution is determined by

$$\widehat{\boldsymbol{x}} = \arg\max_{\boldsymbol{x}} \{p(\boldsymbol{x}|\boldsymbol{y})\} = \arg\min_{\boldsymbol{x}} \{J(\boldsymbol{x})\}, \quad \text{with } J(\boldsymbol{x}) = Q(\boldsymbol{x}) + \lambda_1 H(\boldsymbol{x}) + \lambda_2 S(\boldsymbol{x}).$$

  Note here also that, for the cases where one of the functions $\phi_1(x)$ or $\phi_2(x)$ is a logarithmic function of x, we have to constrain its range to the positive real axis, and we have to solve the following optimization problem

$$\widehat{\boldsymbol{x}} = \arg\max_{\boldsymbol{x}>0} \{p(\boldsymbol{x}|\boldsymbol{y})\} = \arg\min_{\boldsymbol{x}>0} \{J(\boldsymbol{x})\}.$$

  This optimization is achieved by a modified conjugate gradients method.

- The choice of the functions $\phi_i(x)$ and the determination of the parameters (λ_1, λ_2) in the first step is still an open problem.

  In imaging applications we propose to make this choice using our prior knowledge on the nature of the quantity of interest (physics of the application). For example, if the object $\boldsymbol{x}$ is a real quantity equally distributed on the positive and the negative reals, a Gaussian prior, *i.e.*; $(\phi_1(x) = x, \phi_2(x) = x^2)$ is convenient. But, if the object $\boldsymbol{x}$ is a positive quantity or if we know that it represents small extent, bright and sharp objects on a nearly black background (images in radio astronomy, for example), we may choose $(\phi_1(x) = x, \phi_2(x) = \ln x)$, or $(\phi_1(x) = x, \phi_2(x) = x \ln x)$ which are the priors with longer tails than the Gaussian or truncated Gaussian one.

  When the choice of the functions $(\phi_1(x), \phi_2(x))$ is made, we still have to determine the hyperparameters (λ_1, λ_2). For this two main approaches have been proposed. The first is based on the generalized maximum likelihood (GML) which tries to estimate simultaneously the parameters $\boldsymbol{x}$ and the hyperparameters $\boldsymbol{\theta} = (\lambda_1, \lambda_2)$ by

$$(\widehat{\boldsymbol{x}}, \widehat{\boldsymbol{\theta}}) = \arg\max_{(\boldsymbol{x},\boldsymbol{\theta})} \{p(\boldsymbol{x}, \boldsymbol{y}; \boldsymbol{\theta})\} = \arg\max_{(\boldsymbol{x},\boldsymbol{\theta})} \{p(\boldsymbol{y}|\boldsymbol{x})\, p(\boldsymbol{x}; \boldsymbol{\theta})\}, \tag{25}$$

  and the second is based on the marginalization (MML), in which the hyperparameters $\boldsymbol{\theta}$ are estimated first by

$$\widehat{\boldsymbol{\theta}} = \arg\max_{\boldsymbol{\theta}} \left\{p(\boldsymbol{y}; \boldsymbol{\theta}) = \int p(\boldsymbol{x}, \boldsymbol{y}; \boldsymbol{\theta})\, \mathrm{d}\boldsymbol{x}\right\} = \arg\max_{\boldsymbol{\theta}} \left\{\int p(\boldsymbol{y}|\boldsymbol{x})\, p(\boldsymbol{x}; \boldsymbol{\theta})\, \mathrm{d}\boldsymbol{x}\right\}, \tag{26}$$

  and then used for the estimation of $\boldsymbol{x}$ in

$$\widehat{\boldsymbol{x}} = \arg\max_{\boldsymbol{x}} \left\{p(\boldsymbol{x}|\boldsymbol{y}; \widehat{\boldsymbol{\theta}})\right\} = \arg\max_{\boldsymbol{x}} \left\{p(\boldsymbol{y}|\boldsymbol{x})\, p(\boldsymbol{x}|\widehat{\boldsymbol{\theta}})\right\}. \tag{27}$$

  What is important here is that both methods preserve the scale invariant property. For practical applications we have recently proposed and used a method based on the generalized maximum likelihood [8, 9] which has been successfully used in many signal and image reconstruction and restoration problems as we mentioned in the introduction [10].

5. Conclusions

Except for the Gaussian case where all the Bayesian estimators are linear functions of the observed data, in general, the Bayesian estimators are *nonlinear* functions of the data. When dealing with linear inverse problems linearity is sometimes a too strong property, while *scale invariance* often remains a desirable property. In this paper we discussed and proposed a family of generalized exponential probability distributions for the direct probabilities (the prior $p(\boldsymbol{x})$ and the likelihood $p(\boldsymbol{y}|\boldsymbol{x})$), for which the posterior $p(\boldsymbol{x}|\boldsymbol{y})$, and, consequently, the main posterior estimators are scale invariant. Among many properties, generalized exponential can be considered as the maximum entropy probability distributions subject to the knowledge of a finite set of expectation values of some known functions.

A Appendix: General case

We want to find the solutions of the following equation:

$$\forall k > 0, \forall x, \quad \sum_{i=1}^{r} \lambda_i(k)\phi_i(kx) = \sum_{i=1}^{r} \lambda_i(1)\phi_i(x) + \beta(k). \tag{A.1}$$

Making the following changes of variables and notations

$$1/k = \tilde{k},\ kx = \tilde{x},\ \lambda_i(k) = \tilde{\lambda}_i(\tilde{k}), \quad \text{and} \quad \beta_i(k) = \tilde{\beta}_i(\tilde{k}),$$

equation (A.1) becomes

$$\sum_{i=1}^{r} \tilde{\lambda}_i(\tilde{k})\phi_i(\tilde{x}) = \sum_{i=1}^{r} \tilde{\lambda}_i(1)\phi_i(\tilde{k}\tilde{x}) + \tilde{\beta}(\tilde{k})$$

For convenience sake, we will drop the tilde $\tilde{}$, and note $\lambda_i(1) = \lambda_i$, so that we can write

$$\sum_{i=1}^{r} \lambda_i(k)\phi_i(x) = \sum_{i=1}^{r} \lambda_i\phi_i(kx) + \beta(k).$$

Noting

$$S(x) = \sum_{i=1}^{r} \lambda_i\phi_i(x), \quad \text{and} \quad S(kx) = \sum_{i=1}^{r} \lambda_i\phi_i(kx)$$

we have

$$\sum_{i=1}^{r} \lambda_i(k)\phi_i(x) = S(kx) + \beta(k). \tag{A.2}$$

Taking the first $r-1$ derivatives of this equation with respect to k, we obtain

$$\begin{aligned}
\sum_{i=1}^{r} \lambda_i'(k)\phi_i(x) &= x\,S'(kx) + \beta'(k) \\
\sum_{i=1}^{r} \lambda_i''(k)\phi_i(x) &= x^2\,S''(kx) + \beta''(k) \\
\vdots \qquad\qquad & \qquad \vdots \\
\sum_{i=1}^{r} \lambda_i^{(r-1)}(k)\phi_i(x) &= x^{r-1} S^{(r-1)}(kx) + \beta^{(r-1)}(k).
\end{aligned} \tag{A.3}$$

Combining equations (A.2) and (A.3) in matrix form we have

$$\begin{pmatrix} \lambda_1(k) & \cdots & \lambda_r(k) \\ \lambda_1'(k) & \cdots & \lambda_r'(k) \\ \lambda_1''(k) & \cdots & \lambda_r''(k) \\ \vdots & \cdots & \vdots \\ \lambda_1^{(r-1)}(k) & \cdots & \lambda_r^{(r-1)}(k) \end{pmatrix} \begin{pmatrix} \phi_1(x) \\ \phi_2(x) \\ \phi_3(x) \\ \vdots \\ \phi_r(x) \end{pmatrix} = \begin{pmatrix} S(kx)+\beta(k) \\ xS'(kx)+\beta'(k) \\ x^2S''(kx)+\beta''(k) \\ \vdots \\ x^{r-1}S^{(r-1)}(kx)+\beta^{(r-1)}(k) \end{pmatrix} \tag{A.4}$$

If this matrix equation can be inverted, it follows that any function $\phi_i(x)$ is a linear combination of $S(kx)+\beta(k)$ and its $(r-1)$ derivatives with respect to k:

$$\phi_i(x) = \sum_{i=0}^{r} \eta_i(k) \left[x^{(i-1)} S^{(i-1)}(kx) + \beta^{(i-1)}(k)\right]. \tag{A.5}$$

If this is not the case, there exists an interval for k, for which some of the functions $\lambda_i(k)$ are linear combinations of the others [2]. In this case let us show that we will go back to the situation of the problem of lower order r. Let us assume that the last column of the matrix is a linear combination of the others, *i.e.*,

$$\lambda_r(k) = \sum_{i=1}^{r-1} \gamma_i \lambda_i(k).$$

Putting this in the equation (A.1) will give

$$\sum_{i=1}^{r-1} \lambda_i(k)\phi_i(kx) + \left[\sum_{i=1}^{r-1} \gamma_i\lambda_i(k)\right] \phi_r(kx) = \sum_{i=1}^{r-1} \lambda_i(1)\phi_i(x) + \beta(k) + \left[\sum_{i=1}^{r-1} \gamma_i\lambda_i(1)\right] \phi_r(x),$$

and noting $\psi_i(x) = \phi_i(x) + \gamma_i\phi_r(x)$ and $\psi_i(kx) = \phi_i(kx) + \gamma_i\phi_r(kx)$, we obtain

$$\sum_{i=1}^{r-1} \lambda_i(k)\psi_i(kx) = \sum_{i=1}^{r} \lambda_i(1)\psi_i(x) + \beta(k),$$

which is an equation in the same form as (A.1), but of lower order.

Now taking derivatives of both parts of the equation (A.5) with respect to k and noting $kx = u$ we obtain

$$\sum_{i=0}^{r} a_i\, u^i S^i(u) = a \tag{A.6}$$

This is the general expression of a rth order Euler–Cauchy differential equation [1, 2] which is classically solved through the change of variable $u = e^x$. One can find the general expression of its solution in the following form:

$$S(x) = \sum_{m=1}^{M} \left(\sum_{n=0}^{N_m-1} c_{mn}(\ln x)^n\right) x^{\alpha_m} + \sum_{n=0}^{N_0} c_{0n}(\ln x)^n \quad \text{with } M = 0,\cdots r, \text{ and } \sum_{m=0}^{M} N_m = r, \tag{A.7}$$

where M and N_m are integer numbers, and c_{mn}, c_{0n} and α_m are real numbers. In fact the most general solution also incorporates terms of the form

$$\left[\sum_n (\ln x)^n \left(\alpha_n \cos(\ln x) + \beta_n \sin(\ln x)\right)\right] x^d$$

derived from complex α_m and c_{mn}. But we will not consider these terms because the resulting *pdf*'s have oscillatory behavior around zero.

One can give a geometric interpretation of the solutions given in (A.7). For any given order r make a $(r+1) \times (r+1)$ table in the form

$\ln^r x$					
$\vdots$					
$\ln^2 x$					
$\ln x$					
1	$\times$				
	1	x^{α_1}	x^{α_2}	$\cdots$	x^{α_r}

and let r mass points fall down into the columns. To each filled box is assigned a function $\phi_i(x)$ by multiplying the corresponding powers of x and $\ln x$ on the same line and the same column. To illustrate this, we give in the following the first three cases:

Case $r = 1$:

$\ln x$	b	
1	$\times$	a
	1	x^{α_1}

	$\phi(x)$
a	x^{α_1}
b	$\ln x$

Case $r = 2$:

$\ln^2 x$	d		
$\ln x$	bd	c	
1	$\times$	abc	a
	1	x^{α_1}	x^{α_2}

	$\phi_1(x)$	$\phi_2(x)$
a	x^{α_1}	x^{α_2}
b	x^{α_1}	$\ln x$
c	x^{α_1}	$x^{\alpha_1} \ln x$
d	$\ln x$	$\ln^2 x$

Case $r = 3$:

$\ln^3 x$	g			
$\ln^2 x$	fg	c		
$\ln x$	$bdfg$	dc	e	
1	$\times$	$abcdef$	abe	a
	1	x^{α_1}	x^{α_2}	x^{α_3}

	$\phi_1(x)$	$\phi_2(x)$	$\phi_3(x)$
a	x^{α_1}	x^{α_2}	x^{α_3}
b	x^{α_1}	x^{α_2}	$\ln x$
c	x^{α_1}	$x^{\alpha_1} \ln x$	$x^{\alpha_1} \ln^2 x$
d	x^{α_1}	$x^{\alpha_1} \ln x$	$\ln x$
e	x^{α_1}	x^{α_2}	$x^{\alpha_2} \ln x$
f	x^{α_1}	$\ln x$	$\ln^2 x$
g	$\ln x$	$\ln^2 x$	$\ln^3 x$

References

[1] Angot A., "Compléments de mathématiques," Masson ed., Sixième Édition, Paris, 1982.

[2] Bass J., "Cours de mathématiques," Masson ed., Tome II, Quatrième Édition, Paris, 1968.

[3] Demoment G., "Image Reconstruction and Restoration: Overview of Common Estimation Structure and Problems," *IEEE Trans. on Acoustics, Speech, and Signal Processing*, Vol. 37, pp:2024-2036, (1989).

[4] Gull S. F. and Skilling J., "Maximum entropy method in image processing," *IEE Proc.*, **131-F**, pp. 646-659, 1984.

[5] Mohammad-Djafari A. and Idier J., "Maximum entropy prior laws of images and estimation of their parameters," in *W.T. Grandy, Jr. (ed.), Maximum-entropy and Bayesian methods,* Kluwer Academic Publishers, Netherlands, 1990.

[6] Mohammad-Djafari A. and Demoment G., "Maximum entropy Fourier synthesis with application to diffraction tomography," *Applied Optics*, Vol.26, No. 10, pp:1745-1754, (1987).

[7] Mohammad-Djafari A. and Demoment G., "Maximum entropy reconstruction in X ray and diffraction tomography," *IEEE Trans. on Medical Imaging*, Vol. 7, No. 4 pp:345-354, (1988).

[8] Mohammad-Djafari A., "Bayesian Approach with Maximum Entropy Priors to Imaging Inverse Problems, Part I: Foundations," *submitted to IEEE Trans. on Image Processing*, (August, 1993).

[9] Mohammad-Djafari A., "Bayesian Approach with Maximum Entropy Priors to Imaging Inverse Problems, Part II: Applications," *submitted to IEEE Trans. on Image Processing*, (August, 1993).

[10] Nguyen M.K. and Mohammad-Djafari A., "Bayesian Maximum Entropy Image Reconstruction from the Microwave Scattered Field Data," in *A. Mohammad-Djafari and G. Demoment(ed.), Maximum Entropy and Bayesian Methods,* Kluwer Academic Publishers, the Netherlands, 1993.

[11] Skilling J., "Maximum-Entropy and Bayesian Methods," J. Skilling ed., Kluwer Academic Publishers, Dordrecht, 1988.

MAXIMUM ENTROPY SIGNAL TRANSMISSION

Enders A. Robinson
Henry Krumb School of Mines
Columbia University
New York, NY 10027 USA

ABSTRACT. There are two means at our disposal to understand the behavior of physical systems: observation and experimentation. Observation becomes increasingly difficult as an object becomes more remote or obscure. Experimentation is impossible for objects that cannot be manipulated or directly contacted. In such cases it is necessary to use numerical simulation, drawing upon perceived virtual systems expressed through models. Two main approaches to deal with remote or obscure objects come under the headings of the "inverse source problem" and the "inverse medium problem." In the typical inverse source problem, the source of energy is remote, the medium transmits the source signal to an accessible receiver, and information about the source is required. An example of an inverse source problem is classical earthquake seismology where received seismic data are used to determine locations of remote earthquakes. Another example is passive sonar where engine noise from a hidden submarine is used to locate its position. In the typical inverse medium problem the source of energy (usually man-made) is local, the signal penetrates an inaccessible medium that reflects energy back to accessible points, and information about the internal structure of the medium is required. Examples are reflection seismology, radar, and active sonar. The usual approach to either type of inverse problem is first to devise a theoretical model that admits a solution from the available data. Implementation then involves using the theoretical model to find the required solution, often through an iterative improvement method. One of the most basic models is a system with parallel plane layers (the so-called layer-cake model of geophysics). An important characteristic of the layer-cake model is that it yields a transmitted signal that has maximum entropy. Because of this property, a signal from a distant source can be deconvolved to remove the unwanted reverberations that occurred during transmission. Thus it is possible to obtain a good representation of the unknown source signal, and so the inverse source problem for the layer-cake model has an effective computer solution. The layer-cake model also has the characteristic that it yields a reflected signal with both feedforward and feedback components, where the feedback component has maximum entropy. In practical terms, this maximum entropy property means that the received reflected signal preserves the information about the structure of the medium. As a result, the received reflected signal can be deconvolved to give a picture of the internal structure of the medium, and so the inverse medium problem for the layer-cake model also has a computer solution.

1. Introduction.

In 1842, Augusta Ada, Countess of Lovelace, is reputed to have exclaimed: "We must say most aptly, that the analytic engine weaves algebraic patterns just as the Jacquard loom weaves flowers and leaves." The analytic engine, of course, was a very early version of the computer. A science that depends upon computers, not only in routine calculations but also in the most advanced scientific investigations, is geophysics. Data processing in geophysics began in 1952 when the first seismic data analysis was done on the MIT Whirlwind digital computer. Geophysics is data-intensive. In increasing its observational powers, geophysics

G. R. Heidbreder (ed.), Maximum Entropy and Bayesian Methods, 135–147.

has developed mathematical and digital methods that have found applications in many other fields. In turn, geophysics has freely drawn upon concepts and techniques developed in other scientific disciplines: physics, mathematics, electrical engineering, and mechanical engineering. One of the most fruitful areas has been digital signal processing and spectral analysis [3] [5] and especially the field of maximum entropy and Bayesian analysis [4].

Today computers provide a seismic visualization environment that lets the geophysicist interactively explore actual as well as simulated data. In a larger sense, the Earth itself can be considered as a great computing machine, the ultimate computing machine for geophysical studies. As a seismic wave propagates through the subsurface, the internal structure of the Earth determines the actual path of the wave, essentially by means of the mechanism embodied in Huygens' principle. This determination of the physical path within the Earth's layers amounts to real-time physical (analog) computation at each stage of propagation. Digital computer programs try to mimic this physical process so that the wave path can be simulated within the computer. At the present time, the digital computer is separate from the Earth. In time, geophysicists will make the critical linkage. Future seismic experiments will connect digital computers to the Earth in an essential way so that they work together as a unit. In such an ideal situation, the physical process and the simulated model of wave propagation will blend together and become one integrated whole.

A major problem in geophysics is the determination of the internal structure of the Earth. For example, in the study of plate tectonics, the mechanisms producing the internal movements of the Earth lie deep within the crust, mantle, and core of the Earth, out of direct reach of human observers. Except for the limited amount of the shallow subsurface that can be reached by shafts and drill holes, the interior of the Earth lies out of reach and cannot be directly contacted. The only means to study the interior of the Earth is through the use of wave motion and other geophysical phenomena. We are fortunate in that we have available the naturally occurring seismic waves that are generated by earthquakes. Seismic waves travel outward from the origin of an earthquake, and travel to all depths within the Earth. Some of the waves penetrate to deep within the Earth and then travel onward to reach the surface again at distant points. Seismometers at these points record the waves as seismograms. These recorded waves contain vital information about the crust, mantle, and core of the Earth. The waves can be described as messengers that convey the information that can be used to form images of geological features deep within the Earth. For good image quality, a large number of receivers (seismometers) must be used.

As we have seen, earthquake seismology is essentially a transmission problem, with the waves traveling to receivers often located at great distances from the earthquake's source. In contrast, reflection seismology as used in petroleum exploration has both sources and receivers at approximately the same locations on the Earth's surface. In seismic exploration, a man-made source at or near the surface sends seismic waves down into the ground. The waves are reflected at various subsurface rock interfaces and make their way back to the surface where they are recorded. From the received data the geophysicist wants to determine the structure of the subsurface medium. From this knowledge one can infer the presence of possible oil bearing reservoirs. Reflection seismic data represent the single largest class of scientific data collected in digital form. The processing of these data produces subsurface images of unprecedented geological detail. In some cases, the seismic data is integrated with well-logging data, which consist of continuously collected information from devices lowered into drill holes. When circumstances warrant, such data are processed to yield a description of the type of rock and its porosity, and the presence or absence of oil. In addition to petroleum exploration, seismic reflection methods are presently being used on a larger scale to study the internal structure of the continents.

Both in earthquake seismology and in petroleum exploration seismology, great advances have been made in instrumentation. Emphasis now is on imaging, the presentation of the data in a form that is visually accurate and instructive. This new perspective is the main scientific frontier that is being developed today. Computers provide a visualization environment that lets the scientist interactively explore propagation patterns. The ultimate objective is to make a critical linkage between the computers and the transmission medium. In this way the data processing schemes can be driven by the fine structure of the medium itself. Image processing and mapping are then used to create detailed three-dimensional pictures depicting the interior structure. Small-scale heterogeneities within the Earth are revealed with remarkable resolution.

2. Inverse Source Problem and Inverse Medium Problem.

Figure 1 [left] depicts the inverse source problem with the following characteristics: the source of energy is remote; the medium transmits and distorts the source signal; the received data is the transmitted signal corrupted by noise and reverberations produced by the medium; the desired information is the source itself. Figure 1 [right] depicts the inverse medium problem with the following characteristics: the source of energy is local, usually man-made; the medium reflects and distorts the source signal; the received data is the reflected signal corrupted by noise and reverberations; the desired information is the medium.

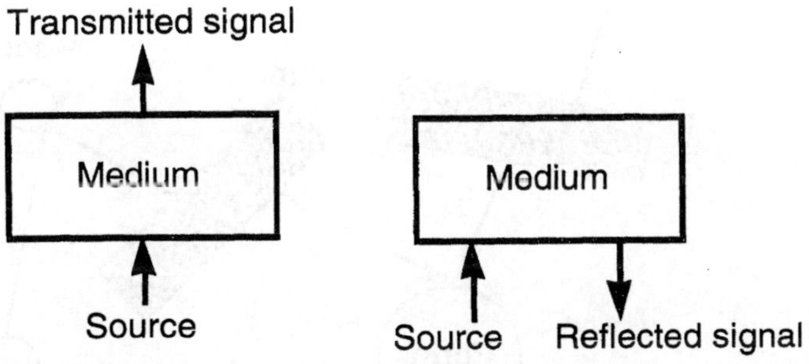

Figure 1: *[Left]: In the inverse source problem, it is required to find the (unknown) source from the (known) received transmitted signal. [Right]: In the inverse medium problem, it is required to find the (unknown) medium from the (known) received reflected signal.*

Astronomy provides examples of both kinds of inverse problems. Let us look at two examples, light from a star and light from the moon. Starlight represents received data that tell us mostly about the distant source (the star), and little about the medium (interstellar space). Spectral analysis of starlight represents a solution to an inverse source problem. Moonlight represents received data that tell us mostly about the medium (in this case, the moon, which is the dominant reflector) and less about the source (the sun). Analysis of moonlight represents the solution of an inverse medium problem, one made famous by Galileo, who wrote in 1610: "It is a very beautiful thing, and most gratifying to the sight, to behold the body of the moon, distant from us by almost sixty earthly radii, as if it were no farther away than two such measures."

In each case light goes through a heterogeneous medium. The star generates its own light. As the light travels to the Earth, the path admits reverberations in the interstellar medium. Each time there is a backward reflection there must be an offsetting forward reflection in order for the light to eventually reach the Earth. As a result, any star-to-Earth transmission path necessarily has an even number of bounces. The Moon reflects light that originates at the Sun. As a result, any Sun-to-Moon-to-Earth reflection path always has an odd number of bounces, made up of the even number due to transmission and the extra bounce upon reflection from the Moon. In Figure 2, two of the many possible multiple raypaths are depicted. One of the depicted paths is a Star-to-Earth first-order multiple path (that is, a transmission path with two bounces). The other depicted path is a Sun-to-Moon-to-Earth first-order multiple path (that is, a reflection path with three bounces). In general, transmission paths have an even number of bounces, and reflection paths have an odd number. As a result, transmission paths represent pure feedback systems, whereas reflection paths represent systems with both feedforward and feedback components.

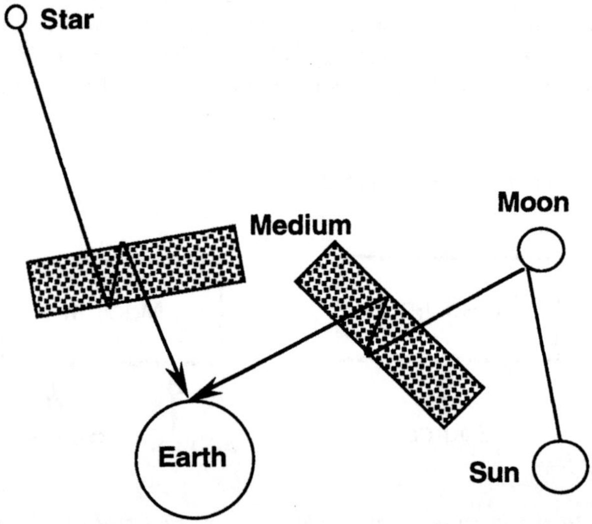

Figure 2: *The raypath from the star depicts a first-order multiple path for transmission. The raypath from the Sun depicts a first-order multiple path reflected from the Moon.*

The classic problem of reflection seismology concerns the determination of the properties of the interior of the Earth from recorded waves that have been reflected from the subsurface layers. As a first step in the mathematical analysis, the problem is usually simplified by assuming that the crust of the Earth is made up of a sequence of horizontal parallel plane layers, each of which is homogeneous, isotropic, and nonabsorptive. This is the classic layer-cake model of reflection seismology. Using this model, we can characterize the medium (from source to receiver) by a series of Fresnel reflection coefficients, each associated with a reflecting subsurface interface. A reflection coefficient is a ratio: the ratio of the amplitude of the reflected wave to that of the incident wave. As a result, a reflection coefficient must have magnitude less than or equal to one. Suppose that the Fresnel reflection coefficient of the first interface is 0.8 and that of the second interface is 0.4. See Figure 3. (In our mathematical analysis, normal incidence is assumed, but the figures are drawn with offset for visual clarity.)

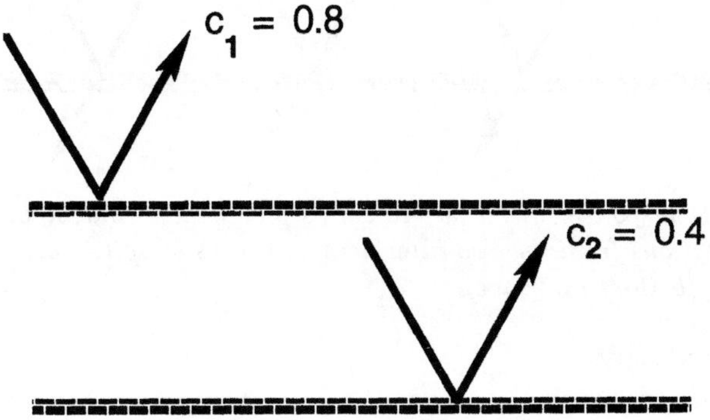

Figure 3: *Two interfaces and their Fresnel reflection coefficients: 0.8 (top interface) and 0.4 (bottom interface).*

A downgoing unit incident wave strikes the system. What is the combined reflectivity of the two interfaces together? Clearly it is not the sum of 0.8 and 0.4, which is 1.2, a number greater than one in magnitude. A reflectivity greater than one in magnitude indicates that reflected energy is greater than the incident energy, an impossible situation. We must therefore look for an explanation of what occurs. The answer is that a reverberation wavetrain is generated by a feedback mechanism in the layer between the two interfaces [2]. The reflected response from the two interfaces is the upgoing reverberation wavetrain escaping from the top interface; this response is given by the series 0.8, 0.144, -0.046, 0.015,..., which is depicted in Figure 4. The transmitted response from the two interfaces is the downgoing reverberation wavetrain escaping from the bottom interface. In (infinite) time, all the energy escapes from the layer, some going up and the rest going down. Thus the total energy from the downgoing incident wave is divided into the energy contained in each of these two escaping wavetrains, namely the reflected response and the transmitted response. As a result, the ratio of the energy of the reflected response to the energy of the incident wave is necessarily less than one. The key idea is that the addition theorem for reflection coefficients combines the two Fresnel reflection coefficients, each of which is less than one in magnitude, to yield a third quantity, a reflected response given by a reverberation wavetrain 0.8, 0.144, -0.046, 0.015,..., with energy ratio less than one.

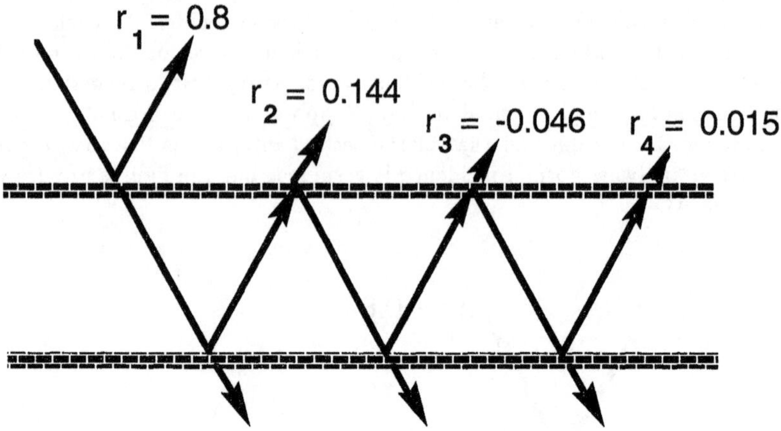

Figure 4: *Reflected response, given by the series 0.8, 0.144, -0.046, 0.015, ..., resulting from internal reflections from the two interfaces with Fresnel reflection coefficients 0.8 (top interface) and 0.4 (bottom interface).*

3. Maximum Entropy.

Figure 5 depicts the pure feedback system that generates the transmitted wave in the layer-cake model. Because the received transmitted signal is generated by a pure feedback system, it must be of the autoregressive type. According to the results of Burg [1], a signal of the autoregressive type has maximum entropy. As a result, the received transmitted signal has maximum entropy, from which it follows that it can be deconvolved to yield the unknown source signal. Thus, for the layer-cake model, the inverse source problem has a computer solution [7].

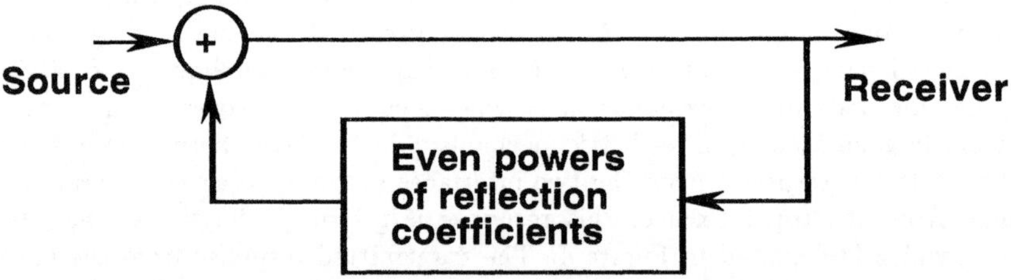

Figure 5: *The pure feedback system describing the transmission problem shown in Figure 1 [left].*

Figure 6 depicts the feedforward-feedback system that generates the reflected wave in the layer-cake model. Because the feedback component is of the autoregressive type, it follows from Burg [1] that the feedback component has maximum entropy. As a result, the reflected signal preserves the information about the individual Fresnel reflection coefficients, and hence the received reflected signal can be deconvolved to yield the layered structure of the medium. Dynamic deconvolution [5] involves mathematically peeling off the layers of the sedimentary column. We start with the recorded reflection seismogram at the surface. At any given interface we must remove the effect of the reflection coefficient from the reflected wave motion in that layer so as to obtain the reflected wave motion in the next deeper layer. In this way, layer by layer, the medium can be reconstructed. Thus, for the layer-cake model, the inverse medium problem also has a computer solution.

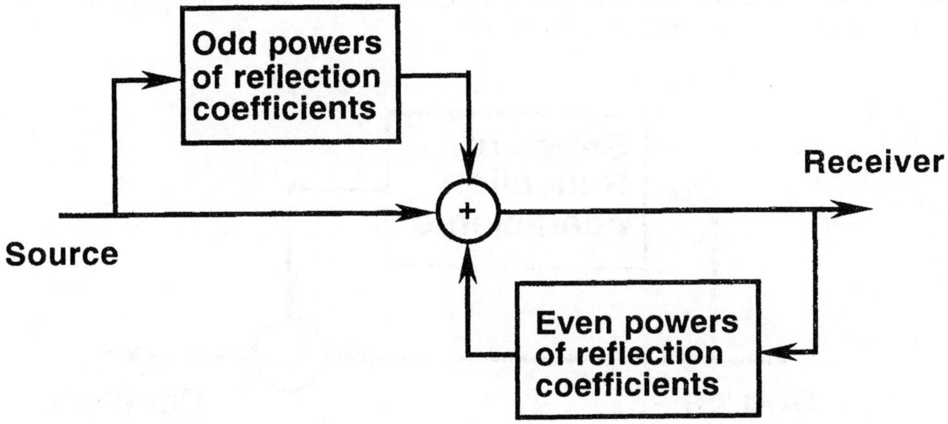

Figure 6: *The feed forward-feedback system of the reflection problem shown in Figure 1 [right].*

Nature often provides media that are weakly inhomogeneous, at least locally. For such media, the reflection coefficients of the interfaces or scatterers are small and random. Weakly inhomogeneous media provide good transmission characteristics for a remote source and good reflection characteristics for a local source. For such cases minimal computer processing such as deconvolution is required. If the mathematics of nature had been otherwise, that is, if smallness and randomness provided poor transmission characteristics for a remote source and poor reflection characteristics for a local source, we would live in a world of obscurity.

Consider the transmission problem in the case when the medium is weakly inhomogeneous (that is, the reflection coefficients are small and random). For this case, the feedback loop reduces approximately to zero. Hence the pure feedback system of Figure 5 reduces (approximately) to a distortionless transmission system (that is, a system with no feedback loop) as shown in Figure 7.

Figure 7: *Distortionless transmission system from source to receiver (feedback loop absent).*

Let us now consider the reflection problem in the case when the medium is weakly inhomogeneous. As above, the feedback loop reduces (approximately) to zero. Furthermore, the higher order terms in the feedforward loop reduce (approximately) to zero, so the feedforward loop essentially contains only the first-order terms. The first order terms, in fact, are the series of the reflection coefficients of the subsurface layers. Hence the feedforward-feedback system of Figure 6 reduces to the pure feedforward system shown in Figure 8. This series of reflection coefficients appears at the output in the form of the received reflected wave. As a result, the received wave mimics the structure of the underground layers, and thus this system produces (within the given approximations) a faithful depiction of the medium at the output.

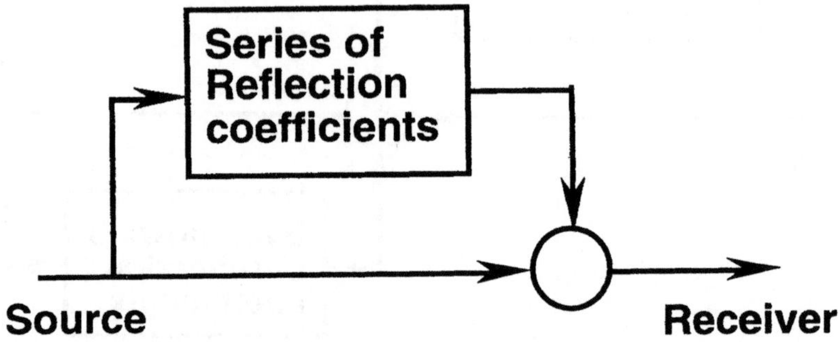

Figure 8: *Pure feedforward system with no feedback loops.*

The above result can be verified every time we look around us. The air is a weakly inhomogeneous medium. As a result, we see a remarkably clear representation of our surroundings. In effect, we see only the important primary reflections despite the fact that the many reflecting objects present produce a myriad of multiple reflections. When we analyze all of the raypaths involved, we immediately realize that the actual raypaths are extremely complex involving many high-order multiple paths. However many of the objects, although numerous, are unimportant in that their reflection coefficients are small and random. According to the mathematics, these small and random reflection coefficients effectively cancel themselves out and we are left with a clear image of the important objects. That is, we clearly see those objects that have large (in magnitude) reflection coefficients. In this way, we perceive the world about us. Shakespeare expressed this idea when he wrote:

For the eye sees not itself
But by reflection, by some other things.

A wavetrain transmitted in the Earth consists of the downgoing direct wave and the myriad of downgoing multiple waves that follow the direct wave in time. Figure 9 is a schematic depiction of a medium showing downgoing waves only. In physical space, all the wavepaths represent energy traveling vertically into the Earth. The figure represents a space-time diagram. Time is measured to the right on the horizontal axis, and vertical depth into the Earth is measured downward on the vertical axis.

Time and depth units are chosen so as to make the velocity equal to one, so a wave travels a unit of depth in a unit of time. As a result, in the space-time diagrams, the raypaths appear as 45 degree lines (negative 45 degree lines for downgoing waves and positive 45 degree lines for upgoing waves). The position of the source is always in the upper left-hand corner (time 0 and depth 0). Referring to Figure 9, we see the raypath of the direct (or primary) wave as the negative 45 degree line on the left, and the reverberations (multiple waves) as the parallel lines that occur to the right of the direct raypath.

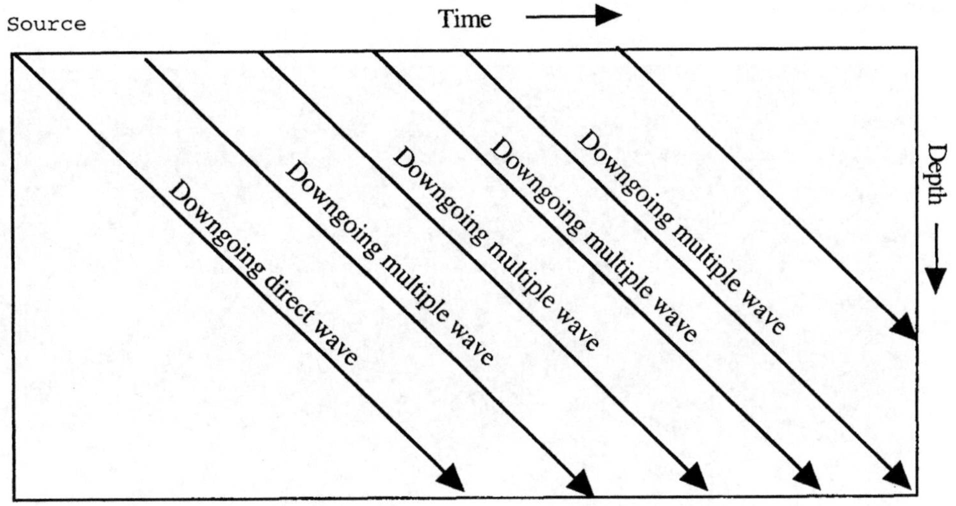

Figure 9: *Transmitted wave motion in a weakly inhomogeneous medium.*

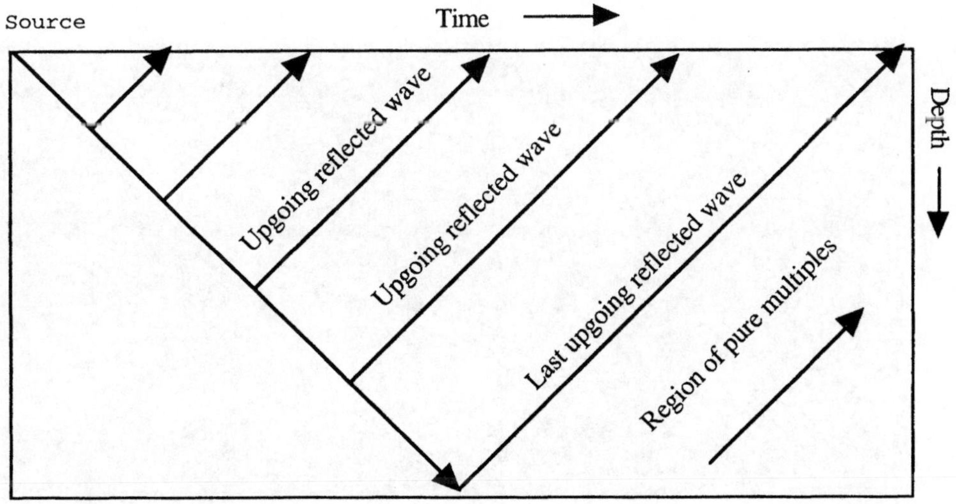

Figure 10: *Transmitted wave motion in a weakly inhomogeneous medium.*

Figure 10 is the corresponding schematic representation of the same medium showing upgoing waves only. These upgoing waves are the waves reflected from the various reflecting horizons as well as their multiples. The upgoing waves appear in this space-time domain as positive 45 degree lines. The last reflecting horizon is assumed to be at the bottom of the figure, giving rise to the so-called last true reflection indicated by the 45 degree line with the label "Last upgoing reflected wave." All rays to the right of this last true reflection are purely multiples.

Figure 11: *Reflected wave motion in a weakly inhomogeneous medium.*

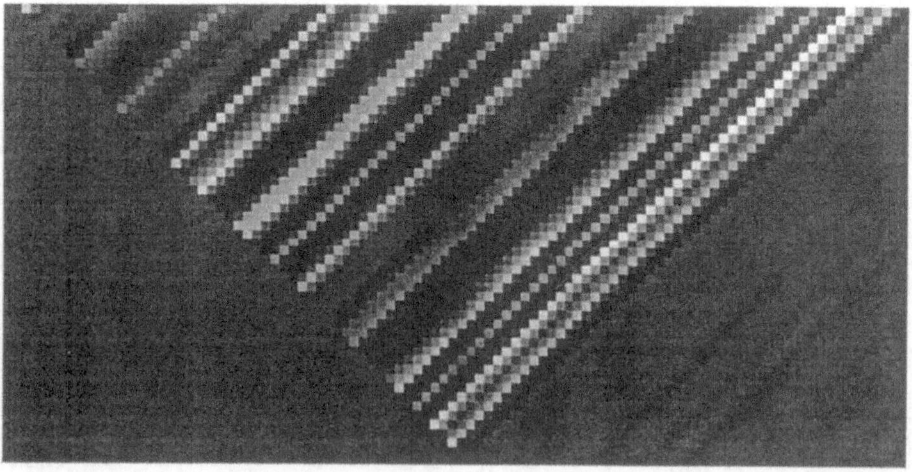

Figure 12: *Reflected wave motion in a weakly inhomogeneous medium.*

Figures 11, 12, 13, and 14 show actual examples. Figure 11 depicts the transmitted wavetrain in a weakly inhomogeneous medium (that is, one with small and random reflection coefficients). The raypath of the direct (or primary) wave is the strong 45 degree line on the left, and the reverberations are the parallel lines to the right of the direct raypath.

The amplitudes of the reverberations are much less than those of the direct wave, so that the direct wave carries most of the transmitted energy. Figure 12 depicts the same situation, but now in the case of reflected waves. The reflected waves faithfully portray the

Figure 13: *Transmitted wave motion in a medium that is not weakly inhomogeneous.*

Figure 14: *Reflected wave motion in a medium that is not weakly inhomogeneous.*

internal structure of the medium, and the region of pure multiples is noticeably weak. These two figures depict the orderliness of wave motion in a weakly inhomogeneous medium.

Let us now consider a different medium, one that is not weakly inhomogeneous (in this case, one with large and random reflection coefficients). In order to make this situation comparable to the situation depicted in Figures 11 and 12, we use a medium with the same internal structure; that is we use the same reflection coefficients but amplify them by a factor of 15 to make them larger in magnitude. Figure 13 depicts the transmitted wavetrain, and Figure 14 the reflected wave train. Both figures show the chaotic behavior that masks the true nature of the medium. Such cases can be decoded by computer processing, but, in more extreme instances, even computer processing may not be adequate to produce order out of chaos.

4. Conclusion.

In a heterogeneous medium, most waves do not follow the direct route but bounce back and forth at internal inhomogeneities on their way to the receiver. The wavetrains made up of these bounces appear in the received signal in the form of reverberations. The general principle is that transmission paths have an even number of bounces, whereas reflection paths have an odd number. In the case of the layer-cake model, transmission paths represent pure feedback systems, whereas reflection paths represent feedforward-feedback systems. As a result, a received transmitted signal can be decoded (by signal processing) to yield a depiction of the source. Generally a transmission process allows us to perceive a distant source but obscures the structure of the medium. For example, we can see a star because it is a source of light. Except for the solar system, all of the mass so far seen in the universe are sources of light or other electromagnetic energy. The remaining mass of the universe, about 90 percent of the whole, has not yet been seen. If this unseen mass contains no sources of energy, and if the reflections from neighboring stars or galaxies are not above noise level, this mass will never be seen.

In contrast, a reflected signal from a local source is produced by a feedforward-feedback system. The received reflected signal can be readily decoded (by signal processing) to produce a depiction of the medium. That is, a reflection process allows us easily to see the structure of the medium, but hides the nature of the source. For example, we can clearly see the Moon and other objects in the solar system because of reflection of the Sun's light, but moonlight give us little knowledge about the nature of the sun (the source). Often we encounter media that are weakly inhomogeneous, at least within certain ranges. The Fresnel reflection coefficients of the interfaces or scatterers of such media are small and random, and thus these media provide good transmission characteristics for a remote source and good reflection characteristics for a local source. In such cases, minimal computer processing is required.

References

[1] J. P. Burg, "Maximum Entropy Spectral Analysis", Ph.D. thesis, Stanford University, 1975.

[2] K. E. Burg, M. Ewing, F. Press, and E. J. Stulken, "A seismic wave guide phenomenon," *Geophysics*, **16**, 594, 1951.

[3] W. A. Gardner, *Statistical Spectral Analysis, A Nonprobabilistic Theory*, Prentice Hall, 1988.

[4] E. T. Jaynes, "On the rationale of maximum entropy methods", *Proc. IEEE*, **70**, 1039, 1982.

[5] S. L. Marple, *Digital Spectral Analysis with Applications*, Prentice Hall, 1987.

[6] E. A. Robinson, "Dynamic Predictive Deconvolution", *Geophysical Prospecting*, vol. 23, pp. 779–797, 1975.

[7] E. A. Robinson, "Spectral approach to geophysical inversion by Lorentz, Fourier, and Radon transforms," *Proc. IEEE* **70**, 1039, 1982.

MAXIMUM QUANTUM ENTROPY FOR CLASSICAL DENSITY FUNCTIONS

Timothy C. Wallstrom
Theoretical Division, MS-B213
Los Alamos National Laboratory
Los Alamos, New Mexico 87545

ABSTRACT. Maximum Quantum Entropy (MQE), recently introduced by Richard Silver and also known as Quantum Statistical Inference (QSI), is a method of estimating smooth, non-quantum-mechanical ("classical") densities, given information about those densities. It is formally analogous to Maximum Entropy (ME), the difference being that the Shannon entropy is replaced by the quantum entropy. We present a concise description of MQE from a mathematical perspective, not relying on physical analogy. We introduce density matrices and the quantum entropy, compare MQE with ME, discuss the nature of constraints in MQE and show how these constraints influence the density estimate. We conclude with a discussion of the status of MQE as a maximum entropy method.

1. Introduction

Maximum entropy methods have been found to be useful not only in statistical physics, but also in many statistical applications which have nothing to do with physics [19, 6]. This is presumably because they encapsulate principles of probabilistic reasoning which do not depend on physics [8, 13, 21]. Such methods have also been applied to quantum statistical mechanics, where the entropy to be maximized is not the usual Shannon entropy, but instead the so-called "quantum entropy" [10, 2]. This generalization is normally thought to be a requirement of the application: the quantum entropy is just the Shannon entropy of the subjective part of the uncertainty in the physical situation; it automatically ignores the inherent quantum-mechanical uncertainty. Recently, however, Richard Silver has suggested that the quantum entropy, like the Shannon entropy, might be useful outside the context of physics, as part of the apparatus of statistical analysis [16, 17, 14].

Intuitively, the reason this might be helpful is as follows. Suppose that we have a continuous probability density f, which we know to possess some local smoothness, and we wish to define its entropy. If we discretize it coarsely into components f_i, we might expect that these components are independent, in which case a reasonable expression for the entropy might be the Shannon entropy [18] $S_S(f) = -\sum f_i \log f_i$. At some finer level of description, however, nearby f_i are correlated, so this is not the correct expression for the entropy. The quantum entropy does take into account the correlations between nearby points, because it enables one to constrain, in effect, the derivatives of the function. This leads to estimates which are locally smooth. The nature of this smoothing is determined by the choice of smoothing operator $\mathbf{L}$, and the width by a hyperparameter β.

The purpose of this paper is to give a concise introduction to Maximum Quantum Entropy (MQE) from a primarily mathematical perspective. This affords a significantly

G. R. Heidbreder (ed.), Maximum Entropy and Bayesian Methods, 149–155.

new perspective from previous work, which relied on physical analogy, and clarifies the logical foundations of the method. After introducing density matrices in §2, we introduce MQE by comparing it with the formally similar Maximum Entropy (ME) method (§3). In §4, we discuss the nature of the MQE constraints, and in §5 and §6, we show how these constraints affect the estimate. After briefly discussing the use of the relative quantum entropy in §7, we examine the status of MQE as a maximum entropy method in §8.

2. Density matrices

The quantum entropy is defined in terms of a symmetric, positive matrix D, the density matrix. The density function is given by the diagonal elements of D: $f_x = D_{xx}$. We do not insist that f be normalized to one. For purposes of exposition and implementation, and to avoid mathematical complications, it is sometimes easier to deal with a discretization: $f_i = D_{ii}$. This is a good approximation if the discretization scale is small compared to the scale of the smallest features, *cf.* [12]. Obviously, there is an infinite number of density matrices corresponding to each f, since the off-diagonal elements remain unspecified. We will see that the off-diagonal elements provide local smoothing between neighboring values of f. The matrix D, and thus the off-diagonal elements, are determined through a maximum quantum entropy variational principle.

3. Maximum Quantum Entropy

MQE is most easily understood by comparing it to the better-known maximum entropy method (ME) [7, 9].

In ME, we calculate f by maximizing the Shannon entropy, $S_S(f) = \sum f_i - 1 - \sum f_i \log f_i$, subject to constraints $A_m(f) = a_m$ [9]. (Here $S_S(f)$ has been generalized to non-normalized f.) By the method of Lagrange multipliers, this is equivalent to maximizing $S_S(f) - \sum \lambda_m A_m(f)$. Assuming the A_m are linear, the solution to this problem is $f_i = \exp(-\sum_m \lambda_m A_{mi})$, where $A_m(f) = \sum A_{mi} f_i$, and where the λ_m are to be determined by the constraints. For comparison with MQE, we write f as the diagonal elements of a density matrix D. Let $\mathcal{F}_i$ be the matrix with a 1 in the ith diagonal position and zeros everywhere else: $(\mathcal{F}_i)_{jk} = \delta_{ji}\delta_{ik}$. Define $\mathcal{A}_m = \sum_i A_{mi}\mathcal{F}_i$. Then $f_i = D_{ii} = \mathrm{Tr}(D\mathcal{F}_i)$, where

$$D = \exp(-\sum_m \lambda_m \mathcal{A}_m). \tag{1}$$

D is just a diagonal matrix with elements $D_{ii} = \exp(-\sum_m \lambda_m A_{mi})$.

In MQE, we calculate D by maximizing the quantum entropy

$$S(D) = \mathrm{Tr}D - 1 - \mathrm{Tr}(D \log D), \tag{2}$$

subject to the constraints $\mathrm{Tr}(D\mathcal{A}_m) = a_m$ and $\mathrm{Tr}(D\mathbf{L}) = b$ where $\mathbf{L}$ is some smoothing operator. Note that $\mathrm{Tr}(D\mathcal{A}_m) = A_m(f)$. This maximization problem is thus the same as ME, except that we use the quantum entropy instead of the Shannon entropy, and we add a constraint on $\mathrm{Tr}(D\mathbf{L})$, the "expectation" of $\mathbf{L}$. The solution to this problem is

$$D = \exp(-\sum_m \lambda_m \mathcal{A}_m - \beta\mathbf{L}), \tag{3}$$

where the λ_m and β are determined by the constraints. This expression for D is exactly the same as ME, except that there is now a non-diagonal smoothing operator $\mathbf{L}$ in the exponential. $\mathbf{L}$ correlates the diagonal elements of D. If $\mathbf{L}_{ij} \to 0$ as $|i-j|$ increases, these correlations will be local. $\sum_m \lambda_m \mathcal{A}_m$ and $\beta\mathbf{L}$ do not commute, so sophisticated mathematical methods are required to analyze the dependence of D and f on the constraints. These methods are taken from quantum statistical mechanics.

We have assumed that the λ_m and β are unique, given a_m and b. This follows from the concavity of the entropy in the constrained parameters. Define the *relevant* entropy $S(a)$ by $S(a_1,\ldots,a_n) = S(D)$, where D maximizes the entropy under the constraints $\mathrm{Tr}(\mathcal{O}_m D) = a_m$ ($\mathcal{O}_m$ is $\mathcal{A}_m$ or $\mathbf{L}$.) $S(D)$ is concave [23]; $S(a)$ inherits this concavity[3]. If we define the Legendre transform of S as $T(\lambda) = S(a) - \lambda \cdot a$, where $\lambda_i = \partial S(a)/\partial a_i$, there is thus at most one λ for any a. Therefore, $S(a)$ is well-defined. Conversely, any λ maps into exactly one a. Therefore, there is a bijective correspondence between λ and the set $a = \mathrm{Tr}(D(\lambda)\mathcal{O})$; this is sometimes called duality. This implies that if f and b are known, D is uniquely defined; we write $S(f,b)$ for $S(D)$. Eq. (3) defines a manifold of density matrices. This can be parameterized by (f,b); a set of constraints (a) parameterizes a submanifold.

4. Constraints

In MQE, our constraints are linear in D; these can be divided into two types. The first depends only on the diagonal elements of D, which is to say, on f. Such constraints are used to constrain certain values of f, or to embody *a priori* information that we might have about f. Examples are f_j, $\sum f_i$, $\sum f_i x_i^n$, $\sum_{a<x_i<b} f_i$, or $\sum_j K(j-i)f_j$, to constrain f_j, the normalization of f, its nth moment, a confidence interval, or a convolution around f_i.

The second type of constraint involves off-diagonal elements of D as well. This is only really used in our smoothness constraint, $\mathrm{Tr}(D\mathbf{L}) = b$. Initially, it seems peculiar to constrain $\mathrm{Tr}(D\mathbf{L})$ rather than the value of $\mathbf{L}$ on f itself. If we tried to represent this constraint directly in terms of f, however, we would wind up constraining something like $\sum_{ij} f_i \mathbf{L}_{ij} f_j$, which is not linear in f, and would thus not be amenable to the maximum-entropy formalism. Use of the density matrix thus enables us to incorporate smoothness constraints into the Maximum Entropy formalism, and this constitutes the central novelty of MQE.

5. Properties of the density matrix

As noted, the density matrix is in a form which arises frequently in physics; therefore techniques from physics can be exploited to calculate its properties. We write

$$D = \exp(-U(\lambda) - \beta\mathbf{L}), \tag{4}$$

where $U(\lambda) = \sum \lambda_m \mathcal{A}_m$. We go to a continuous basis in this section. This expression for D has the following properties as a function of β. First, as $\beta \to 0$, $f(x) \to \exp(-U(\lambda)(x))$ and we recover ME. For intermediate values of β, we diagonalize the operator $\beta H = U + \beta\mathbf{L}$. Let ϕ_n be the eigenfunctions of H, and let ϵ_n be the eigenvalues. Then

$$f(x) = \sum e^{-\beta\epsilon_n} |\phi_n(x)|^2. \tag{5}$$

Finally, as $\beta \to \infty$, only the lowest eigenvector contributes, and we get (for normalized f) $f(x) = |\phi_0(x)|^2$. Note that the λ will change to maintain the constraints; as $\beta \to \infty$, $\lambda \to \infty$ also. This is somewhat different from the usual approach in physics, in which the "potential" U would be fixed and subsumed into the "Hamiltonian" H, where it would also get multiplied by β; also, we are not assuming that $\mathbf{L}$ is second-order.

$D(x, x')$ can be very usefully represented as the solution to a partial differential equation [5]. In fact, it is $D(x, x'; 1)$, where $D(x, x'; t)$ is the solution of the equation

$$\partial_t D = (-\beta \mathbf{L} - U(\lambda))D, \tag{6}$$

with initial condition $D(x, x'; 0) = \delta(x - x')$. Suppose $U = 0$. If $\mathbf{L}$ is quadratic, this is just the kernel of the heat equation: a Gaussian of width $\sqrt{\beta t}$ about x'. If $\mathbf{L}$ is of nth order, $\mathbf{L}$ is a function of $x/\beta^{1/n}$.

6. The dependence of f on the constraints

Suppose we add a constraint $\delta\lambda \mathrm{Tr}(DJ)$. This will add some matrix $(\delta\lambda)J$ to the exponential in our density matrix. If $D = \exp(A)$ and $D' = \exp(A + B)$, then to first order in B [4, 11],

$$\delta D = \int_0^1 dt e^{(1-t)A} B e^{tA}. \tag{7}$$

If J is $\mathcal{F}_{x'}$, the constraint on $f(x')$, then $J(y - z) = \delta(y - x')\delta(x' - z)$, and we get

$$\delta f(x) = -\delta\lambda_{x'} \int_0^1 dt D(x, x'; 1 - t) D(x', x; t), \tag{8}$$

where $\delta\lambda_x$ is the associated Lagrange multiplier and where we have written $D(x, x)$ as $f(x)$. This formula describes the effect on $f(x)$ of changing $f(x')$ (by changing its Lagrange multiplier). Note that the effect is *local*: if x is too far from x', both terms in the integrand will tend to zero.

Define $G^{-1}(x_i, x_j) = -\delta f(x_i)/\delta\lambda(x_j)$; then

$$f'(x) = f(x) - \sum \delta\lambda_j G^{-1}(x, x_j) + O(\delta\lambda^2). \tag{9}$$

Define G to be minus the second derivative of the entropy with respect to f_i and f_j; G^{-1} is clearly its inverse. As a second derivative, G is symmetric and since the entropy is concave, G is positive. Therefore, G^{-1} is also a positive symmetric matrix. Intuitively, G^{-1}, which might be called the MQE linear response function, measures how much f_i changes when f_j is changed (by changing λ_j). The shape of f' (and G^{-1}) as a function of the order $2m$ of $\mathbf{L}$ and the dimension d is studied in a companion paper [22]. The result is that f' will have l continuous derivatives if $2m \geq d + l$.

7. Relative Quantum Entropy

The relative quantum entropy is a generalization of the relative Shannon entropy, and is defined in terms of the density matrices D and D^0 as follows [3]:

$$S(D; D^0) = -\mathrm{Tr}D + \mathrm{Tr}D^0 + \mathrm{Tr}(D(\log D - \log D^0)). \tag{10}$$

This can also be written in terms of the constrained variables. We write Eq. (3) as $D = \exp(-\sum \lambda_i \mathcal{O}_i)$; then $\log D = -\sum \lambda_i \mathcal{O}_i$. Therefore $S(a; a^0) = -S(a) + T(\lambda^0) + \lambda^0 \cdot a$. Then $\partial S/\partial a_i = -(\lambda_i - \lambda_i^0)$, and the second derivative is $-\partial \lambda_i/\partial a_j = -G_{ij}$, which is negative by concavity. Considered as a function of a, $S(a; a^0)$ is zero when $a = a^0$, positive elsewhere, and convex. Both the relative quantum entropy and the relative Shannon entropy (Kullback-Liebler divergence) are information divergences in a precise technical sense. (Define as in [1, p.84], with $S(a)$ and $T(\lambda)$ the conjugate potentials.) If we want to have a single parameter characterizing the smoothness, we can define $I_Q^{(\beta)}(f; f^0) = S(f, b; f^0, b^0)$, where b is fixed by the requirement that $\partial S/\partial b = 0$; this amounts to fixing β at the value corresponding to (f^0, b^0), rather than fixing b^0, and is usually easier for technical reasons.

The exponential of the relative quantum entropy can be used as a prior for Bayesian inference [16, 17, 14]. Specifically, one assumes a prior probability of the form

$$P^{(\alpha,\beta)}(f; f^0) \propto \exp(-\alpha I_Q^{(\beta)}(f; f^0)), \tag{11}$$

where α and β are hyperparameters to be determined. One then generally takes as one's estimate the mode of the posterior distribution. This is identical to Quantified Maximum Entropy [20], except that it uses the quantum entropy instead of the Shannon entropy.

Generally, this technique negotiates a tradeoff, parameterized by α, between minimizing the likelihood and minimizing the relative quantum entropy. In the limit in which the likelihood enforces the constraints rigidly, we recover the technique discussed in §3. These techniques are logically distinct, but to avoid a proliferation of terminology, we extend the use of the term MQE to cover the more general method; this is also known as Quantum Statistical Inference (QSI) [16]. Although the method of choosing the Lagrange multipliers is generalized, the form of the density matrix is the same, so §4-§7 still apply.

The exact form of the posterior distribution will depend on the likelihood, which will depend in turn on the type of problem being considered. The two primary applications which have been worked out are non-parametric regression problems, including inverse problems [14], and density function estimation (DFE) [15]. I_Q is convex, and the second derivative of the log-likelihood is negative in both of these cases, so the posterior probability has a unique maximum, which may be found numerically.

8. Maximum Quantum Entropy for non-quantum data

To the writer's knowledge, MQE is the first instance in which the quantum entropy has been proposed as a part of the statistical apparatus, as opposed to being a part of the description of the object under study (*i.e.*, a quantum-mechanical system). Can one justify the maximization of the quantum entropy of a non-quantum density in terms of the usual arguments for maximizing the entropy?

For clarity, I will use the term MaxEnt to refer to the logical framework in which the maximization of the entropy has been justified [8], ME to refer to the application of MaxEnt to images using the entropy $-\sum f_i \log f_i$ [18], and MQE to refer to the procedure described in this paper.

Normally, in MaxEnt, we start with some set of *fixed*, *mutually exclusive* and exhaustive events. The entropy is defined as the Shannon entropy of the probabilities of these events. Given constraints on these probabilities, MaxEnt estimates these probabilities as those which maximize the entropy subject to the constraints. In the absence of constraints we recover equal probabilities, in accordance with Laplace's Principle of Indifference.

In the Brandeis dice problem [8], for example, the events are that face i will turn up when the die is rolled; the constraint might be that the average value of a toss is some number, say 4.5. In ME, where the entropy is defined as $-\sum f_i \log f_i$, the event is that some unit of intensity will show up in the ith pixel; we might have constraints on some linear functionals of f, as in §3.

MQE does not appear to satisfy the assumptions of the MaxEnt framework, when applied to classical densities. If we wish to write the quantum entropy in the form of a Shannon entropy, $-\sum w_i \log w_i$, we must diagonalize the density matrix, which can be done by diagonalizing the operator $\sum \lambda_j \mathcal{A}_j + \beta\mathbf{L}$. In this basis, the eigenstates are $\{\phi_i\}$, the diagonal elements of the density matrix are w_i, the entropy is simply the Shannon entropy of the w_i, and the estimate for the function is $\sum w_i |\phi_i(x)|^2$. In the framework of MaxEnt, therefore, the events are that a unit of intensity $|\phi_i(x)|^2$ will contribute to the estimate $f(x)$, and the posterior probability of this event is w_i. Other justifications for MQE may be proposed, but this appears to be the unique way of fitting it into the MaxEnt framework, as described above.

Note, however, that the "events" are neither fixed nor mutually exclusive. They are not fixed because they are eigenstates of an operator which depends on the λ_j. In effect, if we try to define the entropy to square with the MaxEnt formalism, the very definition of the entropy changes as we change the λ_j. Furthermore, the events are not mutually exclusive because the ϕ_i overlap. In general, f can be composed of the ϕ_i in many different ways. Our algorithm picks out one of these uniquely, but it is not clear how to justify this in terms of MaxEnt.

For these reasons, I believe that MQE is not, strictly speaking, a maximum entropy technique, when applied outside of quantum mechanics. (This discussion does not apply to the use of quantum entropy in quantum mechanics, which involves important additional considerations beyond the scope of this paper.) At the same time, the more conventional ME approach, in which the individual pixels are considered independent, cannot, strictly speaking, be justified in terms of MaxEnt either, when the density is known to be smooth. This is merely a reflection of the profound difficulty in relating real-world problems to abstract principles of reasoning.

MQE smooths the density estimate because the smoothing operator in the exponent mixes adjacent values of the density. It may be a very useful approach to the estimation of density functions, and it gives rise to some very beautiful mathematics. It provides an intriguing and very novel statistical model. It should be investigated further and compared to existing techniques in non-trivial real-world applications.

ACKNOWLEDGMENTS. I would like to thank Harry Martz, David Wolpert, and especially Richard Silver for useful conversations.

References

[1] Shun-ichi Amari. *Differential-Geometrical Methods in Statistics.* Springer-Verlag, Berlin, 1985.

[2] R. Balian and N. L. Balazs. 'Equiprobability, inference, and entropy in quantum theory'. *Ann. Phys.*, 179:97–144, 1987.

[3] Roger Balian, Yoram Alhassid, and Hugo Reinhardt. 'Dissipation in many-body systems: A geometric approach based on information theory.' *Phys. Rep.*, 131:1–146, 1986.

[4] Richard P. Feynman. 'An operator calculus having applications in quantum electrodynamics.' *Phys. Rev.*, 84:108–128, 1951.

[5] Richard P. Feynman. *Statistical Mechanics*. Benjamin/Cummings, Reading, 1972.

[6] Paul F. Fougère, editor. *Maximum Entropy and Bayesian Methods: Dartmouth, U.S.A., 1989*. Kluwer, Dordrecht, 1990.

[7] Stephen F. Gull. 'Developments in maximum entropy data analysis.' In J. Skilling, editor, *Maximum Entropy and Bayesian Methods*, pages 53–71, Dordrecht, 1989. Kluwer.

[8] E. T. Jaynes. 'Where do we stand on maximum entropy?' In R. D. Levine and M. Tribus, editors, *The Maximum Entropy Formalism*. M.I.T. press, Cambridge, 1978.

[9] Edwin T. Jaynes. 'Information theory and statistical mechanics, I.' *Phys. Rev.*, 106:620–630, 1957.

[10] Edwin T. Jaynes. 'Information theory and statistical mechanics, II.' *Phys. Rev.*, 108:171–190, 1957.

[11] Robert Karplus and Julian Schwinger. 'A note on saturation in microwave spectroscopy.' *Phys. Rev.*, 73:1020–1026, 1948.

[12] D. W. Scott, R. A. Tapia, and J. R. Thompson. 'Nonparametric probability density estimation by discrete maximum penalized-likelihood criteria.' *Ann. Statist.*, 8:820–832, 1980.

[13] John E. Shore and Rodney W. Johnson. 'Axiomatic derivation of the principle of maximum entropy and the principle of minimum cross-entropy.' *IEEE Trans. Info. Th.*, pages 26–37, 1980.

[14] Richard Silver and Harry Martz. 'Quantum statistical inference.' *to be published*, 1993.

[15] Richard Silver, Timothy C. Wallstrom, and Harry Martz. 'Maximum quantum entropy for non-parametric density function estimation.' *to be published*, 1993.

[16] Richard N. Silver. 'Quantum statistical inference.' In M. Djafari and G. Demoment, editors, *Maximum Entropy and Bayesian Methods*, pages 167–182. Kluwer, Dordrecht, 1993.

[17] Richard N. Silver. 'Quantum statistical inference.' In *Physics and Probability: Essays in Honor of E.T.Jaynes*, pages 223–238. Cambridge University Press, Cambridge, 1993.

[18] John Skilling. 'Classic maximum entropy.' In J. Skilling, editor, *Maximum Entropy and Bayesian Methods*, pages 45–52, Dordrecht, 1989. Kluwer.

[19] John Skilling, editor. *Maximum Entropy and Bayesian Methods: Cambridge, England, 1988*. Kluwer, Dordrecht, 1989.

[20] John Skilling. 'Quantified maximum entropy.' In P. F. Fougére, editor, *Maximum Entropy and Bayesian Methods*, pages 341–350, Dordrecht, 1990. Kluwer.

[21] Y. Tikochinsky, N. Z. Tishby, and R. D. Levine. 'Alternative approach to maximum-entropy inference.' *Phys. Rev. A*, 30(5):2638–2644, 1984.

[22] Timothy C. Wallstrom. 'Smoothing in maximum quantum entropy.' *These proceedings*.

[23] Alfred Wehrl. 'General properties of entropy.' *Rev. Mod. Phys.*, 50:221–260, 1978.

SMOOTHING IN MAXIMUM QUANTUM ENTROPY

Timothy C. Wallstrom
Theoretical Division, MS-B213
Los Alamos National Laboratory
Los Alamos, New Mexico 87545

ABSTRACT. The method of Maximum Quantum Entropy (MQE) has been described in a companion paper. Here we give criteria for the smoothness properties of the MQE estimate, as a function of the smoothing operator and the dimension. With point constraints, the MQE estimate will have continuous derivatives of order l in d dimensions if and only if the elliptic smoothing operator is of order $2m$, where $2m > l+d$. It is thus impossible to constrain individual points unless $2m > d$, and the derivative will have discontinuities unless $2m > d+1$.

The method of Maximum Quantum Entropy (MQE) has been described in a companion paper [3]. We recall that our density estimate is of the form

$$f(x) = \sum e^{-\beta\epsilon_i} |\phi_i(x)|^2, \tag{1}$$

where $\phi_i(x)$ and ϵ_i are the eigenfunctions and eigenvalues, respectively, of the operator H, defined by

$$\beta H = \beta \mathbf{L} + U(\lambda). \tag{2}$$

$\mathbf{L}$ is a differential operator, and U is a multiplication operator.

We wish to determine the smoothness properties of f as a function of U, $\mathbf{L}$, and the dimension d. Our result is that if $\mathbf{L}$ is an elliptic operator of order $2m$, and if U constrains the density at a finite number of points so that it is the sum of delta functions, then in d dimensions f will be of class C^l if and only if $l < 2m - d$. (We say that f is of class C^l if all partial derivatives of order l or less exist and are continuous.)

What does this result mean for MQE? If $2m \leq d$ we obtain singularities at the points which are constrained; this is catastrophic because the whole point of introducing the constraint is to adjust f to some finite value. This implies, in particular, that we must use higher-order smoothers to implement MQE in 2 or more dimensions. If $2m \leq d+1$, our estimate is continuous but its derivative is not. This is the case for a second-order smoother in one-dimension; the visual consequence is kinks in the estimate. These frequently detract considerably from the visual appeal; see [2]. This implies that we are usually better off using a higher-order smoother even in one dimension. Finally, if $2m > d+1$, we get an estimate which is continuous and has a continuous derivative.

We develop this result heuristically; a rigorous discussion will appear elsewhere [4]. In a continuous basis, our constraints on the density are of the form

$$\mathcal{A}_m = \int dx\, A_m(x) \mathcal{F}_x, \tag{3}$$

G. R. Heidbreder (ed.), Maximum Entropy and Bayesian Methods, 157–159.

where $(\mathcal{F}_x)_{yz} = \delta(y-x)\delta(x-z)$ (these are now Dirac delta functions); therefore,

$$U(\lambda)\psi(x) = \sum \lambda_m A_m(x)\psi(x). \tag{4}$$

Eq. (1) expresses $f(x)$ as an infinite sum of solutions to the eigenvalue equation

$$(\beta \mathbf{L} + U(\lambda))\psi = E\psi. \tag{5}$$

U is the sum of the individual A_m's. If $A_m(x)$ constrains the value of f at x_i, it will be a Dirac delta function: $A_m(x) = \delta(x - x_i)$. We will specialize to this case in this paper, although other cases arise in applications.

We simplify the problem by a series of observations: (1) The smoothness of f is the same as that of the individual eigenfunctions ψ. (2) The smoothness of each ψ is the same as that of ϕ, where ϕ solves the inhomogeneous equation with one delta function:

$$(\beta \mathbf{L} - E)\phi = \mu\delta(x) \tag{6}$$

(3) We can replace $\beta\mathbf{L}$ by $(-\Delta)^m$, if $\mathbf{L}$ is elliptic of order $2m$ with analytic coefficients, without affecting the smoothness of ϕ. All of these assertions, though plausible, require an argument; see [4] for details.

We are thus reduced to an examination of the fundamental solution (or Green's function) of the equation

$$((-\Delta)^m - E)\phi = 0. \tag{7}$$

If $d > 2m$ ($d = 2m$), this is well-known to go as $1/r^{d-2m}$ ($\log r$). Otherwise, the answer can be obtained by a Fourier transform:

$$\phi(x) = \frac{1}{(2\pi)^d}\int \frac{d^d k e^{ik\cdot x}}{|k|^{2m} - E}; \tag{8}$$

this expression can also be used to calculate derivatives. (We take the principal value over the singularity.) It is not too difficult to show that all derivatives of order $l < 2m - d$ are continuous, but that $D_i^l\phi$ diverges at $x = 0$ if $l \geq 2m - d$. (If l is even this is straightforward. If l is odd, calculate $D_i^l\phi(\epsilon) - D_i^l\phi(-\epsilon)$ and replace $k\epsilon$ by k'; asymptotically the integral scales as ϵ^{2m-d-l}. If l is odd and $l = 2m - d$, then $D_i^l\phi$ is merely discontinous at $x = 0$. D_i means the derivative is taken in the direction x_i.) Therefore $\phi \in C^l$ and $f \in C^l$ if and only if $l < 2m - d$.

Remarks: (1) Alternatively, we can invoke results of John [1] which state, in the spherically symmetric case, that if $2m \geq d$, the singular part of the kernel goes asymptotically as r^{2m-d} (d odd), or $(\log r)r^{2m-d}$ (d even) for small r. (For nonsymmetric kernels there is an analytic prefactor.) Recalling that $\partial/\partial x = (r/x)(\partial/\partial r)$, we easily recover the above results. (2) For non-integral m, we can define pseudo-differential operators by means of the Fourier transform:

$$(-\Delta)^m + E : \hat{f} \mapsto (|k|^{2m} + E)\hat{f}. \tag{9}$$

The continuity results (and the above argument) hold in this case also. (3) If U is of a more general form, the inhomogeneous equation can of course be solved with the help of the Green's functions, and if f is approximately constant where U is non-zero, this will provide a useful approximation to the homogeneous equation (5).

In Fig. (1), we plot $G_0^{-1}(x)$, the linear approximation to $\delta f/\delta\lambda$, computed numerically for $d = 1$ and $m = 1/2$, 1 and 2. Our analysis implies that the corresponding f should be discontinous, continuous but with no continuous derivatives, and continuous with continuous first and second derivatives, respectively. Because of the linear approximation, $G_0^{-1}(x)$ for $m = 2$ dips very slightly below the axis. $\delta f/\delta\lambda$ is strictly positive.

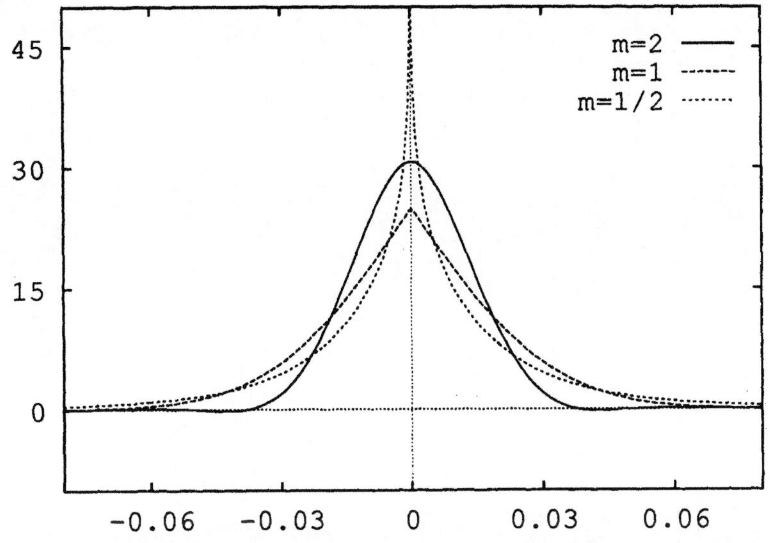

Figure 1: G_0^{-1} for $m = .5$, 1, and 2. We assume f is defined on $[0, 1]$. The scale of the y-axis is determined by the condition that $\int_0^1 G_0^{-1}(x - 0.5)dx = 1$.

ACKNOWLEDGMENTS. I would like to thank Richard Silver for useful conversations.

References

[1] Fritz John. 'The fundamental solution of linear elliptic differential equations with analytic coefficients.' *Comm. Pure and Appl. Math*, 3:273–304, 1950.

[2] Richard N. Silver, Harry Martz, and Timothy C. Wallstrom. 'Quantum statistical inference for density estimation.' In G. Heidbreder, editor, *Maximum Entropy and Bayesian Methods*, Dordrecht, 1994. Kluwer.

[3] Timothy C. Wallstrom. 'Maximum quantum entropy for classical density functions.' These proceedings.

[4] Timothy C. Wallstrom. 'Smoothing operators in maximum quantum entropy.' *to appear*, 1994.

DENSITY ESTIMATION BY MAXIMUM QUANTUM ENTROPY

R. N. Silver, T. Wallstrom, H. F. Martz
MS B262
Los Alamos National Laboratary
Los Alamos, NM 87545

ABSTRACT. A new Bayesian method for non-parametric density estimation is proposed, based on a mathematical analogy to quantum statistical physics. The mathematical procedure is related to maximum entropy methods for inverse problems and image reconstruction. The information divergence enforces global smoothing toward default models, convexity, positivity, extensivity and normalization. The novel feature is the replacement of classical entropy by quantum entropy, so that local smoothing is enforced by constraints on differential operators. The linear response of the estimate is proportional to the covariance. The hyperparameters are estimated by type-II maximum likelihood (evidence). The method is demonstrated on textbook data sets.

1. Introduction.

Non-parametric density estimation has been studied extensively by statisticians. If a set of N_s observations, $\{x_i\}$, is identically and independently drawn from a probability density function $f(x)$, the problem is to estimate f when no parametric form is known. A variety of non-Bayesian methods, such as histograms and kernel density estimators, have been developed and applied to density estimation [for reviews, see Silverman, 1986; Izenman, 1991]. There has been comparatively little work on maximum penalized likelihood methods [see, e.g., Good and Gaskins, 1980], despite their potential advantages such as a Bayesian interpretation, the ability to combine explicit prior knowledge with the data, the ability to combine data from different sources, etc. The situation in density estimation contrasts sharply with that of inverse problems and image reconstruction where maximum penalized likelihood methods are dominant [see, e.g., Titterington, 1985; Demoment, 1989].

In a maximum penalized likelihood (MPL) framework, the density estimate is determined from the maximum of

$$Q(f) \equiv \sum_{i=1}^{N_s} \ln\left(f(x_i)\right) - \alpha I(f; f_o, \beta) \tag{1}$$

as a functional of f. The first term in (1) is the log-likelihood function and the second term is the penalty function (alternative terms are *regularization functional*, or *information divergence*). Here f_o is a default model in the absence of data, $I(f; f_o, \beta)$ is zero when $f = f_o$ and monotonically increases as f diverges from f_o, α is a global smoothing hyperparameter (or *statistical regularization parameter*), and β is a local smoothing hyperparameter.

We propose a *Maximum Quantum Entropy* (MQE) method for density estimation, which corresponds to a choice for the penalty function, I_Q. The mathematical structures we use originated in *quantum statistical mechanics*; hence, an alternative name is *Quantum Statistical Inference - QSI* [Silver, 1993]. MQE is a variation upon the maximum entropy (ME)

G. R. Heidbreder (ed.), Maximum Entropy and Bayesian Methods, 161–174.

methods [Skilling and Gull, 1989] that have been applied extensively to inverse problems and image reconstruction. The penalty functions used in ME are various modifications of the Shannon entropy of information theory, which in fact originated in the 19th century development of classical statistical physics. The penalty function for MQE is the more general concept of *relative quantum entropy*, which was developed [von Neumann, 1927] for applications to quantum statistical physics.

Both ME and MQE enforce desirable properties of density estimators such as global smoothing toward a default model, positivity, normalization, extensivity, and convex optimization. But, in addition, MQE enforces local smoothing by constraining the expectation values of differential operators. The maximum local smoothing limit of MQE is traditional penalized likelihood [Good and Gaskins, 1980] which does not enforce extensivity. The zero local smoothing limit of MQE is classic ME. MQE was applied previously to inverse problems [Silver, 1993], where it was shown to improve upon ME wherever local smoothing is important. MQE may be compared to an alternative proposal [Skilling and Gull, 1989; Robinson, 1991] to smooth ME using 'intrinsic correlation functions' and 'hidden images', which does not incorporate local smoothing in the penalty function.

The purpose of the present paper is adapt MQE to density estimation. The theory will be developed within a Bayesian framework referring to [Silver and Martz, 1993] for mathematical details. The method is illustrated using textbook data sets [Scott, 1993].

2. MQE Density Functions.

In MQE, the density function, the constraints and the entropy are all expressed in terms of a new concept in statistics, the *density matrix*. $D(x, x')$ is an $\infty \times \infty$ matrix which is real symmetric and positive semidefinite. The density function f is equal to the diagonal elements of $\mathbf{D}$,

$$f(x) = D(x, x) \quad . \tag{2}$$

$\mathbf{D}$ will be determined uniquely by the combination of constraints on f and a maximum entropy principle. Without loss of generality, we assume $0 \leq x \leq 1$ and impose appropriate boundary conditions.

The density matrix, $\mathbf{D}$, can be diagonalized by a unitary transformation,

$$D(x, x') = \sum_{n=0}^{\infty} \psi_n(x) w_n \psi_n(x') \quad . \tag{3}$$

The ψ_n are orthonormal and complete forming a Hilbert space. The *weights* satisfy $w_n \geq 0$. Hence,

$$f(x) = \sum_{n=0}^{\infty} w_n \psi_n^2(x) \geq 0 \quad , \tag{4}$$

and

$$\sum_{n=0}^{\infty} w_n = 1 \quad . \tag{5}$$

(For practical calculations, we will show below that the ψ_n may be obtained as eigenfunctions of a linear differential operator, and the w_n are related to the eigenvalues.)

Linear Lagrange constraints on f may be written in terms of $\mathbf{D}$. Data constraints are

$$\int_0^1 U(x)f(x)dx = E(\mathbf{U}) = Tr\{\mathbf{UD}\} \quad , \tag{6}$$

where $(\mathbf{U})_{x,x'} = U(x)\delta(x - x')$. For example, if the constraints consist of a set of $E(\mathbf{O}_i) = \int_0^1 O_i(x)f(x)dx$, then $U(x) = \sum_i \lambda_i O_i(x)$ for Lagrange multipliers λ_i. The normalization constraint on f is

$$E(\mathbf{1}) = Tr\{\mathbf{D}\} = 1 \quad . \tag{7}$$

The key constraint is local smoothing, which is defined by the choice of an Hermitian differential operator $\mathbf{L}$ whose expectation value is the local smoothing constraint,

$$E(\mathbf{L}) = Tr\{\mathbf{LD}\} = \sum_{n=0}^{\infty} w_n \int_0^1 \psi_n(x)\mathbf{L}\psi_n(x)dx \quad . \tag{8}$$

We have used quadratic, $\mathbf{L}_2 \equiv -\partial^2/\partial x^2$, and quartic, $\mathbf{L}_4 \equiv \partial^4/\partial x^4$, differential operators. (We note that there are many other possible choices including x-dependent forms.) This explicit constraint applied to $\mathbf{D}$ is an implicit local smoothing constraint on f. In ME f will have the same singularity structure as U, whereas in MQE f will have smoother singularities than U depending on the choice of $\mathbf{L}$. The singularity structure of U is determined by the nature of the data analysis problem. For example, for inverse problems U consists of a sum of Lagrange multipliers times point spread functions which are most often already locally smooth. However, we shall see that for density estimation U consists of a sum of δ-functions. Then, ME produces an f with δ-function singularities, a MQE constraint on $\mathbf{L}_2$ requires f to be continuous, and a MQE constraint on $\mathbf{L}_4$ requires f to have continuous first derivatives. For a more comprehensive discussion see [Wallstrom, 1993].

These constraints are still not sufficient to uniquely specify $\mathbf{D}$, so now we invoke a maximum entropy principle. The *quantum entropy* of a density matrix is

$$S_Q \equiv -Tr\{\mathbf{D}\ln(\mathbf{D})\} = -\sum_{n=0}^{\infty} [w_n \ln(w_n)] \quad . \tag{9}$$

S_Q is invariant to unitary transformations of the Hilbert space. It is not a relative entropy, so that in the absence of constraints all eigenfunctions are equally likely. One can prove that S_Q is a concave function of $\mathbf{D}$ [Wehrl, 1978]. The maximum entropy principle is to maximize S_Q subject to the constraints of the problem. Using the method of Lagrange multipliers, maximize

$$Q(\mathbf{D}) \equiv S_Q - \beta E(\mathbf{L}) - E(\mathbf{U}) + (\mu + 1)E(\mathbf{1}) \quad , \tag{10}$$

where the Lagrange multipliers are chosen so that the constraints are satisfied. The local smoothing constraint on $E(\mathbf{L})$ has Lagrange multiplier β, the data constraint has Lagrange multiplier U, and the normalization constraint has Lagrange multiplier $\mu + 1$.

The maximum of (10) is found at

$$\mathbf{D} = \exp\left(-\mathbf{H} + \mu\mathbf{1}\right) \quad , \tag{11}$$

where

$$\mathbf{H} \equiv \beta \mathbf{L} + \mathbf{U} \quad . \tag{12}$$

This constitutes an exponential family of density matrices parameterized by U, β, and μ. Within this family, there is a one-to-one correspondence between a choice of density function, f, and a corresponding density matrix, $\mathbf{D}$.

Diagonalizing $\mathbf{D}$, we find that the ψ_n in (3) are eigenfunctions of $\mathbf{H}$, i.e.

$$\mathbf{H}\psi_n(x) = \varepsilon_n \psi_n(x) \quad . \tag{13}$$

The weights are

$$w_n = \exp\left(-\varepsilon_n + \mu\right) \quad . \tag{14}$$

For example, for $\mathbf{L}_2$ (12) reads

$$-\beta \frac{\partial^2 \psi_n(x)}{\partial x^2} + U(x)\psi_n(x) = \varepsilon_n \psi_n(x) \quad , \tag{15}$$

which is analogous to the time-independent Schrödinger equation. Such eigenvalue equations may alternatively be derived from variational principles as developed in Sturm-Liouville theory.

The local smoothness of f is adjusted by tuning β. For reasonable choices of $\mathbf{L}$ (such as the quadratic and quartic), the ε_n increase monotonically with n and with β. The number of nodes in $\psi_n(x)$ also increase monotonically with n, so that small n corresponds to smoother $\psi_n^2(x)$. For $\beta = 0$ (ME) there is no local smoothing. As β is increased fewer eigenfunctions contribute to (4) resulting in smoother f.

The normalization of f is maintained by choosing

$$\mu = -\ln\left(\sum_{n=0}^{\infty} e^{-\varepsilon_n}\right) \quad . \tag{16}$$

We are now ready to identify the penalty function, I_Q, in (1). The penalty function is a *relative quantum entropy*,

$$I_Q = Tr\{\mathbf{D}\ln(\mathbf{D}) - \mathbf{D}\ln(\mathbf{D}_o)\} \quad , \tag{17}$$

where $\mathbf{D}_o$ is the density matrix corresponding to the default model f_o. This may be regarded as a straightforward generalization of the Kullback-Liebler entropy used in ME methods from density functions to density matrices. In the limit of no local smoothing, $\beta \rightarrow 0$, MQE reduces to ME. Alternatively, let $Q_o(\mathbf{D})$ be the entropy variational functional similar to (10) whose maximum is at $\mathbf{D}_o$. Then

$$I_Q = Q_o(\mathbf{D}_o) - Q_o(\mathbf{D}) \quad . \tag{18}$$

It follows that $I_Q \geq 0$. We summarize the mathematical properties satisfied by I_Q which are critical to its relevance to statistics.

The concavity property of S_Q means that $\mathbf{G}$ defined by

$$\delta^2 S_Q = -\frac{1}{2}\int G(x,x')\delta f(x)\delta f(x')dxdx' \tag{19}$$

is positive semidefinite (no negative eigenvalues). The consequence is that one can prove *duality* properties between I_Q and its Legendre transform,

$$C_Q(U; U_o, \beta) \equiv I_Q(f; f_o, \beta) + \int f(x)U(x)dx \quad , \tag{20}$$

which is a cumulant generating functional. First order variations may be shown to be

$$\delta C_Q = \int f(x)\delta U(x)dx \qquad \delta I_Q = \int [-U(x) + U_o(x)]\, \delta f(x)dx \quad . \tag{21}$$

Second order variations are

$$\delta^2 I_Q = \frac{1}{2}\int G(x,x')\delta f(x)\delta f(x')dxdx' \quad \delta^2 C_Q = -\frac{1}{2}\int G^{-1}(x,x')\delta U(x)\delta U(x')dxdx' \quad . \tag{22}$$

Notice the dual symmetry between f and U in these relations, which is analogous to the dual symmetry between observables and Lagrange multipliers in traditional ME methods.

Legendre transform dual mathematical structures in statistics of this form may be given a differential geometry interpretation [Amari, 1985]. From (17) $I_Q(f_o; f_o, \beta) = 0$, and from (21) $dI_Q(f; f_o, \beta)/df = 0$ at $f = f_o$. Hence, I_Q is an *information divergence*, and $\mathbf{G}$ is a *Riemann metric* in the manifold of f.

The concavity property ensures a dual (one-to-one) relation between conjugate variables, f and U,

$$\delta f(x) = -\int G^{-1}(x,x')\delta U(x')dx' \quad . \tag{23}$$

Because of this relation, $\mathbf{G}^{-1}$ may be termed a *linear response function.* For typical choices of local smoothing operator, $\mathbf{L}$, (including the quadratic and quartic) one can demonstrate that $G^{-1}(x,x')$ peaks at $x - x' = 0$ and falls off faster than a power law as $\mid x - x' \mid$ increases, a property we term *locality.* The characteristic width of $G^{-1}(x,x')$ is termed the *correlation length*, γ. For $\mathbf{L}_2$, $\gamma \propto (\beta)^{1/2}$. For $\mathbf{L}_4$, $\gamma \propto (\beta)^{1/4}$. For example, let $\mathbf{G}_o^{-1}$ be the linear response function for no data constraints and a flat default model, i.e. $U = 0$. Then for $\mathbf{L}_2$, one can prove $G_o^{-1}(x,x') \propto (1 - erf(\mid x - x' \mid /\gamma))/\gamma$. Figure 1 illustrates the behavior of $\mathbf{G}_o^{-1}$ for quadratic and quartic local smoothing. Note that for quadratic smoothing $\mathbf{G}^{-1}$ is strictly positive, whereas for higher order smoothing $\mathbf{G}^{-1}$ can have negative components at large $\mid x - x' \mid$. The non-linearity of MQE guarantees that $f \geq 0$ regardless of the choice of local smoothing.

Readers familiar with density estimation may be tempted to identify $\mathbf{G}^{-1}$ with the kernel in a kernel density estimation procedure. Readers familiar with ME may be tempted to identify $\mathbf{G}^{-1}$ with the *intrinsic correlation function* used in the [Skilling and Gull, 1989] proposal to correct ME for local smoothing using *hidden ME images.* However, there are significant differences. For example, in both these methods no structure in f can be narrower than the width of the kernel or intrinsic correlation functions, whereas in MQE the non-linearity permits structure in f which is much narrower than the width of $\mathbf{G}^{-1}$.

ME ($\beta = 0$) satisfies *local extensivity*, which means that the penalty function is an additive function of the $f(x)$ at each point. However, we often have prior knowledge or evidence in the data that f is locally smooth, which violates the local extensivity property. MQE relaxes this condition to *non-local extensivity*, defined as follows. Let δI_Q^i be a change

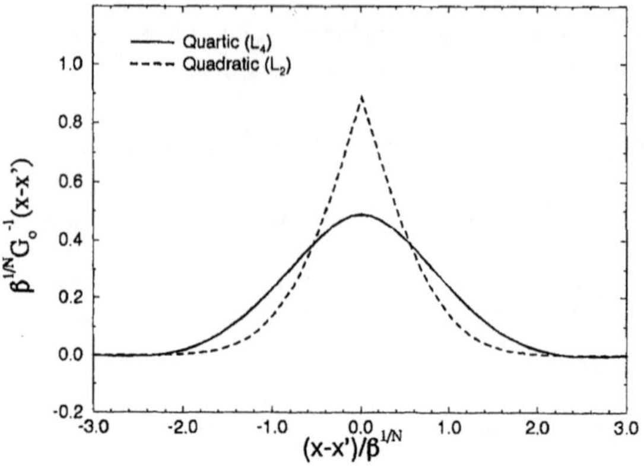

Figure 1: **Linear Response Functions** - $\mathbf{G}_o^{-1}$ *for local smoothing constraints of the form* $\mathbf{L}_N = \partial^N/\partial x^N$, *no data constraints, and a flat default model, i.e.* $U = 0$. β *is the Lagrange multiplier for the local smoothing constraint on the density matrix. Results are shown for quadratic (dashed) and quartic (solid) local smoothing.*

in I_Q corresponding to a change δf^i in f. Let the δf^i have compact and disjoint supports separated by much more than γ. Then non-local extensivity means $\delta I_Q \simeq \sum_i \delta I_Q^i$ for $\delta f = \sum_i \delta f^i$. This may be shown by combining the locality properties of $\mathbf{G}^{-1}$ with (22). In comparison the MPL method of [Good and Gaskins, 1980] does not obey any form of extensivity, because it is equivalent to $\gamma \to \infty$.

These *convexity* and *non-local extensivity* properties of I_Q satisfy important desiderata for both image reconstruction and density estimation. In the latter case non-local extensivity is compromised only by the added constraint on the normalization of f. Many other mathematical properties of I_Q have been established in physics contexts [for reviews, see Wehrl, 1978; Balian, 1991].

3. Application to Density Estimation.

We apply these properties of I_Q to the MPL problem defined by (1). From (21), the first order variation of $Q(f)$ requires that the MPL estimate satisfies

$$\sum_{i=1}^{N_s} \frac{\delta(x - x_i)}{f(x)} + \alpha(U(x) - U_o(x)) = 0 \quad . \tag{24}$$

From (22) the second order variation (Hessian matrix) is positive semi-definite, so that solution of (24) is a problem for convex non-linear optimization methods [Skilling, 1993].

A variety of interpretations of MPL methods exist including ways to estimate hyperparameters and quantify error estimates for any choice of penalty function [Thompson, 1991]. We specialize to the Bayesian interpretation of MQE. Bayes theorem is

$$P[f \mid \{x_i\}; f_o, \alpha, \beta] \times P[\{x_i\}; f_o, \alpha, \beta] = P[\{x_i\} \mid f] \times P[f; f_o, \alpha, \beta] \quad . \tag{25}$$

The *likelihood function* is

$$P[\{x_i\} \mid f] = \prod_{i=1}^{N_s} f(x_i) \quad . \tag{26}$$

The *prior probability* for f is taken to be

$$P[f; f_o, \alpha, \beta] \propto \exp\left[-\alpha I_Q(f; f_o, \beta)\right] \quad . \tag{27}$$

Then $P[f \mid \{x_i\}; f_o, \alpha, \beta]$ is the *posterior probability* for f, and $P[\{x_i\}; f_o, \alpha, \beta]$ is the *marginal likelihood*, or *evidence.* We take the best estimate, $\hat{f}$, from the maximum of the posterior probability which is equivalent to maximizing (1). Thus, the MPL estimate is equivalent to a Maximum A Posteriori (MAP) estimate in the Bayesian interpretation.

The hyperparameters α and β are estimated from the maximum of the evidence. This method is termed *type-II maximum likelihood* (ML-II) in the statistics literature [Good, 1983; Berger, 1985], and the *evidence procedure* in the ME literature. The marginal likelihood is obtained by integrating Bayes theorem (25) over f. A metric must be used in this integration over f in order to enforce invariance to coordinate transformations. The appropriate choice is the Jeffrey's prior $\sqrt{\det(\alpha\mathbf{G})}$, which is equivalent to a *Riemann volume* factor for the f-manifold in differential geometry. We evaluate the integral in a Gaussian approximation to the expansion of $Q(f)$ in $\ln(f/\hat{f})$ about $Q(\hat{f})$. The resulting marginal likelihood is

$$P[\{x_i\}; f_o, \alpha, \beta] \propto \frac{1}{\sqrt{\det\left(\mathbf{1} + \frac{\mathbf{M}}{\alpha}\right)}} \times \exp Q(\hat{f}) \quad , \tag{28}$$

where the $N_s \times N_s$ matrix $\mathbf{M}$ is

$$M_{ij} \equiv \frac{\hat{G}^{-1}(x_i, x_j)}{\hat{f}(x_i)\hat{f}(x_j)} \quad .$$

The first term on the r.h.s. of (28) favors the simpler f of large α and β, so that it may be termed an *Ockham factor.* The second term, $\exp Q(\hat{f})$, favors the more complicated f of small α and β, and it is termed the *data factor.* The balance between the Ockham factor and data factor determines the optimal hyperparameters, $\hat{\alpha}$ and $\hat{\beta}$. We find empirically that the ML-II optimization of hyperparameters is convex for all data sets studied so far.

The covariance of the MAP estimate can be calculated using the same Gaussian approximations employed in the calculation of the marginal likelihood. The result is

$$Cov[f(x), f(x')] = \frac{\hat{G}^{-1}(x, x')}{\hat{\alpha}} - \sum_{i,j=1}^{N_s} \frac{\hat{G}^{-1}(x, x_i)}{\hat{\alpha}\hat{f}(x_i)} \left(\mathbf{1} + \frac{\mathbf{M}}{\hat{\alpha}}\right)^{-1}_{ij} \frac{\hat{G}^{-1}(x_j, x')}{\hat{f}(x_j)\hat{\alpha}} \quad . \tag{29}$$

We interpret

$$N_d \equiv \alpha \int \frac{Cov[f(x), f(x)]}{\hat{f}(x)} dx \tag{30}$$

as the *number of degrees of freedom* in $\hat{f}$. One can prove $N_d \geq 0$. In the absence of data, the prior $N_d^o = Tr\{\mathbf{G}_o^{-1}\}$ is proportional to $1/\gamma$. This provides a simple interpretation of the local smoothing hyperparameter β, because it determines the correlation length scale γ which is inversely proportional to N_d^o. ME ($\beta = 0$) corresponds to an infinite N_d, which is why ME has infinite error bars on individual points of the MAP estimate, $\hat{f}$. MQE ($\beta \neq 0$) has a finite N_d and finite error bars on individual points.

Convergence of density estimation can be monitored using N_d, because the effect of the data is to reduce it toward zero. Let f_t be the true density function. As N_s becomes large one may use the property

$$E\left(\sum_{i=1}^{N_s} O(x_i)\right) = N_s \int O(x) f_t(x) dx \quad , \tag{31}$$

to approximate the integral in (30). The result is

$$N_d \approx \frac{\alpha}{N_s} \sum_{i=1}^{N_s} \frac{\hat{f}(x_i)}{f_t(x_i)} \left(\mathbf{M}(\alpha \mathbf{1} + \mathbf{M})^{-1}\right)_{i,i} \quad . \tag{32}$$

In analogy with developments in ME [Skilling and Gull, 1989], we define

$$N_g \equiv Tr\{\mathbf{M}(\alpha\mathbf{1} + \mathbf{M})^{-1}\} \quad , \tag{33}$$

as the *number of good measurements.* Manifestly, $N_s \geq N_g \geq 0$. Then, to the extent that $\hat{f}$ has converged to f_t, (32) and (33) imply that $N_d \rightarrow \alpha N_g / N_s \leq \alpha$.

One can also derive a fundamental relation between the linear response of the MQE MAP estimate to perturbations and the covariance matrix,

$$\delta \hat{f}(x) = -\alpha \int Cov[f(x), f(x')] \delta U_p(x') dx' \quad . \tag{34}$$

Here δU_p is an infinitesimal perturbation in U which may be due to changes in the default model, changes in the data, changes in other constraints, etc. For example, an infinitesimal change in the default model corresponds to $\delta U_p(x) = -\int G_o(x, x') \delta f_o(x') dx'$. Putting (34) in words, the covariance matrix also describes the sensitivity of the MAP estimate to changes in prior knowledge or data. Large errors on the MAP estimate correspond to high sensitivity to input information, and small errors correspond to low sensitivity.

4. Examples.

We apply MQE to three textbook examples of density estimation problems: the duration of eruptions of the Old Faithful Geyser; the amount of annual snowfall in Buffalo; and the Lawrence Radiation Lab (LRL) particle physics data. For each data set, we urge readers to examine the corresponding sections of [Scott, 1993] to compare the performance of MQE with other approaches to density estimation.

To obtain the numerical results presented in this paper, we used Newton-Raphson for the non-linear optimization and matrix diagonalization of a discrete approximation to MQE to calculate f from knowledge of U. The number of pixels (bins) used is indicated directly on the figures for each data set. In other words, the raw data were histogrammed prior to applying MQE. We chose pixels widths which were much narrower than any structure in f, so the discretization should not significantly affect the estimate. All the MQE calculations used a flat default model, f_o, normalized to unit integral over the range of x. The values for the hyperparameters, α and β, are quoted for data scaled to the range $0 \leq x \leq 1$. The term, *optimal estimate*, means that the hyperparameters were chosen to maximize the marginal likelihood.

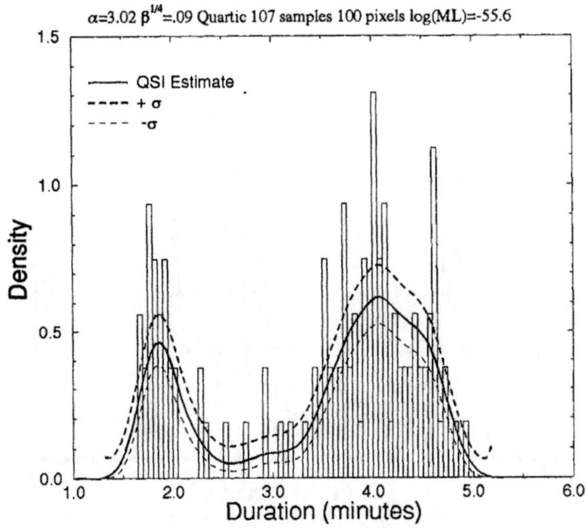

Figure 2: **Old Faithful Eruptions** - *107 measurements of the duration of geyser eruptions are displayed as a histogram with 100 bins. The solid line is the optimal MQE estimate obtained with quartic local smoothing,* $\mathbf{L}_4$. *The dashed lines indicate* $\pm$ *one standard deviation errors on the MQE point estimate.*

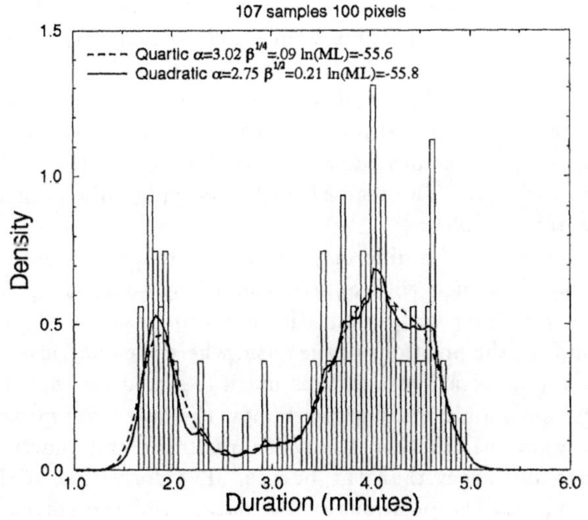

Figure 3: **Old Faithful Eruptions** - *Comparison of optimal MQE estimates for quadratic (solid) and quartic (dashed) local smoothing. The marginal likelihoods (ML) and correlation lengths are nearly identical. The quadratic estimate is unsatisfactory because it shows bumps at the positions of the data. The bumps are smaller than the error bars in Fig. 2 and not statistically significant. Nevertheless, the higher order smoothing of the quartic estimate is preferred.*

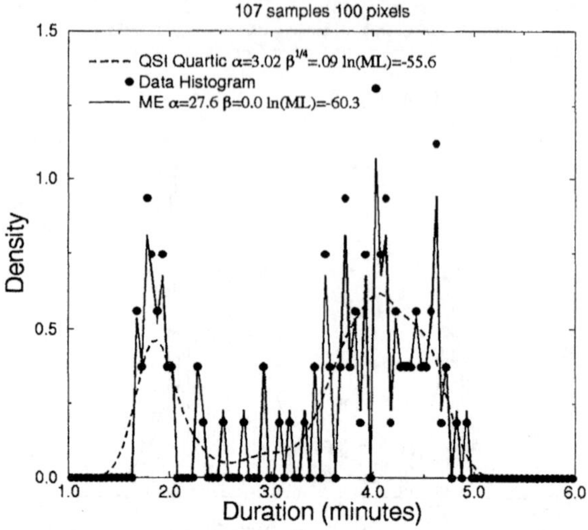

Figure 4: **Old Faithful Eruptions** - *Comparison of optimal MQE (dashed) with quartic smoothing and maximum entropy (solid) which has no local smoothing. Dots are the data histogram. The ratio of marginal likelihoods (ML) favoring MQE over ME is* 110.

Figures 2-4 show results for the duration of eruptions of the Old Faithful Geyser. The raw data from 107 eruptions are displayed as a histogram using 100 bins. Note that this histogram is not an optimal histogram estimate of f, which would use a much smaller number of bins. Rather, this histogram is simply a convenient way to display the raw data. In Fig. 2 the solid line is the optimal MQE estimate obtained for $\alpha = 3.02$ and $\beta^{1/4} = 0.09$ with quartic local smoothing. The dashed curve shows $\pm$ one standard deviation point estimates of errors on the MQE estimate, which are calculated from (29) according to $\sigma(x) = \sqrt{Cov[f(x), f(x)]}$. These provide only a partial representation of the full covariance matrix for the MQE estimate. The reader can be the judge of whether the optimal MQE estimate and errors are credible.

Figure 3 shows the effect of a different choice for the local smoothing constraint. The optimal MQE estimate obtained with quartic smoothing (dashed) is compared to the optimal MQE estimate obtained with quadratic smoothing (solid). The quadratic estimate appears to have bumps at the positions of the data, where the quartic estimate appears to be smooth. Therefore, the quadratic estimate is much less credible than the quartic, because the true density function should not depend on how the data were measured. However, one may argue that the apparent differences between quadratic and quartic are not significant. The MQE error bars are larger than the bumps. The derivative of the quartic estimate would also show bumps at the positions of the data. And the correlation lengths, γ, for the two estimates are nearly identical. (Let γ be defined as the half-width-half-maximum of $\mathbf{G}^{-1}$. Then from Fig. 1 and the values of β in Fig. 3 we find $\gamma_2 = .105$ and $\gamma_4 = .099$.) Indeed, there does not appear to be any Bayesian preference for the type of local smoothing, and the marginal likelihoods for the quadratic and quartic estimates are nearly identical. Nevertheless, we prefer, and we will use, quartic local smoothing f@ΔȷPhe rest of the figures in this paper. In Bayesian language, a strong *hyperprior* favors higher order smoothing.

Figure 4 compares the optimal MQE estimate (dashed) with the optimal ME estimate (solid) which has no local smoothing. The ME estimate consists of spikes at the positions of the data, and it is not credible. In this case there is a strong Bayesian preference; the marginal likelihood of the optimal MQE estimate is 110 times larger than the marginal likelihood of the ME estimate. This observation poses a question: Why does ME often work extremely well for inverse problems? As discussed earlier, the smoothness of f is determined by a combination of the smoothness of U and the local smoothing. The U's for inverse problems consist of a sum of Lagrange multipliers multiplying point spread functions (or kernels), whereas the U's for density estimation are sums of δ-functions. Typical point spread functions are already locally smooth, so that additional local smoothing is much less important. However, MQE would still be preferred over ME for most inverse problems because it provides point estimates of errors on f.

The data in Figure 5 are measurements of the annual snowfall in Buffalo over a period of 63 years. The data are displayed as a histogram with 100 bins. The optimal MQE estimate (solid) consists of a single bump. This data set has been studied using almost all available density estimation methods, and the results are displayed in [Scott, 1993]. Almost all methods, with the exception of a cross validation kernel method, produce density estimates showing three bumps. Figure 6 shows a non-optimal MQE estimate (dashed) with three bumps obtained by tuning the local smoothing hyperparameter down from large $\beta^{1/4}$ to $\beta^{1/4} = 0.1$. The parameter α is still adjusted to maximize the marginal likelihood. However, the optimal MQE estimate with one bump is 23 times more likely (judged by the ratio of marginal likelihoods) than the non-optimal MQE estimate with three bumps. And the error estimates are as large as the bumps, so they have no statistical significance.

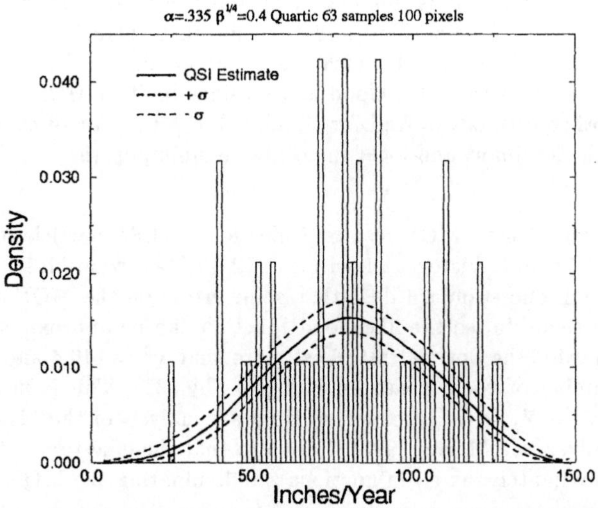

Figure 5: **Buffalo Snowfall** - *63 measurements of the annual snowfall in Buffalo are displayed as a histogram with 100 bins. The solid line is the optimal MQE estimate obtained with quartic local smoothing. Dashed lines are the ± one standard deviation error bars on the MQE estimate.*

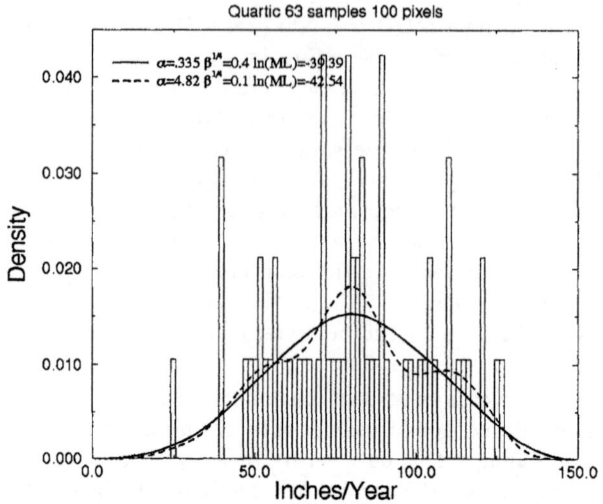

Figure 6: **Buffalo Snowfall** - *The solid line with a single bump is the optimal MQE estimate obtained at large β. The dashed line with three bumps is a non-optimal MQE estimate obtained by reducing $\beta^{1/4}$ to* 0.1. *The single bump estimate is 23 times more likely than the three bump estimate.*

The Buffalo snowfall is the only one of our three examples where optimal MQE agrees with the penalized likelihood method of Good and Gaskins using quartic smoothing. The equivalence means that Eq. (4) is dominated by the lowest ε_n eigenfunction. The only operative constraint is local smoothing and the quantum entropy is almost zero. This corresponds to a marginal likelihood which has a flat maximum for $0.4 \leq \beta \leq \infty$. For our other data sets, we find that this Good and Gaskins limit of MQE is not optimal and produces oversmoothed estimates. And for simulated f with a lot of sharp structure, the entropy constraint is dominant and local smoothing is unimportant.

Finally, Figs. 7 and 8 show MQE results (solid) for the LRL particle physics data. The data consist of 25752 counts histogrammed into 172 10 MeV wide bins. The gray area in Fig. 8 indicates the ± one standard deviation point errors on the MQE estimates. There are many counts in each bin, so the likelihood function can be approximately related to a χ^2 statistic. We find for the optimal MQE estimate that $\chi^2 = 146.4$ and that $N_g = 43.3$, where N_g is the number of good measurements given by (31). This is in rough agreement with the relation, $\chi^2 + N_g \approx N_{bins}$, expected from an analysis of the ML-II procedure for inverse problems [Silver and Martz, 1993]. Note also that the quantum entropy, $S_Q = 1.78$, indicates that approximately six eigenfunctions are dominating the MQE estimate in (4).

We regard these maximum quantum entropy (or quantum statistical inference) results for density estimation as very encouraging. The introduction of quantum entropy dramatically expands the potential applications for maximum entropy methods. Considerable further testing and development will be needed to realize the full potential of quantum methods for statistics, inverse problems, and image reconstruction.

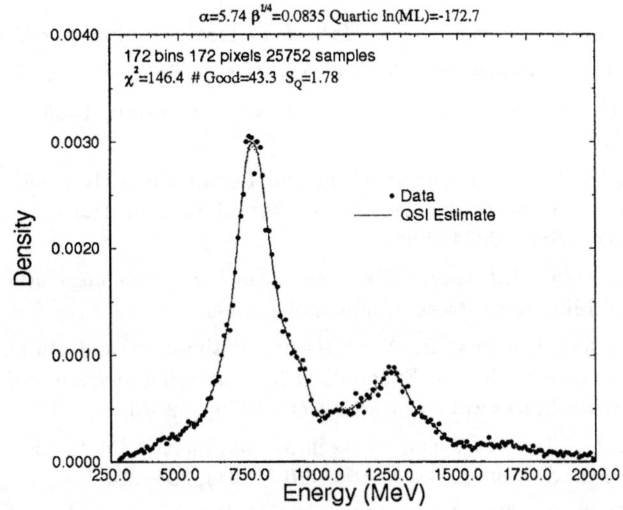

Figure 7: **LRL Particle Physics Data** - *Data consist of 25752 counts histogrammed into 172 10 MeV wide bins. The solid line is the optimal MQE estimate.*

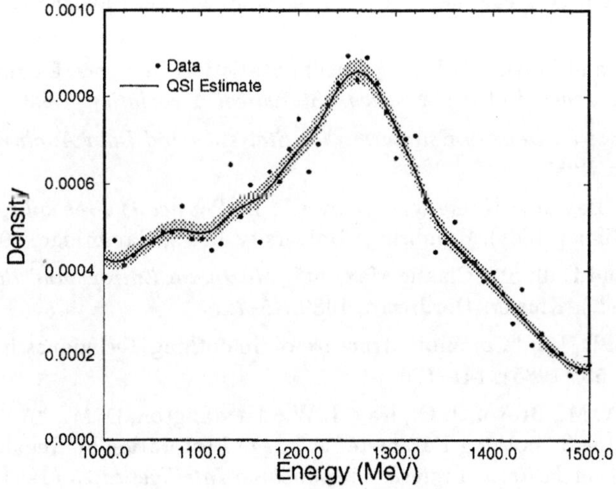

Figure 8: **LRL Particle Physics Data** - *Detail of Figure 7. The boundaries of the gray area are the ± one standard deviation errors on the MQE estimate.*

References

[1] Amari, S., *Differential-geometrical Methods in Statistics,* Springer-Verlag, Berlin, 1985.

[2] Balian, R., *From Microphysics to Macrophysics,* Springer-Verlag, Berlin, 1991.

[3] Berger, J. O., *Statistical Decision Theory and Bayesian Analysis,* Springer-Verlag, Berlin, 1985.

[4] Demoment, G., "Image Reconstruction and Restoration: Overview of Common Estimation Structures and Problems", *IEEE Transactions on Acoustics, Speech and Signal Processing* **37** (1989), 2024–2036.

[5] Good, I. J., *Good Thinking: The Foundations of Probability and Its Applications,* University of Minnesota Press, Minneapolis, 1985.

[6] Good, I. J., and Gaskins, R. A., "Density Estimation and Bump Hunting by the Penalized Likelihood Method Exemplified by Scattering and Meteorite Data'], *Journal of the American Statistical Association* **75** (1980), 42–73.

[7] Izenman, A. J., "Recent Developments in Nonparametric Density Estimation", *Journal of the American Statistical Association* **86** (1991), 205–224.

[8] Robinson, D. R. T., "Maximum Entropy with Poisson Statistics", *Maximum Entropy and Bayesian Methods* W. T. Grandy, L. H. Schick (eds.) Kluwer, Dordrecht (1991), 337–341.

[9] Scott, D. W., *Multivariate Density Estimation,* John Wiley & Sons, Inc., New York, 1993.

[10] Silver, R. N., "Quantum Statistical Inference", *Maximum Entropy and Bayesian Methods,* A. Djafari, G. Demoment, (eds), Kluwer Academic Publishers, Dordrecht, 1993, 167–182.

[11] Silver, R. N. and Martz, H. F., "Quantum Statistical Inference for Inverse Problems", submitted to *Journal of the American Statistical Association*, 1993.

[12] Silverman, B. W., *Density Estimation for Statistics and Data Analysis,* Chapman and Hall, London, 1986.

[13] Skilling, J., "Bayesian Numerical Analysis", In *Physics & Probability,* W. T. Grandy, Jr., P. W. Milonni (eds), Cambridge University Press, Cambridge, 1993, p. 207–222.

[14] Skilling, J. and Gull, S., "Classic MaxEnt", *Maximum Entropy and Bayesian Methods,* J. Skilling (ed.), Kluwer, Dordrecht, 1989, 45–71.

[15] Titterington, D. M., "Common Structure of Smoothing Techniques in Statistics", *Int. Statist. Rev.* **53** (1985), 141–170.

[16] Thompson, A. M., Brown, J. C., Kay, J. W., Titterington, D. M., "A Study of Methods of Choosing the Smoothing Parameter in Image Restoration by Regularization", *IEEE Transactions on Pattern Analysis and Machine Intelligence* **13** (1991), 326–339.

[17] von Neumann, J., *Gött. Nachr.* **273**, 1927.

[18] Wallstrom, T., "Generalized Quantum Statistical Inference", to be published, 1993.

[19] Wehrl, A., "General Properties of Entropy", *Reviews of Modern Physics* **50** (1978), 221–260.

BELIEF AND DESIRE

Anthony J. M. Garrett
Byron's Lodge, 63 High Street
Grantchester
Cambridge CB3 9NF
England

ABSTRACT. The sum and product rules are consequences of associating a single number with a conditioned proposition: the Boolean algebra of the propositions induces an algebra for the numbers, which with a few additional assumptions gives the sum and product rules. We call the associated number *probability*, and can interpret it as representing strength of belief that the proposition be true. The extra assumptions are empirical and cannot be derived from logical argument alone; consequently, probabilistic logic has a semi-empirical basis, and is not a pure consequence of consistency requirements. In addition to the capacity for belief we also have the capacity for desire: but numerical measure of the strength of desire that a conditioned proposition be true does not obey the same laws, because the additional assumptions which led for probability to the sum and product rules do not hold for desirability. The equations, whose solution gives a calculus of desirabilities, are derived. The notion of desirability clarifies the relative status of probabilistic inference and decision theory, in which probability and desirability are combined when the desirability that a proposition be true is conditioned on an action of our choosing. Desirability is then known – among other names – as *loss function*; however, it remains a valid concept when there is no choice of action. It is a valuable clarifying notion.

1. Propositions and Numbers Associated With Them.

Define a proposition as something which we perceive as either TRUE or FALSE. This is the bottom line, since if we try instead to think of something as taking either of two values (0 or 1, say, or Heads or Tails), it is still TRUE or FALSE that the variable takes one value and not the other; consequently, associated with the proposition is a *truth variable* whose value is either TRUE or FALSE (though we might not be certain which). Even if we are foolhardy enough to generalise the notion of a proposition to something we perceive as TRUE or FALSE or SOMETHING ELSE, whatever that may be, then we still see it as uniquely TRUE or FALSE that the generalised proposition is TRUE, or FALSE, or SOMETHING ELSE. To go any deeper we must investigate the nature of truth, which is not the aim here. Now, the strength of belief that a proposition is TRUE, conditional on the truth of other propositions, is what Bayesians mean by its *probability*. (Objectors to the Bayesian view should replace the word 'probability' throughout by 'belief-strength'; nothing else will change.) Since we can conceive of one strength of belief being greater than another, and since we require these to be transitive, an ordering exists: probabilities can be placed along a line, and brought into correspondence with real numbers ordered along that line. Probabilities are representable by real numbers.

The same argument, and real-number representation, applies to desirabilities – the strengths with which we *want* propositions to be true, conditioned on the truth of other

G. R. Heidbreder (ed.), Maximum Entropy and Bayesian Methods, 175–186.

propositions. In particular, if we are about to make one of several possible actions, the conditioning information might include the action we take. This is the first step in a decision theory, of which action to take; the concept of desirability makes clear the historically confused relationship between inference and decision theory.

Setting aside the interpretations which we, as beings capable of belief or desire, give to these numbers, the idea is to associate real numbers with conditioned propositions. The truth values of the propositions obey Boolean algebra, known to Aristotle and the ancient Greeks, and familiar today in the design of logic circuitry. This algebra of the propositions induces an algebra for the numbers associated with the propositions. In a robot whose circuitry is designed to simulate human action in uncertain conditions, these numbers will have some physical representation, such as voltage, and that voltage is a *number*. The first worker to explore the connection between the Boolean calculus of propositions, and the corresponding calculus of the numbers, was R.T. Cox, in 1946 [1]. We shall investigate this connection in detail here.

Cox has also advocated a calculus of *questions* [2], to which the truth of the propositions are the answers; the questions have a quantitative, numerical *bearing* on each other. This is not helpful, because we are interested in inference from given information, not given questions.

We divert briefly to discuss non-Boolean calculi of propositions. Much has recently been written, for example, on 'quantum logic', which focusses on 'both ... and' rather than 'either ... or'. Logic is a mode by which human beings perform reasoning and, in the absence of evidence that any particular logic is hard-wired into the brain (which operationally is just a neural net), it is perhaps culturally assigned. *I* have no idea what it means to say that something is both TRUE *and* FALSE -- that I am in both Cambridge and Oxford at the same instant, or that an electron is both spin-up *and* spin-down (these statements are neither more nor less unreasonable, since logic does not distinguish size) -- but it is possible that the statements would convey meaning to, say, a Martian. Of course, how someone brought up in a quantum culture communicates with someone brought up in an Aristotelian one remains a problem; hence 'quantum logic' will never be a satisfying solution to quantum paradoxes in our Aristotelian culture. Aristotelian cultures are perhaps better placed to study systematically the correlations, or patterns, occurring in the natural world ('science'). It seems, though, that logic is a *practical* subject: to see which logic is in use on any planet, we have to observe its logicians. This is in fact perfectly natural, since logic is concerned not with nature, but with the language we use to describe it: with epistemology, not ontology.

In analysing the differing logics, we perforce use our *own* logic; a process I call introspection. It might be that Aristotelian logic, uniquely, is singled out through introspection. The earlier comment about TRUE, FALSE and SOMETHING ELSE is a start on analysing this. If so, Aristotelian logic is compulsory, and has nothing to do with culture.

Ending this speculative diversion, we return to Boolean logic, and examine the corresponding numerical calculus. For a proposition X, the truth variable, which takes values either TRUE (T) or FALSE (F), will be denoted v_X. We begin by associating a single number with the truth of a proposition, conditioned on the truth of another; for proposition X, conditioned on the truth of proposition Z, denote this number $n[v_X = T|v_Z = T]$. The conditioning solidus is to be read as "supposing that", not "given that". We now employ the conventional shorthand of writing 'X' to mean $v_X = T$; it follows that $\overline{X}$ means $v_{\overline{X}} = T$, or, since $\overline{X}$ is the negative of proposition X, that $v_X = F$. Accordingly, our number is written as $n_{X|Z}$; it should be borne in mind that its dependence on X is parametric, through v_X. For joint propositions, expressed through the logical product, we take it that $n_{XY|Z}$ is expressible in terms of the four further numbers $n_{X|Z}$, $n_{Y|Z}$, $n_{X|YZ}$ and $n_{Y|XZ}$:

$$n_{XY|Z} = \mathcal{F}(n_{X|YZ}, n_{Y|Z}, n_{Y|XZ}, n_{X|Z}). \tag{1}$$

The commutativity and associativity properties of the truth values of propositions now induce constraints on the function $\mathcal{F}$. Commutativity ($XY = YX$, in shorthand) implies that $n_{XY|Z} = n_{YX|Z}$, so that, from (1),

$$\mathcal{F}(q, r, s, t) = \mathcal{F}(s, t, q, r). \tag{2}$$

Associativity, the condition that $A(BC) = (AB)C \equiv ABC$, implies that if we decompose the number corresponding to the triple product ABC in differing ways, using (1), the results must coincide. (Products of greater than three propositions are then automatically associative.) In decomposing $n_{ABC|D}$, define numbers for the twelve single conditioned propositions as

$$\begin{array}{rclcrcl}
\alpha & = & n_{A|BCD} & \quad & \tau & = & n_{A|D} \\
\beta & = & n_{B|ACD} & & \theta & = & n_{B|D} \\
\gamma & = & n_{C|ABD} & & \chi & = & n_{C|D} \\
\lambda & = & n_{A|BD} & & \eta & = & n_{A|CD} \\
\mu & = & n_{B|CD} & & \rho & = & n_{B|AD} \\
\nu & = & n_{C|AD} & & \xi & = & n_{C|BD}
\end{array} \tag{3}$$

One decomposition is

$$n_{ABC|D} = \mathcal{F}(n_{AB|CD}, n_{C|D}, n_{C|ABD}, n_{AB|D}), \tag{4}$$

which, on further decomposing the double products on the RHS, gives

$$n_{ABC|D} = \mathcal{F}(\mathcal{F}(\alpha, \mu, \beta, \eta), \chi, \gamma, \mathcal{F}(\lambda, \theta, \rho, \tau)). \tag{5}$$

An alternative decomposition is

$$\begin{aligned}
n_{ABC|D} &= \mathcal{F}(n_{A|BCD}, n_{BC|D}, n_{BC|AD}, n_{A|D}) && (6) \\
&= \mathcal{F}(\alpha, \mathcal{F}(\mu, \chi, \xi, \theta), \mathcal{F}(\beta, \nu, \gamma, \rho), \tau). && (7)
\end{aligned}$$

By equating the RHS's of (5) and (7) we obtain a functional equation for $\mathcal{F}$, involving the twelve variables (3). These variables are not independent of each other: the equation must be solved jointly with the functional equations obtained by permuting A, B and C in this analysis, and with the functional equations of commutativity, $n_{AB|D} = n_{BA|D}$ (and two permutations), and $n_{AB|CD} = n_{BA|CD}$ (and two permutations). There are no further simultaneous equations, although any generality in the solution for $\mathcal{F}$ must conform to certain special-case constraints; for example, if D tells us that C is true (the implication $D \to C$), which in truth values is $v_{CD} = v_D$, then CD can be replaced by D throughout, and $n_{C|D...}$ is the number corresponding to TRUTH of C.

The analysis of Cox [1] (which is in a specific context) tells us that a solution for $\mathcal{F}$ is

$$\mathcal{F}(q,r,s,t) = \Phi^{-1}(\Phi(q)\Phi(r)), \tag{8}$$

provided that Φ and its inverse Φ^{-1} are unique functions of their arguments, so that the corresponding mapping is 1:1. (Uniqueness is often wrongly held to mean that Φ is strictly monotonic; in fact, specific branches of non-monotonic functions suffice.) Direct verification readily indicates that (8) is a solution when n and Φ are multi-component entities, provided that the mapping remains 1:1. By analogy with (8), we recognise that further solutions are

$$\mathcal{F}(q,r,s,t) = \Phi^{-1}(\Phi(q)\Phi(s)) \tag{9}$$

and

$$\mathcal{F}(q,r,s,t) = \Phi^{-1}(\Phi(r)\Phi(t)). \tag{10}$$

The commutativity condition (2) gives a new equation relating q, r, s and t when the solution (8) is substituted into it. Solutions (9) and (10) are invariant under the symmetry (2).

We now exploit the commutativity relation (2) to write one of the four arguments of $\mathcal{F}$ in terms of the others. We can eliminate either of the 'correlated' pair $n_{X|YZ}$, $n_{Y|XZ}$, to give, for example,

$$n_{XY|Z} = \mathcal{F}^*(n_{X|YZ}, n_{Y|Z}, n_{X|Z}), \tag{11}$$

or either of the 'uncorrelated' pair $n_{X|Z}$, $n_{Y|Z}$, to give, for example

$$n_{XY|Z} = \mathcal{F}^{**}(n_{X|YZ}, n_{Y|Z}, n_{Y|XZ}). \tag{12}$$

Although this procedure breaks the $X \leftrightarrow Y$ symmetry, there are now only three arguments on the RHS's of (11) and (12). However, the functional equations for $\mathcal{F}^*$ or $\mathcal{F}^{**}$ are no simpler than for $\mathcal{F}$.

2. The Calculus of Probability.

Now let us be specific and interpret the number n as representing strength of belief; henceforth we denote it p. Let proposition Z include the implication "X is TRUE if Y is TRUE", or $Y \to X$; written in truth values, this is $v_{XY} = v_Y$. Then $v_X = T|v_{YZ} = T$, and $p_{X|YZ}$ represents the strength of belief one has in the truth of something that is certainly TRUE. This is clearly a unique number, independent of all details of proposition X, and we denote it p_T. The decomposition (11) becomes

$$p_{Y|Z} = \mathcal{F}^*(p_T, p_{Y|Z}, p_{X|Z}). \tag{13}$$

Since nothing is stated about Y – only about $X|Y$ – relation (13) must hold for arbitrary $p_{Y|Z}$, so that the RHS of (13) is independent of its third argument, and

$$\mathcal{F}^*(p_T, r, s) = r. \tag{14}$$

This relation constrains the function $\mathcal{F}^*$. There is no simplification or constraint upon taking Z to contain the converse implication "Y is TRUE if X is TRUE". By taking Z to

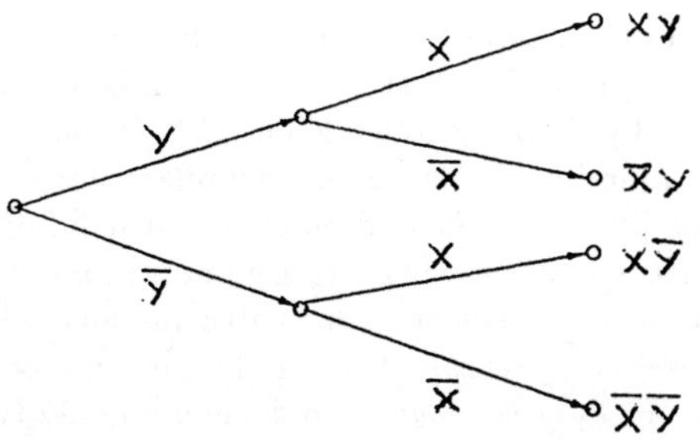

Figure 1: Interdependencies Among Logical Propositions.

tell us that Y is TRUE, it follows that $v_{XY} = v_X$, and that $v_{YZ} = v_Z$, so that (11) reduces to

$$\mathcal{F}^*(q, p_T, q) = q. \tag{15}$$

Upon taking Z to tell us that X is TRUE, a special case of (14) is all that results. We cannot take Z to tell us that Y is FALSE, since this would contradict the conditioning statement that $v_{YZ} = T$ in the first argument of $\mathcal{F}^*$ in (11). By taking Z to tell us that X is FALSE, (11) reduces to

$$\mathcal{F}^*(p_F, r, p_F) = p_F, \tag{16}$$

where p_F is the number assigned to FALSEHOOD, which like p_T is unique. No further constraints on the function $\mathcal{F}^*$ can be generated from special cases.

We now make an assumption: that $\mathcal{F}^*(q, r, s)$ is fully independent of its third argument, whether or not the first argument takes the special value p_T. This means that

$$p_{XY|Z} = \mathcal{F}^\dagger(p_{X|YZ}, p_{Y|Z}). \tag{17}$$

The ultimate justification for this step is practical: simplicity is preferable, and the logic system arising from (17), programmed into a robot, proves sufficient to emulate human logical reasoning. There is in addition a persuasive, albeit not conclusive, theoretical argument, which is illustrated using the tree diagram of Figure 1. In order that XY be TRUE, given Z, it is first necessary that Y be TRUE, given Z; consequently the probability $p(Y|Z)$ is required. This gets us halfway along the top path of Figure 1. In order to proceed to XY, we require now that X be TRUE, given Y (and Z), so that $p(X|YZ)$ is needed. If, however, Y is FALSE, then XY is FALSE whatever the status of X, so that $p(X|Z)$ is not needed. Figure 1 also indicates why we did not include negations, such as $p(X|\overline{Y}Z)$, in the RHS of (1).

Purported proofs of (17) by logic alone have been given by Tribus [3], and by Smith and Erickson [4] (from which Figure 1 is drawn). However, Tribus calls $\mathcal{F}^*$ by the names F_7

and F_9, and on examination simply gives no reason to discount either. Smith and Erickson call it F_5 (or F_{10}) and present two arguments: the first is that the constraint (14) causes the third argument of $\mathcal{F}^*$ to drop out whatever the value of the first argument, which is trivial to disconfirm by counter-example; the second is based on an assumption that the p's of two arbitrary conditioned propositions are equal, which is unjustified.

Nothing is made simpler by choosing to work with $\mathcal{F}^{**}$ (from (12)) instead of $\mathcal{F}^*$. Smith and Erickson [4] call this function F_6 and give a (similar) flawed argument for dropping the third argument; Tribus [3] calls it F_8 and objects that it becomes ill-defined in special circumstances which, in fact, condition one of its arguments upon both truth and falsehood of the same proposition at once. This is enough to give any algorithm indigestion: that the error is immediately exposed is a valuable warning flag, not a vice.

It is easy to show that, if $p_{XY|Z}$ is *assumed* to depend on only two of $p_{X|YZ}, p_{Y|Z}, p_{X|Z}$, then (17) is the only choice which makes any sense: for example, dependence on only $p_{Y|Z}$ and $p_{X|Z}$ fails to allow for correlations between Y and X, so that any interpretation of the numbers in which we employ correlations – such as probability – cannot be described. This argument is not a special case like (14), but is, again, *practical*.

Our assumption that $p_{XY|Z}$ depends only on $p_{X|YZ}$ and $p_{Y|Z}$ (or equally on the 'other' pair $p_{Y|XZ}$ and $p_{X|Z}$) is where Cox [1] came in. The assumption causes the function $\mathcal{F}^\dagger$ to decouple from $\mathcal{F}$: by equating (5) and (7) and suppressing the last two arguments of $\mathcal{F}$, we have

$$\mathcal{F}^\dagger(\mathcal{F}^\dagger(\alpha,\mu),\chi) = \mathcal{F}^\dagger(\alpha,\mathcal{F}^\dagger(\mu,\chi)) \tag{18}$$

where α, μ and χ can be varied independently. This equation, known appropriately in the lore of functional equations as the associativity equation, has a long history [5]. Its general solution was quoted at (8):

$$\mathcal{F}^\dagger(q,r) = \Phi^{-1}(\Phi(q)\Phi(r)), \tag{19}$$

where Φ and its inverse, Φ^{-1}, are unique but otherwise arbitrary functions. The solution is exchangeable with respect to q and r. The easiest derivation of (19) proceeds by some elegant sleight of hand involving differentiation with respect to the arguments of $\mathcal{F}^\dagger$, and is given by Cox [1] and Tribus [3]. Direct substitution immediately verifies, however, that either side of (18) is equal to $\Phi^{-1}(\Phi(\alpha)\Phi(\mu)\Phi(\chi))$; since this procedure does not involve differentiation, (19) is a solution – though not necessarily the general solution – irrespective of differentiability. (A tentative solution can always be made rigorous by successful verification!) That (19) is fully general is proved by Aczél [5], who constructs it as the general solution using a standard iteration technique from the theory of functional equations; a tutorial exposition of this analysis, eschewing the crabbed language of modern pure mathematics, is given by Smith and Erickson [4].

In summary, we write our solution of (17) in the form

$$\Phi(p_{XY|Z}) = \Phi(p_{X|YZ})\Phi(p_{Y|Z}). \tag{20}$$

By deleting the third argument of $\mathcal{F}^*$ in the special results (14) and (15), we have the further condition

$$\mathcal{F}^\dagger(p_T,r) = \Phi^{-1}(\Phi(p_T)\Phi(r)) = r, \tag{21}$$

from which it follows, operating with Φ, that $(\Phi(p_T)-1)\Phi(r)=0$ for all r, so that

$$\Phi(p_T) = 1. \tag{22}$$

Likewise, (16) tells us that

$$\Phi^{-1}(\Phi(p_F)\Phi(r)) = p_F, \tag{23}$$

and by operating on both sides with Φ we have $\Phi(p_F)(\Phi(r) - 1) = 0$, so that

$$\Phi(p_F) = 0. \tag{24}$$

Since our results involve p and Φ only in the combination $\Phi(p\ldots)$ we can, without any change of content, absorb the arbitrary function Φ into the number p. For example, suppose that $\Phi(r) = \frac{1}{100}r$; then (20) becomes $100p_{XY|Z} = p_{X|YZ}p_{Y|Z}$, and we have $p_T = 100$, $p_F = 0$. Here, p represents probability on a scale from 0 to 100, as in percentages; but the content of the theory is clearly the same as on a scale from 0 to 1. Henceforth we denote $\Phi(p\ldots)$ by P, so that (20), (22) and (24) become

$$P_{XY|Z} = P_{X|YZ}P_{Y|Z}, \tag{25}$$

$$P_T = 1, \qquad P_F = 0. \tag{26}$$

What has happened is that the arbitrariness in the representation of strength of belief - if p represents it, so does any function of p - has been reduced: we work with that representation which satisfies (25) and (26). Arbitrariness has not been completely removed, however, for (25) and (26) are invariant under $p \to p^\ell$, and this lesser ambiguity persists further in the analysis.

We recognise (25) as the product rule, and (26) as the standard values for probabilities of propositions known to be TRUE or FALSE; these values arise from the analysis, and are not merely conventions. Bayes' theorem follows on interchanging X and Y in (25) and equating the results:

$$\frac{P_{X|YZ}}{P_{X|Z}} = \frac{P_{Y|XZ}}{P_{Y|Z}}. \tag{27}$$

This is in fact the relation corresponding to (2) above. It states that if TRUTH of Y makes more probable the TRUTH of X, then the converse holds. Bayes' theorem corresponds to a great deal of intuitive reasoning [3].

We now make a further assumption: that the probability corresponding to TRUTH of a proposition can be extracted if we know the probability corresponding to TRUTH of its negation. This is clearly a desirable feature of our probability theory:

$$P_{X|Z} = G(P_{\overline{X}|Z}). \tag{28}$$

Since double negation returns a proposition to itself, the double application of G is the unit operator: write $X = \overline{W}$ in (28) and make the temporary change in notation $P_{X|Z} \to P(X)$, so that $P(\overline{W}) = G(P(\overline{\overline{W}})) = G(P(W)) = G(G(P(\overline{W})))$, or $G(G(r)) = r$, or $G(r) = G^{-1}(r)$: G is its own inverse. This condition is satisfied by any single-valued function $G(x)$ which is unchanged on reflection in the line $y = x$. We also require, from (26), that $G(1) = 0$, $G(0) = 1$.

Further conditions on the function G follow from the requirement that the logical *sum* of propositions, which is related to negation through the logical relation $\overline{A} + \overline{B} = \overline{AB}$, be commutative and associative. First, commutativity: we use (28), and (25) and (27), to express $P_{X+Y|Z}$ in terms of $P_{X|Z}$, $P_{Y|Z}$ and $P_{XY|Z}$, and demand that the result be exchangeable in X and Y. Using our revised notation, we have

$$\begin{aligned} P(X+Y) &= P(\overline{\overline{X}\,\overline{Y}}) && (29)\\ &= G(P(\overline{X}\,\overline{Y})) && (30)\\ &= G(P(\overline{Y})P(\overline{X}|\overline{Y})) && (31)\\ &= G(G(P(Y))\,G(P(X|\overline{Y}))) && (32)\\ &= G\Big(G(P(Y))\,G\Big(\frac{P(X)P(\overline{Y}|X)}{P(\overline{Y})}\Big)\Big) && (33)\\ &= G\Big(G(P(Y))\,G\Big(\frac{P(X)G(P(Y|X))}{G(P(Y))}\Big)\Big) && (34)\\ &= G\Big(G(P(Y))\,G\Big(\frac{P(X)G(\frac{P(XY)}{P(X)})}{G(P(Y))}\Big)\Big). && (35) \end{aligned}$$

Upon defining $x \equiv P(X)$, $y \equiv P(Y)$, $m \equiv P(XY)$, and exchanging X and Y and equating the result to (35), we have the functional equation

$$G(y)G\Big(\frac{xG(\frac{m}{x})}{G(y)}\Big) = G(x)G\Big(\frac{yG(\frac{m}{y})}{G(x)}\Big), \tag{36}$$

where x, y and m are all independent. To solve this we put $m = 0$ (recall that $G(0) = 1$), giving the equation

$$G(y)G\Big(\frac{x}{G(y)}\Big) = G(x)G\Big(\frac{y}{G(x)}\Big). \tag{37}$$

Define further $x' = G(x)$, $y' = G(y)$, so that, on combining (37) with self-reciprocity of G, we have

$$y'G\Big(\frac{G(x')}{y'}\Big) = x'G\Big(\frac{G(y')}{x'}\Big). \tag{38}$$

Cox [1] and subsequently Tribus [3] derived equation (38) in less direct ways. They then showed that the solution which satisfies $G(1) = 0$, $G(0) = 1$ (and self-reciprocity, which follows upon putting one of x' and y' to unity in (38)) is

$$G(r) = (1 - r^{\ell})^{1/\ell} \tag{39}$$

where ℓ is arbitrary. Aczél [5] again derives (39) without assuming differentiability. It is readily verified that this solution satisfies the full functional equation (36) (we would be in deep trouble if it didn't), so that (28) becomes

$$P(X)^{\ell} + P(\overline{X})^{\ell} = 1 \tag{40}$$

and (35) simplifies to

$$P(X+Y)^{\ell} = P(X)^{\ell} + P(Y)^{\ell} - P(XY)^{\ell}. \tag{41}$$

We take $\ell = 1$ and so define uniquely the representation of strength of belief, or probability, with which we work. Written in full, (40) and (41) become

$$P_{X|Z} + P_{\overline{X}|Z} = 1, \tag{42}$$

$$P_{X+Y|Z} + P_{XY|Z} = P_{X|Z} + P_{Y|Z}. \tag{43}$$

Equation (42) is the well-known sum rule of probability. The result (43), which reduces to (42) if $Y = \overline{X}$, automatically ensures associativity of the logical sum, since

$$\begin{aligned} P_{A+(B+C)|D} &= P_{A|D} + P_{B+C|D} - P_{A(B+C)|D} & (44)\\ &= P_{A|D} + P_{B|D} + P_{C|D} - P_{BC|D} - P_{AB+AC|D} & (45)\\ &= P_{A|D} + P_{B|D} + P_{C|D} - P_{BC|D} - P_{AB|D} - P_{AC|D} + P_{ABAC|D} & (46)\\ &= P_{A|D} + P_{B|D} + P_{C|D} - P_{BC|D} - P_{AB|D} - P_{AC|D} + P_{ABC|D}, & (47) \end{aligned}$$

which is exchangeable with respect to A, B and C. The step from (46) to (47) follows from the logical relation $v_{AA} = v_A$.

It is often useful, in Bayes' theorem (27), to use the sum rule to rewrite $P_{Y|Z}$, on the RHS, as

$$\begin{aligned} P_{Y|Z} &= P_{Y|Z}(P_{X|YZ} + P_{\overline{X}|YZ}) & (48)\\ &= P_{XY|Z} + P_{\overline{X}Y|Z} & (49)\\ &= P_{Y|XZ}P_{X|Z} + P_{Y|\overline{X}Z}P_{\overline{X}|Z}. & (50) \end{aligned}$$

Propositions of the type "the tree is between height h and $h + dh$" enable quantitative and continuous parameters to be handled.

We have now derived the product and sum rules of probability. Our derivation has been based on the identification of a probability with the TRUTH of a conditioned proposition: the Boolean algebra of the propositions then induces an algebra for the probabilities, and this gives the product and sum rules. However, some further assumptions based on experience, not logic, had to be made: that $p_{XY|Z}$ need depend only on two of $p_{X|Z}$, $p_{Y|Z}$, $p_{X|YZ}$ and $p_{Y|XZ}$. Also, it was taken that the probabilities corresponding to TRUE and FALSE propositions were independent of the nature of the proposition; and that the probability of a proposition *can* be expressed uniquely in terms of the probability of its negation. All well and good; but now let us turn our attention from probability to *desirability*.

3. Desirability, and Decision Theory.

In fact, desirability does not obey Bayes' theorem. To see this, we compare experimentally Bayes' theorem for desirabilities with practical, human reasoning. Let X denote the proposition "I win a large sweepstake", let Y denote "I have debts to Mr. Jones", and let Z denote "Mr. Jones is a gangster". Does desirability d obey

$$\frac{d_{X|YZ}}{d_{X|Z}} = \frac{d_{Y|XZ}}{d_{Y|Z}} \ ? \tag{51}$$

Clearly desirability can be positive or negative; the negative of desirability is undesirability. Now, it is positively desirable to win a sweepstake; and very much more so when one has debts to a gangster: the LHS of (51) (considerably) exceeds unity. On the RHS, $d_{Y|XZ}$ is weakly negative; it is not pleasant to have debts, but not serious if they can easily be paid off. By contrast, $d_{Y|Z}$ is strongly negative. So the RHS is positive but (considerably) less than unity, and not equal to the LHS. Evidently some of the assumptions specific to probability, made in the passage to Bayes' theorem (27), fail for desirability: there is an asymmetry in the relation between X and Y.

Asymmetry is a warning that desirability is very different from probability. If $d_{XY|Z}$ is taken to obey the counterpart of (1) and be expressible as a function of $d_{X|YZ}$, $d_{Y|Z}$, $d_{Y|XZ}$ and $d_{X|Z}$ then, although commutativity still enables us to eliminate one of these, we cannot make the further assumption – leading to the product rule and Bayes' theorem – that dependence is on $d_{X|YZ}$ and $d_{Y|Z}$ alone. To make progress we must tackle the full set of equations for $\mathcal{F}$, referred to after (7). This is an important task for the future.

Further differences between desirability and probability emerge. It is not the case that the desirability of a proposition known to be TRUE (or FALSE) is independent of what the proposition is, as happens for probability ($P = 1$ or 0). Therefore the solutions of the functional equations must be qualitatively more subtle. They must always be invariant under the arbitrary transform $d \to \Phi(d)$, however, since if d is a numerical representation of strength of desire then so is $\Phi(d)$. To make progress it is best to be aware of these problems, but to confront them only as they arise in seeking solutions of the functional equations for $\mathcal{F}$.

Also, there is no reason why $d_{\overline{X}|Z}$ should be deducible from $d_{X|Z}$, or the converse. Clearly the sum rule itself is violated by desirabilities: intuitively it is highly desirable to win a raffle, but only weakly undesirable not to. For desirabilities we can expect only to write $d_{X+Y|Z}$, like $d_{XY|Z}$, in terms of $d_{X|Z}$, $d_{Y|Z}$, $d_{X|YZ}$ and $d_{Y|XZ}$. Associativity then constrains this dependence; the corresponding functional equation demands study.

So far we have assumed that probability and desirability are mutually decoupled. Practice tells us that probability is indeed decoupled from desirability: it is universally held as an error to let what you want to believe influence what you should believe. Wishful thinking, although as old as the human race, is pathological. But is desirability decoupled from probability? Perhaps the reason why it is highly desirable, but only weakly undesirable, to win a raffle is because it is highly improbable that you win, but highly probable that you don't; as fewer people enter the raffle and your probabilities change, it seems more undesirable to lose. The issue is not clear.

When we have a choice of actions, the desirability that a particular proposition be TRUE often depends on the action we take. For example, choosing to bet on Heads, on a single throw of a coin, makes it more desirable to us that Heads comes up. There accordingly exists a calculus of which choice to make: a theory of *decision*. We take it as axiomatic that our purpose is to maximise the expected payoff, in money or desirability. Clearly probability is involved somewhere, since if we glean secret information that the coin is biased, we bet on the face to which we assign higher probability. The English language includes some single words which express this combination of probability and (negative) desirability: *risk*, for example. By considering the assignment of probability, desirability and risk to a conditioned proposition, it becomes clear that coupling of desirability to probability is permitted by the equations of commutativity and associativity which arise when *two* numbers are attached to each conditioned proposition. We expect to find solutions with one number – the probability – decoupled; but symmetry with respect to the two numbers will be broken.

Using betting to illustrate decision theory has both virtues and vices. The virtue is that it is easy to comprehend that more money is more desirable: the relation, although not necessarily linear, is readily understood to be monotonic. The vice is that one places one's bets with a bookmaker, who chooses what odds to quote; this extra choice detracts from the decision-theoretic problem faced by the punter, because it is game-theoretical. Here, though, we are looking only at the punter's interests when faced with quoted odds. (Further game-theoretical factors apply when taking account of handicapping. Be warned also that the word 'odds' is ambiguous: the bookmaker assigns his own probabilities and his own odds $O \equiv P/(1-P)$; but the 'odds' he quotes are different from these and do not satisfy the corresponding normalisation condition, in order to give him the expectation of profit from equally informed punters and those worse informed. A less fundamental derivation of the sum and product rules has been based on this [6].) A comprehensive – and valuable – formalism exists for distributing a stake over several horses, given your probabilities and the odds quoted at you; here, though, we concentrate on the ideas, rephrasing them using the clarifying concept of desirability. Desirability can also be called utility or, when there is a choice of actions, the loss function. (In search theory, for example, there is a choice of paths, whose desirabilities might be proportional to their lengths.) Like probability, desirability goes by many disguises, and the names depend on whether the inventor is more optimistic or more pessimistic, and more 'objectivist' or 'subjectivist', in outlook. The name *desirability*, like probability, correctly conveys the idea of a number, assigned objectively and interpreted by human consciousness. This is the key to its usefulness as a tutorial concept.

Denote by B_h and B_t the propositions "I bet on Heads" (or Tails), and by R_h and R_t the propositions "the result is Heads" (or Tails). Then, given our further information I, we construct the four desirabilities $d(R_h|B_hI)$, $d(R_h|B_tI)$, $d(R_t|B_hI)$, $d(R_t|B_tI)$, and our probabilities $P(R_h|I)$, $P(R_t|I)$, and set up the expected desirability betting on Heads,

$$< d_h >= d(R_h|B_hI)P(R_h|I) + d(R_t|B_hI)P(R_t|I), \tag{52}$$

and on Tails,

$$< d_t >= d(R_h|B_tI)P(R_h|I) + d(R_t|B_tI)P(R_t|I). \tag{53}$$

We decide to bet on Heads or Tails according to whichever of these is the larger.

Parameter estimation is an important example of decision theory. The end result of any probabilistic calculation of a parameter is always a probability distribution or density, and if we propose to choose a single 'best value' we must specify what we mean by 'best': best for what? By *best* we mean *most desirable*, and desirabilities are assigned according to the task at hand. We choose the value of the parameter, corresponding in (52), (53) to B_h or B_t, according to a calculation of this type; as a simple example, the most probable value – the mode – corresponds to a δ-function desirability peaked at the greatest probability. (This is also what is involved in comparative hypothesis testing.) Finally, whether or not a result is 'significant' is a decision, which must involve desirabilities and the information from which they are assigned; the question is otherwise incomplete.

Conceptually, that is all there is to the decision process. However, we have skated over the matter of how to assign desirability in the first place, having examined only the interrelation of logically related propositions. Assignment of desirability is utterly an unsolved problem; while probability is assigned through a symmetry principle called Maximum Entropy [3,7], no corresponding principle is known for desirability. The usefulness of the concept of desirability rests largely on the fact that it transcends decision theory, remaining useful even when one has no choice of action since there is still a legitimate amount to want something. This in turn clears up the confused relationship between Bayesian probabilistic inference, on the one hand, and decision theory on the other. Following the influential book of Wald [8], published in 1950, probabilistic inference came to be seen by many as a part of decision theory. Others, particularly in the modern 'Maximum Entropy – Bayesian' fraternity, see decision theory as a trivial tack-on to probability theory. By facilitating dialogue, the concept of desirability indicates that the truth lies in between: decision theory *is* additional to probability, but non-trivially: indeed, its internal logic is so much more complicated than probability that it is still hidden.

4. Conclusions.

We have seen that inductive, probabilistic, logic is not a purely theoretical construct, but depends at some points in its construction on observational comparison with human reasoning. The notion, parallel to probability, of desirability that a proposition be TRUE, has been introduced; the structure of the calculus for desirabilities is more complicated than for probabilities, and solution of the full functional equations for $\mathcal{F}$ is a key task for the future. When desirability is conditioned on actions over which we have choice, it corresponds to the loss function of decision theory; decision-making proceeds, as usual, by combining this function with the probabilities. The idea of desirability clears up confusion over the relative standing of probabilistic inference and decision theory.

References

[1] Cox, R.T. 1946. Probability, Frequency and Reasonable Expectation. *American Journal of Physics* **14**, 1-13.

[2] Cox, R.T. 1979. Of Inference and Inquiry. In: R.D. Levine and M. Tribus (eds), *The Maximum Entropy Formalism.* MIT Press, Cambridge, Massachusetts, U.S.A. pp119-167.

[3] Tribus, M. 1969. *Rational Descriptions, Decisions and Designs.* Pergamon, New York, U.S.A.

[4] Smith, C.R. and Erickson, G.J. 1990. Probability Theory and the Associativity Equation. In: P.F. Fougère (ed), *Maximum Entropy and Bayesian Methods, Dartmouth, U.S.A., 1989.* Kluwer, Dordrecht, Netherlands. pp17-30.

[5] Aczél, J. 1966. *Lectures on Functional Equations and Their Applications.* Academic Press, New York, U.S.A.

[6] de Finetti, B. 1974. *Theory of Probability*, Vols. 1 and 2. Wiley, New York, U.S.A. (English translation.)

[7] Jaynes, E.T. 1983. *E.T. Jaynes: Papers on Probability, Statistics and Statistical Physics.* R.D. Rosenkrantz (ed). Synthese Library **158**. Reidel, Dordrecht, Netherlands.

[8] Wald, A. 1950. *Statistical Decision Functions.* Wiley, New York, U.S.A.

A BAYESIAN GENETIC ALGORITHM FOR CALCULATING MAXIMUM ENTROPY DISTRIBUTIONS

Neil Pendock
Department of Computational and Applied Mathematics
University of the Witwatersrand
Johannesburg, South Africa
e-mail : 076neil@witsvma.wits.ac.za

ABSTRACT. The principle of maximum entropy [MaxEnt] is a powerful information-theoretic tool for solving inverse problems. MaxEnt solutions are the most honest answers to ill-posed inverse problems in that they have the least amount of structure not inferred by the data. The practical estimation of maximum entropy distributions is a difficult numerical problem due to the non-linearity of the entropy functional and the large number of parameters to be estimated in most problems. A further complication is that continuous optimization schemes do not necessarily give the correct MaxEnt estimate when the underlying distribution is discrete. We show how a genetic algorithm may be constructed within a framework of Bayesian inference and used to efficiently search the high dimensional parameter distribution space and locate MaxEnt distributions. We illustrate the approach by presenting a genetic algorithm to solve Jaynes' dice problem : if we toss a die many times and count the average number of dots that show, what were the frequencies with which the different faces appeared?

1 The principle of maximum entropy

E.T. Jaynes was the first person to fully realize the power of maximizing entropy subject to data constraints [MaxEnt] as a means of solving ill-posed inverse problems. The method is now a well established technique of data analysis and finds wide-spread application. Loredo (1990) provides an overview of recent MaxEnt applications. There are many ways to justify MaxEnt :

- Shore and Johnson (1980) provide an axiomatic derivation for maximizing entropy from four principles of consistent inference as the only consistent criterion for choosing one solution to an ill-posed inverse problem, from many possibilities.
- Tikochinsky *et al* (1984) argue that entropy is the only appropriate selection algorithm for combining data from reproducible experiments.
- Shannon (1949) showed that entropy is the only consistent information measure for a discrete probability distribution and demonstrated how our knowledge of available information may be combined using mixing entropies and then used to solve the problem at hand.
- Kullback (1959) proposed an entropic method of statistical inference which parallels Jaynes' application of MaxEnt to solving problems in statistical mechanics.

G. R. Heidbreder (ed.), Maximum Entropy and Bayesian Methods, 187–195.

- The Bayesian way to tackle inverse problems is to express the posterior parameter distribution as the product of a parameter prior and data likelihood distribution. Many authors have proposed entropic priors from first (physical) principles for applications ranging from image reconstruction and image restoration [Skilling and Gull (1984)], to geophysics [Rietsch (1977)] and economics [Zellner and Highfield (1988)]. The Bayesian formulation produces a distribution of solutions and a particular one (e.g. the most likely) may be obtained by maximizing the posterior distribution. This is equivalent to MaxEnt if an entropic prior is chosen.

- Jaynes (1989) himself provides a convincing justification by means of his *entropy concentration theorem* :

If a random experiment has M possible outcomes and we perform the experiment N times, the frequencies of occurrence of each outcome $\{f_i\}$ have entropy $-\sum_i^M f_i \log f_i$. If we have m data d_j constraining the $\{f_i\}$ in the form of linearly independent constraints

$$\sum_i a_{ji} f_i = d_j \quad 1 \le j \le m < M$$

then $F\%$ of outcomes will have entropy outside the interval $[H_{max} - \nabla H, H_{max}]$ where $\nabla H = \chi^2_{M-m-1} \frac{F}{2N}$. The interpretation of this theorem is that, for large N, most feasible distributions $\{f_i\}$ will have an entropy close to H_{max}.

2 Jaynes' dice problem

A good illustration of the entropy concentration theorem and the MaxEnt method is Jaynes' Brandeis dice problem [Jaynes (1989)] which may also be regarded as a pathological inverse problem of estimating six parameters from two pieces of data :

If we toss a die N times and observe that the average number of spots up was 4.5, what were the frequencies $\{n_i\}$ with which the six different faces appeared? This statement gives us two expectation constraints on $\{n_i\}$ namely $\sum_i n_i = N$ and $\sum_i i\ n_i = 4.5\ N$ as well as the physical realizability constraints that $0 \le n_i \le N\ \forall i$ and that n_i be an integer.

There are many sets of $\{n_i\}$ which fit these constraints, in particular

26	88	134	185	246	321
34	85	130	178	244	329
54	79	114	166	240	347
29	94	128	177	241	331
39	88	123	172	240	338

all do. Each distribution of outcomes of tossing the die N times may be made in

$$W = \frac{N!}{n_1! \cdots n_6!}$$

ways. As $N \to \infty$, using Stirling's approximation for !, we see that $\frac{\log W}{N} \to -f' \log f$ where f is a 6-element frequency vector with entries $f_i = \frac{n_i}{N}$. Thus if we repeated the whole experiment many times, the distribution with maximum entropy would appear most often.

This is surely a reason for choosing that distribution with maximum entropy from the set of admissible distributions as the most reasonable answer to our ill-posed inverse problem. One way of finding the MaxEnt distribution is to

$$\begin{aligned} &Maximize \quad -f' \log f \\ &\quad subject\ to \quad Af = d \end{aligned}$$

where

$$A = \begin{pmatrix} 1 & 1 & 1 & 1 & 1 & 1 \\ 1 & 2 & 3 & 4 & 5 & 6 \end{pmatrix}$$

$f = (f_1, \cdots, f_6)'$ and $d = (1, 4.5)'$. Introducing two Lagrange multipliers $\lambda = (\lambda_1, \lambda_2)'$ we have to maximize

$$f' \log f + \lambda' (Af - d)$$

which has solution

$$f = e^{-A' \lambda}$$

where λ may be found by solving the nonlinear system of (two) equations

$$Ae^{-A' \lambda} = d$$

Linear systems are usually easier to solve than non-linear ones, so we could start by linearizing (1). Newton's method provides us with a method of doing this : choose a trial λ_0 and expand (1) in Taylor's series around λ_0 ignoring quadratic and higher order terms. The resulting linear system

$$Ae^{-A'\lambda_n} - (\lambda_{n+1} - \lambda_n)' A \left(Ae^{-A'\lambda_n}\right)' = d$$

may then be solved iteratively. Jaynes presents the MaxEnt solution

$$f = (0.0543, 0.0788, 0.1142, 0.1654, 0.2398, 0.3475)$$

with entropy $H_{max} = 1.61358$.

The above formulation may be used to solve the general linear ill-posed inverse problem. There are some drawbacks with our implementation, however. In the dice experiment proposed by Jaynes, the die was tossed $N = 1000$ times and the above solution is not realizable in that we would have had to observe 54.3 occurrences of the side with a single spot. In fact, we may estimate the frequencies f_i to as much accuracy as we like - clearly a ridiculous ability given that N is finite! Of course it is our *knowledge* of f_i given all available data that we may state with as much precision as we like, although in many applications, the optimal discrete realization which gives rise to the $\{f_i\}$ may be important. In fact, there is no guarantee that the MaxEnt distribution for a discrete problem is even close to the MaxEnt distribution for the equivalent continuous problem! For the dice tossed one thousand times, truncating the continuous solution produces an infeasible distribution $(54, 79, 114, 165, 240, 348)$. The correct answer [calculated by the program listed in the Appendix] is $(54, 79, 114, 166, 240, 347)$.

A second problem lies in the numerical estimation of λ. Newton's method is a local procedure and we have no guarantee of achieving a global optimum - in fact in the above implementation, we will climb the closest hill to our starting point λ_0.

The presense of noise in the data is also a reality which should not be ignored for most "real-world" inverse problems. What we really want is the MaxEnt parameter distribution which generates "mock" data consistent with the "real" data. A Bayesian genetic algorithm for estimating MaxEnt distributions may help in all three areas.

3 A genetic algorithm

Genetic algorithms [GAs] are stochastic optimization algorithms which attempt to maximize a fitness (objective) function by finding an optimal model (parameter distribution) or set of models. The essence of a GA is that model space is explored by establishing a population of models and allowing fit models to exchange information between themselves [reproduction]. We shall discuss the workings of a GA in a Bayesian framework which is a general approach to solving inference problems and illustrate their use in solving MaxEnt problems by considering Jaynes' dice problem :

A model corresponds to a particular realization of observed face frequencies $n_1, \cdots, n_6$ of the die. We shall code each model as a binary string of $6 \times k$ bits where k is the number of bits necessary to represent N, the number of times the die was tossed. We decided to only generate feasible models since the unconstrained model space has N^6 members and even the subspace containing all distributions $\{n_i\}$ with $\sum_i n_i = N$ has

$$\binom{N+5}{N}$$

models. A Fortran routine to generate feasible models is listed in the Appendix and was used to generate an initial population of one hundred models.

Models are selected for reproduction according to their fitness. We used the posterior parameter distribution as the fitness function. Bayes theorem states that

$$p(f|d) = \frac{p(f)\, p(d|f)}{p(d)}$$

In view of Jaynes' entropy concentration theorem, the entropic prior

$$p(f) \propto \exp(-f' \log f)$$

is appropriate. The same prior is arrived at by adopting the criterion that a prior should be as uninformative as possible. With no prior information, Bernoulli's principle of insufficient reason would assign a uniform prior likelihood $p_i = \frac{1}{6}$ to the expected frequencies. We know that f cannot be uniform but an uninformative prior for f could be achieved by minimizing the distance between f and p. Kullback (1959) showed that the appropriate metric between two distributions which are frequencies of reproducible trials is the minimum information discrimination statistic or cross entropy

$$\sum f_i \log \frac{f_i}{p_i}$$

which is equivalent to an entropic prior for uniform $\{p_i\}$.

The data likelihood distribution is unity for our initial model population since they all fit the data exactly although their offspring may stray from feasibility and must be punished, as is the case in natural populations! To achieve this we made an *ad hoc* decision to use the n-norm euclidean distance with associated likelihood distribution

$$p(d|f) \propto \exp(-|\sum_i i\, f_i - 4.5|^n)\ \exp(-|\sum_i f_i - 1|^n)$$

This generalized Gaussian distribution weights both data constraints the same. The normalization constants have been ignored for the above two distributions as they will be incorporated into the constant $p(d)$.

The model population fitness function was chosen to be the logarithm of the posterior parameter distribution, i.e.

$$-f' \log f - |\sum_i i\, f_i - 4.5|^n - |\sum_i f_i - 1|^n$$

n may be interpreted as a weighting factor for the data likelihood $p(d|f)$ against the prior $p(f)$. The data d define a point in $m-$dimensional data space. Each model f generates another point in data space with a likelihood $p(d|f)$. The relative importance of this point as compared to its *a priori* likelihood $p(f)$ depends on n. For most real (noisy) data the feasible region consists of an $m-$dimensional cloud containing d. The radius of this cloud may be modulated by n. In the Jaynes' dice problem, if we choose a less robust metric (i.e. n large) the requirement that $\sum f_i = 1$ and $\sum i\, f_i = 4.5$ exactly is relaxed and we explore a larger model space, looking for MaxEnt distributions.

The GA evolves as follows : two models are selected by drawing random samples from $p(f|d)$. A cross-over point is selected at random and the two parent models exchange right-most bit strings. The two least fittest models are removed from the population and the process continues. Mutation, i.e. the flipping of a randomly chosen bit from a randomly chosen model occurs with a probability of 0.01 and prevents stagnation of the model population around a local maximum of the fitness function. The mutation process is akin to a random walk through model space. Figure 1 shows how the overall performance [maximum model fitness] of the GA degrades as the process of generating new models becomes less evolutionary and more random.

The above GA has a simple form and performs remarkably well. We assumed the die was tossed one thousand times ($N = 1000$) and the GA was allowed to run for one hundred thousand generations. Results for two different data likelihood functions (generalized Gaussians with n=1 and n=2, respectively) are presented in tables 1 and 2. As the data constraint is relaxed (table 2), higher entropy models are located.

4 Conclusions

GAs have several advantages over deterministic procedures for the estimation of MaxEnt distributions :

- GAs are global optimization procedures which efficiently search high dimensional parameter space for optimal distributions.

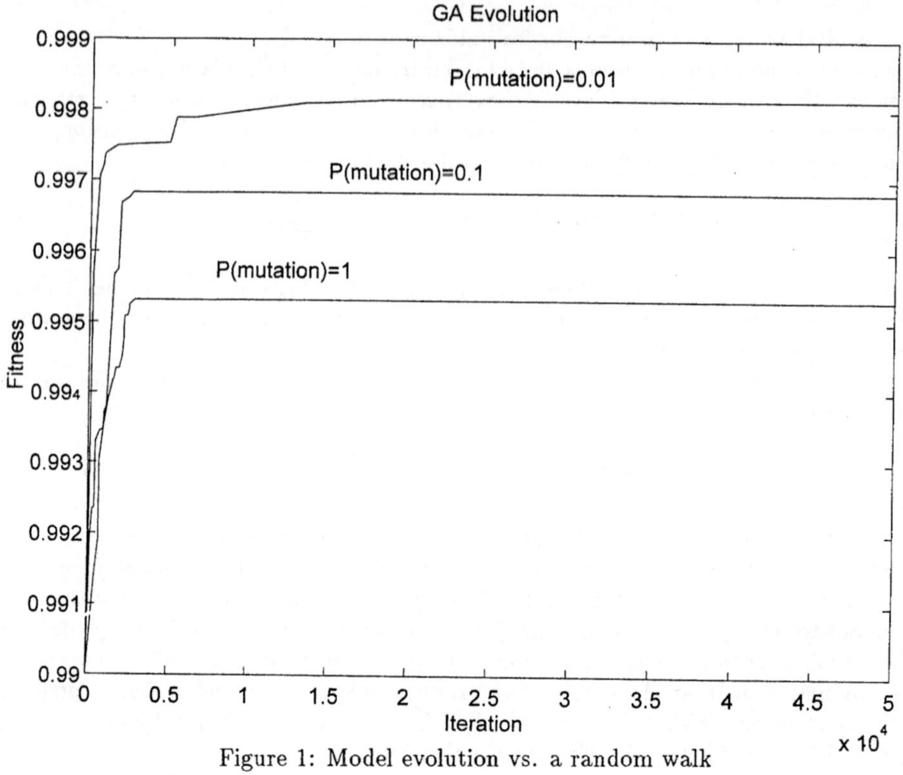

Figure 1: Model evolution vs. a random walk

Generation	entropy	$\sum_i n_i$	$\sum_i in_i$	n_1	n_2	n_3	n_4	n_5	n_6
1305	1.6010	1001	4503	57	72	85	196	283	308
25931	1.6135	1001	4492	58	87	106	166	226	358
35291	1.6155	1001	4495	55	90	106	166	226	358

Table 1: Fit models for $n = 1$

Generation	entropy	$\sum_i n_i$	$\sum_i in_i$	n_1	n_2	n_3	n_4	n_5	n_6
91	1.6129	1004	4464	68	120	85	129	227	375
164	1.6147	1008	4488	76	82	85	201	195	369
27028	1.6159	1006	4507	66	88	101	142	260	349
89584	1.6304	1009	4488	64	88	101	172	247	337

Table 2: Fit models for $n = 2$

- The computer code to implement a GA is simple. Computer storage is modest and numerical instability is not an issue.
- The presense of noise in the data may be easily incorporated using a Bayesian formulation of the fitness function in terms of a data likelihood and parameter prior distribution.

MaxEnt is a powerful technique for data analysis which may be used to produce *honest* answers to ill-posed inverse problems utilizing all available prior information and data while remaining uncommited about structure in the solution about which nothing is known. The Bayesian formulation is a convenient way of designing the fitness function and model selection criteria for a genetic algorithm.

References

[1] Jaynes, E.T. (1989). Papers on probability, statistics and statistical physics, R.D. Rosenkrantz (ed.), Kluwer.

[2] Kullback, S. (1959). Information theory and statistics, Wiley, 1959.

[3] Loredo, T.J., (1990). From Laplace to supernova SN 1987A : Bayesian inference in astrophysics, in *Maximum Entropy and Bayesian Methods*, P F Fougère (ed.), Kluwer.

[4] Rietsch, E. (1977). The maximum entropy approach to inverse problems, *J. Geophysics*, **42**, 489-506.

[5] Shannon, C.& W. Weaver, (1949). The mathematical theory of communication, Univ. Illinois Press.

[6] Shore, J.E. & R.W. Johnson, (1980). Axiomatic derivation of the principle of maximum entropy and the principle of minimum cross-entropy, *IEEE Trans. Info. Th.* **IT-26**, 26-37.

[7] Skilling, J. & S. F. Gull, (1984). Maximum entropy method in image processing, *Proc. IEE* **131**.

[8] Tikochinsky, Y., Tishby, N.Z. & R.D. Levine, (1984). Consistent inference of probabilities for reproducible experiments, *Phys. Rev. Let.* **52**, 1357-1360.

[9] Zellner, A. & R. Highfield (1988). Calculation of maximum entropy distributions and approximation of marginal posterior distributions. *J. Econometrics* **37**, 195-209.

5 Appendix

```
      subroutine pick(m1, m2, n)
c
c     Fortran routine to generate consistent samples n
c     subject to the constraints
c     n(1) + n(2) + n(3) + n(4) + n(5) + n(6) = m1
c     n(1) + 2*n(2) + 3*n(3) + 4*n(4) + 5*n(5) + 6*n(6) = m2
c
      integer n(6), seed
      data seed/3171/

1     k1 = m1
      k2 = m2
      do i=1,4
            n(i) = int(RAN(seed)*k1) + 1
            k1 = k1 - n(i)
            k2 = k2 - i*n(i)
            if (k1 .lt. 0 .or. k2. lt. 0) goto 1
      enddo
      n(5) = 6*k1 - k2
      n(6) = k2 - 5*k1
      if (n(5) .lt. 0 .or. n(6) .lt. 0) goto 1
      return
      end
```

```
      parameter (nthrow = 1000, nspot = 4500, emax = 1.6)
      integer n(6)
c
c     Fortran program to exhaustively search for the MaxEnt
c     solution to Jaynes' Brandeis dice problem
c
      do 4 i1 = 0,nthrow
              n(1) = i1
              k2 = nthrow - i1
              l2 = nspot - i1
              do 3 i2 = 0, k2
                      n(2) = i2
                      k3 = k2 - i2
                      l3 = l2 - 2*i2
                      if (l3 .lt. 0) goto 4
                      do 2 i3 = 0, k3
                              n(3) = i3
                              k4 = k3 - i3
                              l4 = l3 - 3*i3
                              if (l4 .lt. 0) goto 3
                              do i4 = 0, k4
                                      n(4) = i4
                                      k5 = k4 - i4
                                      l5 = l4 - 4*i4
                                      if (l5 .lt. 0) goto 2
                                      n(5) = 6*k5 - l5
                                      if (n(5) .lt. 0) goto 2
                                      n(6) = l5 - 5*k5
                                      if (n(6) .lt. 0) goto 2
                                      ent = 0.0
                                      do 1 i=1, 6
                                              x = n(i)/1000.0
                                              if (x .le. 0.0) goto 1
                                              ent = ent - x*alog(x)
1                                     continue
                                      if (ent .gt. emax) then
                                              emax = ent
                                              write(6,*)ent, n
                                      endif
                              enddo
2                     continue
3             continue
4      continue
       stop
       end
```

A *MATHEMATICA*™ PACKAGE FOR SYMBOLIC BAYESIAN CALCULATIONS

Paul Desmedt*, Ignace Lemahieu†
Department of Electronics and Information Systems,
University of Ghent, St.-Pietersnieuwstraat 41,
B-9000 Ghent, Belgium.

K. Thielemans
Katholieke Universiteit Leuven
Instituut voor Theoretische Fysica
Celestijnenlaan 200 D
B- 3001 Leuven, Belgium

ABSTRACT. A package, `BayesCalc`, is presented that extends the standard possibilities of *Mathematica.* The implemented extensions allow the automatic, symbolic calculation of many operations needed in the daily application of Bayesian theory. The main feature of the package is the symbolic calculation of posterior probabilities.

Some examples are given to illustrate the proposed package.

1. Introduction

Mathematica [1] is a program for doing symbolic mathematical manipulations by computer. These manipulations are performed according to some built-in mathematical rules. *Mathematica* allows the user to extend these rules. Here the package `BayesCalc` is presented. This package implements most of the rules needed for the application of Bayesian probability theory.

Because Bayesian probability theory uses only a restricted number of rules, it forms an excellent subject for implementation in *Mathematica.* Furthermore the application of the rules of Bayesian probability theory is straightforward. This elegant property further limits the complexity of the implementation.

The main purpose of the presented package `BayesCalc` is the automatic calculation of posterior probabilities. These posterior probabilities are calculated from a number of probabilistic relations (prior probabilities, sampling distributions, ...) and parameter ranges. These probabilistic relations and parameters embody the relevant information I for a particular problem. The principal concern of the user is the correct specification of this information.

In the next section a summary of the most important rules of Bayesian probability is given. Section 3 presents the main features offered by the implemented *Mathematica*

* supported by a grant from IWONL, Brussels, Belgium
† research associate with the NFWO, Brussels, Belgium
[1] *Mathematica* is a trademark of Wolfram Research Inc. For details, see [1].

G. R. Heidbreder (ed.), Maximum Entropy and Bayesian Methods, 197–203.

package `BayesCalc`. No implementation details are discussed in this paper. The technical features are considered in [2], which is available from the authors. In section 4 the package is illustrated with some examples.

Notation:
Mathematica input and output is written in `typeset` font. The *Mathematica* user interface is simulated by preceding the input lines by "*In[n] :=*", and output statements by "*Out[n]=*", where n is the *Mathematica* line number. If *n=1*, no previous inputs to the *Mathematica* package are required.

2. Basic rules of Bayesian probability theory

Bayesian probability theory is based on the application of only two rules [3, 4, 5]: the sum rule,

$$bp(A+B|I) = bp(A|I) + bp(B|I) - bp(AB|I), \tag{1}$$

and the product rule,

$$bp(AB|I) = bp(A|I)\, bp(B|AI) = bp(B|I)\, bp(A|BI). \tag{2}$$

Here A and B are two hypotheses and I embodies the available information. Hypotheses A and B may also represent sets of hypotheses. Examples of hypotheses are "parameter x has value 10", "it will rain tomorrow", ... The information I includes prior probability laws, logic dependencies,...

Here hypotheses (single or sets) are represented by capital letters (A, B). Parameters will be represented by small letters (x, y).

The basic principle of Bayesian probability theory is to calculate the probability of all the unknown hypotheses A conditional on all the supposed known hypotheses B, i.e., $bp(A|BI)$. The resulting probabilistic relation $bp(A|BI)$ can constitute a starting point for parameter estimation, decision theory, hypothesis testing, etc.

Another important tool in Bayesian probability theory is the marginalization procedure. This marginalization procedure is applied when the probability $bp(A|BI)$ is needed independently of an unknown parameter n, called nuisance parameter. The appropriate procedure consists in calculating the joint probability $bp(An|BI)$ of the unknown hypotheses A and the nuisance parameter n. The nuisance parameter n is then eliminated (marginalized) by integrating the joint probability over the total range of the parameter n, i.e.,

$$bp(A|BI) = \int_{n_{min}}^{n_{max}} bp(A\;n|B\;I)\,\mathrm{d}n, \tag{3}$$

where n_{min} and n_{max} define the range over which the parameter n can vary.

3. General features of the `BayesCalc` package

The general notation used by the package `BayesCalc` for a conditional probability is `bp[{A}, {B}]` which stands for $bp(A \mid B\; I)$. Note that the curly brackets "{}" determine which hypotheses are on the right or the left of the conditional sign. If a hypothesis consists of a set of hypotheses, the distinct hypotheses are separated by commas.

Before any posterior probabilities can be calculated, we need the specification of the proper information I. Consider first the specification of probabilistic relations. The general procedure to specify probabilistic relations has the syntax:

$$\texttt{definebp}[\{\texttt{A}\},\{\texttt{B}\},\mathit{userFunc}[A,B]] \tag{4}$$

In conventional notation this means: $bp(A|BI) = userFunc(A,B)$. For instance

$$\texttt{definebp}[\{\texttt{x}\},\{\mu,\sigma\},\texttt{gauss}[\texttt{x},\mu,\sigma]] \tag{5}$$

specifies that the measurement `x` has a Gauss probability distribution with mean μ and variance σ^2. (`gauss` is a function defined in `BayesCalc`, but the explicit formula would do equally well.) Unconditional probabilistic relations are defined by:

$$\texttt{definebp}[\{\texttt{A}\},\mathit{userFunc}(A)] \tag{6}$$

Note that if a probabilistic relation exists between two hypotheses `A` and `B`, these hypotheses are logically dependent. The package `BayesCalc` uses this property to determine the logical dependencies between hypotheses. It assumes that two hypotheses are independent unless a probabilistic relation of the form (4) links them. Dependencies may be declared explicitly by[2]:

$$\texttt{dependent}[\{\texttt{A}\},\{\texttt{B}\}] \tag{7}$$

The second part of the prior information consists of the definition of the ranges of the parameters. The input

$$\texttt{defineRange}[\texttt{a},\{\texttt{b},\texttt{c}\}] \tag{8}$$

specifies that the parameter `a` can have values between `b` and `c`.

Once the proper information I concerning the particular problem is specified, the requested probability relation can be calculated by

$$\texttt{compute}[\texttt{bp}[\{\texttt{A}\},\{\texttt{B}\}]]. \tag{9}$$

The previous form is immediately transformed with the product rule by the package to

$$\frac{1}{\texttt{bp}[\{\texttt{B}\}]}\texttt{bp}[\{\texttt{A},\texttt{B}\}] \tag{10}$$

The denominator of this expression is a normalization constant and will in general not be expandable in a product of user specified probabilistic relations. As a consequence, the denominator will usually be returned unchanged. The nominator will be expanded by recursive use of the product rule. This expansion will ultimately result in a product of user defined probabilistic relations.

If the package was unable to find an explicit expression for the joint probability, equation (10) is returned. The recursive expansion procedure is explained in [2].

[2] All hypotheses can be made dependent on each other by default with `allDependent[]`. Independency of hypotheses should then explicitly be declared by: `independent[{A},{B}]`

In numerous applications, the result returned by `compute` will be sufficient to solve the problem at hand. However, sometimes the normalized version of this expression is desired. The procedure to obtain the normalized version is:

$$\texttt{normalize}[\mathit{probFunc}[A, B], \{\texttt{A}\}] \tag{11}$$

where *probFunc[A,B]* is a general probabilistic relation and {`A`} is the set of parameters for which the normalization is performed[3].

An implementation of the sum rule is also provided by the logical "`or`" operation. Thus

$$\mathit{In[1]} := \texttt{compute[bp[\{or[a,b]\}]]} \tag{12}$$

yields

$$\mathit{Out[1]} = \texttt{bp[\{a\}]} + \texttt{bp[\{b\}]} - \texttt{bp[\{a,b\}]} \tag{13}$$

Whenever a logical "`or`" is detected by the package, the sum rule will first be applied before any product rule expansion is attempted.

A full list of the available functions and the corresponding conventional meaning is given in table 1.

Of course, once the required probabilistic relation is obtained from the package `BayesCalc`, all the built-in *Mathematica* routines are available for further manipulations. For instance one can obtain plots, perform differentiations,...

4. Examples

The routines of `BayesCalc` are illustrated by two examples taken from an introductory text written by Loredo (section 6 of [6]).

The first problem is the estimation of the Poisson rate `b` when a number of counts `nb` were observed during a measurement time `T`. The experiment can be performed to estimate the background activity rate `b` of the sky. In this example, `nb` is the number of events measured from an "empty" part of the sky. The user defines this information by

In[1] := (14)

```
Needs["BayesCalc`"]
definebp[{b}, 1/b]
definebp[{b}, {nb}, poisson[nb, T b]]
defineRange[b, {0, Infinity}]
```

The posterior probability for the rate is obtained by

$$\mathit{In[2]} := \texttt{compute[bp[\{b\},\{nb\}]]} \tag{15}$$

$$\mathit{Out[2]} = \frac{1}{\texttt{bp[\{nb\}] nb! Exp[T b]}}\ \texttt{T}^{\texttt{nb}}\ \texttt{b}^{\texttt{nb}-1} \tag{16}$$

[3]However, the *Mathematica* integration routines may fail to find the integrals needed for normalization. Therefore, some frequently occurring integrals were precomputed and stored in `BayesCalc`. This also speeds up most calculations.

Mathematica **rule**	**Conventional meaning**
Information specification	
`reset[]`	clears all previous user specified probabilistic settings
`allDependent[]`	make all hypotheses dependent on each other
`dependent[{A},{B}]`	explicitly make hypotheses A dependent on hypotheses B
`independent[{A},{B}]`	explicitly make hypotheses A independent on hypotheses B
`definebp[{A}, {B},` *probFunc[A,B]*`]`	$bp(A \mid BI) = probFunc(A, B)$
`definebp[{A},` *probFunc[A]*`]`	$bp(A \mid I) = probFunc(A)$
`defineRange[a,{b,c}]`	$a \in [b, c]$
Probability relations	
`gauss[x,mu,sigma]`	$\frac{1}{\sqrt{2\pi sigma^2}} \exp(-\frac{1}{2}\frac{(x-mu)^2}{sigma^2})$
`poisson[x,rate]`	$\exp(-rate)\ rate^x\ \frac{1}{x!}$
`uniform[a]`	uniform prior for parameter a
`jeffreys[a]`	Jeffreys prior for parameter a
Utilities	
`compute[bp[{A},{B}]]`	calculate probability $bp(A \mid BI)$
`normCompute[bp[{A},{B}]]`	calculate normalized posterior probability $bp(A \mid BI)$
`normalize[`*probFunc[A,B]*`,{C} ]`	normalize *probFunc[A,B]* for hypotheses C
`marginalize[` *probFunc[A,B]*`,{C} ]`	marginalize *probFunc[A,B]* for hypotheses C
`mean[`*probFunc[A]*`,c]`	calculate mean
`stdev[`*probFunc[A]*`,c]`	calculate standard deviation
`moment[`*probFunc[A]*`,n, c]`	calculate n-th order moment
Logical "or" and logical "and"	
`bp[{or[a,b]}]`	$bp(a + b \mid I)$
`bp[{or[a,and[b,c]]}]`	$bp(a + bc \mid I)$
`bp[{a,b}]`	$bp(ab \mid I)$

Table 1: Overview of routines provided by `BayesCalc`

The normalized version is obtained by [4]:

$$In[3] := \texttt{normalize}[\%, \texttt{b}] \tag{17}$$

The result of this operation is

$$Out[3] = \frac{\texttt{T}^{\texttt{nb}}\ \texttt{b}^{\texttt{nb}-1}}{\texttt{Exp[T b] Gamma[nb]}} \tag{18}$$

which equals formula (41) of [6].

The next example is an extension of the previous one. A supplementary measurement is made from a radiation source in the sky with unknown rate **s**. The number of counts observed is **n** and the measurement time is **t**. The requested posterior probability is $bp(s\ b \mid n\ nb\ I)$. Again we define the available information:

$$\begin{array}{l} In[1] := \\ \quad \texttt{reset}[] \\ \quad \texttt{definebp}[\{\texttt{n}\}, \{\texttt{s},\texttt{b}\}, \texttt{poisson}[\texttt{n}, (\texttt{s}+\texttt{b})\texttt{t}]] \\ \quad \texttt{definebp}[\{\texttt{s}\}, \{\texttt{b}\}, 1/(\texttt{s}+\texttt{b})] \\ \quad \texttt{defineRange}[\texttt{b}, \{0, \texttt{Infinity}\}] \\ \quad \texttt{definebp}[\{\texttt{nb}\}, \{\texttt{b}\}, \texttt{poisson}[\texttt{nb}, \texttt{T b}]] \\ \quad \texttt{definebp}[\{\texttt{b}\}, 1/\texttt{b}] \end{array} \tag{19}$$

Simply typing

$$In[2] := \texttt{compute}[\texttt{bp}[\{\texttt{s},\texttt{b}\}, \{\texttt{n},\texttt{nb}\}]] \tag{20}$$

will immediately give a result equivalent to equation (45) of [6]:

$$Out[2] = \frac{\texttt{b}^{(-1+\texttt{nb})}\ \texttt{E}^{-\texttt{T b}-(\texttt{b}+\texttt{s})\texttt{t}}\ (\texttt{b}+\texttt{s})^{-1+\texttt{n}}\ \texttt{t}^{\texttt{n}}\ \texttt{T}^{\texttt{nb}}}{\texttt{bp}[\{\texttt{n},\texttt{nb}\}]\ \texttt{n}!\ \texttt{nb}!} \tag{21}$$

The function `reset[]` clears all user specified probabilistic relations. It is recommended to start each new problem with a call of the `reset[]` function.

Other examples are given in the `BayesCalc` package. These examples include the use of sum rules, gauss probability relations,...

5. Obtaining the BayesCalc package

The package is freely available from the authors. Any constructive criticisms, suggestions for improvements can be addressed to the authors and will be highly appreciated.

6. Conclusions

The presented `BayesCalc` package allows the automatic calculation of probabilities in the presence of the properly specified prior information. The posterior probabilities are obtained by successive applications of the product rule.

Additional tools are available for the marginalization procedure, evaluation of expected values and many other frequently needed operations.

---

[4] "%" is a *Mathematica* short hand for the last output.

References

[1] S. Wolfram, *Mathematica, A System for Doing Mathematics by Computer*, Addison-Wesley Publishing Company, Inc., 1991

[2] P. Desmedt, K. Thielemans, *Technical aspects of the* Mathematica *package* `BayesCalc`, ELIS Technical Report DG 93-13, 1993

[3] E.T. Jaynes, *Probability theory as logic*, to be published

[4] J.O. Berger, *Statistical decision theory and Bayesian analysis*, Springer-Verlag, 1985

[5] H. Jeffreys, in *Theory of probability,* Oxford University Press, 1939

[6] T.J. Loredo, *From Laplace to supernova SN 1987A : Bayesian inference in astrophysics*, in Maximum entropy and Bayesian methods, Dartmouth, Reidel, 1990, pp. 81-142

A MULTICRITERION EVALUATION OF THE MEMSYS5 PROGRAM FOR PET

Paul Desmedt *, Ignace Lemahieu†, Koen Bastiaens
Department of Electronics and Information Systems,
University of Ghent, St.-Pietersnieuwstraat 41, 9000 Ghent,
Belgium.

ABSTRACT.
In Positron Emission Tomography (PET) images have to be reconstructed from noisy projection data. The noise on the PET data can be modeled by a Poisson distribution.

In this paper, the performance of MemSys5, a maximum entropy based general purpose reconstruction algorithm, on such data is tested. A number of different criteria were applied to evaluate the algorithm: quadratic distance to reference images, edge detection capacity, flatness recovery, etc.

1. Introduction

Positron Emission Tomography (PET) is a tomographic method to display metabolic activity in a slice through a patient's body. The particular construction of the PET scanner and the use of a radioactive tracer entail the modeling of the data by a Poisson distribution.

In PET the most popular reconstruction technique is the Filtered Backprojection (FB) algorithm. This reconstruction algorithm is based on a Fourier Transform technique and it is extremely fast. However, since the FB algorithm does not account for the noise present in the data, the reconstructions suffer from severe noise artifacts.

To deal with the noise, statistical reconstruction techniques are investigated. The most widely used statistical reconstruction technique is the Maximum Likelihood - Expectation Maximization (ML-EM) algorithm. The ML-EM algorithm searches for the image that maximizes the likelihood of the data. Hence, no prior information about the images is used. This ML-EM algorithm is iterative and it succeeds in suppressing the noise. However, when the algorithm is iterated too long, the reconstructed image starts to degrade [1].

One possibility to avoid the image degradation is the introduction of some prior information. As a general purpose prior the entropy type priors are commonly used. An elaborate program that uses an entropic prior is the MemSys5 program developed by Skilling and Gull [2, 3]. MemSys5 is an iterative algorithm that is based on a conjugate gradient algorithm [4].

In this paper the performance of the MemSys5 algorithm is evaluated. A number of different performance criteria are treated. Some of these performance criteria are: quadratic distance of reconstructed image to reference image, edge detection capacity, flatness recovery. This wide range of criteria is evaluated because the reconstructed image will presumably serve a number of very distinct purposes. Indeed, the same reconstructed image may

*supported by a grant from IWONL, Brussels, Belgium
†senior research associate with the NFWO, Brussels, Belgium

G. R. Heidbreder (ed.), Maximum Entropy and Bayesian Methods, 205–211.

in a medical environment serve many different tasks: the visual detection of anomalies, the automatic detection of edges by computer programs, the time evolution of the tracer concentration in a particular organ of the patient, etc.

2. Evaluation criteria

A number of different approaches to evaluate images are available. First, general distance measures such as the quadratic difference between the reconstructed image and the reference image (or scaled quadratic difference) can be used. The antagonist of these simple measurement criteria is the evaluation of the task performance by the human observer [5]. This is a very laborious evaluation method and numerical observers are investigated to replace the human observer criterion [6]. Here we will not investigate the human observer performance nor the related numerical observer performance. We choose to evaluate the general distance functions and some computer task specific numerical evaluators. Of course these procedures can only be applied when the reference image is available. Therefore the tests are performed on simulated data. These data are obtained by applying a projection routine to a known reference image.

The general distance measures are the quadratic distance *dist* and the scaled quadratic distance *sc_dist*

$$dist = \frac{1}{N}\sum_{i=0}^{N}(f_i^{(ref)} - f_i^{(k)})^2 \tag{1}$$

$$sc\_dist = \frac{1}{N}\sum_{i=0}^{N}\left(\frac{1}{(f_i^{(ref)})^2}(f_i^{(k)} - f_i^{(ref)})^2\right), \quad f_i^{(ref)} \neq 0, \tag{2}$$

where $f_i^{(ref)}$ and $f_i^{(k)}$ respectively represent the i-th pixel intensity of the reference image and the image estimate at the k-th iteration. The number of pixels in the image is N.

The first specific task considered, is the ability of the reconstructed image to reproduce edges. To evaluate the performance of the edge detection, the following procedure is employed. An edge detector is applied that returns an image with pixel value 1 for border pixels and pixel value 0 for the other pixels. This binary image will be called the border image. The edge detector consists of the application of a Prewitt gradient operator [7]. The result of this operation is a gradient image. Afterwards pixels in the gradient image with a value that is at least 10% of the maximum gradient pixel value are selected as border pixels. The same edge detector is applied to the reference image and the current image estimation $\mathbf{f}^{(k)}$.

The border image obtained from the image estimation is then compared to the border image of the reference image. The first criterion *border_I* counts the number of edge pixels of the reference image that are missed in the reconstructed image. The second criterion *border_II* counts the number of falsely presumed edge pixels in the reconstructed image. Both numbers are scaled by the number of edge pixels in the border image of the reference image. This results in the criteria

$$border\_I = \frac{1}{true\_borders}\sum_{i=0}^{N} b_i^{(ref)}(1 - b_i^{(k)}) \tag{3}$$

$$border\_II = \frac{1}{true\_borders}\sum_{i=0}^{N}(1 - b_i^{(ref)})b_i^{(k)} \tag{4}$$

where *true_borders* is the number of border pixels of the reference image, i.e.:

$$true\_borders = \sum_{i=0}^{N} b_i^{(ref)}. \tag{5}$$

Where $b_i^{(ref)}$ and $b_i^{(k)}$ are the pixel values of the border images of respectively the reference image and the estimated image.

The next two criteria are concerned with specific regions of the image. Such regions are commonly used in medical practice to delineate structures of interest. They are called Regions Of Interest (ROI). The first criterion, *ROI_1*, is related to the visual appearance of the reconstructed image. In general a noisy data set will result in a rather noisy looking reconstructed image. To evaluate the occurrence of these irregularities a ROI is selected with a uniform activity distribution. For this ROI the standard deviation is calculated. Hence, *ROI_1* will be zero for the reference image. For the image estimations the magnitude of *ROI_1* is an indication of the noise degradation of the ROI.

This error criterion is defined by:

$$ROI\_1 = \frac{1}{\sum_{i\in ROI} f_i^{(k)}}\sqrt{\frac{1}{N_{ROI}-1}\sum_{i\in ROI}(f_i^{(k)} - \bar{f}^{(k)})^2}, \tag{6}$$

where $\bar{f}^{(k)}$ is the mean pixel intensity value of the *ROI* of the image estimation. N_{ROI} is the number of pixels in the ROI.

The next criterion checks the quadratic difference between the estimated ROI mean and the ROI mean of the reference image, i.e.:

$$ROI\_2 = \frac{1}{(\sum_{i\in ROI} f_i^{(ref)})^2}\left(\sum_{i\in ROI} f_i^{(k)} - \sum_{i\in ROI} f_i^{(ref)}\right)^2. \tag{7}$$

3. Experiments and discussion

The software phantom is based on structure intensities obtained in real brain scans. The phantom is rendered in fig. 1. It is digitized on a 64×64 grid. From this image, projection data is created with 64 angular and 64 lateral positions. Then Poisson noise is added to the projection data. To alter the importance of the noise effect, three different intensity levels are used. The three different mean intensities are: intensity 1: 198, intensity 2: 395, intensity 3: 790.

For the three intensities the different error criteria are shown as a function of the iteration number. The stopping criterion of the MemSys5 program corresponds to the classic maximum entropy solution [3]. However, the iteration procedure of MemSys5 is stopped when the number of good degrees of freedom could not be found to acceptable accuracy (error code 6).

A smoothing of the image can be performed by the Intrinsic Correlation Function (ICF) [2]. In this paper the ICF is not used. The effect of the ICF on the reconstructed image will form the subject of further investigations.

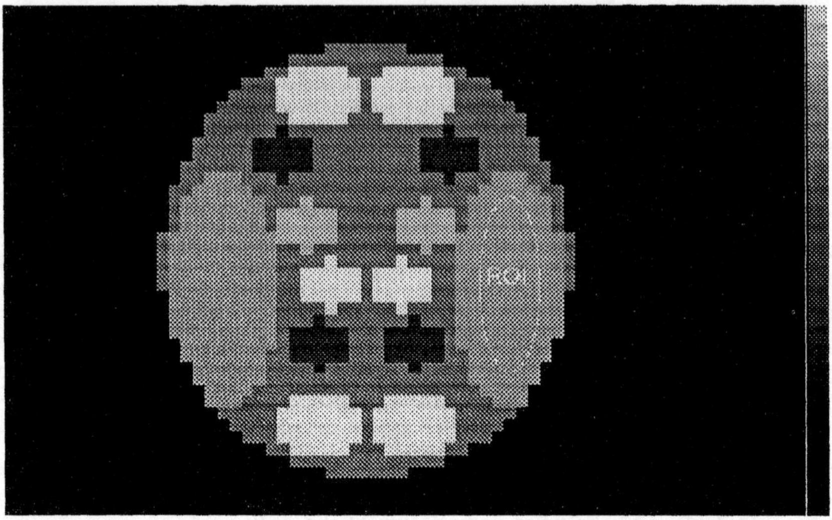

Figure 1: Reference image.

Fig. 2 shows the general distance measures *dist* and *sc_dist*. We observe that in both plots a minimum occurs for all the intensity levels. The minima on corresponding curves are also located at approximately the same iteration number on both plots. For instance, the *dist* and *sc_dist* curves for intensity 3 both attain their minimum at iteration 8. Furthermore the minimum of the scaled distance curve decreases when dealing with higher intensities. This is due to the better signal to noise ratio when the mean intensity level increases. Moreover, the iteration at which the minimum occurs shifts to later iterations with increasing intensities. This can also be observed for the ML-EM algorithm [1].

In the ML-EM algorithm these effects (minimum in the general measure curves, decrease of value of minimum when signal to noise ratio improves,...) are attributed to the noise. The reconstructed image is believed to overfit the data. Since these effects also occur for the reconstructions with the MemSys5 program, the same cause can be stated. The MemSys5 program apparently overfits the data.

Fig. 3 shows the two border detection criteria: *border_I* and *border_II*. For all intensity levels the *border_I* criterion becomes very low ($< 5\%$), indicating that almost all borders of the reference image are retrieved by the image estimation. However when looking at the *border_II* plot, it is observed that up to 40% of supplementary edge pixels are found. Especially, the number of extra border pixels in the first iterations is extremely large.

The region dependent criteria are shown in fig. 4. The delineation of the region is depicted on fig. 1. It is clear that the mean intensity of the region is very well retrieved (*ROI_2*). Also interesting to note is that the standard deviation (*ROI_1*) steadily increases as the iterations proceed. This indicates that the noisy appearance continuously increases with the iteration number. Despite this visual deterioration the mean intensity is still recovered with high accuracy.

We also observe that the different criteria agree on the optimal iteration number. E.g.,

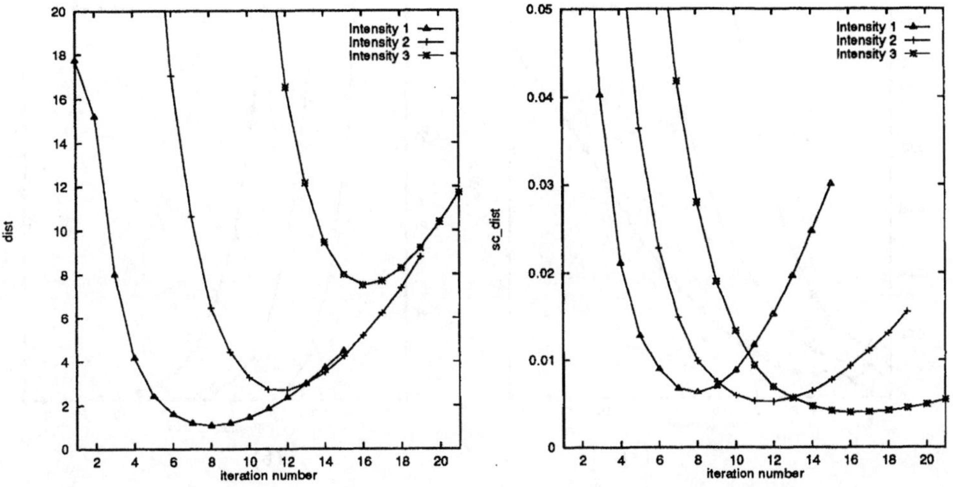

(a) (b)

Figure 2: General measures as a function of iteration number: (a) *dist* (b) *sc_dist*

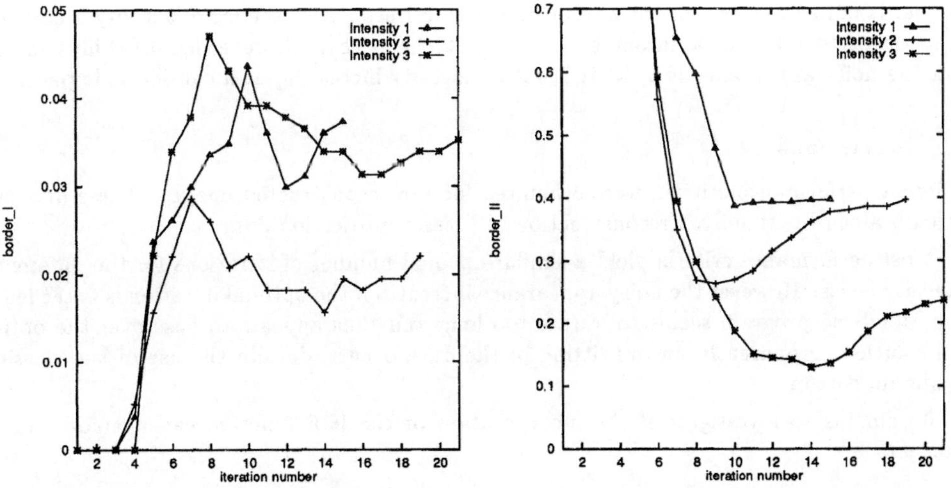

(a) (b)

Figure 3: Border criteria as a function of iteration number: (a) *border_I* (b) *border_II*

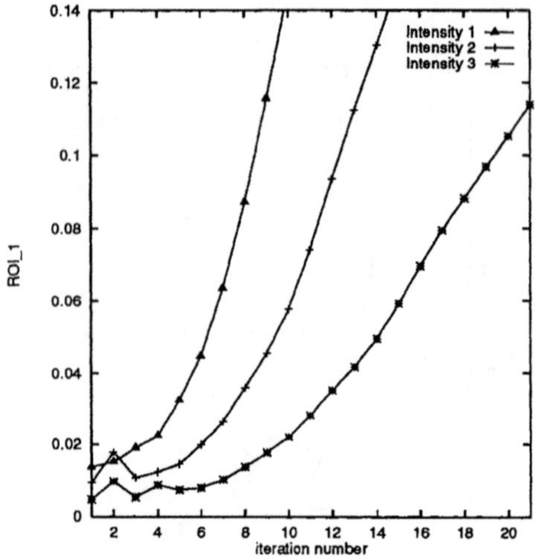

(a)

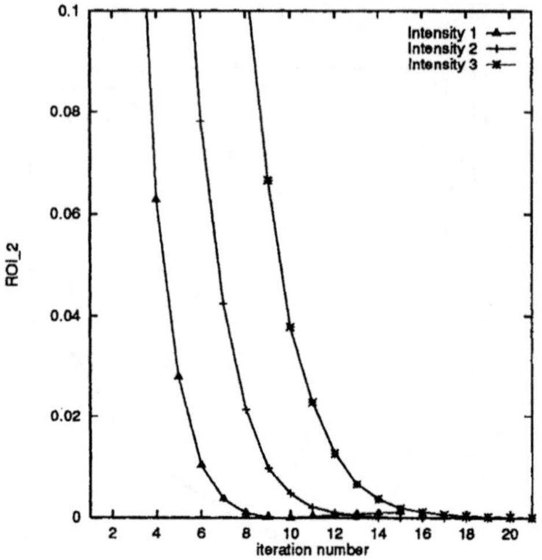

(b)

Figure 4: ROI specific criteria as a function of iteration number: (a) ROI_1 (b) ROI_2

it can be checked that for intensity 2 the optimal iteration number is around 12. This agreement of optimal iteration number was already observed when discussing fig. 2. But this iteration number also approximately corresponds to a stabilization of the different border detection performances (fig. 3). However the occurrence of an optimum iteration number is not so obvious for the border detection criteria. Also the mean activity seems to be retrieved to sufficient accuracy by iteration 12 (fig. 4 (b)). However fig. 4 (a) illustrates that the noisy appearance is by that iteration already increasing at a considerable pace.

4. Conclusions

Different performance criteria were evaluated for the reconstructed images. These images were obtained by MemSys5 reconstruction of Poisson projection data.

Most performance criteria yield a similar optimal number of iterations for the different intensity levels. However the noisy appearance increases if the optimal iteration is exceeded. The MemSys5 program seems to iterate too long and thus appears to pass over the optimal solution. Apparently an overfitting of the data occurs, despite the use of the classic maximum entropy.

It remains to investigate if the incorporation of the ICF function can overcome this problem.

5. Acknowledgements

We are indebted to the Cambridge group for the use of the MemSys5 package. We especially want to thank Steve Gull, Mark Charter and Jason Welch.

References

[1] Snyder D.L., Miller M.I., Thomas L.J., Politte D.G., *Noise and edge artifacts in maximum-likelihood reconstructions for emission tomography*, IEEE Trans. Med. Imaging, Vol. MI-6, 1987, pp. 228-238

[2] Gull S.F., Skilling J., *Quantified Maximum Entropy: MemSys5. Users' Manual*, 1991

[3] Skilling J., *Classic maximum entropy*, in Maximum entropy and Bayesian methods, Cambridge, Reidel, Ed. J. Skilling, Kluwer Academic Publishers, 1989, pp. 45 - 52

[4] Skilling J., Gull S.F., *Algorithms and applications*, in Maximum entropy and Bayesian methods in inverse problems, Reidel, Ed. C.R. Smith and W.T. Grandy, Kluwer Academic Publishers, 1985, pp. 83 - 132

[5] Hanson K.M., *Optimization of the constrained algebraic reconstruction technique for the performance of a variety of visual tasks*, in Information processing in medical imaging, 1991, pp. 45-57

[6] Herman G.T., Yeung K.T.D., *Evaluators of image reconstruction algorithms*, in Technical report No. MIPG#151, University of Pennsylvania, 1989

[7] Pratt W.K., *Digital image processing*, John Wiley & Sons, Inc., 1991

PARALLEL MAXIMUM ENTROPY RECONSTRUCTION OF PET IMAGES

Koen Bastiaens, Paul Desmedt*, Ignace Lemahieu†
Department of Electronics and Information Systems,
University of Gent, St.-Pietersnieuwstraat 41,
9000 Gent, Belgium.
email: kb@inwphys.rug.ac.be

ABSTRACT. The application of a maximum entropy reconstruction method to PET images requires a long computation time. To overcome this problem multiprocessor machines could be used. In this paper we present a parallelization method for the Green expectation maximization method.

1. Introduction

Positron Emission Tomography (PET) is an imaging technique to visualize the distribution of radio-nuclides in an object. It is generally used as a medical diagnostic procedure to evaluate metabolic activity in the human body. Low levels of positron emitting radioactive material are introduced in the organ to be studied. Then, a PET scanner is used to measure the counts of positron-electron annihilation events.

In general two classes of algorithms are used to reconstruct the images from the data recorded by the PET scanner. The first group consists of analytic algorithms, e.g. filtered backprojection algorithms. Other algorithms are based on iterative techniques, e.g. ML-EM (maximum likelihood expectation maximization) and maximum entropy reconstruction methods.

Due to the size of the problem, the execution of all algorithms requires a lot of computing resources. Various efforts have been made to cope with these difficulties. These efforts have led to modern PET scanners with fast implementations of the filtered backprojection algorithm. For this image reconstruction method the image is available only few moments after the patient has left the scanner.

But the execution of the iterative methods is still very time consuming. However, the iterative algorithms have a number of advantages over other faster algorithms. One of the key benefits is the reduction of the statistical noise artifact.

To overcome the problem of the long computation time multiprocessor machines or general purpose supercomputers could be used. It has been shown [4] that an implementation of the maximum likelihood reconstruction method on a general purpose supercomputer is indeed very fast, but unfortunately a supercomputer is not affordable for everyone. The lower cost/performance ratio of most multiprocessors makes them much more interesting for

* supported by a grant from IWONL, Brussels, Belgium
† research associate with the NFWO, Brussels, Belgium

G. R. Heidbreder (ed.), Maximum Entropy and Bayesian Methods, 213–219.

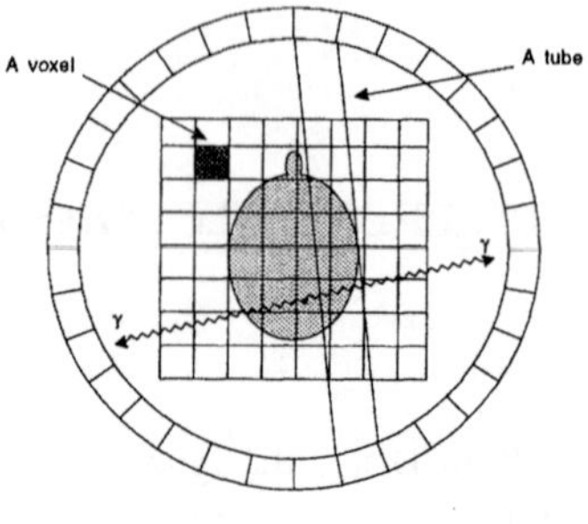

Figure 1: A simplified model of a PET scanner

most applications. Based on the same argument we can exclude the development of a dedicated multiprocessor. Instead, we are interested in commercially available multiprocessor systems.

2. The reconstruction algorithm

When a positron in the radionuclide recombines with an electron, a pair of photons is generated. These photons travel in opposite directions on a straight line. The PET scanner counts the number of detected photon pairs. A pair of scanner detectors on a potential path of two photons is called a *tube*. Not all possible tubes are considered for measurements of coincidences. Only those tubes that connect a detector with a number of t_d detectors around the opposite detector are considered. Figure 1 shows a simplified model of a PET scanner.

We will use the following definitions: the image is composed of V voxels, the number of photon pairs emitted from the voxel with index i is denoted as λ_i, the image is denoted as a vector λ. The number of detected photon pairs are denoted as d_i ($i \in \{1, ..., T\}$), with T the number of tubes. The scanner data is represented as a point $\mathbf{d}$ in a T-dimensional space. $\psi_{j,i}$ is proportional to the probability that a photon pair emitted from voxel i is detected by tube j. The matrix ψ (dimension $T \times V$) is called the transfer matrix.

The solution of the image reconstruction problem is to find the image λ that is most probable when the data $\mathbf{d}$ is observed, i.e. the image for which $p(\lambda \mid \mathbf{d})$ is maximized. Based on the Bayes theorem this probability can be rewritten as:

$$p(\lambda \mid \mathbf{d}I) = \frac{p(\lambda \mid I)p(\mathbf{d} \mid \lambda I)}{p(\mathbf{d} \mid I)}.$$

The solution proposed by the ML-EM algorithm is to find the image for which $p(\mathbf{d} \mid \lambda)$ is maximized, i.e. the probability $p(\lambda)$ is considered independent of λ. However, after a number of iterations increasing noise can considerably deteriorate the image. The introduction of prior information can solve this problem. We will use the entropy prior proposed in [3]:

$$p(\lambda \mid I) \sim \exp\left[\alpha\mathcal{P}(\lambda)\right] = \exp\left[-\alpha \sum_{i=1}^{V} \lambda_i \ln \lambda_i\right],$$

and the modification of the EM (expectation maximization) iteration scheme proposed in [2]:

$$\lambda^{k+1} = \frac{\mathbf{1}_V}{\psi^T \mathbf{1}_T - \alpha \nabla \mathcal{P}(\lambda^k)} . (\psi^T (\frac{\mathbf{d}}{\psi \lambda^k})) . \lambda^k ;$$

λ and $\mathbf{d}$ are column matrices and $\mathbf{1}_V$ is a column matrix with all elements equal to 1; the division and '.' (multiplication) operation represent element wise operations. In the case of the Liang entropy prior the following is valid [3]:

$$-\nabla \mathcal{P}(\lambda) = 1 + \ln(\lambda).$$

The reconstruction starts from an initial guess of the image and the iteration scheme is applied a number of times.

The operation $\psi\lambda$, which is the computation of the projections of the image on the detectors for different angles, requires $O(V \times T)$ operations. The same number of operations is required for $\psi^T \mathbf{d}$, the computation of the backprojection of the data on the image. It is clear that an on the fly computation of the elements of the transfer matrix is not preferable. However in general the tremendous size of the matrix ψ makes it impossible to store it.

Different effects have their contributions to the value of the elements of ψ. These effects are e.g. the geometrical shape of the scanner, the attenuation, the efficiency of the detectors, the positron range, etc. We can rewrite the matrix ψ as a composition of three matrices: $\psi = \mathbf{CAG}$. The elements of the matrix $\mathbf{G}$ $(T \times V)$ depend on the geometry of the scanner. Actually, this is the projection matrix. The matrix $\mathbf{A}$ $(T \times T)$ models the attenuation. The matrix $\mathbf{C}$ $(T \times T)$ models, among other things, the finite size of the detectors which has a broadening effect on the tubes. The contributions of the other effects are neglected. We can now rewrite the iteration scheme as:

$$\lambda^{k+1} = \frac{\mathbf{1}_V}{\mathbf{G}^T \mathbf{A}^T \mathbf{C}^T \mathbf{1}_T + \alpha(\mathbf{1}_V + \ln(\lambda^k))} . (\mathbf{G}^T \mathbf{A}^T \mathbf{C}^T (\frac{\mathbf{d}}{\mathbf{CAG}\lambda^k})) . \lambda^k .$$

The matrix $\mathbf{A}$ is diagonal, therefore the actual storage requirement for $\mathbf{A}$ is the same as for the scanner data. The multiplication with $\mathbf{C}$ can be modeled as a convolution with a rather small convolution kernel. Only this convolution kernel must be stored. Although the attenuation correction and the convolution are not carried out as matrix multiplications they will be treated as such in the mathematical exposition.

This subdivision of ψ does not reduce the storage requirements, but when an additional assumption is made the storage requirements of $\mathbf{G}$ can be reduced significantly. If the detector size and the voxel size are nearly equal, the projection of a voxel on a detector bank over a certain angle can only cover parts of at most two detectors. In that case it is sufficient to store for each voxel-angle pair the detector numbers and the fraction of the projection that covers the detector.

The storage requirements of $\mathbf{G}$ can be reduced further by a factor 8 when the reconstructed area is restricted to the largest circle that fits into the image [1]. Of course this is only applicable when the assumption is made that the object under study is always situated in the corresponding area. With this assumption there exist for each voxel-tube pair another 7 voxel-tube pairs that have the same value for the geometrical factor (the same value of $\mathbf{G}_{j,i}$). Figure 2 shows that a voxel-tube pair rotated over a number of times 90° around

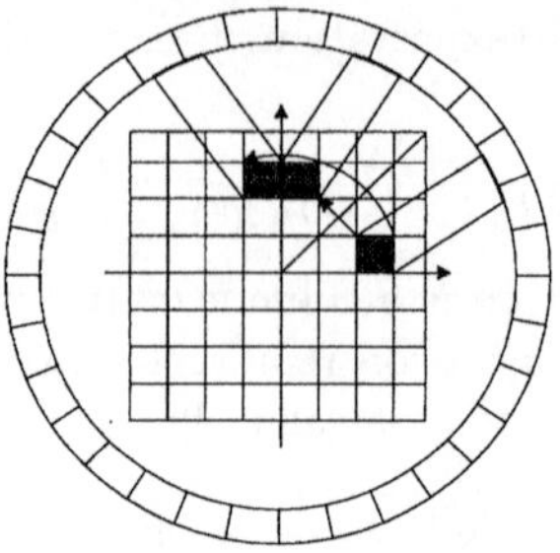

Figure 2: An illustration of the storage reduction

the center of the image has the same geometrical value. A rotation of a voxel-tube pair is the voxel-tube pair found by both rotating the voxel and the tube. Further, a voxel-tube pair derived by mirroring the voxel and the tube with respect to the line $x = y$ also has the same value.

3. Parallelization of the reconstruction method

The maximum entropy reconstruction algorithm is an excellent candidate for a data parallel execution. Data parallelism means that the data space can be partitioned in such a way that the same algorithm can be applied on the different sets of the data. When dealing with data parallelism, the challenge is to partition the data in such a way that the distribution of the workload over the different processors is optimized and that the communication and the synchronization overhead is kept minimal.

A first data partitioning scheme is easily derived from the iteration scheme. The computation of the value of a certain voxel is independent of the computation of values of the other voxels. Thus, the image can be partitioned in a number of independently computable sets of voxels. It is preferable to choose sets of equal size to give each processor the same workload. The computation of the values of a certain set still requires the complete scanner data set, the projection matrix and the whole image computed in the previous iteration step. This way of partitioning is called the *partitioning by voxel* scheme.

It is less obvious, but it is also possible to partition the scanner data in a number of equal sets. With each part of the data set the image is computed and as a final step these individual computed images are added together. This scheme is called the *partitioning by tube* scheme.

The disadvantage of these partitioning schemes is that they require more operations than a sequential execution. E.g. the value $\mathbf{d}/(\mathbf{CAG}\lambda^k)$ is computed at different times in the partitioning by voxel scheme. This can be justified for distributed multiprocessors when the communication cost of this value is higher than the computation cost.

For a shared memory multiprocessor a much more efficient partitioning scheme is proposed which is a mixture of both schemes: a *partitioning by voxel and by tube scheme.* The iteration scheme can be decomposed in three steps: (1) $\gamma = \mathbf{G}^T\mathbf{A}^T\mathbf{C}^T\mathbf{1}_T + \mathbf{1}_V$, (2) $\delta = \mathbf{A}^T\mathbf{C}^T.(\mathbf{d}/(\mathbf{CAG}\lambda^k))$, (3) $\lambda^{k+1} = (1/(\gamma + \alpha \ln(\lambda^k))).(\mathbf{G}^T\delta).\lambda^k$.

The first step must be computed only once because the value of α is the same for every

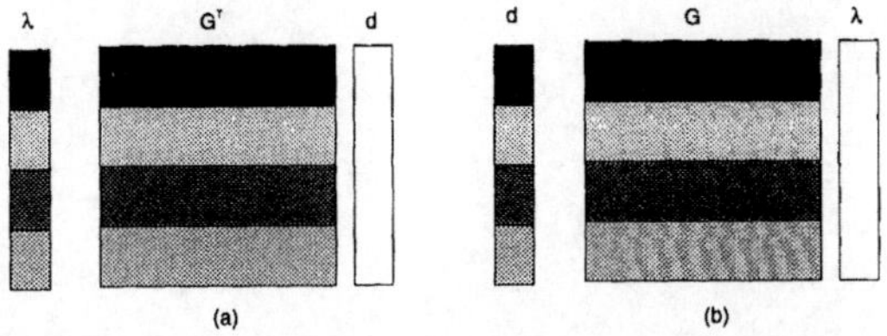

Figure 3: Illustration of the partitioning: (a) by voxel (b) by tube

iteration step. The other two steps must be computed every iteration. In the second step every operation gives a result of dimension T, in the third step of dimension V. So for the second step the partitioning by tube scheme can be applied and for the third step the partitioning by voxel scheme. The first step can further be decomposed in $\zeta = \mathbf{A}^T\mathbf{C}^T\mathbf{1}_T$ and $\gamma = \mathbf{G}^T\zeta + \alpha\mathbf{1}_V$. The computation of γ can be partitioned by tubes, while the second substep can be partitioned by voxels.

In the second step the computation of δ is distributed over the different processors. Therefore **d**, **A**, **C** and **G** are partitioned in sets of rows. In figure 3 the partitioning of the (back)projection operation is given as an illustration. Suppose we want to create x different tasks, then each task has to process T/x tubes. In the third step δ is now taken as a whole, but γ, λ^k and $\mathbf{G}^T$ are now partitioned by rows. Each task has now to process V/x voxels.

The tasks of step three can only start as soon as all the tasks of the second step are finished, because the value of δ is needed. To accomplish this the necessary synchronization must be introduced. The iteration step is finished when all tasks of the third step are finished. So here again synchronization must be introduced. In spite of the use of shared data there is no extra synchronization needed to prevent uncontrolled access, because shared data is only read. Each task will only write in its assigned partition of the image.

For reasons of clarity elementary matrix operations were used in the explanation of the parallelization method. Due to the previously presented storage reduction the (back)projection operations are no longer elementary matrix multiplications.

4. Results

The major goal of the parallelization of sequential algorithms is to minimize the processing time. For a given multiprocessor system one can use the absolute execution time to compare different parallelization schemes. But the absolute execution time does not indicate how efficient the available computing power is utilized. Furthermore, the way the used compiler generates its code and the particular features of the architectures have an effect on the absolute execution time.

The *speedup* (sequential execution time / parallel execution time) is a normalized metric for performance comparisons among the same or different multiprocessor systems. The speedup indicates how much faster a parallel program is executed on a multiprocessor compared to sequential processing. We will also present some of the absolute execution times. Absolute execution times provide an idea of the time between the recording of the data and the availability of the reconstructed images.

The target platform that we are using is a SUN 630MP model 140 running under Solaris

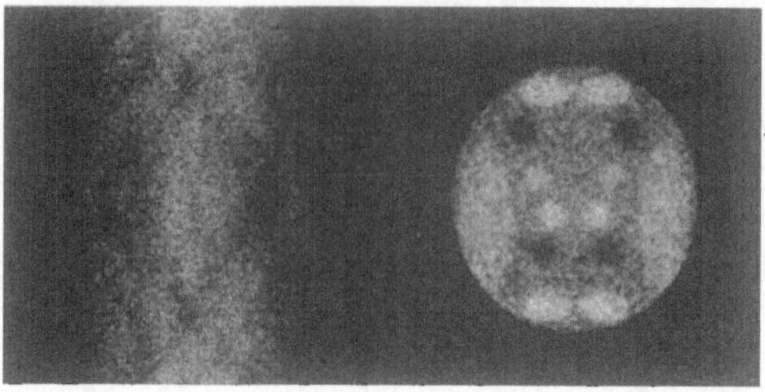

Figure 4: The data recorded by the scanner, and the reconstructed image ($\alpha = 1$) after 40 iteration steps

# of proc.	64×64	256×256	# of proc.	64×64	256×256
2	1.969	1.988	1	16.18	1034
4	3.782	3.927	4	4.279	263.3

Table 1: Comparison of the speedups and the absolute execution times (in s) for 10 iterations

2.2. It is a shared memory multiprocessor with 4 processors and a performance of 28.5 Mips, 4.2 MFlops per processor.

The measured runs consisted of 10 iteration steps. Only the time in the iteration steps was taken into account; the initialization times were not incorporated. The initialization time is very small in comparison with the execution time of 10 iteration steps. In practical situations many more iteration steps are made.

Before the timing measurements are given, an example of a reconstructed image is given. The image consists of 256 by 256 voxels; the scanner data has the same dimensions. It is a phantom which is a model for the human head. Figure 4 shows the data recorded by the scanner and the (in parallel) reconstructed image.

Table 1 gives a summary of the measurements. The timing measurements were carried out on the presented 256 by 256 image as well as on a 64 by 64 image. The number of chosen data partitions was 4, equal to the number of processors. Figure 5 shows clearly that a nearly linear speedup is achieved.

The performance could be further improved if we can make the assumption that the object under study is always situated in the largest circle that fits into the image. In that case only 80% of the pixels must be processed.

5. Conclusion

We have presented a method for parallelizing a maximum entropy reconstruction method. A data partitioning scheme is presented for optimal performance on a shared memory multiprocessor system. The measurements show a (nearly) linear speedup for the different implementations.

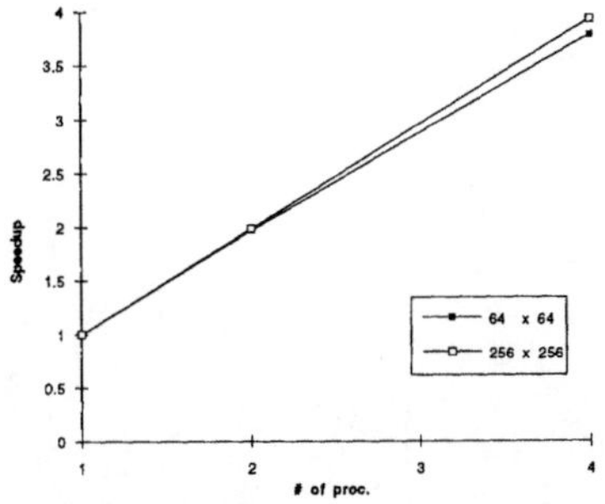

Figure 5: Nearly linear speedup is achieved

References

[1] C.M. Chen, S.-Y. Lee, and Z.H. Cho, "Parallelization of the EM Algorithm for 3-D PET Image Reconstruction," *"IEEE Trans. Med. Imaging*, vol. 10, No. 4, Dec., pp. 513–522, 1991.

[2] P.J. Green,"Bayesian inage reconstruction from emission tomography data using a modified EM algorithm," *IEEE Trans. Med. Imaging*, vol. 9, pp. 84–93, 1990.

[3] Z. Liang, R. Jaszczak, and K. Geer, "On Bayesian image reconstruction from projections: uniform and nonuniform a priori source information," *IEEE trans. med. imaging*, vol. 8, pp. 227–235, 1989.

[4] L. Kaufman, "Implementing and accelerating the EM algorithm for positron emission tomography," *IEEE Trans. Med. Imaging*, vol. MI-5, pp. 16-22, March 1987.

BAYESIAN NON-LINEAR MODELING FOR THE PREDICTION COMPETITION

David J.C. MacKay
Cavendish Laboratory,
Cambridge, CB3 0HE. United Kingdom
mackay@mrao.cam.ac.uk

ABSTRACT. The 1993 energy prediction competition involved the prediction of a series of building energy loads from a series of environmental input variables. Non-linear regression using 'neural networks' is a popular technique for such modeling tasks. Since it is not obvious how large a time-window of inputs is appropriate, or what preprocessing of inputs is best, this can be viewed as a regression problem in which there are many possible input variables, some of which may actually be irrelevant to the prediction of the output variable. Because a finite data set will show random correlations between the irrelevant inputs and the output, any conventional neural network (even with regularisation or 'weight decay') will not set the coefficients for these junk inputs to zero. Thus the irrelevant variables will hurt the model's performance.

The Automatic Relevance Determination (ARD) model puts a prior over the regression parameters which embodies the concept of relevance. This is done in a simple and 'soft' way by introducing multiple regularisation constants, one associated with each input. Using Bayesian methods, the regularisation constants for junk inputs are automatically inferred to be large, preventing those inputs from causing significant overfitting.

An entry using the ARD model won the competition by a significant margin.

1 Overview of Bayesian modeling methods

A practical Bayesian framework for adaptive data modeling has been described in (MacKay 1992). In this framework, the overall aim is to develop probabilistic models that are well matched to the data, and make optimal predictions with those models. Neural network learning, for example, is interpreted as an **inference** of the most probable parameters for a model, given the training data. The search in model space (*i.e.*, the space of architectures, noise models, preprocessings, regularizers and regularisation constants) can then also be treated as an inference problem, where we infer the relative probability of alternative models, given the data. Bayesian model comparison naturally embodies **Occam's razor,** the principle that states a preference for simple models.

Bayesian optimization of model control parameters has four important advantages. (1) No validation set is needed; so all the training data can be devoted to both model fitting and model comparison. (2) Regularisation constants can be optimized on-line, *i.e.* simultaneously with the optimization of ordinary model parameters. (3) The Bayesian objective function is not noisy, as a cross-validation measure is. (4) Because the gradient of the evidence with respect to the control parameters can be evaluated, it is possible to optimism a large number of control parameters simultaneously.

Bayesian inference for neural nets can be implemented numerically by a deterministic method involving Gaussian approximations, the 'evidence' framework (MacKay 1992), or

G. R. Heidbreder (ed.), Maximum Entropy and Bayesian Methods, 221–234.

by Monte Carlo methods (Neal 1993). The former framework is used here.

NEURAL NETWORKS FOR REGRESSION

A supervised neural network is a non-linear parameterized mapping from an input $\mathbf{x}$ to an output $\mathbf{y} = \mathbf{y}(\mathbf{x}; \mathbf{w})$. Here, the parameters of the net are denoted by $\mathbf{w}$. Such networks can be 'trained' to perform regression, binary classification, or multi-class classification tasks.

In the case of a regression problem, the mapping for a 'two-layer network' may have the form:

$$h_j = f^{(1)}\left(\sum_k w_{jk}^{(1)} x_k + \theta_j^{(1)}\right); \quad y_i = f^{(2)}\left(\sum_j w_{ij}^{(2)} h_j + \theta_i^{(2)}\right) \tag{1}$$

where, for example, $f^{(1)}(a) = \tanh(a)$, and $f^{(2)}(a) = a$. The 'weights' w and 'biases' θ together make up the parameter vector $\mathbf{w}$. The non-linearity of $f^{(1)}$ at the 'hidden layer' gives the neural network greater computational flexibility than a standard linear regression. Such a network is trained to fit a data set $D = \{\mathbf{x}^{(m)}, \mathbf{t}^{(m)}\}$ by minimizing an error function, *e.g.*,

$$E_D(\mathbf{w}) = \frac{1}{2}\sum_m \sum_i \left(t_i^{(m)} - y_i(\mathbf{x}^{(m)}; \mathbf{w})\right)^2. \tag{2}$$

This function is minimized using some optimization method that makes use of the gradient of E_D, which can be evaluated using 'backpropagation' (the chain rule) (Rumelhart *et al.* 1986). Often, regularisation or 'weight decay' is included, modifying the objective function to:

$$M(\mathbf{w}) = \beta E_D + \alpha E_W \tag{3}$$

where $E_W = \frac{1}{2}\sum_i w_i^2$. The additional term decreases the tendency of a model to 'overfit' the details of the training data.

NEURAL NETWORK LEARNING AS INFERENCE

The above neural network learning process can be given the following probabilistic interpretation. The error function is interpreted as the log likelihood for a noise model, and the regularizer is interpreted as a prior probability distribution over the parameters:

$$P(D|\mathbf{w}, \beta, \mathcal{H}) = \frac{1}{Z_D(\beta)} \exp(-\beta E_D); \quad P(\mathbf{w}|\alpha, \mathcal{H}) = \frac{1}{Z_W(\alpha)} \exp(-\alpha E_W). \tag{4}$$

The minimization of $M(\mathbf{w})$ then corresponds to the **inference** of the parameters $\mathbf{w}$, given the data:

$$P(\mathbf{w}|D, \alpha, \beta, \mathcal{H}) = \frac{P(D|\mathbf{w}, \beta, \mathcal{H}) P(\mathbf{w}|\alpha, \mathcal{H})}{P(D|\alpha, \beta, \mathcal{H})} = \frac{1}{Z_M} \exp(-M(\mathbf{w})). \tag{5}$$

This interpretation adds little new at this stage. But new ideas emerge when we proceed to higher levels of inference.

SETTING REGULARISATION CONSTANTS α AND β

The control parameters α and β determine the flexibility of the model. Bayesian probability theory can tell us how to set these parameters. All we need to do is write down the inference we wish to make, namely the probability of α and β given the data, and then use Bayes' theorem:

$$P(\alpha,\beta|D,\mathcal{H}) = \frac{P(D|\alpha,\beta,\mathcal{H})P(\alpha,\beta|\mathcal{H})}{P(D|\mathcal{H})} \tag{6}$$

The data-dependent term, $P(D|\alpha,\beta,\mathcal{H})$, is the normalizing constant from our previous inference (5); we call this term the 'evidence' for α and β. This pattern of inference continues if we wish to compare our model $\mathcal{H}$ with other models, using different architectures, regularizers or noise models. Alternative models are ranked by evaluating $P(D|\mathcal{H})$, the normalizing constant of inference (6).

Assuming we have only weak prior knowledge about the noise level and the smoothness of the interpolant, the evidence framework optimizes the constants α and β by finding the maximum of the evidence. If we can approximate the posterior probability distribution by a Gaussian,

$$P(\mathbf{w}|D,\alpha,\beta,\mathcal{H}) \simeq \frac{1}{Z'_M} \exp\left(-M(\mathbf{w}_{\text{MP}}) + \frac{1}{2}(\mathbf{w}-\mathbf{w}_{\text{MP}})^{\text{T}}\mathbf{A}(\mathbf{w}-\mathbf{w}_{\text{MP}})\right), \tag{7}$$

then the maximum of the evidence has elegant properties which allow it to be located on-line. I summarize here the method for the case of a single regularisation constant α. As shown in (MacKay 1992), the maximum evidence α satisfies the following self-consistent equation:

$$1/\alpha = \sum_i w_i^{\text{MP}\,2}/\gamma \tag{8}$$

where $\mathbf{w}^{\text{MP}}$ is the parameter vector which minimizes the objective function $M = \beta E_D + \alpha E_W$ and γ is the 'number of well-determined parameters', given by $\gamma = k - \alpha \text{Trace}(\text{A}^{-1})$, where k is the total number of parameters, and $\mathbf{A} = -\nabla\nabla \log P(\mathbf{w}|D,\mathcal{H})$. The matrix A^{-1} measures the size of the error bars on the parameters $\mathbf{w}$. Thus $\gamma \to k$ when the parameters are all well-determined; otherwise, $0 < \gamma < k$. Noting that $1/\alpha$ corresponds to the variance σ_w^2 of the assumed distribution for $\{w_i\}$, equation (8) specifies an intuitive condition for matching the prior to the data, $\sigma_w^2 = \langle w^2 \rangle$, where the average is over the γ effective parameters; the other $k-\gamma$ effective parameters having been set to zero by the prior.

Equation (8) can be used as a re-estimation formula for α. The computational overhead for these Bayesian calculations is not severe: one only needs evaluate properties of the error bar matrix, A^{-1}. In my work I have evaluated this matrix explicitly; this does not take a significant time if the number of parameters is small (a few hundred). For large problems these calculations can be performed more efficiently (Skilling 1993).

AUTOMATIC RELEVANCE DETERMINATION

The automatic relevance determination (ARD) model (MacKay and Neal 1994) is a Bayesian model which can be implemented with the methods described in (MacKay 1992).

Consider a regression problem in which there are many input variables, some of which are actually irrelevant to the prediction of the output variable. Because a finite data set will show random correlations between the irrelevant inputs and the output, any conventional neural network (even with regularisation) will not set the coefficients for these junk inputs to zero. Thus the irrelevant variables will hurt the model's performance, particularly when the variables are many and the data are few.

What is needed is a model whose prior over the regression parameters embodies the concept of relevance, so that the model is effectively able to infer which variables are relevant and switch the others off. A simple and 'soft' way of doing this is to introduce multiple regularisation constants, one 'α' associated with each input, controlling the weights from that input to the hidden units. Two additional regularisation constants are used to control the biases of the hidden units, and the weights going to the outputs. Thus in the ARD model, the parameters are divided into classes c, with independent scales α_c. Assuming a Gaussian prior for each class, we can define $E_{W(c)} = \sum_{i\in c} w_i^2/2$, so the prior is:

$$P(\{w_i\}|\{\alpha_c\},\mathcal{H}_{\mathrm{ARD}}) = \frac{1}{\prod Z_{W(c)}} \exp(-\sum_c \alpha_c E_{W(c)}), \tag{9}$$

The evidence framework can be used to optimism all the regularisation constants simultaneously by finding their most probable value, *i.e.*, the maximum over $\{\alpha_c\}$ of the evidence, $P(D|\{\alpha_c\},\mathcal{H}_{\mathrm{ARD}})$.[1] We expect the regularisation constants for junk inputs to be inferred to be large, preventing those inputs from causing significant overfitting.

In general, caution should be exercised when simultaneously maximizing the evidence over a large number of hyperparameters; probability maximization in many dimensions can give results that are unrepresentative of the whole probability distribution. In this application, the relevances of the input variables are expected to be approximately independent, so that the joint maximum over $\{\alpha_c\}$ is expected to be representative.

2 Prediction competition: part A

The American Society of Heating, Refrigeration and Air Conditioning Engineers organized a prediction competition which was active from December 1992 to April 1993. Both parts of the competition involved creating an empirical model based on training data (as distinct from a physical model), and making predictions for a test set. Part A involved three target variables, and the test set came from a different time period from the training set, so that extrapolation was involved. Part B had one target variable, and was an interpolation problem.

The task

The training set consisted of hourly measurements from September 1 1989 to December 31 1989 of four input variables (temperature, humidity, solar flux and wind), and three target variables (electricity, cooling water and heating water) — 2926 data points for each target. The testing set consisted of the input variables for the next 54 days — 1282 data points. The organizers requested predictions for the test set; no error bars on these predictions were requested. The performance measures for predictions were the Coefficient of Variation ('CV', a sum squared error measure normalized by the data mean), and the mean bias error ('MBE', the average residual normalized by the data mean).

[1]The quantity equivalent to γ is $\gamma_c = k_c - \mathrm{Trace}_c(\mathbf{A}^{-1})$, where the trace is over the parameters in class c, and k_c is the number of parameters in class c.

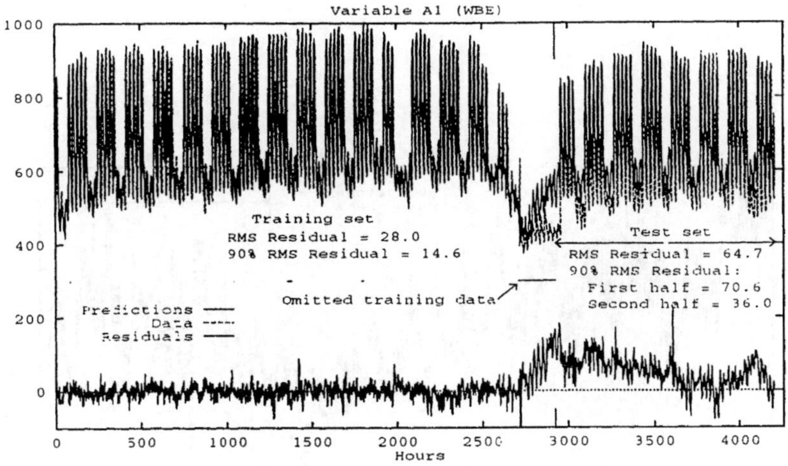

Figure 1: **Target A1 — Electricity**

The three target variables are displayed in their entirety, along with my models' final predictions and residuals, in figures 1–3.

METHOD

A large number of neural nets were trained using the ARD model, for each of the prediction problems. The data seemed to include some substantial glitches. Because I had not yet developed an automatic Bayesian noise model that anticipates outliers (though this certainly could be done (Box and Tiao 1973)), I omitted by hand those data points which gave large residuals relative to the first models that were trained. These omitted periods are indicated on some of the graphs in this paper. 25% of the data was selected at random as training data, the remainder being left out to speed the optimizations, and for use as a validation set. All the networks had a single hidden layer of tanh units, and a single linear output (figure 4). It was found that models with between 4 and 8 hidden units were appropriate for these problems.

A large number of inputs were included: different temporal preprocessings of the environmental inputs, and different representations of time and holidays. All these inputs were controlled by the ARD model. ARD proved a useful guide for decisions concerning preprocessing of the data, in particular, how much time history to include. Moving averages of the environmental variables were created using filters with a variety of exponential time constants. This was thought to be a more appropriate representation than time delays, because (a) filters suppress noise in the input variables, allowing one to use fewer filtered inputs with long time constant; (b) with exponentially filtered inputs it is easy to create (what I believe to be) a natural model, giving equal status to filters having timescales 1, 2, 4, 8, 16, etc..

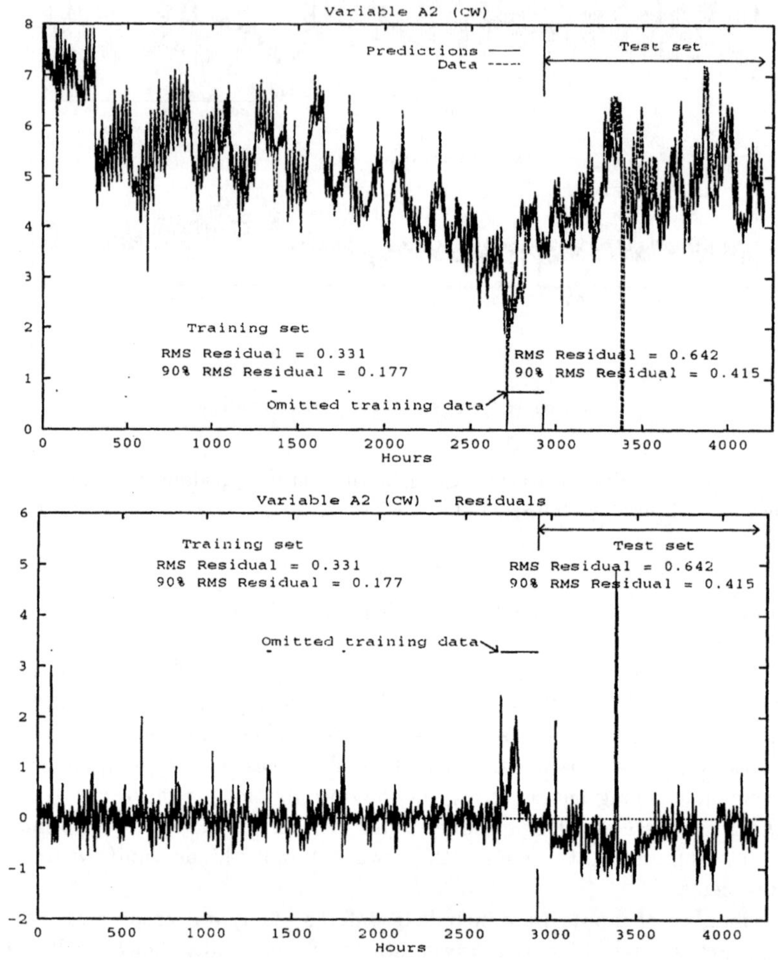

Figure 2: **Target A2 — Cooling water**

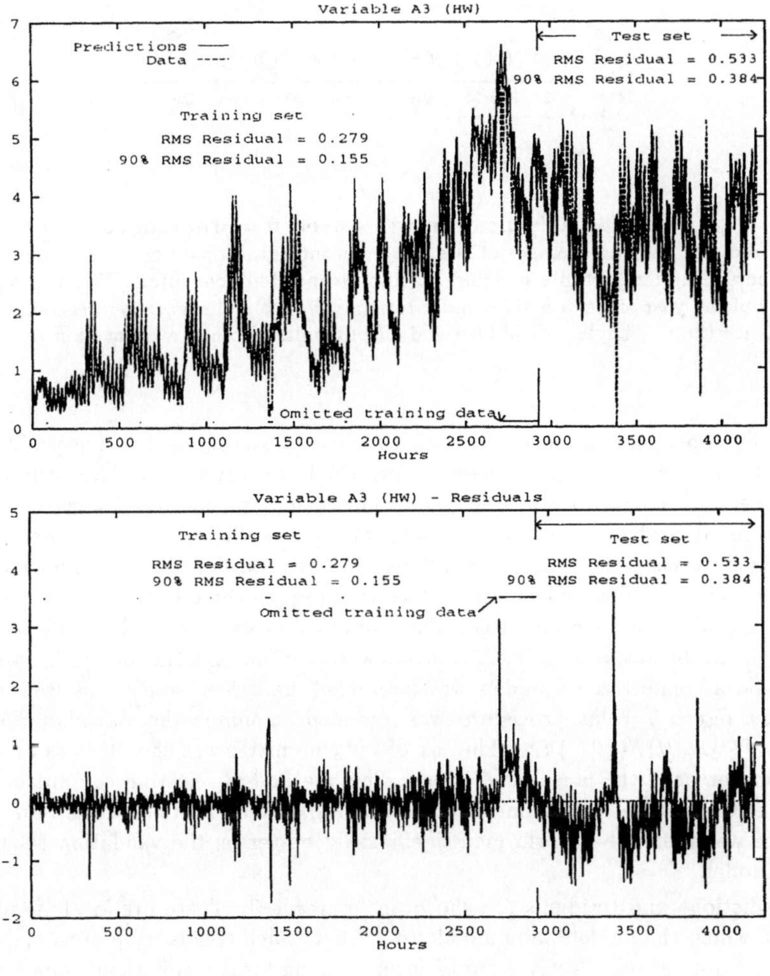

Figure 3: **Target A3 — Heating water**

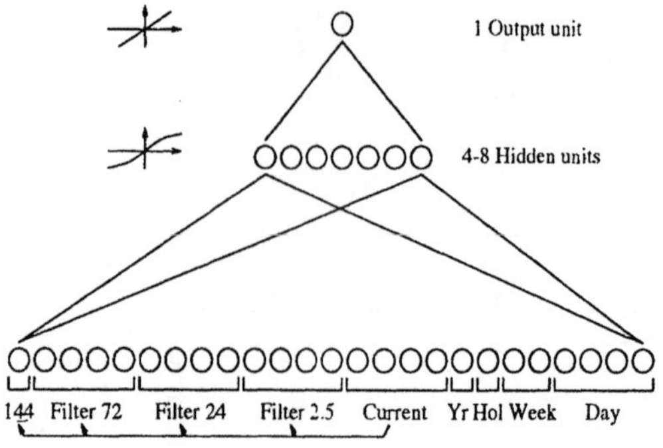

Figure 4: **A typical network used for problem A**
The filters produced moving averages of the four environmental inputs on three time-scales: 2.5, 24 and 72 hours. The temperature variable was also given a 144 hour filter. Time was represented using the cos of the year angle, a holiday indicator, and the cos and sin of: the week angle, the day angle, and twice the day angle. All hidden and output units also had a connection to a bias unit (not shown).

The on–line optimization of regularisation constants was successful. For problem A, 28 such control constants were simultaneously optimized in every model. The optimization of a single model and its control constants took about one day on a Sun 4 workstation, using code which could probably be made substantially more efficient. About twenty models were optimized for each problem, using different initial conditions and different numbers of hidden units. Most models did not show 'overtraining' as the optimization proceeded, so 'early stopping' was not generally used. The numerical evaluation of the 'evidence' for the models proved problematic, so validation errors were used to rank the models for prediction. For each task, a committee of models was assembled, and their predictions were averaged together (see figure 5); this procedure was intended to mimic the Bayesian predictions $P(\mathbf{t}|D) = \int P(\mathbf{t}|D,\mathcal{H})P(\mathcal{H}|D)\, d\mathcal{H}$. The size of the committee was chosen so as to minimize the validation error of the mean predictions. This method of selecting committee size has also been described under the name 'stacked generalization' (Breiman 1992). In all cases, a committee was found that performed significantly better on the validation set than any individual model.

The predictions and residuals are shown in figures 1–3. There are local trends in the testing data which the models were unable to predict. Such trends were presumably 'overfitted' in the training set. Clearly a model incorporating local correlations among residuals is called for. Such a model would not perform much better by the competition criteria, but its on-line predictive performance would be greatly enhanced.

In the competition rules, it was suggested that scatter plots of the model predictions versus temperature should be made. The scatter plot for problem A3 is particularly interesting. Target A3 showed a strong correlation with temperature in the training set (dots in figure 6b). When I examined my models' predictions for the testing set, I was surprised to find that, for target A3, a significantly offset correlation was predicted ('+'s in figure 6a).

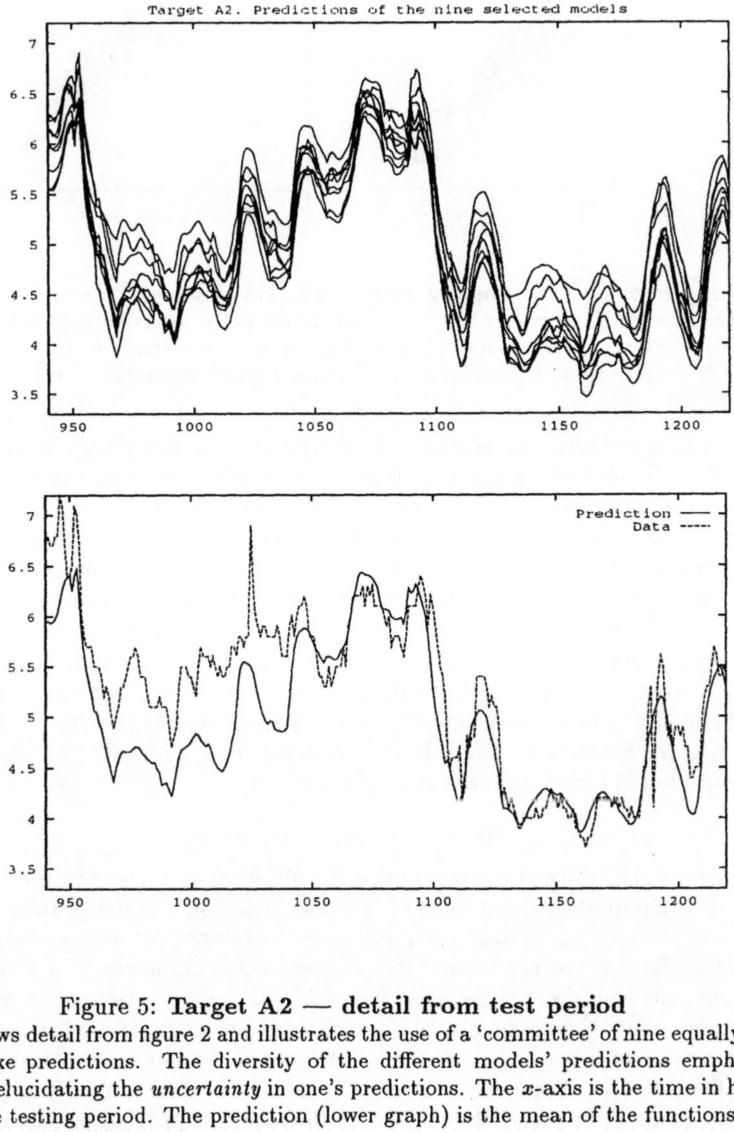

Figure 5: **Target A2 — detail from test period**

This figure shows detail from figure 2 and illustrates the use of a 'committee' of nine equally weighted models to make predictions. The diversity of the different models' predictions emphasizes the importance of elucidating the *uncertainty* in one's predictions. The x-axis is the time in hours from the start of the testing period. The prediction (lower graph) is the mean of the functions produced by the nine models (upper graph).

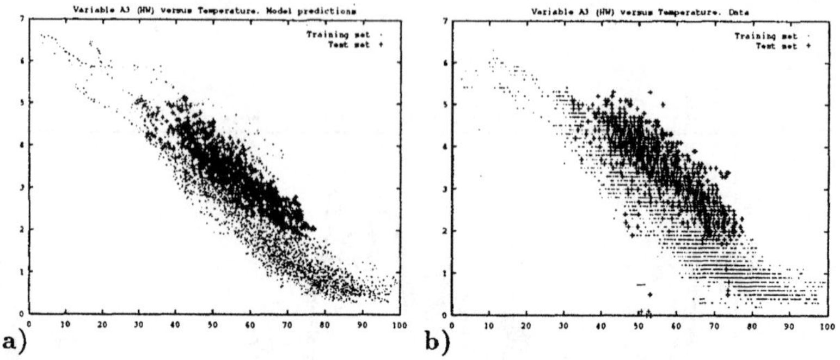

Figure 6: **Predictions for target A3 (HW) versus temperature**
a) Model predictions. This graph shows that my model predicted a substantially different correlation between target A3 and temperature (+) from that shown in the training set (·). **b)** Data. This predicted offset was correct. Units: hot water /10^6Btu versus temperature / F.

This change in correlation turned out to be correct ('+'s in figure 6b). This indicates that these non-linear models controlled with Bayesian methods discovered non-trivial underlying structure in the data. Most other entrants' predictions for target A3 showed a large bias; presumably none of their models extracted the same structure from the data.

In the models used for problem A3, I have examined the values of the parameters $\{\alpha_c, \gamma_c\}$, which give at least a qualitative indication of the inferred 'relevance' of the inputs. For prediction of the hot water consumption, the time of year and the current temperature were the most relevant variables. Also highly relevant were the holiday indicator, the time of day, the current solar and wind speed, and the moving average of the temperature over the last 144 hours. The current humidity was not relevant, but the moving average of the humidity over 72 hours was. The solar was relevant on a timescale of 24 hours. None of the 2.5 hour filtered inputs seemed especially relevant.

How much did ARD help?

An indication of the utility of the ARD prior was obtained by taking the *final* weights of the networks in the optimal committees as a starting point, and training them further using the standard model's regularizer (*i.e.*, just three regularisation constants). The dotted lines in figure 7 show the validation error of these networks before and after adaptation. As a control, the solid lines show what happened to the validation error when the same networks were used as a starting point for continued optimization under the ARD model. The validation error is a noisy performance measure, but the trend is clear: the standard models suffer between 5% and 30% increase in error because of overfitting by the parameters of the less relevant inputs; the ARD models, on the other hand, do not overfit with continued training. The validation errors for the ARD model in some cases change with continued training, because my restarting procedure set the α_i to default values, which displaced the model parameters into a new optimum.

On the competition test data, the performance difference between these two sets of

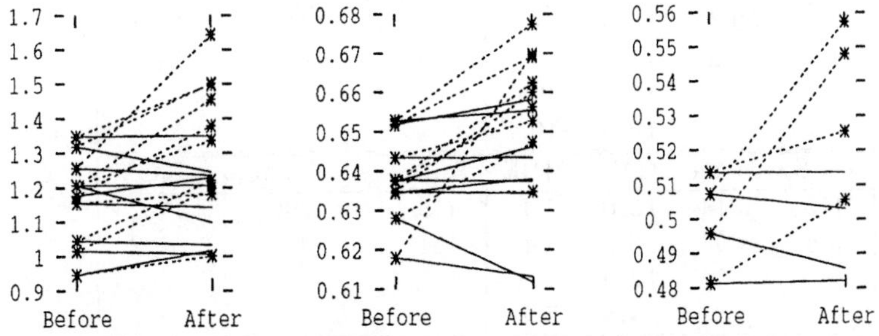

A1) A2) A3)

Figure 7: **Change in validation error when the ARD prior is suspended**
The solid lines without stars show the performance of ARD models. The dotted lines with stars show the models with ARD suspended. In most cases, these standard ('ARD off') models get significantly worse.

models is not so pronounced, because the residuals are dominated by other effects. Maybe the greatest contribution of the ARD method to this problem was that it guided the choice of input variables to include large time-delays.

After the competition, it was revealed that the building in this study was a large university engineering center in Texas. Some of the glitches in the data were caused by the bursting of cold water pipes during a frost — a rare event apparently not anticipated by Texan architects!

The holiday period for staff ended on January 1st, but the student population did not return to the building for a couple of weeks. This may account for the significant bias error in the predictions of electricity usage (figure 1). Another factor which changed between the training period and the test period is that the Computer Science department moved to another building. This too will have caused a reduction in electricity usage. The reduction in electricity consumption may also account for some fraction of the biases in the cold and/or hot water supplies: one might expect less cooling water to be used, or more heating water, to make up the missing energy. The observed average electrical power deficit (according to my model) of 50kW corresponds to an expected decrease in CW or increase in HW consumption of 0.17 $\times 10^6$Btu (assuming that the CW and HW figures measure the actual energy delivered to the building). This is only about a fifth of the overall shift in correlation between HW and temperature shown in figure 6b. In fact, relative to my models, both CW and HW showed an increase of about 0.2 $\times 10^6$Btu.

3 Prediction competition: part B

The data for part B consisted of 3344 measurements of four input variables at hourly intervals during daylight hours over about 300 days. Quasi–random chunks of this data set had been extracted to serve as a test set of 900. The other 2444 examples were accompanied by a single target variable. The physical source of the data were measurements of solar

Problem A1	RMS	Mean	CV	MBE	$RMS_{90\%}$	$Mean_{90\%}$	RCV
ARD	64.7	50.3	**10.3**	8.1	54.1	42.2	**11.1**
ARD off	71.2	56.2	**11.4**	9.0	59.3	47.3	**12.2**
Entrant 6			**11.8**	10.5			
Median			**16.9**	-10.4			
Problem A2	RMS	Mean	CV	MBE	$RMS_{90\%}$	$Mean_{90\%}$	RCV
ARD	.642	-.314	**13.0**	-6.4	.415	-.296	**11.2**
ARD off	.668	-.367	**13.5**	-7.4	.451	-.349	**12.2**
Entrant 6			**13.0**	-5.9			
Median			**14.8**	-7.6			
Problem A3	RMS	Mean	CV	MBE	$RMS_{90\%}$	$Mean_{90\%}$	RCV
ARD	.532	-.204	**15.2**	-5.8	.384	-.167	**9.15**
ARD off	.495	-.121	**14.2**	-3.5	.339	-.094	**8.08**
Entrant 6			**30.6**	-27.3			
Median			**31.0**	-27.0			

Problem B	RMS	Mean	CV	MBE	$RMS_{90\%}$	$Mean_{90\%}$	RCV
ARD	11.2	1.1	**3.20**	0.32	6.55	0.67	**.710**
Entrant 6			**2.75**	0.17			
Median			**6.19**	0.17			

Key:

My models:

ARD — The predictions entered in the competition using the ARD model.

ARD off — Predictions obtained using derived models with the standard regularizer.

Other entries:

Entrant 6 — The entry which came 2nd by the competition's average CV score.

Median — Median (by magnitude) of scores of all entries in competition.

Raw Performance measures:

RMS — Root mean square residual.

Mean — Mean residual.

CV — Coefficient of variation (percentage). The competition performance measure.

MBE — Mean Bias Error (percentage).

Robust Performance measures:

$RMS_{90\%}$ — Root mean square of the smallest 90% of the residuals.

$Mean_{90\%}$ — Mean of those residuals.

RCV — $RMS_{90\%}$/(90% data range).

Normalizing constants:

Problem	Mean of test data	90% data range
A1	624.77	486.79
A2	4.933	3.7
A3	3.495	4.2
B	350.8	923

Table 1: **Performances of different methods on test sets**

flux from five outdoor devices. Four of the devices had a fixed attitude. The fifth, whose output was to be predicted, was driven by motors so that it pointed at the sun. The aim is to enable four cheap fixed devices to substitute for one expensive moving one. Clearly, information such as the day of the week and past history of the input variables was not expected to be relevant. However, I did not realize this, and I spent some time exploring different temporal preprocessings of the input. Satisfyingly, all time-delayed inputs, and the time of the week, were correctly found to be irrelevant by the ARD model, and I pruned these inputs from the final models used for making predictions — without physical comprehension of the problem.

The inputs used in the final models were the four sensor measurements, and a five dimensional continuous encoding of the time of day and the time of year. For training, one third of the training set was selected at random, and the remaining two thirds were reserved as a validation set. This random selection of the training set was later regretted, because it leaves randomly distributed holes where there are no training data. This caused my models' predictions to become unnecessarily poor on a small fraction of the testing data. As in part A, a committee of networks was formed. Each network had between 5 and 10 hidden units.

Results

Problem B was a much easier prediction problem. This is partly due to the fact that it was an interpolation problem, with test data extracted in small chunks from the training set. Typical residuals were less than 1% of the data range, and contrasts between different methods were not great. Most of the sum-squared error of my models' predictions is due to a few outliers.

4 Discussion

The ARD prior was a success because it made it possible to include a large number of inputs without fear of overfitting.

Further work could be well spent on improving the noise model, which assumes the residuals are Gaussian and uncorrelated from frame to frame. A better predictive model for the residuals shown in figures 1–3 might represent the data as the sum of the neural net prediction and an unpredictable, but auto-correlated, additional disturbance. Also, a robust Bayesian noise model is needed which captures the concept of outliers.

In conclusion, the winning entry in this competition was created using the following data modeling philosophy: use huge flexible models, including all possibilities that you can imagine might be appropriate; control the flexibility of these models using sophisticated priors: and use Bayes as a helmsman to guide the search in this model space.

Acknowledgments

I am grateful to Radford Neal for invaluable discussions. I thank the Hopfield group, Caltech, and the members of the Radioastronomy lab, Cambridge, for generous sharing of computer resources. This work was supported by a Royal Society research fellowship, and by the Defense Research Agency, Malvern.

References

BOX, G. E. P., and TIAO, G. C. (1973) *Bayesian inference in statistical analysis.* Addison–Wesley.

BREIMAN, L. (1992) Stacked regressions. Technical Report 367, Dept. of Stat., Univ. of Cal. Berkeley.

MACKAY, D. J. C. (1992) A practical Bayesian framework for backpropagation networks. *Neural Computation* 4 (3): 448–472.

MACKAY, D. J. C., and NEAL, R. M. (1994) Automatic relevance determination for neural networks. Technical Report in preparation, Cambridge University.

NEAL, R. M. (1993) Bayesian learning via stochastic dynamics. In *Advances in Neural Information Processing Systems 5*, ed. by C. L. Giles, S. J. Hanson, and J. D. Cowan, pp. 475–482, San Mateo, California. Morgan Kaufmann.

RUMELHART, D., HINTON, G. E., and WILLIAMS, R. (1986) Learning representations by back–propagating errors. *Nature* **323**: 533–536.

SKILLING, J. (1993) Bayesian numerical analysis. In *Physics and Probability*, ed. by W. T. Grandy, Jr. and P. Milonni, Cambridge. C.U.P.

BAYESIAN MODELING AND CLASSIFICATION OF NEURAL SIGNALS

Michael S. Lewicki

Computation and Neural Systems Program
California Institute of Technology 216–76
Pasadena, CA 91125

lewicki@cns.caltech.edu

ABSTRACT. Identifying and classifying action potential shapes in extracellular neural waveforms has long been the subject of research, and although several algorithms for this purpose have been successfully applied, their use has been limited by some outstanding problems. The first is how to determine shapes of the action potentials in the waveform and, second, how to decide how many shapes are distinct. A harder problem is that action potentials frequently overlap making difficult both the determination of the shapes and the classification of the spikes. In this report, a solution to each of these problems is obtained by applying Bayesian probability theory. By defining a probabilistic model of the waveform, the probability of both the form and number of spike shapes can be quantified. In addition, this framework is used to obtain an efficient algorithm for the decomposition of arbitrarily complex overlap sequences. This algorithm can extract many times more information than previous methods and facilitates the extracellular investigation of neuronal classes and of interactions within neuronal circuits.

1 Introduction

Waveforms of extracellular neural recordings often contain action potentials (APs) from several different neurons. Each voltage spike in the waveform shown in figure 1 is the result of APs from one or more neurons. An individual AP typically has a fast positive component and a fast negative component and may have additional slower components depending on the type of neuron and where the electrode is positioned with respect to the cell. Determining what cell fired when is a difficult, ill-posed problem and is compounded by the fact that cells frequently spike simultaneously which results in large variations in the observed shapes.

Identifying and classifying the APs in a waveform, which is commonly referred to as "spike sorting", has three major difficulties. The first is determining the AP shapes, the second is deciding the number of distinct shapes, and the third is decomposing overlapping spikes into their component parts. In general, these cannot be solved independently since the solution of one will affect the solution of the others. Algorithms for identifying and classifying APs (see Schmidt, 1984 for a review) fall into two main categories: feature clustering and template matching.

Feature clustering involves describing features of APs, such as the peak value, spike width, slope, *etc.*, and using a clustering algorithm to determine distinct classes in the

G. R. Heidbreder (ed.), Maximum Entropy and Bayesian Methods, 235–253.

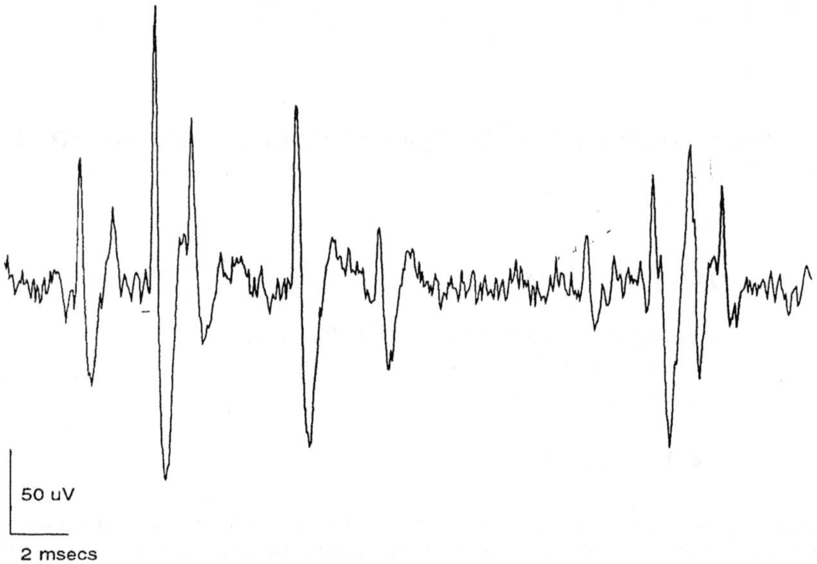

The extracellular waveform shows several different action potentials generated by an unknown number of neurons. Note the frequent presence of overlapping APs which can, in the case of the right most group, completely obscure individual spikes. The waveform was recorded with a glass-coated platinum iridium electrode in zebra finch nucleus MAN (courtesy of Allison Doupe, Caltech).

set of features. Using a small set of features, although computationally efficient, is often sufficient only to discriminate the cells with the largest APs. Increasing the number of features in the clustering often yields better discrimination, but there still remains the problem of how to choose the features, and it is difficult with such techniques to handle overlapping spikes.

In template matching algorithms, typical action potential shapes are determined, either by an automatic process or by the user. The waveform is then scanned and each event classified according to how well it fits each template. Template matching algorithms are better suited for classifying overlaps since some underlying APs can be correctly classified if the template is subtracted from the waveform each time a fit is found. The main difficulty in template matching algorithms is in choosing the templates and in decomposing complex overlap sequences.

The approach demonstrated in this paper is to model the waveform directly, obtaining a probabilistic description of each action potential and, in turn, of the whole waveform. This method allows us to compute the class conditional probabilities of each AP which quantifies the certainty with which an AP is assigned to a given class. In addition, it will be possible to quantify the certainty of both the form and number of spike shapes. Finally, we can use this description to decompose overlapping APs efficiently and to assign probabilities to alternative spike model sequences.

2 Modeling Action Potentials

First we consider the problem of fitting a model to events from a single cell. Let us assume that the data from the event we observe (at time zero) is a result of a fixed underlying spike function, $s(t)$, plus noise:

$$d_i = s(t_i) + \eta_i. \tag{1}$$

A computationally convenient form for $s(t)$ is a continuous piece-wise linear function:

$$s(t) = y_j + \frac{v_j}{h}(t - x_j), \qquad x_j \leq t < x_{j+1}, \tag{2}$$

where $h = x_{j+1} - x_j, j = 1 \ldots R$, and $v_j = y_{j+1} - y_j$. We will treat R and the x_j's as known. The noise, η, is modeled as a Gaussian process with zero mean and standard deviation σ_η.

2.1 The posterior for the model parameters

From the Bayesian perspective, the task is to infer the posterior distribution of the parameters, $\mathbf{v} = \{v_1, \ldots, v_R\}$, given the data from the observed events, D, and our prior assumptions of the spike model, M. Applying Bayes' rule we have

$$P(\mathbf{v}|D, \sigma_\eta, \sigma_w, M) = \frac{P(D|\mathbf{v}, \sigma_\eta, M)\, P(\mathbf{v}|\sigma_w, M)}{P(D|\sigma_\eta, \sigma_w, M)}. \tag{3}$$

$P(D|\mathbf{v}, \sigma_\eta, M)$ is the probability of the data for the model given in (2) and is assumed to be Gaussian:

$$P(D|\mathbf{v}, \sigma_\eta, M) = \frac{1}{Z_D(\sigma_\eta)} \exp\left[-\frac{1}{2\sigma_\eta^2}\sum_{i=1}^{I}(d_i - s(t_i))^2\right], \tag{4}$$

where $Z_D(\sigma_\eta) = 1/(2\pi\sigma_\eta^2)^{I/2}$. The time of the ith data point, d_i, is taken to be relative to the corresponding event, *i.e.*$t_i = t_i^{(n)} - \tau^{(n)}$. By convention, $\tau^{(n)}$ is the time of the inferred AP peak. The data range over the predetermined extent of the action potential.[1]

$P(\mathbf{v}|\sigma_w, M)$ specifies prior assumptions of the structure of $s(t)$. Ideally, we want a distribution over $\mathbf{v}$ from which typical samples result only in shapes that are plausible APs. Conversely, this space should not be so restrictive that legitimate AP shapes are excluded. We adopt a simple approach and use a prior of the form

$$P(s(t)|\sigma_w, M) \propto \exp\left[-\int du\, s^{(m)}(u)^2/\sigma_w^2\right], \tag{5}$$

where the superscript (m) denotes differentiation. $m = 1$ corresponds to linear splines, $m = 3$ corresponds to cubic splines, *etc.* The smoothness of $s(t)$ is controlled through the parameter σ_w with small values of σ_w penalizing large fluctuations. A prior simply favoring smoothness ensures minimal restrictions on the kinds of functions we can interpolate, but it doesn't buy us anything either. If we had a more informative prior, we would require less data to reach the same conclusions about the form of $s(t)$. Any reasonable prior should have little effect on the shape of the final spike function if there are abundant data. Even though the prior may have little effect on the shape, it still plays an important role in model comparison which will be discussed in section 4.

[1] For the examples shown here, this range is from 1 msec before the spike peak to 4 msec after the peak.

The components of the posterior distribution for $\mathbf{v}$ are now defined. There still remains, however, the problem of determining σ_η and σ_w. An exact Bayesian analysis requires that we eliminate the dependence of the posterior on σ_η and σ_w by integrating them out:

$$P(\mathbf{v}|D, M) = \int d\sigma_\eta \, d\sigma_w \, P(\mathbf{v}|D, \sigma_\eta, \sigma_w, M) \, P(\sigma_w, \sigma_\eta|M). \tag{6}$$

In this paper, we use the approximation $P(\mathbf{v}|D, M) \approx P(\mathbf{v}|D, \sigma_w^{\mathrm{MP}}, \sigma_\eta^{\mathrm{MP}}, M)$. The most probable values of $\mathbf{v}$, σ_w, and σ_η were obtained using the methods of MacKay (1992). Note that at this point, we could use probability theory to compare alternative spike models, in essence to determine the *most probable* spike model given the data. For example, we might choose cubic splines instead of piece-wise linear functions or choose priors that better represented our knowledge about spike shapes. The piece-wise linear spike models discussed here can be made to fit any fixed shape, since they can contain arbitrarily many segments. With 75 segments, the spike models have been descriptively sufficient for the all the data we have observed. Figure 1a shows the result of fitting one spike model to data consisting of 40 APs.

2.2 Checking the assumptions

Before proceeding to the more complicated cases of multiple spike models and overlapping spikes, we must check our assumptions on real data. Equation (1) assumes that the noise process is invariant throughout the duration of the AP, but in principle this need not be the case. For example, the noise might show larger variation at the extremes. The spike model residuals, $\eta_i = d_i - s(t_i)$, shown in figure 1a, give no indication of an amplitude-dependent noise process.

A second assumption we have made is that the noise is Gaussian. Figure 1b shows a Gaussian distribution with the inferred width σ_η overlaid on a normalized histogram of the residuals from figure 1a. The most significant deviation is in the tails of the distribution which reflects the presence of overlapping spikes. In this case, the overlaps are evenly distributed over the range of the fitted event so they have little effect on the model's form in the limit of large amounts of data. The model would be poorly inferred, however, if the overlaps were not uniformly distributed over the interval, for example if one cell tended to fire within a few milliseconds of another. This is a common problem in practice and will be addressed in section 5.

An assumption which has not been tested is whether the residuals are independent. Figures 1c and 1d show that the noise in these data is slightly correlated. This has little effect on the fit of the models but does affect the accuracy of the probabilities discussed in the later sections. A convenient way of reducing the correlation is to sample close to the Nyquist rate to avoid correlation introduced by the amplifier filters.

3 Multiple Spike Shapes

When an event contains multiple APs, determining the component spike shapes is more difficult because the classes are not known *a priori*. We cannot infer the parameters for one spike model if we don't know what data is representative of its class. Furthermore, if two spike models are similar, it is possible that an observed event could have come from

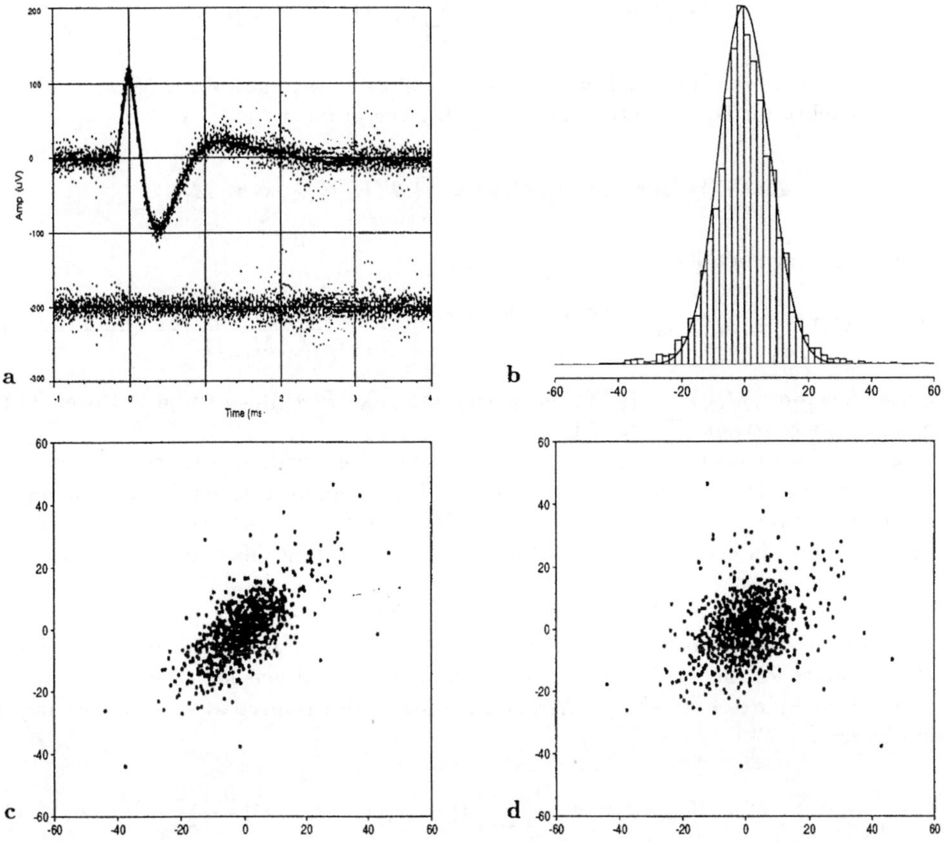

Figure 1: (a) Spike model fit to data consisting of 40 APs. The solid line is a 75 segment piece-wise linear model. Each AP is aligned with respect to the inferred spike peak. Each dot is one sample point. The residual error for each sample, $\eta_i = d_i - s(t_i)$, is offset by -200μV and plotted below. The flat residuals indicate that the data is well-fit by the model. (b) Normalized histogram of the residuals from **a**. The curve is the Gaussian inferred with the methods discussed in the text. The outliers result from overlapping APs which can be seen in the data in **a**. (**c** and **d**) Lagged scatter plot of a sample of the residuals in **a**. (**c**) η_i vs η_{i+1}. (**d**) η_i vs η_{i+2}. These graphs indicate that there is some correlation between η_i and η_{i+1} (**c**), but little between η_i and η_{i+2} (**d**). This is expected for these data because the sampling rate (20kHz) was higher than the Nyquist rate (14kHz).

either class with equal probability. The uncertainty of which class an event belongs to can be incorporated with a mixture distribution (Duda and Hart, 1973).

The probability of a particular event, $\mathbf{D}_n$, given all spike models, $M_{1:K}$, is

$$P(\mathbf{D}_n|\mathbf{v}_{1:K}, \boldsymbol{\pi}, \sigma_\eta, M_{1:K}) = \sum_{k=1}^{K} \pi_k P(\mathbf{D}_n|\mathbf{v}_k, \sigma_\eta, M_k), \tag{7}$$

where π_k is the *a priori* probability that a spike will be an instance of M_k ($\sum \pi_k = 1$). The joint probability for $\mathbf{D}_{1:N} = \{\mathbf{D}_1 \ldots \mathbf{D}_N\}$ is simply the product

$$\mathcal{L} = P(\mathbf{D}_{1:N}|\mathbf{v}_{1:K}, \boldsymbol{\pi}, \sigma_\eta, M_{1:K}) = \prod_{n=1}^{N} P(\mathbf{D}_n|\mathbf{v}_{1:K}, \boldsymbol{\pi}, \sigma_\eta, M_{1:K}). \tag{8}$$

The posterior for multiple spike models is then

$$P(\mathbf{v}_{1:K}, \boldsymbol{\pi}|\mathbf{D}_{1:N}, \sigma_\eta, \boldsymbol{\sigma}_w, M_{1:K}) = \frac{P(\mathbf{D}_{1:N}|\mathbf{v}_{1:K}, \boldsymbol{\pi}, \sigma_\eta, M_{1:K})\, P(\mathbf{v}_{1:K}|\boldsymbol{\sigma}_w, M_{1:K})\, P(\boldsymbol{\pi}|M_{1:K})}{P(\mathbf{D}_n|\sigma_\eta, \boldsymbol{\sigma}_w, M_{1:K})}. \tag{9}$$

We use $P(\mathbf{v}_{1:K}|\boldsymbol{\sigma}_w, M_{1:K}) = \prod_k P(\mathbf{v}_k|\sigma_{wk}, M_k)$ and take $P(\boldsymbol{\pi}|M_{1:K})$ to be flat over $[0,1]^K$ subject to the constraint $\sum_k \pi_k = 1$.

Note that we have implicitly assumed that the spike occurrence times are Poisson in nature with mean firing rates proportional to π_k. This assumes as little as possible about the temporal structure of the spikes. A more powerful description, *e.g.*modeling the distribution of the inter-spike interval, would be obtained by incorporating this information into (8).

3.1 Maximizing the posterior

We proceed as before to find the maxima of the posterior which will give us the most probable values for the whole set of spike models. The conditions satisfied at the maxima of $\mathcal{L}$ given in (8) are obtained by differentiating $\log \mathcal{L}$ with respect to $\mathbf{v}_k$ and equating the result to zero,

$$\frac{\partial \log \mathcal{L}}{\partial \mathbf{v}_k} = \sum_{n=1}^{N} P(M_k|\mathbf{D}_n, \mathbf{v}_k, \boldsymbol{\pi}, \sigma_\eta) \frac{1}{\sigma_\eta^2} \sum_i [d_{n,i} - s_k(t_i - \tau_n\, ; \mathbf{v}_k)] \frac{\partial s_k(t_i\, ; \mathbf{v}_k)}{\partial \mathbf{v}_k} = 0, \tag{10}$$

where τ_n is the occurrence time of $\mathbf{D}_n$. Thus we obtain a soft clustering procedure in which the error for each event, $\mathbf{D}_n$, is weighted by the probability that it is an instance of M_k:

$$P(M_k|\mathbf{D}_n, \mathbf{v}_k, \boldsymbol{\pi}, \sigma_\eta) = \frac{\pi_k P(\mathbf{D}_n|\mathbf{v}_k, \sigma_\eta, M_k)}{\sum_k \pi_k P(\mathbf{D}_n|\mathbf{v}_k, \sigma_\eta, M_k)}. \tag{11}$$

Although (10) can be solved exactly, it is still expensive to compute, because it uses all of the data. We adopt the approach of estimating each $\mathbf{v}_k$ by fitting each model to a reduced event list allowing the possibility of an event being in the lists of multiple models. These lists are obtained by sampling events from the whole data set and including an event in a model's reduced event list with probability proportional to $P(M_k|\mathbf{D}_n, \mathbf{v}_k, \boldsymbol{\pi}, \sigma_\eta)$. We apply the techniques used in the previous section to determine the values for $\boldsymbol{\sigma}_w$, and in turn the most probable values of $\mathbf{v}_{1:K}$.

Differentiating (8) and finding the condition satisfied at the maximum, we obtain the re-estimation formula

$$\pi_k = \frac{1}{N}\sum_n P(M_k|\mathbf{D}_n, \mathbf{v}_k, \boldsymbol{\pi}, \sigma_\eta). \tag{12}$$

For each model, σ_η can be estimated using the methods of the previous section. The mixture model estimate for σ_η is obtained by a weighted average of the individual estimates using weight π_k.

3.2 Selecting events from the data

For these demonstrations, any peak in the waveform that deviated from DC level by more than 4 times the estimated RMS noise level was labeled as an event, $\mathbf{D}_n$. Once an event is located, it is important to obtain accurate estimates of the occurrence time (with each spike model) by maximizing (4) over τ_n. For the largest models, deviations from the optimal value as little as one-tenth the sampling period will introduce misfit errors greater than σ_η. The τ_n's must be re-estimated as the spike models change for optimal results. An efficient way to perform this optimization is to use the k-d trees discussed in section 5.

3.3 Initial conditions

Since the re-estimation formulas derived here will find *local* maxima, it is critical to use good initial conditions for the spike models. Poor fits will result if there are too few spike models representing what are in fact several distinct APs. Conversely, if there are more spike models than distinct APs, not only will there be excess computational overhead, but there is no guarantee that each AP will be represented, since some spike functions may converge to represent the same AP class. Ideally, we want all potential spike shapes to be represented in the initial spike function set, $s_{1:K}(t)$. One approach toward obtaining an even representation of the AP shapes is to initialize each spike function to single events so that $\max_t s(t) - \min_t s(t)$ is evenly distributed with a separation proportional to the estimated waveform RMS noise. This approach works well for present purposes, because the height of an AP captures much of the variability among classes. By erring on the side of starting with too many spike models, we can obtain a good initial representation of the AP shapes. There is still a need to decide if two different models should be combined and if one class should be split into two. How to choose the number of spike models objectively will be demonstrated in the next section.

4 Determining the Number of Spike Models

If we were to choose a set of spike models which best fit the data, we would wind up with a model for each event in the waveform. We might think of heuristics which would tell us when two spike models are distinct and when they are not, but *ad hoc* criteria are notoriously dependent on particular circumstances, and it is difficult to state precisely what information the rules take into account. A solution to this dilemma is provided by probability theory (Jeffreys, 1939; Jaynes, 1979; Gull, 1988).

To determine the *most probable* number of spike models, we need to derive the probability of a set of spike models, denoted by $\{S_j = M^{(j)}_{1:K}\}$, conditioned only on the data and information known *a priori*, which we denote by H. From Bayes' rule, we obtain

$$P(S_j|\mathbf{D}_{1:N}, H) = \frac{P(S_j|H)P(\mathbf{D}_{1:N}|S_j, H)}{P(\mathbf{D}_{1:N}|H)}. \tag{13}$$

The only data dependent term is $P(\mathbf{D}_{1:N}|S_j, H)$ which is called the *evidence* for S_j. If we assume all the hypotheses $S_{1:J}$ under consideration are equally probable, $P(\mathbf{D}_{1:N}|S_j, H)$ ranks alternative spike sets, since it is proportional to $P(S_j|\mathbf{D}_{1:N}, H)$. With equal priors, the ratio $P(\mathbf{D}|S_i, H)/P(\mathbf{D}|S_j, H)$ is equal to the Bayes factor in favor of hypothesis S_i over hypothesis S_j which is the standard way to compare hypotheses in the Bayesian literature.

The evidence for S_j is obtained by integrating out the nuisance parameters in (9):

$$P(\mathbf{D}_{1:N}|S_j) = \int d\mathbf{v}_{1:K} d\pi d\sigma_\eta d\boldsymbol{\sigma}_w \, P(\mathbf{D}_{1:N}|\mathbf{v}_{1:K}, \pi, \sigma_\eta, S_j)\, P(\mathbf{v}_{1:K}|\boldsymbol{\sigma}_w, S_j)\, P(\pi|S_j)\, P(\sigma_\eta, \boldsymbol{\sigma}_w|S_j). \tag{14}$$

This integral is analytically intractable, but it is often well-approximated with a Gaussian integral which for a function $f(\mathbf{w})$ is given by

$$\int d\mathbf{w} f(\mathbf{w}) \approx f(\hat{\mathbf{w}})\,(2\pi)^{d/2}\,|-\nabla\nabla \log f(\mathbf{w})|^{-1/2}, \tag{15}$$

where d is dimension of $\mathbf{w}$, $\hat{\mathbf{w}}$ is a (local) maximum of $f(\mathbf{w})$, $|\mathbf{A}|$ denotes the determinant of $\mathbf{A}$, and the derivatives are evaluated at $\hat{\mathbf{w}}$. With this we obtain the evidence for spike set S_j,

$$\begin{aligned} P(\mathbf{D}_{1:N}|S_j, H) &= P(\mathbf{D}_{1:N}|\hat{\mathbf{v}}_{1:K}, \hat{\pi}, \hat{\sigma}_\eta, S_j)\, P(\hat{\mathbf{v}}_{1:K}|\boldsymbol{\sigma}_w, S_j)\, P(\hat{\pi}|S_j)\, P(\hat{\boldsymbol{\sigma}}_w, \hat{\sigma}_\eta|S_j) \\ &\quad \cdot (2\pi)^{d/2}\,|-\nabla\nabla \log P(\mathbf{D}_{1:N}|\mathbf{v}_{1:K}, \pi, \sigma_\eta, S_j)|^{-1/2}\, \Delta \log \hat{\boldsymbol{\sigma}}_w\, \Delta \log \hat{\sigma}_\eta . \end{aligned} \tag{16}$$

where $\Delta \log \hat{\boldsymbol{\sigma}}_w = \prod_k \sqrt{2/\gamma_k}$, $\Delta \log \hat{\sigma}_\eta = \sqrt{2/(NI-\gamma)}$, and $d = KR + K + 1$. γ_k is the number of good degrees of freedom for M_k (MacKay, 1992) which can be thought of as the number of parameters that are well-determined by the data. $\gamma = \sum_k \gamma_k$. $P(\sigma_w, \sigma_\eta|S_j)$ is assumed to be separable and flat over $\log \sigma_w$ and $\log \sigma_\eta$. Since the labeling of the models is arbitrary, an additional factor of $1/K!$ must be included to estimate the posterior volume accurately. The Hessian $-\nabla\nabla \log P(\mathbf{D}_{1:N}|\mathbf{v}_{1:K}, \pi, \sigma_\eta, S_j)$ (with respect to $\mathbf{v}_{1:K}$ and π) was evaluated both analytically and using a diagonal approximation. Both methods produced similar results, and the latter, being much faster to compute, was used for these demonstrations. Notice that the approximation for the evidence decomposes into the best-fit likelihood for the best fit parameters times the other terms which collectively constitute a complexity penalty called the Ockham factor (MacKay, 1992). Since this factor is the ratio of the posterior accessible volume in parameter space to the prior accessible volume, it is smaller for more complicated models. Overly broad priors will introduce a bias toward simpler models. Unless the best-fit likelihood for complex models is sufficiently larger than the likelihood for simple ones, the simple models will be more probable.

A convenient way of collapsing the spike set is to compare spike models pairwise. Two models in the spike set are selected along with a sampled set of events fit by each model. We then evaluate $P(\mathbf{D}|S_1)$ and $P(\mathbf{D}|S_2)$. S_1 is the hypothesis that the data is modeled by a single spike shape, S_2 says there are two spike shapes. Included in the list of spike models should be a "null" model which is simply a flat line at DC. This hypothesis says that there

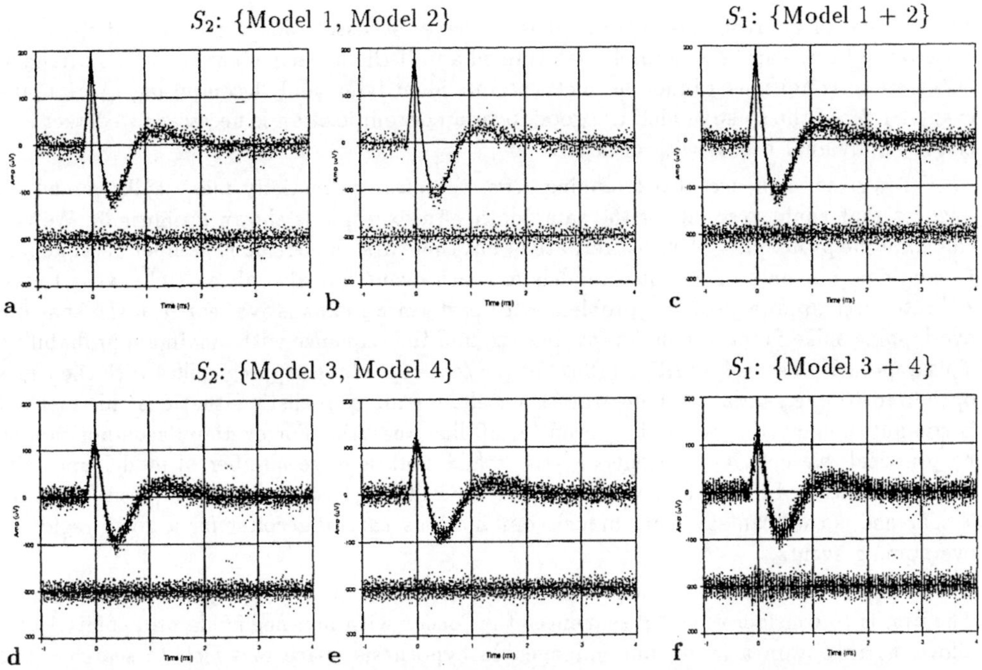

Figure 2: The most probable number of distinct spike models is determined by evaluating the evidence for alternative hypotheses for a given set of data. Simple hypotheses are generated by selecting similar shapes in a spike set. S_2 is the hypothesis that there are two distinct spike models; the fits of two such models to a sampled set of data are shown in **a** and **b**. S_1 is the hypothesis that there is only one spike model; the fit of this model is shown in **c**. In this case, even though the total misfit is less for S_2, the simpler hypothesis, S_1, is more probable by exp(111) to 1. In the second row, S_2 (**d** and **e**) is more probable than S_1 (**f**) by exp(343) to 1. Note the increase in residual error in the model shown in **f**. The difference between model 3 and model 4 is shown in figure 8 (where they are labeled M1 and M2 respectively). The large probability ratios reported here result mainly from the non-Gaussian outliers in the noise. A more realistic noise model, such as heavy-tailed Gaussian, would result in more accurate probability ratios.

are no events and that the data is a result of only the noise. Examples of this comparison are illustrated in figure 2. If $P(\mathbf{D}|S_1) > P(\mathbf{D}|S_2)$, we replace both models in S_2 by the one in S_1. The procedure terminates when no more pairs can be combined to increase the evidence.

5 Decomposing Overlapping Events

The method of inferring the spike models we have discussed thus far is valid if the event occurrence times can be accurately determined and if the noise is Gaussian and stationary. Often these conditions cannot be met without identifying and decomposing overlapping events. Even if the spike models are good, overlap decomposition is necessary to detect and classify individual events with accuracy.

For a given sequence of overlapping APs, there are potentially many spike model sequences that could account for the same data. An example is shown in figure 3. We can calculate the probability of each alternative, but there are an enormous number of sequences to consider, not only all possible models for each event but also all possible event times. A brute-force approach to this problem is to perform an exhaustive search of the space of overlapping spike functions and event times to find the sequence with maximum probability. This approach was used by Atiya (1992) in the case of two overlapping spikes with the times optimized to one sample period. Unfortunately, for many realistic situations this method is computationally too demanding even for off-line analysis. For overlap decomposition to be practical, we need an efficient way to fit and rank a large number of model potential spike sequences. In addition, we would like to state precisely what hypothesis subspace is searched, so we can say what model combinations *cannot* account for a given region of overlapping events.

We can obtain a more efficient decomposition algorithm by employing two techniques. The first is to consider only AP sequences that occur with non-negligible probability. This allows us to obtain a large, but manageable hypothesis space in which to search. The second is to make the search itself efficient using appropriate data structures and dynamic programming.

5.1 Restricting the overlap hypothesis space

The main difficulty with overlapping APs is that there is no simple way to determine the event times. For many overlaps, such as the one in figure 4a, the event times can be determined directly, because the APs are separated enough so that the models can be fit independently. As the degree of overlap increases, as in figures 4b and c, accurate classification of one event depends on accurate classification of the surrounding events. In this case, the overlapping models must be fit simultaneously. Moreover, since small misalignments of the model with respect to the event can introduce significant residual error, each model in the overlap sequence must be precisely aligned.

The continuum of possible event times is the major factor contributing to the multitude of potential overlap models. We can reduce this space significantly if we consider to what precision the τ_n's must be optimized. For a given spike model, $s_k(t)$, the maximum error

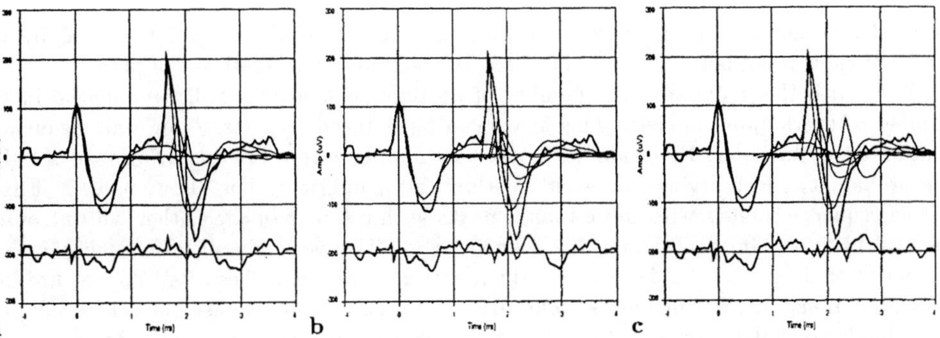

a b c

Figure 3: Over-fitting also occurs in the case of decomposing overlapping events. Shown are 3 of many well-fitting solutions for a single region of data. Thick lines are drawn between the data samples. The thin lines are the spike functions (These examples were taken from the first iteration of the algorithm, so the spike functions are noisy estimates of the underlying AP shapes). The best-fitting solution in this case is not the most probable; the solution with 4 spike functions shown in **a** is more than 8 times more probable than either **b** (5 spike functions) or **c** (6 spike functions) even though these fit the data better. Simply using the *best-fitting* overlap solution results in a dramatic increase in classification error especially in the number of false positives for the smaller models. Finding the *most probable* overlap solution minimizes the classification error.

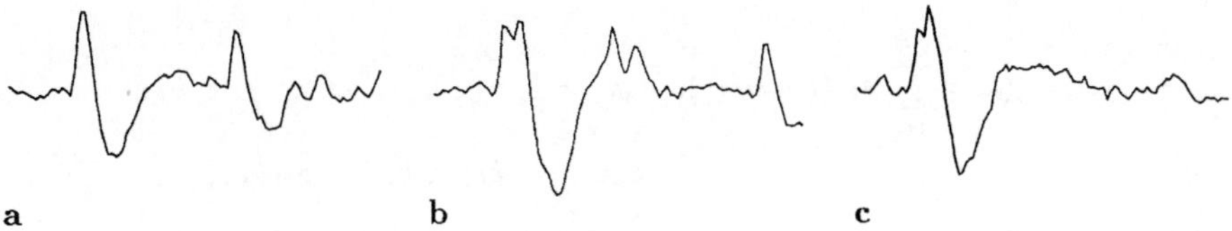

a b c

Figure 4: As the peaks of two action potentials get closer together, it becomes more difficult to classify either one with accuracy. It is necessary in this case (**b** and **c**) to fit multiple models simultaneously.

resulting from a misalignment of δ_k is given by[2]

$$\epsilon = \delta_k \max_t \left| \frac{ds(t)}{dt} \right| . \tag{17}$$

From this we obtain the precision necessary to ensure that the error introduced by the model alone is less than ϵ and only need to choose among a discrete set of points.[3]

Even with this reduction, the number of possible sequences is still exponential in the number of overlapping models. This space can be reduced by considering only sequences that are likely to occur. For example, if there are 5 units with a Poisson firing rate of 20Hz, the probability of observing two events within half a millisecond is about 0.0012. Eliminating sequence models with more than 2 peaks within 0.5ms of each other will introduce about 0.1% error. In this manner, the desired trade-off between classification accuracy and computational cost can be determined. In practice, however, spikes often do not fire in a Poisson manner but fire in bursts. The firing rate model in this case should be adapted accordingly so that the expected number of missed events is estimated accurately.

5.2 SEARCHING THE OVERLAP HYPOTHESIS SPACE

Let us first outline the decomposition algorithm. To fit general model sequences, we use the methodology of dynamic programming. The event data is fit in sections from left to right. At every stage, a list is maintained of all plausible sequences[4] from the restricted hypothesis space determined by the methods described above. The length of data fit is extended by computing for each sequence on the list all plausible models that result by fitting the residual structure in the next region. The probabilities for all sequences are then recomputed, discarding any sequences below the probability threshold. The search terminates when no further overlaps are encountered in the most probable sequence model.

We now discuss each step in more detail. The primary operation in the algorithm is that of determining the most probable sequence models for a region of data. For efficiency, we precompute all possible waveform segments and store the set in a k-d tree (Bently 1975) with which a fixed-radius nearest neighbor search can be performed in time logarithmic in the number of models (Friedman *et al.*, 1977; Ramasubramanian and Paliwal, 1992). $O(N \log N)$ time is required to construct the tree, but once it is set up, each nearest-neighbor search is very fast. The set of overlap functions for a region from a to b around the spike peak is defined by

$$\Lambda_{k_{1:L},n}(t) = \sum_{j=1}^{L} s_{k_j}(t - n\delta_{k_j}), \quad k_j = 1, \ldots, K, \quad k_1 < \ldots < k_L, \tag{18}$$
$$a < t - n\delta_{k_j} < b, \quad n \text{ integer},$$

where L is the maximum number of overlapping spike function segments in the peak region $[a, b]$, and δ_{k_j} is the τ-resolution for $s_k(t)$ defined in (17). The size of the peak region is somewhat arbitrary; the larger the region, the larger the number of waveform segments

[2] We ignore the discontinuities in the derivative of the piece-wise linear model.

[3] For these demonstrations we use $\epsilon = 0.5\sigma_\eta$ which results in δ_k's ranging from 0.05 to 0.3 sampling periods.

[4] By plausible sequences we mean sequences with probability greater than a specified threshold.

that must be considered, but the smaller the number of plausible overlap sequences found. In practice, the size of the peak region is largely limited by the memory required for the k-d tree. For these demonstrations, we take $L = 2$ (up to two overlapping spike functions segments) with a peak region of 0.25ms and include a "noise" model Λ_0 which has constant value equal to the DC voltage level. The number of waveform segments in the set can be reduced by eliminating overlapping spike functions for which the peak would have been (with high probability) detected at a sample position other than that of the data. Even with this reduction, an 11-model spike set results in about 50,000 waveform segments.

Once the best-fitting waveform segments for the first peak region are obtained, each segment is extended until the next peak in the residuals for that segment. This peak is then fit using the k-d tree which generates additional overlap sequences. As long as the introduction of new waveform segments does not alter our conclusions about the ordering sequence list, for example by fitting structure in a preceding region, we ensure either that one of the overlap sequences is true or that the sequences we are considering cannot account for the data.

After each sequence from the original list has been extended, the probability of each sequence model, $\mathbf{c}_i$, is recomputed. The exact relation is given by

$$P(\mathbf{c}_i|\mathbf{D},\mathbf{S}) = \int d\boldsymbol{\tau}_i \frac{P(\mathbf{D}|\mathbf{c}_i,\boldsymbol{\tau}_i,\mathbf{S})P(\mathbf{c}_i,\boldsymbol{\tau}_i|\mathbf{S})}{P(\mathbf{D}|\mathbf{S})}, \tag{19}$$

where $\mathbf{D}$ is the subset of data common to all sequences, and $\mathbf{S} = \{\mathbf{v}_{1:K}, \boldsymbol{\pi}, \sigma_\eta, M_{1:K}\}$. The form of the probability density function, $P(\mathbf{D}|\mathbf{c}_i,\boldsymbol{\tau}_i)$, is the same as (4). Equation (19) can be approximated with a Gaussian integral by treating each peak region as a separable component,

$$P(\mathbf{c}_i|\mathbf{D},\mathbf{S}) \approx \frac{P(\mathbf{D}|\mathbf{c}_i,\hat{\boldsymbol{\tau}}_i,\mathbf{S})\,(2\pi)^{C/2}\prod_j d_j^{-1/2}\,P(\mathbf{c}_i|\mathbf{S})\,P(\boldsymbol{\tau}_i|\mathbf{S})}{P(\mathbf{D}|\mathbf{S})}, \tag{20}$$

where C is the total number of spike functions in the sequence, and d_j is the determinant of Hessian of the τ's for the jth peak region. The values needed to compute the Hessians can be obtained directly from the k-d tree. Note that integrating over $\boldsymbol{\tau}_i$ performs the function of Ockham's Razor by penalizing sequences with many spike models. Omitting this would reduce the solution to one of maximum likelihood which chooses the sequence that best fits the data. For example, the solutions shown in figure 3b and 3c both fit the data better than in 3a, but by (20), 3a is more than 8 times more probable than the others.

$P(\mathbf{c}_i,\boldsymbol{\tau}_i|\mathbf{S})$ describes the *a priori* probability of the sequence of models in $\mathbf{c}_i$ with associated occurrence times $\boldsymbol{\tau}_i$. For this discussion, we assume $P(\mathbf{c}_i|\mathbf{S})$ to be Poisson with rate proportional to $\langle\pi_k\rangle$ and $P(\boldsymbol{\tau}_i|\mathbf{S})$ to be proportional to $1/\langle\pi_k\rangle$. Useful alternatives for $P(\mathbf{c}_i,\boldsymbol{\tau}_i|\mathbf{S})$ include models which take into account a refractory period or describe different types of spiking patterns.

Once the probabilities for the sequence models have been computed, the improbable models are discarded. The decomposition algorithm iterates until no overlapping structure is found in the most probable model. The search can fail if an outlier is encountered or if the true sequence is outside the hypothesis space. Otherwise, upon termination the search results in a list of all plausible sequence models of the given data along with their associated probabilities. Example decompositions are shown in figure 5.

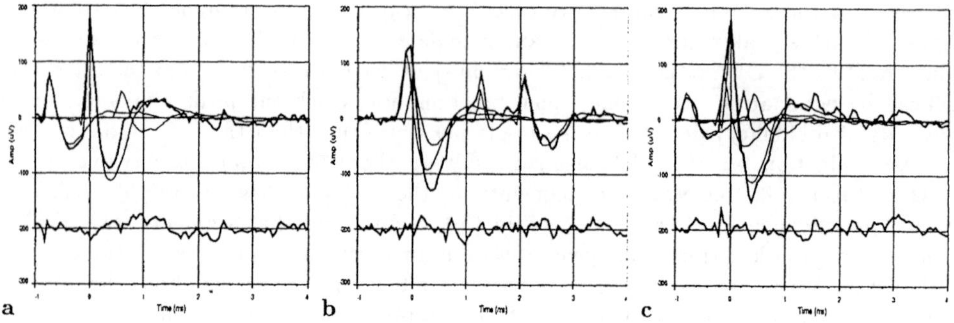

Figure 5: Example overlap solutions. Thick lines are drawn between the data samples. The thin lines are the spike models. The overlap sequence in **a** has 3 spike functions, **b** contains 4 spike functions, and **c** contains 5 spike functions.

6 Performance on Real Data

The algorithm was first tested on real data, a section of which was shown in figure 1. The whole waveform consisted of 40secs of data, filtered from 300 to 7000Hz and sampled at 20kHz. Three iterations of the algorithm were performed with overlap decomposition after the second (with $L = 1$) and third (with $L = 2$) iterations. Spike models which occurred less than 10 times were discarded for efficiency, and the remaining events were reclassified. The inferred spike models are shown in figure 6. The residuals indicate that these spike models account for almost all events in the 40sec waveform. Out of about 1500 total events, only 6 were not fit to within $5\sigma_\eta$. By eye, these events looked very noisy and had no obvious composition in terms of the spike models. One possibility is that they resulted from animal movement. Such events were not present in the synthesized data set described in section 7 where all the events were fit with the inferred spike models.

By eye, all the models look distinct except perhaps for M_2 and M_3. One way to see the difference between these two models is to fit the data from model 3 with model 2 as shown in figure 7. With a single electrode it is difficult to determine whether or not these two shapes result from different neurons, but they are clearly two types of events. One possibility is that these are different states of the same neuron; another is that the shape in model 3 results from a tight coupling between two neurons. Recording with multiple electrodes from a local region of tissue would help resolve issues like this.

In spite of all the math, the algorithm is fast. Inferring the spike set with overlap decomposition takes a few minutes on a Sun Microsystems Sparc IPX. Classification of the 40 second test waveform with overlap decomposition (using $L = 1$) takes about 10 seconds.

7 Performance on Test Data

The accuracy of the algorithm was tested by generating an artificial data set composed of the six inferred shapes shown in figure 6. The event times were Poisson distributed with

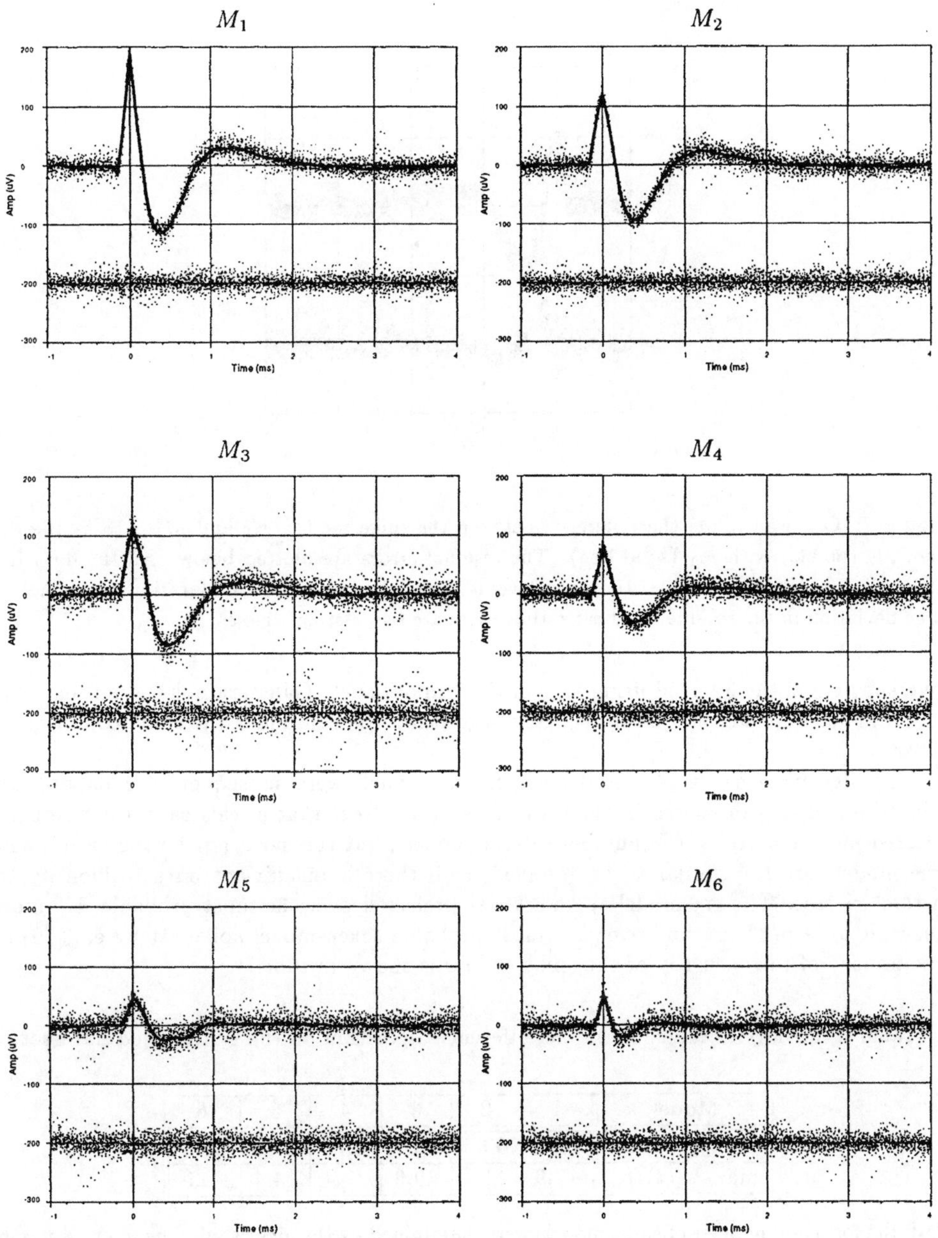

Figure 6: The solid lines are the inferred spike models. The data overlying each model is a sample of at most 40 events. The residual errors are plotted below each model.

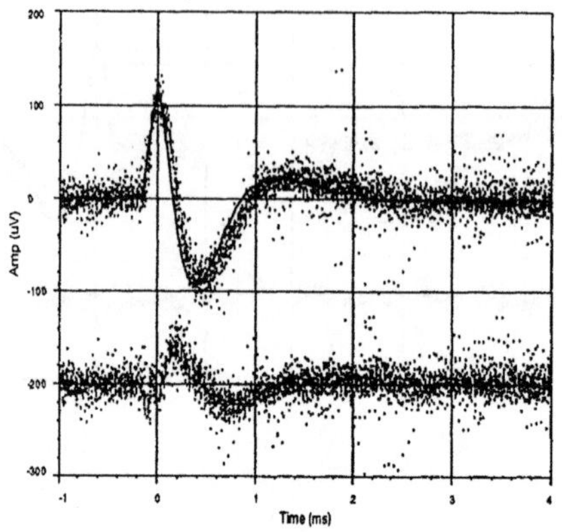

Figure 7: One way to see the difference between the spike models M_1 and M_2 is to fit the data from M_2 (points) with M_1 (solid line). The residual errors are plotted below. All the data from both spike models is plotted. If the noise level is constant throughout the duration of the AP, the large deviation in the residuals indicates that there are two distinct classes.

frequency equal the inferred firing rate of the real data set. Gaussian noise was then added with standard deviation equal to σ_η. The algorithm was run under the same conditions as above.

The algorithm chose 14 initial spike models which were subsequently collapsed to 6 using the methods discussed in the previous section. Note that in this case, the number of inferred models matches the number of true models, but this need not be the case if some true models are too similar to be resolved, or if there is insufficient data to identify two distinct classes. The six model spike set was preferred over the most probable five-model spike set by exp(34) : 1 and over the most probable seven-model spike set by exp(19) : 1. A summary of the accuracy of the spike shapes is shown in table 1.

Table 1: Results of the spike model inference algorithm on the synthesized data set.

Model	1	2	3	4	5	6
$\Delta_{\max}/\sigma_\eta$	0.44	0.36	1.07	0.78	0.84	0.40
$\max_t s_k(t)/\sigma_\eta$	17.9	11.1	10.6	7.4	4.4	5.0

Both the form and number of spike models were determined by the algorithm. The inferred number of spike models matched the true number (6 models). The middle row is the maximum absolute difference between the true spike model and the inferred model normalized by σ_η. The last row is the normalized peak of the inferred spike models which is an indication of how far each type of AP is above the noise level.

Table 2: Classification results for the non-overlapping events of the synthesized data set.

True Models	Inferred Models 1	2	3	4	5	6	Missed Events	Total Events
1	17	0	0	0	0	0	0	17
2	0	25	1	0	0	0	0	26
3	0	0	15	0	0	0	0	15
4	0	0	0	116	0	0	1	117
5	0	0	0	0	56	0	17	73
6	0	0	0	0	0	393	254	647

Table 3: Classification results for the overlapping events of the synthesized data set.

True Models	Inferred Models 1	2	3	4	5	6	Missed Events	Total Events
1	22	0	0	0	0	0	0	22
2	0	36	1	0	0	0	0	37
3	0	0	20	0	0	0	0	20
4	0	1	0	116	0	1	3	121
5	0	0	0	1	61	1	19	82
6	0	0	0	3	2	243	160	408

Tables 2 and 3: Each matrix component indicates the number of times true model i was classified as inferred model j. Events were missed if the true spikes were not detected in an overlap sequence or if all sample values for the spike fell below the event detection threshold ($4\sigma_\eta$). There was 1 false positive for M_5 and 7 for M_6. See text for additional comments.

The results of inferring and classifying the synthesized data set are shown for the non-overlapping spikes in table 2 and for the overlapping spikes in table 3. An event was considered an overlap if the extent[5] overlapped the extent of another event. Perfect performance would have all zeros in the off-diagonal entries and no undetected events. An event can be missed if it is not detected in an overlap sequence or if all its sample values fall below the threshold for event detection ($4\sigma_\eta$). The tables indicate that for the largest four spikes, the performance is nearly perfect, even including the overlapping cases.

Performance is worst in the smallest two spike models where there are a large number of missed events. For these models, there are typically only two or three samples that would be expected to exceed the noise level. As the threshold for event detection is lowered, there is a tradeoff between the number of real spikes missed and the number of false positives, since random fluctuations in the noise can easily produce small spike – like shapes which get misclassified as one of the small spike models. The number of below threshold missed events

[5]The extent of a event is defined as the minimum and maximum values in time at which the best-fitting spike function differs from DC by more than $0.5\sigma_\eta$.

can be minimized (with additional computational expense) by computing the probabilities at every sample point instead of only those that cross threshold. It is worth noting that this situation often does not pose a problem in practice, since observed spikes just above the noise level frequently correspond to many different neurons.

8 Discussion

Formulating the task as the inference of a probabilistic model made clear what was necessary to obtain accurate spike models. Optimizing the τ_n's is crucial for both inference and classification, but this step is commonly ignored by algorithms which cluster the sample points or derive spike shapes from principal components. The soft clustering procedure makes it possible to determine the spike shapes with accuracy even when they are highly overlapping. Unless the spike shapes are well-separated, hard clustering procedures such as k-means will lead to inaccurate estimates of the spike shapes.

Probability theory also provided an objective means of determining the number of spike models which is an essential reason for the success of this algorithm. With the incorrect number of spike models overlap decomposition becomes especially difficult. If there are too few spike models, the overlap data cannot be fit. If there are too many, decomposition becomes a very expensive computation. The evidence has proved to be a sensitive indicator of when two classes are distinct, as was shown in figure 7. Previous approaches have relied on *ad hoc* criteria or the user to make this decision, but such approaches cannot be relied upon to work under varying circumstances since their inherent assumptions are not explicit. An advantage of probability theory is that the assumptions are explicit, and given those assumptions, the answer provided by the evidence is optimal.

One might wonder if the user, having much more information than has been incorporated into the model, can make better decisions than the evidence about what constitutes distinct spike models. Probability theory provides a calculus for stating precisely what can be inferred from the data given the model. When the conclusions reached through probability theory do not fit our expectations, it is due to a failure of the model or a failure of the approximations (if approximations are made). From the performance on the synthesized data, however, the approximations appear to be reasonable. Thus when the conclusions reached through the evidence are at variance with the user's, information is at hand about possible shortcomings of the current model. In this manner, new models can be constructed, and moreover, they can be compared objectively using the evidence.

Probability theory is also essential for accurate overlap decomposition. It is not sufficient just to fit data with compositions of spike models. That leads to the same over-fitting problem encountered in determining the number of spike models and in determining the spike shapes. The Ockham penalty introduced by integrating out the τ's was key to finding the most probable fits and consequently for achieving accurate classification. Previous approaches have been able to handle only a limited class of overlaps, mainly due to the difficulty in making the fit efficient. The algorithm we have described can fit an overlap sequence of virtually arbitrary complexity in milliseconds.

In practice, the algorithm we have described allows us to extract much more information from an experiment than with previous methods. Moreover, this information is qualitatively different from a simple list of spike times. Having reliable estimates of the action potential shapes makes it possible to study the properties of these classes, since distinct neuronal types can have distinct neuronal spikes (Connors and Gutnick 1990). With stereotrodes this advantage would be amplified, since it is then possible to estimate somatic size which is another distinguishing characteristic of cell type. Finally, accurate overlap decomposition makes it possible to investigate interactions among local neurons which were previously very difficult to observe.

Acknowledgements
I thank David MacKay for helpful discussions and encouragement during the early stages of this work and Jamie Mazer for many conversations and extensive help with the development of the software. Thanks also to Allison Doupe and Ken Miller for helpful feedback on the manuscript. This work was supported by Caltech fellowships and an NIH Research Training Grant.

References

[1] Atiya, A.F. (1992). Recognition of multiunit neural signals. *IEEE Trans. on Biomed. Eng.* **39**(7), 723–729.

[2] Bently, J.L. (1975). Multidimensional binary search trees used for associative searching. *Comm. ACM* **18**(9), 509–517.

[3] Connors, B.W. and Gutnick, M.J. (1990). Intrinsic firing patters of diverse neocortical neurons. *TINS* **13**(3), 99–104.

[4] Duda, R.O. and Hart, P.E. (1973). *Pattern classification and scene analysis,* Wiley–Interscience.

[5] Friedman, J.H., Bently, J.L., and Finkel, R.A. (1977). An algorithm for finding best matches in logarithmic expected time. *ACM Trans. Math. Software* **3**(3), 209–226.

[6] Gull, S.F. (1988). Bayesian inductive inference and maximum entropy, in *Maximum Entropy and Bayesian Methods in Science and Engineering, vol. 1: Foundations*, Erickson, G.J. and Smith, C.R. (eds.), Kluwer.

[7] Jaynes, E.T. (1979). Review of *Inference, Method, and Decision* (Rosenkrantz, R.D.). *J. Am. Stat.Assoc.* **74**, 140.

[8] Jeffreys, H. (1939). *Theory of Probability,* Oxford University Press (3rd revised ed 1961).

[9] MacKay, D.J.C. (1992). Bayesian Interpolation. *Neural Comp.* **43**, 415–445.

[10] Schmidt, E.M. (1984). Computer separation of multi-unit neuroelectric data: a review. *J. Neurosci. Methods* **12**, 95–111.

[11] Ramasubramanian, V. and Paliwal, K.K. (1992). Fast k-dimensional tree algorithms for nearest neighbor search with application to vector quantization encoding. *IEEE Trans. Signal Proc.* **40**(3), 518–531.

[12] Wahba, G. (1990). *Spline Models for Observational Data,* SIAM.

ESTIMATORS FOR THE CAUCHY DISTRIBUTION

K. M. Hanson and D. R. Wolf
Los Alamos National Laboratory, MS P940
Los Alamos, New Mexico 87545 USA
email: kmh@lanl.gov and wolf@lanl.gov

ABSTRACT. We discuss the properties of various estimators of the central position of the Cauchy distribution. The performance of these estimators is evaluated for a set of simulated experiments. Estimators based on the maximum and mean of the posterior probability density function are empirically found to be well behaved when more than two measurements are available. On the contrary, because of the infinite variance of the Cauchy distribution, the average of the measured positions is an extremely poor estimator of the central position. However, the median of the measured positions is well behaved. The rms errors for the various estimators are compared to the Fisher-Cramér-Rao lower bound. We find that the square root of the variance of the posterior density function is predictive of the rms error in the mean posterior estimator.

1. Introduction

We explore the properties of various estimators of the central position of the Cauchy distribution, which is notorious for the divergent nature of its first and higher moments. The results of using different kinds of estimators are evaluated by simulating a series of experiments using a Monte Carlo procedure. Investigation of the Cauchy distribution is profitable because its peculiar properties illustrate some interesting aspects of parameter estimation based on Bayesian analysis. It provides us with an example of how to properly deal with data outliers. Some aspects of this paper have been presented in [1].

2. The Cauchy Distribution

2.1. The problem

Suppose that a radioactive source, located at the position (x_0, y_0), emits gamma rays. A position-sensitive linear detector, colinear with the x axis and extending to infinity in both directions, measures the position x_i that the ith gamma ray hits the detector. The data consist of the values x_i, $i = 1, ..., N$, which we designate by the vector $\mathbf{x}$. The problem is to estimate the location of the source x_0, assuming that y_0 is known. This problem is Gull's lighthouse example [2] cast in another setting.

Assume that the gamma rays are confined to the x-y plane and are emitted uniformly in the angle θ at which they leave the source. From the relation $\tan(\theta) = -y_0/(x_i - x_0)$, which holds for $-\pi < \theta < 0$, the probability density function of a measurement x_i is obtained by using the Jacobian determinant to transform the density function dependence from θ to x_i

$$p(x_i|x_0, y_0) = \frac{y_0}{\pi\left[y_0^2 + (x_0 - x_i)^2\right]} \quad . \tag{1}$$

G. R. Heidbreder (ed.), Maximum Entropy and Bayesian Methods, 255–263.

We call $p(x_i|x_0, y_0)$ the likelihood. It is a properly normalized Cauchy distribution, which is notorious for having an undefined mean and an infinite variance. The width of this distribution may be characterized by its full width at half maximum (FWHM), which is $2y_0$.

In a Bayesian analysis the posterior probability density function for the x_0 position of the source summarizes the state of knowledge concerning x_0 by providing the probability density of every possible value of x_0. The posterior of x_0, given the data $\mathbf{x}$ and the position parameter y_0, is given by Bayes's law

$$p(x_0|\mathbf{x}, y_0) \propto p(\mathbf{x}|x_0, y_0)\, p(x_0|y_0) = p(\mathbf{x}|x_0, y_0)\, p(x_0) \ , \tag{2}$$

where we have assumed that the prior on x_0 is independent of y_0. Proportionality constants are always determined by normalization - the requirement that the probability that some event occurs is unity. If we suppose we have no prior information about the x_0 location of the source, then for the prior $p(x_0)$ we should use a constant over whatever sized region is required. Such a prior is noncommittal about the location of the source. Assume that the photons are emitted independently. Each measured x_i follows the likelihood, Eq. (1), and the assumption of independence simply means that $p(\mathbf{x}|x_0, y_0)$ is the product of the single measurement likelihoods $p(x_i|x_0, y_0)$. Using Bayes law, Eq. (2), and the uniform prior assumption, the full posterior probability is

$$p(x_0|\mathbf{x}, y_0) \propto p(\mathbf{x}|x_0, y_0) = \prod_{i=1}^{N} p(x_i|x_0, y_0) \propto \prod_{i=1}^{N} \left[\frac{y_0}{y_0^2 + (x_0 - x_i)^2}\right] \ . \tag{3}$$

Again, all proportionality constants are determined by normalization. Here this is the requirement that the integral of $p(x_0|\mathbf{x}, y_0)$ over x_0 is unity. From here on, we will often drop explicit mention of y_0 and write the posterior as $p(x_0|\mathbf{x})$.

In the above derivation the posterior probability is proportional to the likelihood because the prior is assumed to be a constant. The likelihood expresses the probability of obtaining the specific set of measurements, given a particular x_0. We emphasize that Bayes's law is necessary to gain information about x_0 from the likelihood [2].

If it were known that the source position was limited to a specific region, an appropriate prior would consist of a function that is a positive constant inside the region and zero outside. This prior would have the effect of eliminating the tails of the posterior probability in (2) outside the legitimate region. This prior would alleviate any problem that might exist with the normalization of the prior.

2.2. Monte Carlo simulation

To numerically test how well various estimators of x_0 perform, we need to generate measurements that simulate a series of experiments. The cumulative probably (also called the distribution function), the probability of a measurement $x_i < u$, is given by

$$P(x_i < u) = \int_{-\infty}^{u} p(x|x_0, y_0)\, dx = \frac{1}{\pi} \tan^{-1}\left(\frac{u - x_0}{y_0}\right) + \frac{1}{2} \ . \tag{4}$$

To generate measurements from the Cauchy distribution, one uses a pseudorandom number generator that provides a number r_i in the interval (0,1) and then maps the result into the x_i value using the inverse of (4), $x_i = x_0 + y_0 \tan[\pi(r_i - \frac{1}{2})]$.

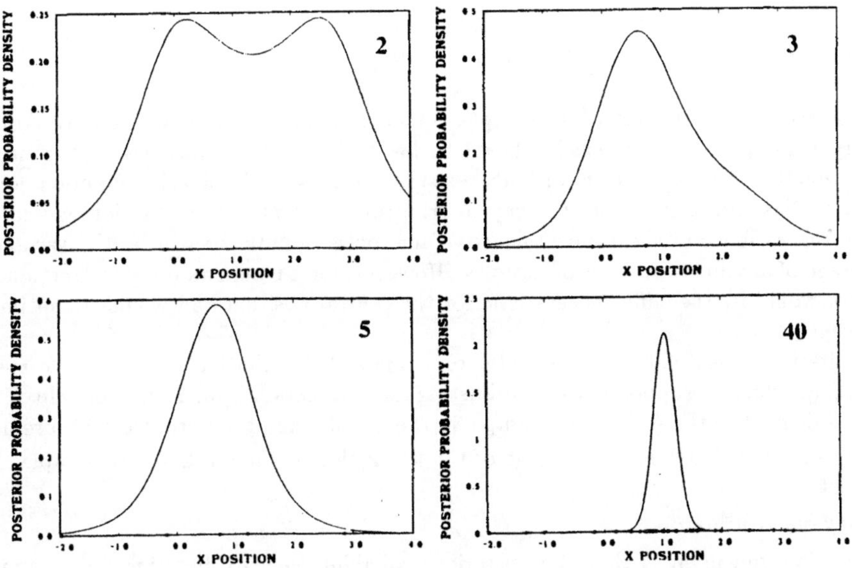

Figure 1: *The posterior probability density function for the x_0 position of the radioactive source assuming that the correct y_0 is known. Each of these plots is shown for one simulated experiment; the measured x_i are displayed on the horizontal axis. The number of measurements is noted in the upper-right corner of each plot.*

Note that when $u - x_0 \gg y_0$, $P(x_i \geq u) = 1 - P(x_i < u) \approx y_0/\pi(u - x_0)$. The probability of getting an x_i value that is greater than 1000 times the FWHM of the distribution ($2y_0$) is roughly 1/1000. Thus the Cauchy distribution offers a superb example of a data distribution with outliers.

The posterior probability given by Eq. (3) is plotted in Fig. 2.2. for specific measurements generated using the Monte Carlo technique described above. The plot for two measurements is bimodal, making it ambiguous to use the maximum posterior probability (see Sect. 3.2) to estimate x_0. As the number of measurements increases, the width of the posterior density function decreases, indicating less uncertainty in the knowledge of x_0. The broad tail of the Cauchy likelihood is increasingly suppressed as the number of measurements increases because the posterior probability involves a product of likelihoods of the individual measurements.

3. Estimation of Location

3.1. Mean and median of the measurements

The average of the x_i measurements (or samples) is often used to estimate the central position of their distribution:

$$\hat{x}_{0(\text{samp mean})} = \frac{1}{N}\sum_{i=1}^{N} x_i \ . \tag{5}$$

The variance of the average of N samples taken randomly and independently from an arbitrary density function is easily shown to be N^{-1} times the variance of the original density function, provided such variance exists. Curiously, the density function for the average of N samples from the Cauchy distribution is identical to that for one sample. Because the variance of the Cauchy density function is infinite, so will be the variance of the average of any finite number of samples. However, for a Gaussian density function, this estimator would be the sufficient statistic for the central position and would be optimal in many ways.

An alternative estimator of the center of a sampled distribution is the sample median $\hat{x}_{0(\text{samp med})}$, which is supposed to be robust against outliers [3, pp. 232]. For odd N, the median is defined as the $\frac{1}{2}(N+1)$th sample in the list of magnitude-ordered measurements; for even N, it is defined as the average of the $(N/2)$th and the $(N/2+1)$th samples from such a list.

3.2. Bayesian estimators

The Bayesian viewpoint is that the posterior probability density function for x_0 summarizes our state of knowledge of x_0 in probabilistic terms. Various types of estimators can be formed from the posterior. The choice of estimator can be based on how the cost of making an error in the estimated quantity depends on the size of the error [1]. The most commonly used estimator in Bayesian analysis is the x_0 value at the maximum of the posterior probability, which we designate by $\hat{x}_{0(\text{MAP})}$, because it is usually called the maximum a posteriori estimator. The MAP estimator minimizes a cost function that is zero for no error and a positive constant for any finite error.

The estimate $\hat{x}_0$ that minimizes the expected mean-square error, i.e. $\int (\hat{x}_0 - x_0)^2\, p(x_0|\mathbf{x})\, dx_0$, is the mean of the posterior density function:

$$\hat{x}_{0(\text{post mean})} = \int x_0\, p(x_0|\mathbf{x})\, dx_0 \ . \tag{6}$$

Defining an integral that is proportional to the kth moment of the posterior given in Eq. (3)

$$I_k(\mathbf{x}) = \frac{y_0}{\pi}\int_{-\infty}^{+\infty} x_0^k \prod_{i=1}^{N} \frac{1}{[y_0^2 + (x_i - x_0)^2]}\, dx_0 \ , \tag{7}$$

the mean (or first moment[1]) of the posterior is

$$\hat{x}_{0(\text{post mean})} = \frac{I_1(\mathbf{x})}{I_0(\mathbf{x})} \ . \tag{8}$$

The integrand in Eq. (7) has simple[2] poles at $x_i^{\pm} \equiv x_i \pm \imath y_0$. By interpreting the integral as one along the real axis in the complex plane and closing the contour at ∞ in the upper

[1]The kth moment of the posterior is $I_k(\mathbf{x})/I_0(\mathbf{x})$.

[2]The procedure described here must be trivially modified when $x_i = x_j$ for $i \neq j$. However, we need not consider such coincident measurements because they represent a set of zero probability.

half plane (which contributes nothing provided the integrand falls off faster than x^{-1}), the desired result is found using the Cauchy residue theorem

$$I_k(\mathbf{x}) = \sum_{i=1}^{N}(x_i^+)^k \prod_{j\neq i} \frac{1}{(x_i^+ - x_j^+)(x_i^+ - x_j^-)}, \quad 0 \leq k < 2N-1 \tag{9}$$

$$= \sum_{i=1}^{N}(x_i^+)^k \prod_{j\neq i} \left[\frac{1}{(x_i - x_j)^2 + 4y_0^2}\right]\left[1 - \frac{2\imath y_0}{x_i - x_j}\right], \quad 0 \leq k < 2N-1, \tag{10}$$

where the second expression is obtained by simply rearranging the product.

Note that by its definition (7), I_k is real for all allowed k. In particular for $k = 1$, the factor $(x_i^+)^k = x_i + \imath y_0$ in Eq. (10) contributes to two summations, one summation with factor x_i and the other summation with factor $\imath y_0$. The summation with factor $\imath y_0$ is identically $\imath y_0 I_0$. Because I_0 is real, $\imath y_0 I_0$ is imaginary. Thus the $\imath y_0$-factor summation must be exactly cancelled by the imaginary part of the x_i-factor summation and we may write

$$I_1(\mathbf{x}) = \Re\left\{\sum_{i=1}^{N} x_i \prod_{j\neq i}\left[\frac{1}{(x_i - x_j)^2 + 4y_0^2}\right]\left[1 - \frac{2\imath y_0}{(x_i - x_j)}\right]\right\}, \quad 0 \leq k < 2N-1. \tag{11}$$

Therefore, the posterior mean estimator (8) has the form of a weighted average of the x_i, $\hat{x}_{0(\text{post mean})} = \sum w_i x_i$, where the sum of the weights is unity. Although this expression looks like a simple variation on the sample average (5), the weights possess a very complicated behaviour. The net effect of the first factor in the product in (11) leads to a diminished contribution from an outlier. But it is very difficult to conceptually grasp the effect of the second factor owing to its complex nature.

Figure 3.2. shows the behavior of the various estimators when a new measurement is combined with five existing measurements. As the value of the new measurement moves away from the other measurements, its net effect on $\hat{x}_{0(\text{post mean})}$ goes to zero. Thus the estimator minimizes the contribution of any measurement that lies far from a cluster of other measurements, which seems to be an ideal treatment of outliers. Because the posterior is independent of the order of the measurements, the same behaviour is expected for any measurement. The posterior maximum estimator behaves similarly to the posterior mean. A new measurement affects the sample mean in a linear fashion because it is just a linear combination of all measurements. The outlier sample can drastically affect the sample mean. The sample median behaves quite differently. The change in the median remains constant as long as the $(N+1)$th sample lies outside the central-most two or three samples, depending on whether N is even or odd, respectively. The estimators based on the posterior are the only ones for which the effect of a single disparate measurement decreases as its descrepancy from the others increases.

The variance of the posterior density function of x_0 for a particular data vector $\mathbf{x}$ is

$$var\{p(x_0|\mathbf{x})\} = \int [x_0 - \hat{x}_{0(\text{post mean})}]^2 p(x_0|\mathbf{x})\, dx_0 = \frac{I_2(\mathbf{x})}{I_0(\mathbf{x})} - \left[\frac{I_1(\mathbf{x})}{I_0(\mathbf{x})}\right]^2, N \geq 2\ . \tag{12}$$

An interesting property of the posterior probability is that its shape depends on $\mathbf{x}$ and is hence usually different for different experiments. See Sect. 4 for its relationship to rms error for $\hat{x}_{0(\text{post mean})}$.

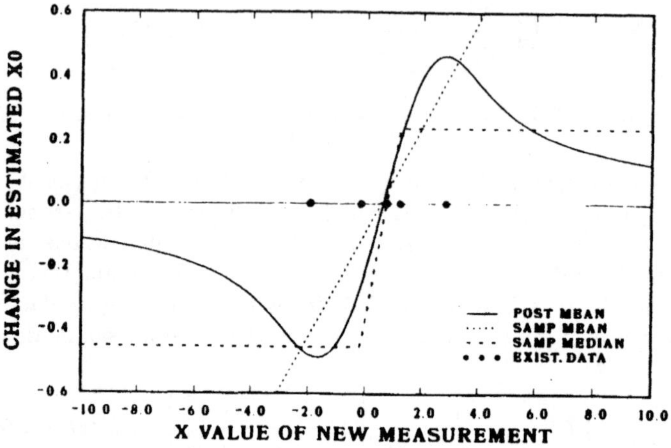

Figure 2: *The change in the estimated position of the Cauchy distribution caused by adding a new measurement x_6 to five existing measurements as a function of the value of the new measurement. The results are shown for several kinds of estimators.*

3.3. Fisher-Cramér-Rao lower bound and Fisher information

The Fisher-Cramér-Rao bound[3] places a lower bound on the variance of any unbiased estimator $\hat{x}(\mathbf{x})$, $var(\hat{x}) \geq \mathcal{I}_N^{-1}$, where $\mathcal{I}_N$ is the Fisher information

$$\mathcal{I}_N \equiv \int \frac{\partial^2 \log[p(x_0|\mathbf{x})]}{\partial x_0^2} p(x_0|\mathbf{x})\, d\mathbf{x} \ , \tag{13}$$

and where N, the number of measurements, is the dimension of $\mathbf{x}$. Because the posterior (3) factors, we have $\mathcal{I}_N = N\mathcal{I}_1$, where $\mathcal{I}_1$ is the single-sample Fisher information given by

$$\mathcal{I}_1 = \frac{4y_0}{\pi} \int_{-\infty}^{+\infty} \frac{(x - x_0)^2}{[y_0^2 + (x - x_0)^2]^3} dx = \frac{1}{2y_0^2} \ . \tag{14}$$

In the last step the integral is evaluated by applying the Cauchy residue theorem (as in Sect. 3.2). Thus the Fisher-Cramér-Rao lower bound on the variance of any unbiased estimator of x_0 is

$$var(\hat{x}) \geq [N\mathcal{I}_1]^{-1} = \frac{2y_0^2}{N} \ . \tag{15}$$

It is important to note that this lower bound is valid only for unbiased estimators, i.e. when averaged over all possible data, it yields the correct result $\int \hat{x}_0(\mathbf{x})\, p(\mathbf{x}|x_0)\, d\mathbf{x} = x_0$. We have confirmed through subsidiary calculations that both the sample mean and posterior mean are unbiased for $N \geq 3$ and that their variances exist for $N \geq 4$.

[3]Fisher stated this lower bound many years before Cramér and Rao [4, p. 66].

Table 1: *Summary of the performance of several estimators of the central position of a Cauchy distribution observed in* 10^5 *trials for a fixed number of samples per trial* N. *The estimators used are the mean and the median of the samples, and the maximum and mean of the posterior probability density function. The last two columns give the Fisher-Cramér-Rao lower bound on the rms error and the rms width of the posterior probability.*

	rms error in estimated position					
N	samp mean	samp median	post max	post mean	FCR	rms post
1	2.85×10^{10}	2.85×10^{10}	2.85×10^{10}	2.85×10^{10}	1.414	∞
2	1.43×10^{10}	1.43×10^{10}	--	1.43×10^{10}	1.000	1.43×10^{10}
3	9.52×10^9	2.828	2.825	2.768	0.816	2.616
5	5.71×10^9	1.103	1.070	0.958	0.632	0.963
10	2.86×10^9	0.578	0.538	0.522	0.447	0.523
20	1.43×10^9	0.373	0.341	0.339	0.316	0.339
40	7.14×10^8	0.256	0.236	0.232	0.224	0.232

4. Simulation Results

The performance of the above estimators for x_0 is tested by simulating 10^5 experiments, each involving a fixed number of measurements N, which are independently drawn from a Cauchy distribution as indicated in Sect. 2.2. The parameters are held fixed at $x_0 = 1$ and $y_0 = 1$ throughout. The results are summarized in Table 1. In these numerical experiments, except for the sample mean, the bias is always observed to be consistent with zero to within its statistical uncertainty, i.e. on the order of the [rms error of the estimator] $/\sqrt{T}$, where T is the number of trials, or experiments. We observe that the average value of the measurements performs terribly! This poor performance was anticipated, owing to the infinite variance of the Cauchy distribution. The only reason that the rms error in $\hat{x}_{0(\text{samp mean})}$ is not infinite, as mentioned, is that only a finite number of trials are included. The largest x_i in the particular sequence of pseudorandom numbers used to generate the 4×10^6 measurements for the $N = 40$ test is 9.03×10^{12}. Because of the symmetry of the likelihood (1) for one and two measurements, all the estimators are identical for $N = 1$ and 2. The posterior mean and maximum perform much better than the sample average for three or more measurements.

The estimators based on the sample median and the maximum of the posterior probability density function perform only slightly worse than the one based on the posterior mean. Just as they demonstrate the weakness of the sample mean estimator, these results underscore the value of the sample median as a simple estimator that is robust against outliers. Empirically, $rms(\hat{x}_{0(\text{samp med})}) > rms(\hat{x}_{0(\text{post max})}) > rms(\hat{x}_{0(\text{post mean})})$, where rms indicates the rms error from Table 1.

The Fisher-Cramér-Rao lower bound on the rms error is seen to be a valid lower bound for the estimators summarized in the table, which only begin to approach the lower bound for $N \geq 20$.

It is natural to ask whether the posterior is predictive of the uncertainty in an estimator. The rms width of the posterior for our Cauchy problem $rms\{p(x_0|\mathbf{x})\}$ may be calculated by taking the square root of the variance in x_0, calculated using Eq. (12). The results for the simulated experiments, shown in the last column of the table, indicate that this calculation does predict the rms error in the $\hat{x}_{0(\text{post mean})}$ estimator.

We note that the shape of the posterior depends on the measured data, as inferred from Fig. 2.2.. This behavior is different for a Gaussian likelihood with uniform prior, for which the width and shape of the posterior for a fixed number of data samples does not depend on the actual data values. We find in 10^5 trials for $N = 5$ that when the trials are selected on basis of $rms\{p(x_0|\mathbf{x})\}$, the rms error in the estimator $\hat{x}_{0(\text{post max})}$ for those trials closely reproduces the chosen $rms\{p(x_0|\mathbf{x})\}$. This result indicates that the posterior probability density function derived for each experiment provides information about the certainty of inferences made on the basis of that experiment. It is intriguing to consider how such information might be used to make decisions about whether more data should be taken to achieve a desired accuracy of interpretation.

5. Discussion

The prior used in the Bayesian analysis was the uniform prior. Because the uniform prior on the real number line is not normalizable, this analysis must be viewed as a limit over normalized priors [5]. In practice, the prior should reflect the state of prior knowledge.

The Bayesian analysis also assumed that the measurement interval was the complete x axis. When the measurement interval is finite, and assuming that a fixed number of measurements are made, the posterior has asymptotically nonzero constant tails. The reason for this is that the probability of the measurements is then simply the product of the probabilities $p(x_i|x_0, y_0)$ of Eq. (1), with each normalized to unity over the finite measurement inteval. For large x_0 the normalization constant is effectively the width of the measurement interval times the value of the Cauchy distribution tail in that interval (Eq. (2) may be used to establish the precise relationship). Thus, the normalization constant has the same large x_0 behavior as the Cauchy distribution that it normalizes, and this gives rise to the asymptotically nonzero constant tail.

For a source of fixed intensity the assumption of fixing the number of measurements corresponds to varying the time interval that the measurements are taken within, measuring until the specified number of photons is gathered. A more reasonable assumption is that the time interval that measurements are taken within is fixed. When this is done the intensity of the source must be taken into account, which translates into a different likelihood for the measurements. In particular, the number of measurements follows a Poisson distribution, so that the likelihood for N measurements discussed in the last paragraph is modified by the factor $P(N|x_0, y_0) = e^{-\lambda}\lambda^N/N!$ where $\lambda = \lambda(x_0, y_0)$ is the source intensity times the total probability that a measurement will occur in the measurement interval.

Acknowledgments

We acknowledge many helpful discussions with Gregory S. Cunningham who suggested using the sample-median estimator. This work was supported by the United States Department of Energy under contract number W-7405-ENG-36.

References

[1] K. M. Hanson. Introduction to Bayesian image analysis. *Proc. SPIE*, vol. 1898:716–731, 1993.

[2] S. F. Gull. Bayesian inductive inference and maximum entropy. In G. J. Erickson and C. R. Smith, editors, *Maximum Entropy and Bayesian Methods in Science and Engineering (Vol. 1)*, pages 53–74. Kluwer Academic, 1989.

[3] J. L. Devore. *Probability and Statistics for Engineering and the Sciences.* Brooks/Cole, Monterey, 1987.

[4] H. L. Van Trees. *Detection, Estimation, and Modulation Theory - Part I.* John Wiley and Sons, New York, 1968.

[5] G. E. P. Box and G. C. Tiao. *Inference in Statistical Analysis.* John Wiley and Sons, New York, 1973 (reprinted 1992).

PROBABILITY THEORY AND MULTIEXPONENTIAL SIGNALS, HOW ACCURATELY CAN THE PARAMETERS BE DETERMINED?

Anand Ramaswami
Department of Physics,
Washington University
One Brookings Drive
St. Louis, Missouri 63130

G. Larry Bretthorst
Department of Chemistry,
Washington University
St. Louis, Missouri 63130

ABSTRACT. Estimating the amplitudes and decay rate constants of exponentially decaying signals is an important problem in science. Understanding how the uncertainty in the parameter estimates depends on the experimental parameters is important as an aid in understanding how to improve the reliability of the parameter estimates. In this paper, probability theory has been applied to this problem with the intent of understanding the relevant experimental parameters. In the case of a single exponential, the uncertainty in the estimated decay rate depends directly on the three halves power of the true decay rate constant, inversely on the signal-to-noise ratio, and inversely on the square root of the number of data values. The uncertainty in the amplitude estimate depends directly on the square root of the true decay rate constant, directly on the noise level, and inversely on the square root of the number of data values. The case of two exponentials has also been analyzed with similar results. However, here the presence of the second signal introduces interference effects which make the estimate more uncertain.

1 Introduction

Exponentially decaying signals occur in many branches of science and engineering. In chemistry the concentrations of the reactants in first order reactions decay exponentially. The same is true in physics for the radioactive nuclear decay and in Nuclear Magnetic Resonance (NMR) for the magnetization of an excited nucleus. Exponentials are also used to model both storage and release of drugs and other exogenous substances, like metabolic tracers, from compartments within the body. In all of these examples, the value of the amplitudes and decay rate constants, contain the information about the dynamics of interest. Various methods have been used to estimate these parameters. Some of these include curve stripping (measuring the slope of the line in a semi-log plot), nonlinear least squares, Prony's method, linear prediction, and Bayesian probability theory. Of these methods, Bayesian probability theory offers a unique opportunity to understand how the parameter estimates depend on experimental parameters because it provides an estimate of the uncertainty in the parameter estimates. The techniques and procedures used in

G. R. Heidbreder (ed.), Maximum Entropy and Bayesian Methods, 265–273.

this paper have been applied previously to exponentially decaying sinusoidal models [1], to chirped sinusoids [2], and to Gaussian point spread functions [3]. When the results of the Bayesian analysis have been compared to more traditional techniques, it has been observed that the traditional methods depend on the experimental parameters in much the same way as the parameter estimates from probability theory [4] and [5]. Consequently, the results derived from probability theory should also be indicative of the behaviour of the estimate from more traditional techniques.

This paper addresses questions of the form "How does the uncertainty in the decay rate constant depend on the sampling time, the number of data points, the signal-to-noise ratio and the true decay rate constant of the signal?" In sections 2 and 3 the uncertainty in the parameter estimates for the decay rate constant and amplitude are determined for the single exponential model. Similarly, in sections 4 and 5 the uncertainty in the parameter estimates are determined for the two (or bi-) exponential model. A different calculation is needed to determine the uncertainty in the parameter estimates for each parameter, though each of these calculations follows the same general outline. First, a data set $D \equiv \{d_1, \ldots, d_N\}$ is postulated. This data has been sampled from a time series $y(t)$ at discrete times t_i $(1 \leq i \leq N)$. The time series $y(t)$ is assumed to be the sum of two terms, a signal plus noise:

$$d_i = y(t_i) = f(t_i) + e_i \quad (1 \leq i \leq N), \tag{1}$$

where e_i represents the noise at time t_i. The signal $f(t)$ is of the form

$$f(t_i) = \sum_{j=1}^{m} B_j G_j(t_i) = \sum_{j=1}^{m} B_j \exp\{-\alpha_j t_i\} \tag{2}$$

where B_j is the amplitude of the jth signal, α_j is the decay rate constant, and m is the number of exponentials. The cases of m=1 and m=2 will be considered in this paper. In the second step of the calculation, the posterior probability density for the parameter of interest (the decay rate constant or amplitude) is computed. Next, a functional form of the data is postulated, and finally the parameter estimates are derived in the (mean $\pm$ standard deviation) form. These estimates explicitly demonstrate the dependence on the experimental parameters and what must be done to make more precise estimates.

2 Estimating Decay Rate Constant: One Exponential Case

The first calculation involves determining how the uncertainty in the estimated decay rate constant depends on the experimental parameters. From Eq. (2), the model equation for the single exponential, $m = 1$, is

$$d_i = B_1 \exp\{-\alpha_1 t_i\} + e_i \quad (1 \leq i \leq N). \tag{3}$$

The posterior probability has been derived previously [6] and the results will simply be given here. Using bounded uniform priors, for both the amplitude and decay rate constant, the posterior probability density for the decay rate constant is given by

$$P(\alpha_1|\sigma, D, I) \propto \exp\left\{\frac{(d \cdot G)^2}{2\sigma^2 G \cdot G}\right\}, \tag{4}$$

where a number of irrelevant constants have been dropped and "·" means sum over discrete times:

$$d \cdot G \equiv \sum_{i=1}^{N} d_i G(t_i) = \sum_{i=1}^{N} d_i \exp\{-\alpha_1 t_i\}, \tag{5}$$

and

$$G \cdot G \equiv \sum_{i=1}^{N} G_i^2 = \sum_{i=1}^{N} \exp(-2\alpha_1 t_i) \approx \frac{1}{2\alpha_1}. \tag{6}$$

The approximation in this equation assumes that $2\alpha_1 N$ is large compared to one; i.e., the signal decays away over the total sampling time, and that the sum may be approximated by an integral. To motivate the integral approximation, suppose the dimensionless decay rate constant is 0.01, that uniform sampling is used, and the time series is sampled for three e-folding times to obtain 300 data values; then the exact sum gives 50.3764, while the approximation gives 50.0. The approximation introduces an error of 0.75%.

The next step in the calculation is to postulate a functional form for the data. If the true values of the amplitude and decay rate constant are $\hat{B}_1$ and $\hat{\alpha}_1$ respectively, then the data are given by

$$d(t_i) = \hat{B}_1 \exp\{-\hat{\alpha}_1 t_i\} + e_i \quad (1 \leq i \leq N) \tag{7}$$

and

$$\begin{aligned} d \cdot G &= \sum_{i=1}^{N} d_i \exp\{-\alpha_1 t_i\} \\ &= \sum_{i=1}^{N} \hat{B}_1 \exp\{-(\hat{\alpha}_1 + \alpha_1)t_i\} + \sum_{i=1}^{N} e_i \exp\{-\alpha_1 t_i\} \\ &\approx \frac{\hat{B}_1}{\alpha_1 + \hat{\alpha}_1}, \end{aligned} \tag{8}$$

where the projection of the model onto the noise was assumed small compared to the projection of the model onto the signal (effectively the high signal-to-noise case). For the postulated data, the posterior probability density for the decay rate constant becomes

$$P(\alpha_1|\sigma, D, I) \propto \exp\left\{\frac{2\alpha_1 \hat{B}_1^2}{2\sigma^2(\alpha_1 + \hat{\alpha}_1)^2}\right\}. \tag{9}$$

The maximum of the posterior probability occurs at $\alpha_1 = \hat{\alpha}_1$; the true value of the parameter. Taylor expanding the exponent about this maximum gives

$$P(\alpha_1|\sigma, D, I) \propto \exp\left\{-\frac{(\alpha_1 - \hat{\alpha}_1)^2 \hat{B}_1^2}{16\sigma^2 \hat{\alpha}_1^3}\right\}, \tag{10}$$

from which one obtains

$$(\alpha_1)_{est} = \hat{\alpha}_1 \pm \sqrt{8}\frac{\sigma}{\hat{B}_1}(\hat{\alpha}_1)^{3/2} \tag{11}$$

as the (mean ± standard deviation) estimate of the decay rate constant.

First note that dimensionless units have been used. The conversion to dimensional units is given by

$$\alpha_1{}' = \frac{\alpha_1 N}{\pi \Delta t} \tag{12}$$

where α_1' is the corresponding dimensional decay rate constant and Δt is the sampling time. Converting the (mean $\pm$ standard deviation) estimate to dimensional quantities one obtains

$$(\alpha_1')_{est} = \hat{\alpha}_1' \pm \frac{\sigma}{\hat{B}_1}\sqrt{\frac{8\pi\Delta t}{N}}(\hat{\alpha}_1')^{3/2}. \tag{13}$$

A number of conclusions can be drawn from this expression:

1. The estimated decay rate constant is equal to the true decay rate constant i.e., as the noise goes to zero, the parameter estimate goes smoothly to the true parameter value.
2. Increasing the signal-to-noise ratio $(\sigma/\hat{B}_1)$ reduces the uncertainty in the estimated decay rate constant.
3. For a fixed acquisition time, increasing the sampling time decreases precision of the estimate: sampling fewer data values over the region where the signal is large decreases the precision of the estimate.
4. Conversely, increasing the number of data points (decreasing the sampling time) improves the precision of the estimate for the decay rate constant.
5. The more rapidly a signal decays the worse the precision of the estimate. Rapidly decaying signals effectively reduce the number of relevant data.

To make the decay rate estimate more precise one can either improve the signal-to-noise ratio, or gather more data values over the region where the signal is large. However, these comments apply only to the decay rate constant. They do not necessarily apply to the amplitude of the signal. To determine if they apply, the same type of calculation must be repeated for the amplitude of the signal, a task to which we now turn our attention.

3 Estimating Amplitude: One Exponential Case

To determine how the amplitude estimate depends on the experimental parameters, the calculation presented in the previous section must be repeated using the posterior probability for the amplitude as the starting point. Given the model, Eq. (3). the posterior probability for the amplitude is given by

$$P(B_1|\sigma, D, I) \propto \int d\alpha_1 \exp\left\{-\frac{B_1^2 G\cdot G - 2B_1 d\cdot G}{2\sigma^2}\right\} \tag{14}$$

where some irrelevant constants have been dropped, and no closed form solution for the integral is known to the authors. For the single exponential model and the postulated data, the posterior probability density for the amplitude becomes

$$P(B_1|\sigma, D, I) \propto \exp\left\{-\frac{(B_1 - \hat{B}_1)^2}{4\sigma^2\hat{\alpha}_1}\right\} \tag{15}$$

where some irrelevant constants were dropped and a Gaussian approximation was used to evaluate the integral over α_1. Examining Eq. (15), the (mean $\pm$ standard deviation) amplitude estimate is given by:

$$(B_1)_{est} = \hat{B}_1 \pm \sigma\sqrt{2\hat{\alpha}_1}. \tag{16}$$

The conversion to dimensional units was given in Eq. (12), from which one obtains

$$(B_1')_{est} = \hat{B}_1' \pm \sigma\sqrt{\frac{2\pi\Delta t\hat{\alpha}_1'}{N}}. \tag{17}$$

A number of conclusions can be drawn from this expression:

1. The estimated amplitude is equal to the true amplitude. As the noise goes to zero, the parameter estimate goes smoothly to the true parameter value.

2. The uncertainty in the estimate does not depend on the signal-to-noise ratio, but varies directly with the noise standard deviation σ. Once a signal is above the noise level one should be able to detect and estimate its amplitude.

3. The uncertainty in the estimate has the same dependence on the sampling time and the total number of data values as the decay rate constant.

4. The uncertainty in the estimate increases as the true decay rate constant increases; but unlike the uncertainty estimate of the decay rate constant, the dependence is on the square root of the true decay rate constant, instead of the three-halves power. Consequently, the uncertainty in the estimated amplitude does not deteriorate as quickly as the uncertainty in the estimated decay rate constant.

Unlike the uncertainty in the estimated decay rate constant, the uncertainty in the estimated amplitude does not depend on the signal-to-noise level; rather it depends only on the noise standard deviation. Increasing the signal strength will not make the amplitude estimate more precise. Only decreasing the noise level or increasing the sampling rate can do that. However, increasing the signal intensity will result in a smaller fractional error.

The calculations presented in this and the previous section show how the uncertainty in the amplitude and decay rate constant depend on the experimental parameters. These estimates are valid for high signal-to-noise data containing a single exponentially decaying signal that decays away in the acquisition time. They are not valid for truncated signals or for data that contain more than a single exponential. Both of these shortcomings are easily corrected by repeating the calculations and making the appropriate assumptions. For truncated data one would not expect any new phenomena to appear. All that one would expect is a more general formula applicable under wider circumstances. However, for the two exponential case one would expect new phenomena to appear. In particular, one would expect to find interference phenomena. To see how the presence of the second exponential affects the parameter estimates these calculations must be repeated for the two exponential model.

4 Estimating Decay Rate Constant: Two Exponential Case

In the next two sections the analysis presented for the single exponential case is generalized to the two exponential case. The question considered in this section is "How accurately can one of the decay rate constants be estimated given data that contain two exponentially decaying signals?" The model equation for such data may be written as

$$d_i = B_1 \exp\{-\alpha_1 t_i\} + B_2 \exp\{-\alpha_2 t_i\} + e_i \quad (1 \le i \le N). \tag{18}$$

Using uniform priors, for the amplitudes and decay rate constants, and assuming the noise standard deviation (σ) is known, the posterior probability density for one of the decay rate constants independent of the value of the other decay rate constant is given by

$$P(\alpha_1|\sigma, D, I) \propto \int d\alpha_2 \exp\left\{\frac{m\overline{h^2}}{2\sigma^2}\right\} \tag{19}$$

where some irrelevant constants have been dropped. The quantity $\overline{h^2}$ is given by

$$\overline{h^2} = \frac{2\alpha_1\alpha_2(\alpha_1+\alpha_2)^2}{(\alpha_1-\alpha_2)^2}\left\{\frac{S_1^2}{2\alpha_2} - \frac{2S_1S_2}{\alpha_1+\alpha_2} + \frac{S_2^2}{2\alpha_1}\right\} \tag{20}$$

with

$$S_1 = d \cdot \exp\{-\alpha_1 t\} \quad \text{and} \quad S_2 = d \cdot \exp\{-\alpha_2 t\}. \tag{21}$$

Postulating two exponential data given by

$$d(t_i) = \hat{B}_1 \exp\{-\hat{\alpha}_1 t\} + \hat{B}_2 \exp\{-\hat{\alpha}_2 t\} + e_i \quad (1 \le i \le N), \tag{22}$$

where $\hat{B}_1$ and $\hat{\alpha}_1$ are the true amplitude and decay rate constant of the first exponential, and $\hat{B}_2$ and $\hat{\alpha}_2$ are the true amplitude and decay rate constant of the second exponential, then S_1 and S_2 are given approximately by

$$S_1 = \frac{\hat{B}_1}{\alpha_1+\hat{\alpha}_1} + \frac{\hat{B}_2}{\alpha_1+\hat{\alpha}_2} \quad \text{and} \quad S_2 = \frac{\hat{B}_1}{\alpha_2+\hat{\alpha}_1} + \frac{\hat{B}_2}{\alpha_2+\hat{\alpha}_2}. \tag{23}$$

The maximum of the posterior probability density again occurs at $\hat{\alpha}_1$; the true value of the parameter. Taylor expanding the exponent about this maximum gives

$$P(\alpha_1|\sigma, D, I) \propto \exp\left\{-\frac{(\alpha_1-\hat{\alpha}_1)^2\hat{B}_1^2(\hat{\alpha}_1-\hat{\alpha}_2)^4}{8\sigma^2\hat{\alpha}_1^3(\hat{\alpha}_1+\hat{\alpha}_2)^4}\right\} \tag{24}$$

where some irrelevant constants have been dropped, and the integral was evaluated in the Gaussian approximation. From Eq. (24) one obtains

$$(\alpha_1)_{est} = \hat{\alpha}_1 \pm 2\sqrt{\frac{\sigma^2\hat{\alpha}_1^3(\hat{\alpha}_1+\hat{\alpha}_2)^4}{\hat{B}_1^2(\hat{\alpha}_1-\hat{\alpha}_2)^4}} \tag{25}$$

as the (mean $\pm$ standard deviation) estimate of the decay rate constant. The conversion to dimensional units was given in Eq. (12), from which one obtains

$$(\alpha_1')_{est} = \hat{\alpha}_1' \pm 2\frac{\sigma}{\hat{B}_1}\frac{(\hat{\alpha}_1+\hat{\alpha}_2)^2}{(\hat{\alpha}_1-\hat{\alpha}_2)^2}\sqrt{\frac{\pi\Delta t}{N}}(\hat{\alpha}_1')^{3/2}. \tag{26}$$

Similarly, for the other decay rate constant one finds,

$$(\alpha_2')_{est} = \hat{\alpha}_2' \pm 2\frac{\sigma}{\hat{B}_2}\frac{(\hat{\alpha}_1+\hat{\alpha}_2)^2}{(\hat{\alpha}_1-\hat{\alpha}_2)^2}\sqrt{\frac{\pi\Delta t}{N}}(\hat{\alpha}_2')^{3/2}. \tag{27}$$

A number of conclusions can be drawn from this expression:

1. The estimated decay rate constant go smoothly to the true value of this parameter as the noise level goes to zero.
2. The uncertainty in the estimated value of the decay rate constant depends only on the signal-to-noise ratio of the component of interest and is independent of the signal-to-noise ratio of the other component.
3. Except for the quadratic ratio, the uncertainty in the estimated parameters has the same dependence on the experimental parameters as in the single exponential case.
4. As the true values of the decay rate constants become comparable the uncertainty in the estimated values for both decay rate constants increases rapidly.

To gain some insight into this last item, suppose the dimensionless decay rate constants are 0.01 and 0.02. Then the uncertainty in the estimated parameters is a factor of 9 worse than if the decay rate constants were well separated. Additionally, suppose N=300, corresponding to an acquisition time of three e-foldings for the longer component, and both exponentials have the same amplitude, then to resolve each of the two decay rate constants at one standard deviation the signal-to-noise ratio must be approximately 10 for each component, or 20 total. But this assumes $N = 300$ data values. If the number of data values is low, for example $N = 30$, then the uncertainty in the estimated parameters increases $\sqrt{10}$. To compensate, the signal-to-noise ratio must be increased to 35 for each component, or 70 total. The presence of the second experimental signal increases the uncertainty in the parameter estimates. But note that when the decay rate constants are very different the uncertainty in the parameter estimates reduce to the single exponential case. So potentially if one can modify the experiment so that one component in very much different from the other, one can reduce the uncertainty in the estimated value of the longer lived component.

5 Estimating Amplitude: Two Exponential Case

The question considered in this section is "How accurately can one of the amplitudes be estimated given data that are known to contain two exponentially decaying signals?" The model equation for this data was given in Eq. (18). The posterior probability for B_1 is given by

$$P(B_1|\sigma, D, I) \propto \int d\alpha_1 d\alpha_2 \exp\left\{-\frac{(B_1-\mathcal{B})^2}{2\sigma^2}\right\} \tag{28}$$

where some irrelevant constants have been dropped, no closed solution for the integral is known to the authors, and

$$\mathcal{B} = \frac{2\alpha_1(\alpha_1+\alpha_2)^2}{\alpha_1^2+\alpha_2^2}\left[d\cdot\exp\{-\alpha_1 t\} + (\frac{2\alpha_2}{\alpha_1+\alpha_2})d\cdot\exp\{-\alpha_2 t\}\right]. \tag{29}$$

For the two exponential model and the postulated data, Eq. (22), the posterior probability density for the amplitude is given approximately by

$$P(B_1|\sigma, D, I) \propto \exp\left\{-\frac{(B_1 - \hat{B}_1)^2}{4\sigma^2\hat{\alpha}_1}\left[\frac{\hat{\alpha}_1 - \hat{\alpha}_2}{\hat{\alpha}_1 + \hat{\alpha}_2}\right]^2\right\} \tag{30}$$

where the integrals were evaluated using a Gaussian approximation. Examining Eq. (30) the (mean $\pm$ standard deviation) amplitude estimate is given by

$$(B_1)_{est} = \hat{B}_1 \pm \sigma\left|\frac{\hat{\alpha}_1 + \hat{\alpha}_2}{\hat{\alpha}_1 - \hat{\alpha}_2}\right|\sqrt{2\hat{\alpha}_1}. \tag{31}$$

The conversion to dimensional units was given in Eq. (12), from which one obtains

$$(B_1')_{est} = \hat{B}_1' \pm \sigma\left|\frac{\hat{\alpha}_1 + \hat{\alpha}_2}{\hat{\alpha}_1 - \hat{\alpha}_2}\right|\sqrt{\frac{2\pi\Delta t\hat{\alpha}_1'}{N}}. \tag{32}$$

Similarly, for the other amplitude one finds,

$$(B_2')_{est} = \hat{B}_2' \pm \sigma\left|\frac{\hat{\alpha}_1 + \hat{\alpha}_2}{\hat{\alpha}_1 - \hat{\alpha}_2}\right|\sqrt{\frac{2\pi\Delta t\hat{\alpha}_2'}{N}}. \tag{33}$$

A number of conclusions can be drawn about estimating the amplitudes when the signal is known to consist of two exponentials:

1. The estimated amplitudes go smoothly to the true values of the amplitudes as the noise level goes to zero.

2. The uncertainty in the amplitude estimate does not depend on the signal-to-noise ratio but varies directly with the noise standard deviation σ. Again, once a signal is above the noise level one should be able to detect and estimate its parameters.

3. Except for the interference factor in front of the square root, the uncertainty in the estimated parameters have the same dependence on the sampling time, the total number of data values and the true decay rate constant as in the single exponential case.

4. As the true values of both decay rate constants become comparable, the uncertainty in the estimated parameters increase. But, the uncertainty in the estimated amplitudes does not increase as rapidly as the uncertainty in the estimated decay rate constant.

To obtain a better understanding of this last item suppose $\alpha_1 = 0.01$, $\alpha_2 = 0.02$ and $N = 300$, (the same values used previously) then the uncertainty in the amplitude estimate is 3 times larger than for the corresponding single exponential case. As the true values of the decay rate constants approach each other the amplitude uncertainty becomes large. However, the signal-to-noise ratio has not changed. If the individual amplitudes cannot be determined there must be some quantity that is still well determined. That quantity is the sum of the two amplitudes. The difference in amplitudes is undetermined. Conversely, if the two decay rate constants are very different, the uncertainty in the amplitude estimate reduces to that found in the single exponential case.

6 Summary

In this paper, probability theory has been used to obtain an understanding of how the uncertainty in the estimated parameters depends on experimental parameters. For decay rate constants, there are two ways to reduce the uncertainty in the estimated parameters: increase the signal-to-noise ratio or increase the sampling rate while holding the acquisition time constant; e.g., take more data over the time one has a signal. For amplitudes a similar result holds for taking more data, but not for increasing the signal-to-noise level. To reduce the uncertainty in the estimated amplitudes, one must decrease the noise level. A potentially much harder condition to fulfill. In the case of data containing two exponentials, probability theory shows how the uncertainty in the parameter estimates depend on the presence of the other exponential. For such data probability theory indicates that the uncertainty in the parameter estimates may be reduced by modifying the experiment in such a way as to separate the true decay rate constants. However, barring this last alternative, there are only two fundamental ways to reduce the uncertainty in the estimated parameters: decrease the noise level (thereby increasing the signal-to-noise of the data) or take more data over the region where the signal is large.

References

[1] Bretthorst, G. Larry, "Bayesian Analysis. III. Applications to NMR Signal Detection, Model Selection and Parameter Estimation," *J. Magn. Reson.* **88**, pp. 571-595 (1990).

[2] Jaynes, E. T., *Bayesian Spectrum and Chirp Analysis,* in "Maximum-Entropy and Bayesian Spectral Analysis and Estimation Problems" (C. R. Smith and G. J. Erickson *Eds.*), p. 1, Reidel, Dordrecht, The Netherlands, 1987.

[3] Bretthorst, G. Larry, Smith, C. Ray, "Bayesian Analysis of Signals from Closely-Spaced Objects," *SPIE Vol.1050 Infrared System and Components 111* (1989).

[4] Kotyk, J. John, Hoffman, G. Norman, Hutton, C. William, Bretthorst, G. Larry, *and* Ackerman, J.H. Joseph "Comparison of Fourier and Bayesian Analysis of NMR Signals. I. Well-Separated Resonances (The Single-Frequency Case)," *J. Magn. Reson.* **98**, pp. 483-500 (1992).

[5] Neil, J. Jeffrey, Bretthorst, G. Larry, *On the Use of Bayesian Probability Theory for Analysis of Exponential Decay Data: An Example Taken from Intravoxel Incoherent Motion Experiments, Magnetic Resonance in Medicine,* **29**, pp. 642-647 (1993).

[6] Bretthorst, G. Larry, *An Introduction to Parameter Estimation Using Bayesian Probability Theory,* in "Maximum Entropy and Bayesian Methods" (P. Fougère *Ed.*), p. 53, Kluwer Academic, Dordrecht, The Netherlands, 1989.

PIXON-BASED IMAGE RECONSTRUCTION

R. C. Puetter and R. K. Piña
Center for Astrophysics and Space Science
University of California, San Diego
9500 Gilman Drive
La Jolla, CA 92093-0111

ABSTRACT. This paper presents the theory of the pixon, the fundamental unit of picture information, and its application to Bayesian image reconstruction. This naturally leads to a discussion of picture information content and the degrees of freedom necessary to describe the underlying image (i.e. the noise-free, undistorted image) within the accuracy of the data. The implications of these concepts for the formulation of appropriate Goodness-of-Fit criteria (i.e. Maximum Likelihood) are discussed. Finally, examples of the applications of these methods to artificial and real data are presented. These examples demonstrate that pixon-based methods produce results superior to both pure Goodness-of-Fit methods and the best examples of Maximum Entropy methods.

1. Introduction

The act of measurement of physical quantities inevitably introduces artifacts. These artifacts can be associated with the statistical limits of the measurement process, e.g. counting statistics, as well as characteristics of the measurement device, e.g. finite resolution or noise of an instrumental origin. Modern approaches for deducing the underlying, uncorrupted physical quantities from the recorded data often turn to Bayesian estimation in which the measurement process is statistically modeled. This is the approach taken here. More specifically, the problem we shall analyze is the recovery of the spatial, temporal, or spectral resolution lost due to the measurement process. This class of problems can be characterized by the equation

$$D(\vec{x}) \;=\; \int dV_y\, K(\vec{x},\vec{y})I(\vec{y}) + N(\vec{x}) \qquad (1)$$

where $D(\vec{x})$ is the recorded data, $I(\vec{x})$ is the underlying signal that one wishes to recover (or reconstruct) as accurately as possible, $K(\vec{x},\vec{y})$ is a kernel function expressing how the act of measurement blurs the true signal, the integration is over the volume in $\vec{y}$-space, and $N(\vec{x})$ is the noise associated with the measurement. Note that while we have presented this problem as an exercise in estimation of true signal in recorded data, equation (1) can be viewed simply as an integral equation and the methods described here as techniques useful in the inversion of this equation.

In later sections of this paper we will present inversions of equation (1) for specific examples. The field we shall draw on for these examples is astronomical imaging. In this case, the integral in equation (1) can be represented as a convolution of the true signal with

G. R. Heidbreder (ed.), Maximum Entropy and Bayesian Methods, 275–292.

a point spread function (PSF), H, i.e.

$$D(\vec{x}) = \int dV_y \, H(\vec{x} - \vec{y}) I(\vec{y}) + N(\vec{x}) \,. \tag{2}$$

However, the methods we describe here are useful for many other problems as well.

Before proceeding with a description of our methods, it is useful to establish a historical perspective. Not surprisingly, the first approach used to solve equation (2) employed Fourier methods. This method takes advantage of the convolution theorem for Fourier Transforms which states that the Fourier Transform of the convolution of two functions is the product of the Fourier Transforms of the individual functions. Hence the "solution" of equation (2) is given by dividing the Fourier Transform of the data by the Fourier Transform of the PSF and performing an inverse Fourier Transform on the quotient. In the absence of noise, this procedure yields an exact result. Unfortunately, this method is notoriously unforgiving with respect to noise, producing strong "ringing", and making the method unsuitable for quantitative analysis of images.

Today, the most successful methods for inverting equation (2) are non-linear in their approach. The simplest among these define a figure-of-merit, or Goodness-of-Fit (GOF) criterion, for the image and then use multi-dimensional optimization methods to maximize this function. Such methods include the familiar Least-Squares method as well as the Lucy-Richardson method (Lucy 1974). While these pure GOF techniques typically give superior inversions to equation (2) than do Fourier methods, they still have many undesirable properties. A common problem is over-resolution in which the algorithm attempts to fit the noise and introduces features which are unnecessary to fit the data. Lucy-Richardson reconstructions, for example, are typically stopped after an arbitrary number of iterations in an attempt to overcome this difficulty. This leaves the unpleasant (and difficult) task of determining which features are "real" and which are not. More sophisticated approaches such as Maximum Entropy (e.g. Skilling 1989) place additional constraints upon the solution based on prior expectations. These additional constraints greatly improve the quality of the solution by regularizing the problem and controlling over-resolution.

Recently, we have introduced a new image reconstruction method based on the pixon (Piña and Puetter 1993, Puetter and Piña 1993). This method greatly expands upon ME methods by introducing a prior with a variable local scale. In fact, our Uniform Pixon Basis (UPB) method (Piña and Puetter, 1993) results in a "Super-Maximum Entropy" reconstruction in which entropy is maximized exactly. In our use of variable correlation length scales, pixon-based methods are similar in some respects to the multi-channel methods of Weir (1991, 1993a) or the Pyramidal Maximum Entropy techniques of Bontekoe, Koper, and Kester (1993). Unlike these techniques, however, pixon-based methods explicitly determine the appropriate local scale based on various criteria. Our most recent Fractal Pixon Basis (FPB) method (Puetter and Piña 1993) selects the local correlation length based on the local structural scale of the image and represents the highest performance image reconstruction method we are aware of to date.

Each of the techniques described above, e.g. GOF, ME, and Pixon-Based methods, can be formalized in terms of Bayesian estimation theory. In order to understand the basis and merits of pixon-based methods relative to competing techniques, we turn now to a brief summary of this theory.

2. The Pixon and Bayesian Estimation

To statistically model the measurement process, a number of issues must be addressed. First, there is the physics describing the measurement process. This, however, is assumed to be set down in equation (1). Second, a number of details related to the measurement process itself, must be properly incorporated. These include the manner in which the data was collected, e.g. as a rectangular grid of counts as is the case for astronomical imaging with a solid state detector, as well as the characteristics of the noise, etc. Finally, a number of decisions must be made regarding how equation (1) is to be inverted mathematically. Common assumptions here are that the image can be represented by a grid of numbers and that the integral represented in equation (1) can be approximated by a discrete sum over this grid. Each of these aspects of the problem, i.e. the physics, the particulars of the measurement, and the mathematical assumptions are part of the model, M, which links the image and the data. Since the goal of any inversion scheme is to obtain the most accurate image, it is important to realize that every aspect of the model affects the accuracy of the image reconstruction, i.e. the physics must be accurately described, the details of the measurement must be adequately noted, and the numerical assumptions made must be appropriate so as not to compromise the inversion.

2.1. The Bayesian Approach

The Bayesian approach to inverting equation (1) is to use Bayes' Theorem to develop a formula for the most probable value of $I(\vec{x})$. This begins by factoring the joint probability distribution of the triplet, D, I and M, i.e., $p(D,I,M)$, where D,I, and M are the data, image, and model respectively. Bayes' Theorem can then be used to factor $p(D,I,M)$ to give

$$p(D,I,M) = p(D|I,M)p(I,M) = p(D|I,M)p(I|M)p(M)\,, \tag{3}$$

or

$$p(D,I,M) = p(I,M|D)p(D) = p(I|D,M)p(D|M)p(M)\,, \tag{4}$$

where $p(X|Y)$ is the probability of X given that Y is known. (Bayes' Theorem states that $p(X,Y) = p(X|Y)p(Y) = p(Y|X)p(X)$.)

Equating the right-hand sides of equations (3) and (4), we find

$$p(I|D,M) = \frac{p(D|I,M)p(I|M)}{p(D|M)} \tag{5}$$

or

$$p(I,M|D) = \frac{p(D|I,M)p(I,M)}{p(D)} \propto p(D|I,M)p(I,M)\,. \tag{6}$$

From equation (6) we can find the M.A.P. (Maximum *A Posteriori*) image using

$$p(I|D) \;=\; \int dM\, p(I,M|D) = \int dM \frac{p(D|I,M)p(I,M)}{p(D)} \propto p(D|I,M_\circ)p(I,M_\circ)\,, \quad (7)$$

where $M_\circ$ is the model from the M.A.P. image/model pair. (The proportionality assumes that the peak in the probability distribution is representative of a typical sample from the posterior probability distribution. While this is usually the case, it is not absolutely guaranteed.)

Equation (5) is the typical starting point of Bayesian image reconstruction in which one wishes to determine the M.A.P. image, i.e. the image which maximizes $p(I|D,M)$. (The M.A.P. image, of course, is only one of several choices for the "best image". Another sensible choice might be the average image, $\langle I\rangle = \int_{D,M} dMdD\; I\; p(I|D,M)$.)

Equations (6) and (7) are our preferred prescription for Bayesian image reconstruction in which both the image and the model are varied simultaneously to obtain the best combined image/model solution, e.g. the M.A.P. image/model pair. Equation (6) might be used directly to find the optimal image, i.e. the image paired with the model in the M.A.P. image/model pair. Alternatively, the M.A.P. image can be found using equation (7).

The significance of the terms in the above equations are well known. The first term, $p(D|I,M)$, is a goodness-of-fit (GOF) quantity. The standard choice for $p(D|I,M)$ is to use $p(D|I,M) = \exp(-\chi^2/2)$, where χ^2 is the chi-square of the residuals. The terms $p(I|M)$ and $p(I,M)$ are "priors". Since they do not depend on the data they can be decided *a priori*. The first of these, i.e. $p(I|M)$, is normally termed the *image prior* and expresses the *a priori* probability of an image given the model. The second, e.g. $p(I,M)$, we have termed the *image/model prior*, and expresses the *a priori* probability of both I and M. In GOF image reconstruction the prior is ignored or is effectively set equal to unity, i.e. there is no prior bias concerning the image or the model. In Maximum Entropy (ME) image reconstruction, the image prior is based upon "phase space volume" or counting arguments and the prior is expressed as $p(I|M) = \exp(\alpha S)$, where S is the entropy of the image and α is an adjustable constant that is used to weight the relative importance of the GOF and image prior. Many different formulations for S and α appear in the literature (Kikuchi and Soffer 1976, Bryan and Skilling 1980, Narayan and Nityananda 1986, and Adorf, Walsh, and Hook 1990). Recently, however, Skilling (1989) and Gull (1989) have shown that there is a natural, Bayesian choice for the value of α. Indeed, this new, natural Bayesian choice for α is directly related to the number of degrees-of-freedom (DOFs) in the data. In this regard, the Bayesian choice for α is directly related to the pixon approach presented here. In fact, if properly performed, we would expect our pixon-based methods and the Bayesian estimate of α to agree exactly on the number of degrees-of-freedom required to describe the data. The advantage of the pixon-based approach, however, is that it actually determines the spatial locations of these DOFs and presents them to the researcher for accurate determination–see below.

The final quantity, $p(D|M)$, is termed the "Evidence" for the model. (Actually, we normally refer to $p(M|D)$ as the Evidence, but $p(D|M)$ is proportional to this quantity and normally equal to this quantity since it is common to assume that both $p(D) = const$ and $p(M) = const$ since there generally is not an *a priori* manner for choosing between valid data sets nor between "sensible" models.) Choosing models on the basis of the Evidence allows one to "peak-up" on more favorable solutions.

3. The Pixon Concept

Equations (6) and (7) indicate that two quantities, the GOF criterion, $p(D|I,M)$, and the image/model prior, $p(I,M)$, are of key importance in obtaining the most probable (i.e. M.A.P.) values for the image/model. In this paper we shall concentrate on a new proposal for the prior. This is discussed below.

3.1. A New, Pixon-Based Prior

Equation (6) will form the basis of our pixon-based methods, i.e. our goal is to determine the M.A.P. image/model pair. Like ME methods we shall base our image/model prior on counting arguments. Unlike standard ME methods, however, we will allow certain aspects of the model to vary simultaneously with the image. By allowing both the image and model to vary simultaneously, we are optimizing our solution over a considerably larger solution space than methods which hold the model constant. Previous workers (Gull 1989, Sibisi 1990, Skilling 1991, MacKay 1992a,b) have already demonstrated the merits of varying the model and have shown the efficacy of selecting between models by maximizing the Evidence for the model.

Of course certain aspects of the model should not be varied. This includes, for example, the physics that we *know* to be true for the problem at hand. In fact, however, this simply means that the prior for this aspect of the model is very highly peaked around the *true* physics. It is so highly peaked, in fact, that the prior is effectively 1 for the physics that we understand to be true and zero for everything else. The same can be said for our understanding of the details of the experimental set-up. We typically know with near absolute certainty the detailed properties of the noise (e.g. the noise may be Gaussian with a particular value of σ) and the layout of the instrument, etc. What we most uncertain about, however, is how the image, $I(\vec{x})$, should be modelled mathematically. Surprisingly, the method of modelling the image can have profound implications for the quality of the reconstruction.

To show how the selection of image representation (we shall use the word basis) affects the quality of the reconstruction, let us next consider the abstract nature of an image and how a generalized image/model prior might be constructed. To do this we shall follow closely the development of Piña and Puetter (1993) and Puetter and Piña (1993). They pointed out that in an abstract sense, an image is a collection of distinguishable events which occur in distinct cells. Hence the value for the image/model prior can be determined from simple counting arguments. If there are N_i events in cell i, and a total of n cells, then the prior probability of that particular image is:

$$p(\{N_i\}, n, N) \quad = \quad = \frac{N!}{n^N \prod_{cells,i} N_i!} = p(I,M)\,, \tag{8}$$

where $\{N_i\}$ is the set of all numbers of events in cells i, N is the total number of events, i.e. $N = \sum N_i$, and the image is now considered to be made up of these events. [In practical terms, an event is a photon count in a photon counting detector or the number of counts in units of the standard deviation of the noise in non-photon counting systems. Furthermore, these "events" are, in fact, units of information, i.e. the knowledge that something (an event) has occurred is the minimal unit of information–see Piña and Puetter (1993) for a discussion. This sense of the term "information" is somewhat different than the classical definition which is in terms of the logarithm of the number of states, etc.] Also note, that the cells used in equation (8) are quite general. In their definition we have not specified a size, shape, or position for the cells. The cell concept simply serves to localize some collections of events.

Since the goal of our reconstruction is to determine the M.A.P. image/model pair, we must maximize the product of the image/model prior given in equation (8) and the GOF term. Most people have a well developed intuition regarding how to maximize the GOF term, i.e. the residuals, $R(\vec{x}) = D(\vec{x}) - \int dV_y\, H(\vec{x} - \vec{y})I(\vec{y})$, must be comparable to the noise (strictly speaking, they should be exactly equal to the noise). Intuition concerning priors is usually less well developed. Equation (8), however, points out the *a priori* desirable properties of the model for the image. These are that the model should contain the fewest number of cells with each containing the largest number of events consistent with maintaining an adequate GOF. We shall call these generalized cells pixons. The pixon name recognizes the pixel (or cell) heritage, and the "-on" suffix recognizes the fundamental nature of the pixon in that the pixons represent an optimal set of cells. Ideally, an image's pixons represent the smallest number of cells (of arbitrary shape, position, etc.) required to fit the data, and represent the minimum DOFs necessary to specify the image. If properly selected, this set is irreducible to a smaller set. Hence pixons are the fundamental units of information in the image. Using a pixon basis is the fulfillment of Occam's Razor formalized in Bayesian terms–it forces the use of the simplest model consistent with the data.

3.2. Fuzzy Pixons: A Practical Pixon-Basis Choice

The simple counting arguments presented in the section above point out the crucial features of the pixon basis, i.e. there should be the fewest number of pixons consistent with fitting the data within the accuracy allowed by the noise. Furthermore, these pixons should contain the maximum information content. While this prescription is exact, we are still left with considerable uncertainty as to how to carry out this prescription in practice. In our attempts to derive suitable, practical pixon bases for image reconstruction, we finally adopted techniques which are similar to those adopted by other authors, i.e. a correlation length method (c.f. Weir 1991, 1993a). This approach controls the number of DOFs by reducing the independence of different parts of the image through explicit spatial correlation. This also causes the resulting degrees of freedom (or pixons) to be "fuzzy", i.e. to be localized but without hard boundaries. This still allows the use of the pixon prior of equation (8), although it does introduce a few computational complexities and mental hurdles for the intuition of the uninitiated. Nonetheless, the practical and performance merits of this approach seem to warrant these modest burdens.

Explicitly, then, our procedure for reducing the DOFs in the image reconstruction is to define the image in terms of a pseudo-image, $I_{pseudo}(\vec{x})$, convolved with a local correlation length, $\delta(\vec{x})$, i.e.

$$I(\vec{x}) = \int_{V_y} dV_y \, K_{pixon}\left(\frac{\vec{y} - \vec{x}}{\delta(\vec{x})}\right) I_{pseudo}(\vec{y}) , \tag{9}$$

where K_{pixon} is a pixon shape function and $\int_{V_y} dV_y \, K_{pixon}(\vec{y}\,/\delta) = 1$. The pseudo-image is defined on a pseudo-grid which typically has a resolution as fine or finer than the data pixel grid. The image is then also defined on a grid with the resolution of the pseudo-grid. Because of the local correlation in equation (9), however, the number of DOFs in the image can be greatly reduced from the number of pixels in the pseudo-grid. For example, if the local correlation length at position $\vec{x}$ is 10 pseudo-pixels then each 100 pixels (10 by 10 pixels) represent a single DOF at this location. Reduction in the DOFs greatly improves the formal value of the image/model prior expressed in equation (6) and removes many common problems with competing methods, e.g. signal correlated residuals and the production of spurious sources–see below.

3.3. An Iterative Procedure for Pixon-Based Reconstruction

There are many possible ways by which one might attempt to obtain the M.A.P. image/model pair for a given data set even when one has decided to use a fuzzy pixon scheme. There is, of course, the brute force method in which the pseudo-image values and the local correlation scales at each point in the pseudo-grid are considered as free variables and the M.A.P. image/model (in this case the correlation lengths might be considered the model) is calculated directly by maximizing $p(D|I,M)p(I,M)$ with any of a selection of multi-dimensional methods. Historically, however, this is not the procedure we have adopted (although we now feel that this is perhaps the best approach). For the calculations presented in Piña and Puetter (1993), Puetter and Piña (1993), and those presented in the sections that follow, we have used an iterative approach that first calculates the image with a fixed model, then an improved model holding the image fixed. This procedure is then iterated to convergence. The scheme is illustrated in Figure 1.

The iterative scheme for calculating the M.A.P. image/model pair starts with an initial guess for the model, i.e. the spatial correlation lengths. A common starting point is to assume that the scale lengths are all equal to 1 pseudo-pixel. This is equivalent to starting out with the standard ME solution for the image. In other words, for the first estimate for the image the fuzzy pixon prior is essentially the ME prior and the GOF criterion can be chosen to be the standard chi-squared value of the residuals. In practice, however, we typically use a simple GOF solution and ignore the ME prior. This is considerably faster in practice and results in a very good first guess. The next step estimates the new local scales, holding the image fixed. This is done by maximizing

$$p(M|D) = \int dI \frac{p(I,M|D)p(I,M)}{p(D)} \propto p(D|I_\circ,M)p(I_\circ,M) \tag{10}$$

i.e. finding the M.A.P. model given the fixed data and current image estimate, $I_\circ$. In our current implementations, this M.A.P. model is determined in only an approximate manner. We simply note, for example, that the prior term, $p(I_\circ,M)$, will insist on the largest possible

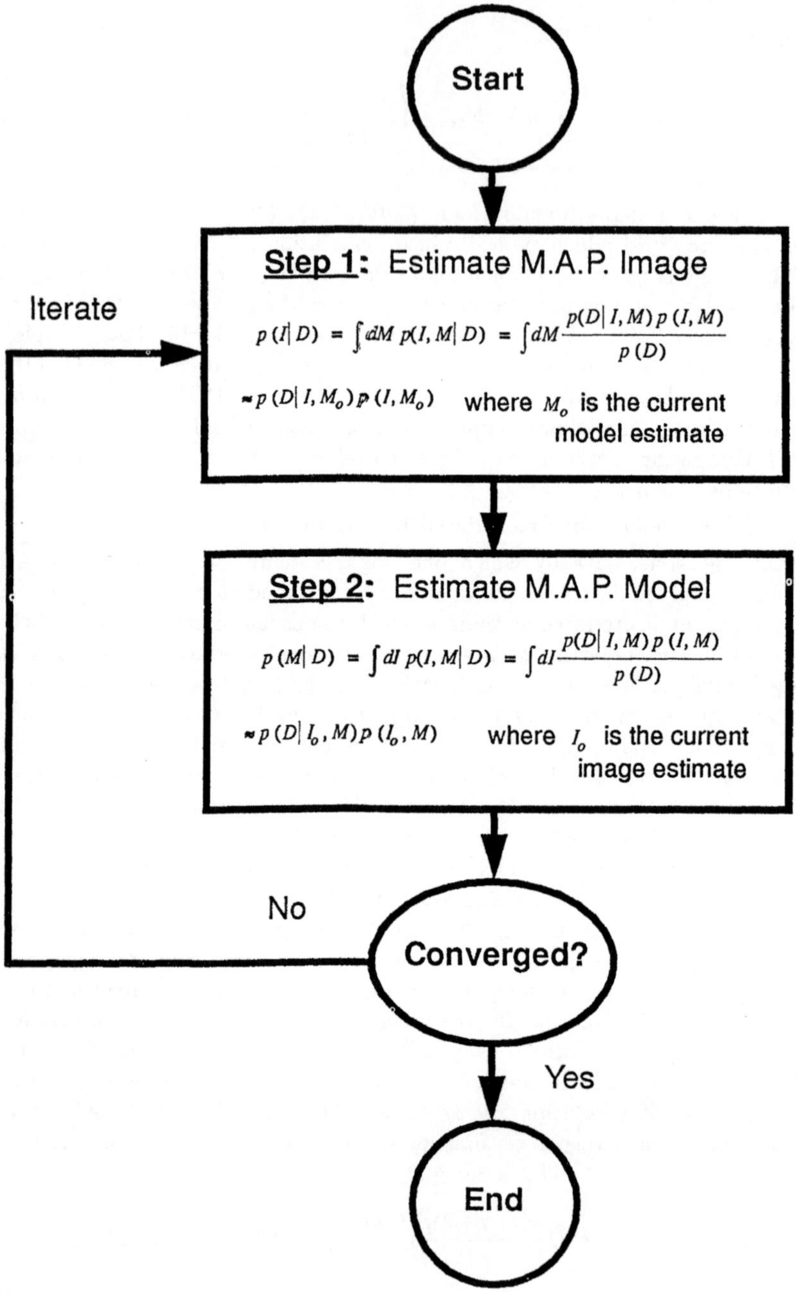

Figure 1: Schematic diagram of iterative scheme for Fractal Pixon Basis (FPB) image reconstruction.

correlation lengths consistent with the GOF, while the GOF term is indifferent to very small correlation lengths since they should always produce acceptable fits. Our procedure is thus simply to find the largest correlation lengths that provide an acceptable fit to the data, i.e., allow a chi-squared value within roughly one standard deviation $\sqrt{2(N-n)}$ of the chi-squared distribution about its modal value $N-n-2$, where N is the number of independent measurements (pixel values) and n is the number of pixons. Once the local scales have been determined, a new image is calculated, etc., and the entire procedure iterated until convergence is obtained.

A bit of intuition into this procedure reveals the fundamental reasons for this method's success. Effectively, this is a fractal technique. Independent of the exact method for obtaining the local correlation lengths (e.g. the brute force method or an iterative method as described above) this procedure seeks the natural local scale present in the underlying image as evidenced by the data. The pixon prior ensures that the procedure takes the largest, i.e. least informative, scale consistent with fitting the data. Our procedure analyzes how a geometric quantity varies as the local scale is varied. In this case we ask how $p(I, M|D)$ varies as the local scale is varied, just as the definition of fractal dimension asks how does the measure (e.g. length, area, etc.) of a geometric object vary as the local scale is changed. For this reason, we have named this entire class of methods Fractal-Pixon methods and the pixon representation of the image the Fractal-Pixon Basis (FPB).

3.4. The Pixon and DOF Maps

In the process of performing a pixon-based image reconstruction all of the appropriate local correlation lengths are determined. If these lengths, $\delta(\vec{x})$, are given in terms of pseudo-pixel sizes, then it is quite easy to determine exactly what fraction of a pixon (or DOF–note, we use the terms DOF and pixon interchangeably) each pseudo-pixel is, i.e. the pseudo-pixel at position $\vec{x}_i$ is

$$\frac{dn_{DOF}}{d(pseudopixel)} = \frac{1}{\int_{V_y} dV_y \, k_{pixon}\left(\frac{\vec{y}}{\delta(\vec{x}_i)}\right)} \tag{11}$$

fraction of a pixon, where $k_{pixon}(x)$ is the pixon shape function normalized to unity at $x=0$. The total number of pixons (or DOFs) in the image is given by

$$N_{pixon} = \sum_{pseudopixels,i} \frac{dn_{DOF}}{d(pseudopixel)} = \sum_{pseudopixels,i} \frac{1}{\int_{V_y} dV_y \, k_{pixon}\left(\frac{\vec{y}}{\delta(\vec{x}_i)}\right)} . \tag{12}$$

We call the "image" formed by the $\delta(\vec{x}_i)'s$ the pixon map, and the image formed by the $dn_{DOF}/d(pseudopixel)$ the DOF map. The pixon map displays the local scale length for pixons while the DOF map displays the density of DOFs in the image. The pixon map gives a lower limit to the spatial scales resolved by the image reconstruction, i.e. a resolution as fine as the local pixon scale (but no finer) has been achieved in the reconstruction. This resolution can be limited either by the quality of the data, e.g. signal-to-noise, or by the lack of real structure at finer scales in the underlying image. Each of these images can be "remapped" into data space by convolution with the PSF. This allows one to see the deduced structural scale and DOF density in data space.

4. Sample Image Restorations

To demonstrate the advances represented by our methods, we present in this section reconstructions from both real and "mock" data sets, i.e. artificially created data. We have done this for several reasons. In the mock data case, the image reconstruction conditions are essentially perfect, i.e. the noise, PSF, and true answer are known *a priori* with arbitrary precision. This leaves no uncertainty in how well each algorithm has performed. However, in imaging situations we rarely encounter such benign conditions. For this reason, we have also included a real data test case in which the noise and PSF characteristics are experimentally determined. Unfortunately, the true answer is also imperfectly known, making validation of the technique more difficult.

Having outlined our testing plan, we still need to decide to what competing techniques we shall compare our methods. In order to make the comparisons as fair as possible, we compare our reconstructions to those performed by other professionals well versed in the competing techniques. This avoids issues of whether the competing reconstructions are the best possible. For the real data test, we have chosen IRAS (Infrared Astronomical Satellite) $60\mu m$ survey scans of the interacting galaxy pair M51 (the "Whirlpool"). This data was used for an international image reconstruction contest at the 1990 MaxEnt Workshop (see Bontekoe 1991), which was attended by leaders in the field of image reconstruction. Hence our reconstruction of M51 will be compared to the best state-of-the-art reconstructions *circa* 1990.

From comparisons like the M51 contest, experts generally agree that ME produces results superior to GOF methods (e.g. Least-Squares and Lucy-Richardson). The reasons are simple. GOF methods do not employ a prior and hence are under-constrained relative to ME methods and typically over-fit the data. As mentioned earlier, this is why Lucy-Richardson (LR) method reconstructions are stopped after an arbitrary number of iterations. This prevents "break-up" of the image into numerous spatially small features. For this reason, we shall concentrate on comparing our reconstructions to ME reconstructions (although we shall present a LR reconstruction of the M51 data set as well, allowing the reader to judge the validity of these claims). The ME code we shall make our comparisons to is MEMSYS, a powerful set of ME algorithms developed by Gull and Skilling (see Gull and Skilling 1991). The MEMSYS algorithms probably represent the best commercial software package available for image reconstruction. The mock data reconstruction example compares our fractal pixon methods with MEMSYS 5, the most current version of the MEMSYS algorithms (see Gull and Skilling 1991). The M51 example compares our results to those of MEMSYS 3, the current version of MEMSYS at the time.

4.1. Example 1: A Mock Data Set Reconstruction

Figure 2 presents FPB and MEMSYS 5 reconstructions of a mock-data set. The MEMSYS 5 reconstructions were performed by Nick Weir of Caltech, a recognized MEMSYS expert, and were supplemented with his multi-channel correlation method which has been shown to enhance the quality of MEMSYS reconstructions (Weir 1991, 1993a). The true, noise-free, unblurred image presented in the top row is constructed from a broad, low-level elliptical Gaussian (i.e. a 2-dimensional Gaussian with different FWHMs in perpendicular directions), and 2 additional narrow, radially symmetric Gaussians. One of these narrow Gaussians is added as a peak on top of the low-level Gaussian. The other is subtracted to

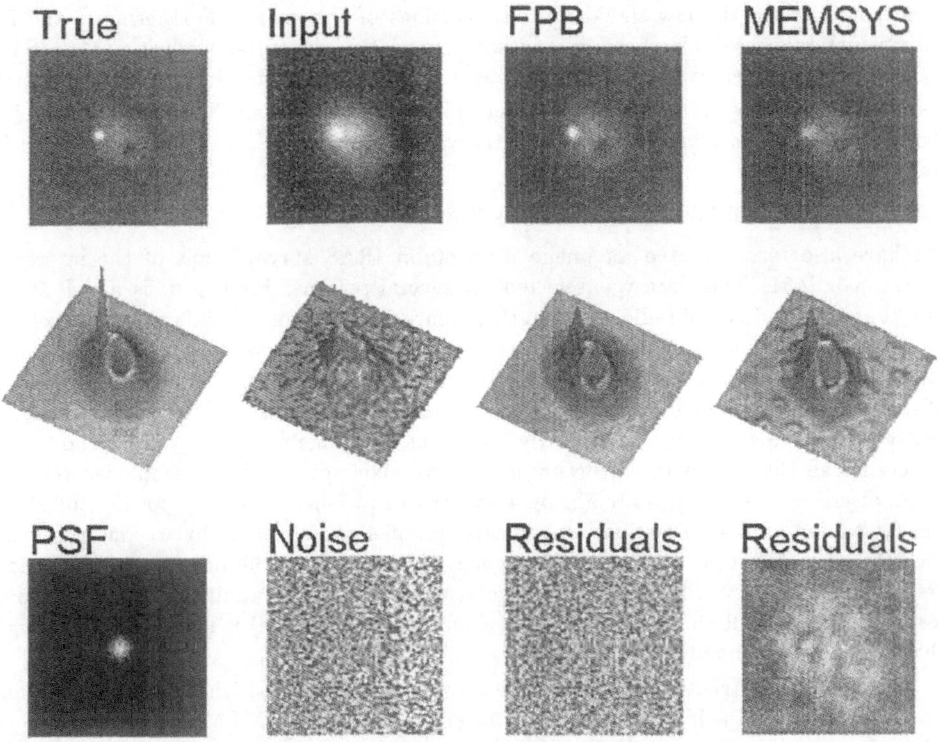

Figure 2: *FPB/MEMSYS 5 Comparison for the mock data set described in the text.*

make a hole. To produce the input image, the true image was convolved with a Gaussian PSF of FWHM=6 pixels, then combined with a Gaussian noise realization. The resulting input image is displayed in the top row. The signal-to-noise ratio on the narrow Gaussian spike is roughly 30. The signal-to-noise on the peak of the low level Gaussian is about 20. The signal-to-noise at the bottom of the Gaussian "hole" is 12.

As can be seen, the FPB reconstruction is superior to the multi-channel MEMSYS result. The FPB reconstruction is free of the low-level spurious sources evident in the MEMSYS 5 reconstruction. These false sources are due to the presence of unconstrained degrees of freedom in the MEMSYS 5 reconstruction and are superimposed over the entire image, not just in the low signal to noise portions of the image. Furthermore, the FPB reconstruction's residuals show no spatially correlated structure, while the MEMSYS 5 reconstruction systematically under-estimates the signal, resulting in biased photometry.

To illustrate the DOF density concept, Figure 3 provides an illustration of the DOF density map, i.e. an image formed from the values of $dn_{DOF}/d(pseudopixel)$ for this first example.

4.2. Example 2: 60 Micron IRAS Survey Scans of M51

We have also reconstructed an image from $60\mu m$ IRAS survey scans of the interacting galaxy pair M51. This data was selected for several reasons. First, M51 is a well studied object at optical, IR, and radio wavelengths. Hence "reality" for this galaxy is relatively well known. Second, as mentioned before, this particular data set was chosen as the basis of an image reconstruction contest. Consequently, there have been a number of serious attempts at performing image reconstruction on this data set by specialists in the field. Finally, the IRAS data for this object is particularly strenuous for image reconstruction methods. This is because all the interesting structure is on "sub-pixel scales" (IRAS employed relatively large, discrete detectors–1.5 arcmin by 4.75 *arcmin* at $60\mu m$) and the position of M51 in the sky caused all scan directions to be nearly parallel. This means that reconstructions in the cross-scan direction (i.e. the 4.75 *arcmin* direction along the detector length) should be significantly more difficult than in the scan direction. In addition, the point source response of the 15 IRAS $60\mu m$ detectors (pixel angular response) is known only to roughly 10% accuracy, and finally, the data is irregularly sampled.

Our FPB reconstruction appears in Figure 4 along with Lucy-Richardson and Maximum Correlation Method (MCM) reconstructions (Rice 1993) and a MEMSYS 3 reconstruction (Bontekoe et al. 1991)–see Gull and Skilling (1991) for a description of the MEMSYS algorithms. The winning entry to the MaxEnt 90 image reconstruction contest was produced by Nick Weir of Caltech and is not presented here since quantitative information concerning this solution has not been published–however, see Bontekoe (1991) for a gray-scale picture of this reconstruction. Nonetheless, Weir's solution is qualitatively similar to Bontekoe's solution (Weir 1993b). Both were made with MEMSYS 3. Weir's solution, however, used a single correlation length channel in the reconstruction. This constrained the minimum correlation length of features in the reconstruction, preventing break-up of the image on smaller size scales. This is probably what resulted in the "winning-edge" for Weir's reconstruction in the MaxEnt 90 contest (Weir 1993b).

As can be seen from Figure 4, our FPB-based reconstruction is superior to those produced by other methods. The Lucy-Richardson and MCM reconstructions fail to signif-

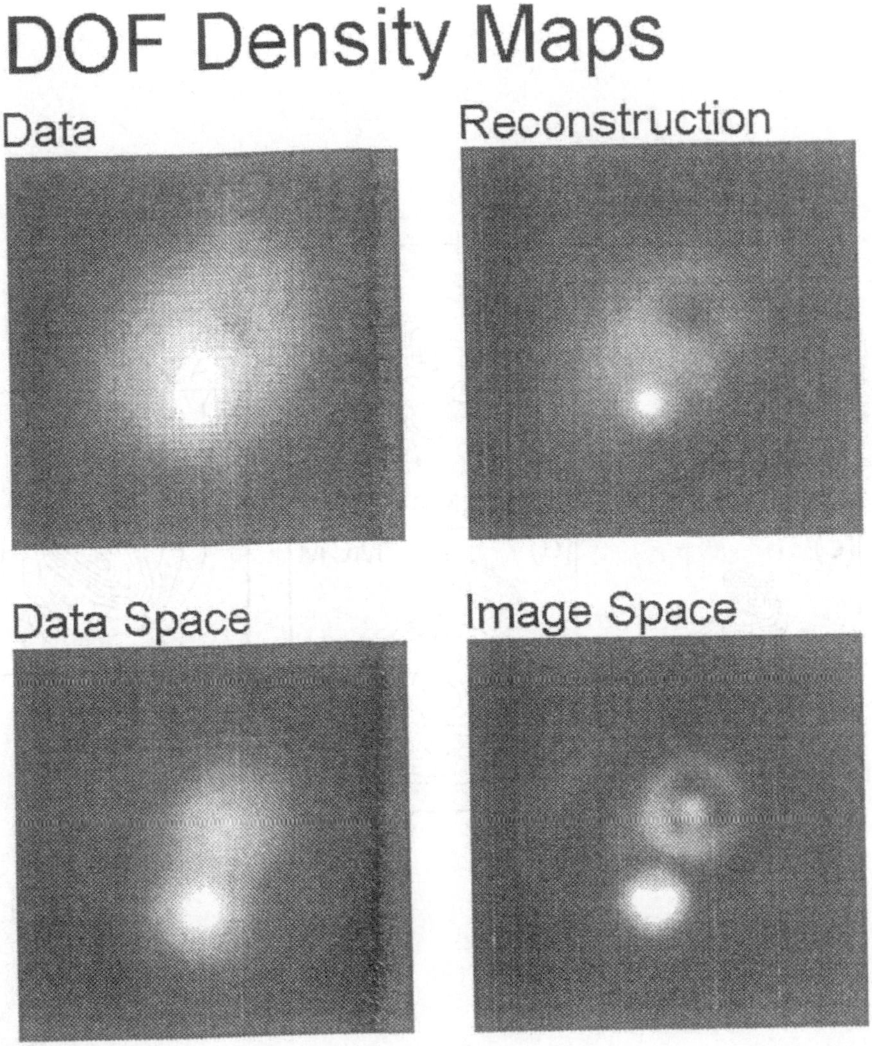

Figure 3: Degree-of-Freedom (DOF) density maps in Data and Image space. The top row shows the input data (top-left) and the reconstructed image (top-right). The deduced DOF density from the reconstructed image is shown below it, i.e. in the lower-right panel, and this DOF density mapped back into Data space is shown in the lower-left panel.

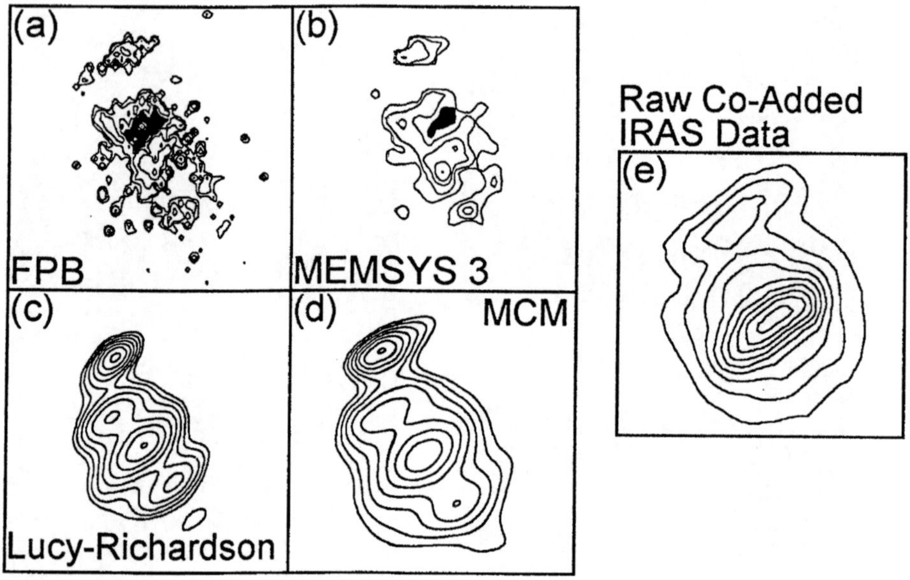

Figure 4: Image reconstruction of the Interacting galaxy M51. (a) FPB-based reconstruction. (b) MEMSYS 3 Reconstruction (c) Lucy-Richardson reconstruction. (d) MCM reconstruction. (e) Raw, co-added IRAS $60\mu m$ data. Figure of panel (b) reproduced from Bontekoe et al. (1991), by permission of the authors. Figures of panel (c) and (d) reproduced from Rice 1993 by permission of the author.

icantly reduce image spread in the cross-scan direction, i.e. the rectangular signature of the 1.5 by 4.75 *arcmin* detectors is still clearly evident, and fail to reconstruct even gross features such as the "hole" (black region) in the emission north of the nucleus–this hole is clearly evident in optical images of M51. The MEMSYS 3 reconstruction by Bontekoe is significantly better. This image clearly recovers the emission "hole" and resolves the north-east and south-west arms of the galaxy into discrete sources. Nonetheless, the level of detail present in the FPB reconstruction is clearly absent, e.g. the weak source centered in the emission hole (again, this feature corresponds to a known optical source).

To assess the significance of the faint sources present in our FPB reconstruction, in Figure 5 we present our reconstruction overlaid with the 5 GHz radio contours of van der Hulst et al. (1988). The radio contours are expected to have significant, although imperfect, correlation with the far infrared emission seen by IRAS. Hence a comparison of the two maps should provide an excellent test of the reality of structures found in our reconstruction. Also identified in Figure 5 are several prominent optical sources and Hα (hydrogen Balmer line emission) knots. As can be seen, the reconstruction indicates excellent correlation with the radio. The central region of the main galaxy and its two brightest arms align remarkably well, and the alignment of the radio emission from the north-east companion and the IRAS emission is excellent. Furthermore, for the most part, whenever there is a source in the reconstruction which is not identifiable with a radio source, it can be identified with either optical or Hα knots. An excellent example is the optical source in the "hole" of emission to the north-east of the nucleus of the primary galaxy or the bright optical source to the north-west of the nucleus (both labeled "Opt" in Figure 5). Because of the excellent correlation with the radio, optical, and Hα images, we are quite confident that all of the features present in our reconstruction are real. Aside from the fact that most of the sources can be identified with emission at other wavelengths, the residual errors in our reconstruction are much smaller than in the MEMSYS 3 reconstruction. As pointed out by Bontekoe et al. (1991) the peak flux in the MEMSYS 3 reconstruction is 2650 units. The residual errors are correlated with the signal and lie between 0 and 430 units. By contrast, the peak value in the FPB reconstruction is 3290 units, the residuals are uncorrelated with the signal, and the residuals lie between -9 and 17 units. (The contour levels for the MEMSYS 3 and FPB reconstructions of Figure 2 are identical and are 150, 300, 600, 1200, and 2400 units.) Furthermore, the large deviation residuals in the FPB reconstruction are due to systematic errors involving incomplete scan coverage of M51. Furthermore, these errors do not lie under the significant flux emitting portions of the M51 image. The residual errors associated with emitting regions in M51 are significantly smaller ($\sigma \approx 1$ unit) and show a roughly Gaussian distribution. Full appreciation of the sensitivity of our technique is only obtained once the reconstruction has been flux calibrated. Formally, the residual error over the majority of the image is 2.7 mJy. This compares with the 280 mJy, 90% completeness limit for the IRAS Faint Source Survey. The largest residual systematic errors associated with incomplete sampling of the M51 region correspond to 50 mJy. This is still a factor of more than 5 times fainter than the IRAS Faint Source Survey limit.

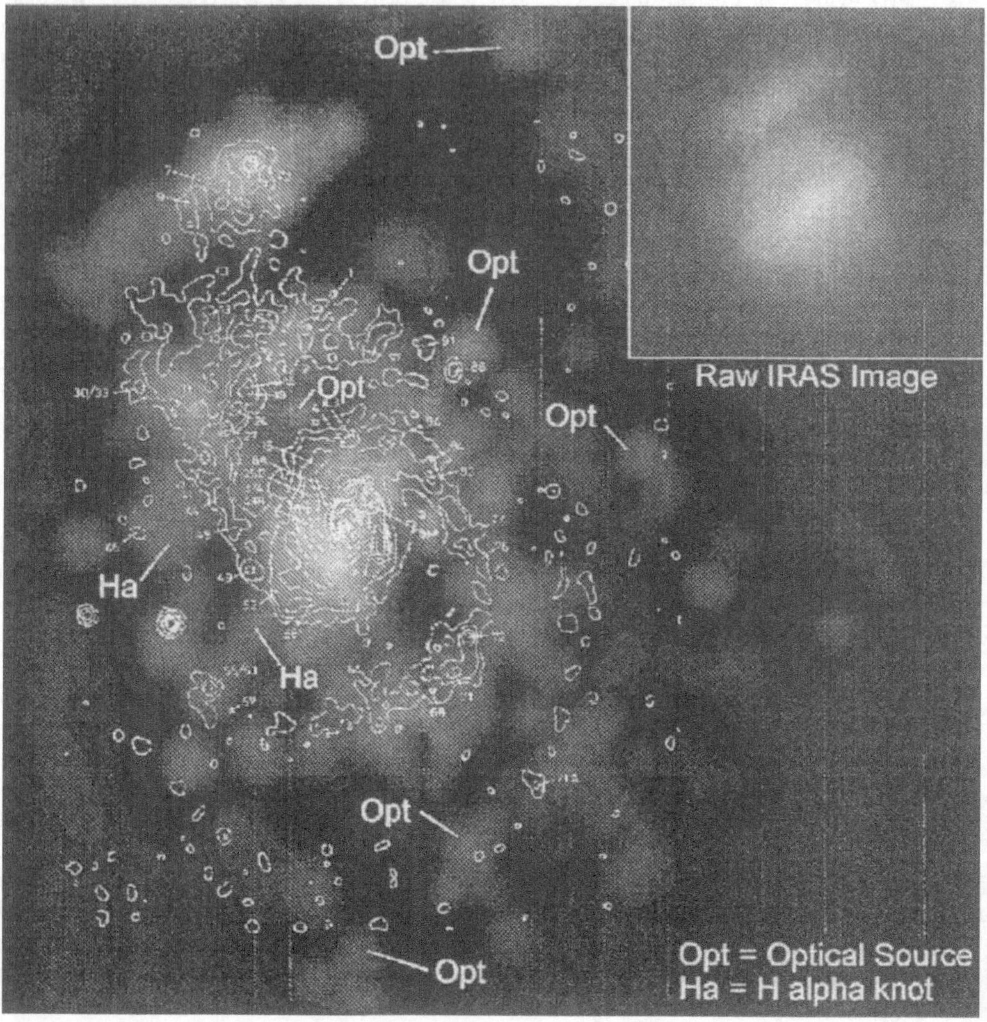

Figure 5: Details of the FPB reconstruction of the IRAS $60\mu m$ survey of M51. The overlaid contours are the 5 GHz data of van der Hulst et al. (1988). Also noted are several of the stronger $60\mu m$ features that can be identified with optical features or $H\alpha$ knots.

5. Conclusions

In conclusion, we have introduced a new concept, the pixon, the use of which provides large improvements in the inversion of a broad class of integral equations, such as those characteristic of the degradation of spectral, spatial, or temporal resolution in the measurement of signals in the presence of noise. The pixon is the fundamental and indivisible unit of information required to describe the underlying signal within the accuracy allowed by the data. In this regard it is an idealized concept. However, this paper also presents a practical and capable approximation to this ideal for a broad range of problems, i.e. the fuzzy, Fractal-Pixon Basis (FPB). This basis uses the local spatial scale relevant to the underlying signal to constrain the inversion of the equations governing the measurement process. In doing so, this method provides performance superior to pure GOF (Maximum Likelihood) and ME methods. Some of the advantages of pixon-based methods are the elimination of signal-correlated residuals and the production of spurious sources typical of other methods. Practical examples from the realm of astronomical image reconstruction show that pixon-based methods can offer large improvements in resolution as well as the detection of extremely weak features in the data.

6. Acknowledgements

The authors would like to thank a number of people for their valuable contributions to this work. We would especially like to thank Nick Weir of Caltech for numerous fruitful discussions regarding image processing and for graciously performing the multi-channel MEMSYS 5 reconstructions presented in this paper. We would also like to thank Romke Bontekoe and Do Kester for providing the M51 IRAS test data set as well as for many helpful conversations. Finally, the authors would like to thank Steven Gull and John Skilling for a number of conversations that greatly expanded our understanding of Bayesian methods in general, and how the pixon fits into Bayesian theory in particular. This work was supported by NASA and the National Science Foundation.

References

[1] Adorf, H.-M., Walsh, J. R., and Hook, R. N., "Restoration Experiments at the ST-ECF", in *The Restoration of HST Images and Spectra*, Proceedings of a Workshop held at the Space Telescope Science Institute, R. L. White and R. J. Allen, Eds., (Baltimore: Space Telescope Science Institute), 121. (1990).

[2] Bontekoe, Tj., R., "The Image Reconstruction Contest", in *Maximum Entropy and Bayesian Methods*, W. T. Grandy, Jr. and L. H. Schick, Eds., (Dordrecht: Kluwer Academic Publishers), 319, 1991.

[3] Bontekoe, Tj., R., Kester, D. J. M., Price, S. D., de Jonge, A. R. W., and Wesselieus, P. R., *Astron. & Astrophysics* 248, 328 1991.

[4] Bontekoe, Tj. R., Koper, K., and Kester, D. J. M. 1993, "Pyramid Maximum Entropy Images of IRAS Survey Data", *Astron. & Astrophys.*, submitted.

[5] Bryan, R. K., and Skilling, J., *Monthly Notices of the Royal Astronomical Society*, 191, 69, 1980.

[6] Gull, S. F., "Developments in Maximum Entropy Data Analysis", in *Maximum Entropy and Bayesian Methods,* J. Skilling, Ed., (Dordrecht, Netherlands: Kluwer Academic Publishers), 53. 1989.

[7] Gull, S. F., and Skilling, J. *MemSys5 Quantified Maximum Entropy User's Manual* 1991.

[8] Kikuchi, R., and Soffer, B. H. in *Image Analysis and Evaluation*, Society of Photographic Scientists and Engineers, Toronto, Canada, July 1976, 95. 1976.

[9] Narayan, R. and Nityananda, R. "Maximum Entropy Image Restoration in Astronomy", *Ann. Rev. Astron. & Astrophys.*, 24, 127, 1986.

[10] Lucy, L. B. 1974, Astron. J., 79, 745.

[11] MacKay, D. J. C. "Bayesian Interpolation", *Neural Computation*, 4,3, 415. 1992a.

[12] MacKay, D. J. C. "The Evidence Framework Applied to Classification Networks", *Neural Computation,* 4,5, 698., 1992b

[13] Piña, R. K., and Puetter, R. C. "Incorporation of Spatial Information in Bayesian Image Reconstruction: The Maximum Residual Likelihood Criterion", *Publications of the Astronomical Society of the Pacific*, 104, 1096, 1992.

[14] Piña, R. K., and Puetter, R. C. "Bayesian Image Reconstruction: The Pixon and Optimal Image Modeling", *Publications of the Astronomical Society of the Pacific*, 105, 630, 1993.

[15] Puetter, R. C., and Piña, R. K. "The Pixon and Bayesian Image Reconstruction", *S.P.I.E. Proceedings,* April 1993, Orlando, FL, in press, 1993.

[16] Rice, W. *Astronomical Journal*, 105, 67, 1993.

[17] Sibisi, S. in *Maximum Entropy and Bayesian Methods,* J. Skilling, Ed. (Dordrecht, Netherlands: Kluwer Academic Publishers), 1991.

[18] Skilling, J. "Classic Maximum Entropy", in *Maximum Entropy and Bayesian Methods,* J. Skilling, Ed., (Dordrecht, Netherlands: Kluwer Academic Publishers), 45, 1989.

[19] Skilling, J. in *Maximum Entropy and Bayesian Methods,* W. T. Grady, Jr. and L. H. Schick, Eds., (Dordrecht, Netherlands: Kluwer Academic Publishers), 1991.

[20] van der Hulst, J. M., Kennicutt, R. C., Crane, P. C., and Rots, A. H., "Radio Properties and Extinction of the H II Regions in M51", *Astron. & Astrophysics,* 195, 38, 1988.

[21] Weir, N, "Applications of Maximum Entropy Techniques to HST Data", in *Proceedings of the ESO/ST-ECF Data Analysis Workshop,* April 1991, P. Grosbo and R. H. Warmels, Eds. (Garching: ESO), 115, 1991.

[22] Weir, N, "A Maximum Entropy-Based Model for Reconstructing Distributions with Correlations at Multiple Scales", *J. Opt. Soc. Am.,* in press, 1993.

SUPER-RESOLVED SURFACE RECONSTRUCTION FROM MULTIPLE IMAGES

Peter Cheeseman, Bob Kanefsky, Richard Kraft,
John Stutz, and Robin Hanson
NASA Ames Research Center
Moffet Field, CA 94035 U.S.A.

ABSTRACT. This paper describes a Bayesian method for constructing a super-resolved surface model by combining information from a set of images of the given surface. We develop the theory and algorithms in detail for the 2-D reconstruction problem, appropriate for the case where all images are taken from roughly the same direction and under similar lighting conditions. We show the results of this 2-D reconstruction on Viking Martian data. These results show dramatic improvements in both spatial and gray-scale resolution. The Bayesian approach uses a neighbor correlation model as well as pixel data from the image set. Some extensions of this method are discussed, including 3-D surface reconstruction and the resolution of diffraction blurred images.

1. Introduction

Consider the problem of how to extract as much information as possible from a set of images, all of the same scene, and of capturing this information in the form of a surface model at maximal resolution. This problem is important in many applications where maximal resolution is paramount. In this paper we focus on space-based remote imaging.

Surface reconstruction from an image set is an example of an inverse problem: if we knew exactly the shape and emittance of the surface, the illumination conditions, the camera angle, etc., we could predict what the camera would observe (the pixels) to within the measurement accuracy. This is the rendering problem addressed by computer graphics. We have the inverse problem: we are given the observed images (pixels) and must use this information to find the most probable surface that could have generated these images. Bayes' theorem provides a formal solution to inverse problems, which we apply here to the surface reconstruction problem.

Because the reconstructed surface can only be determined to within a certain maximum spatial resolution, we represent surfaces by a discrete uniform grid with the surface properties given at each grid point. For the case of a planetary surface, these surface properties could include illumination, albedo, slope, emittance at different wavelengths, etc. We will describe in detail a model using only surface emittance, and then describe how to extend this model. These properties characterize the grid point and describe how it could influence the image pixels once the camera parameters are known. This surface grid is a reconstruction and is *not* what was actually observed. For this reason we call the surface grid elements **mixels** (for **m**odel p**ixels**) to distinguish them from **pixels** which are the *observed* values. Unfortunately, in much of the vision literature, the word pixel is used interchangeably to refer to both inferred and observed values.

We are able to get super-resolved reconstructions from image sets because each pixel of

G. R. Heidbreder (ed.), Maximum Entropy and Bayesian Methods, 293–308.

each image is a new sample of some patch on the observed surface. Two images generated with *exactly* the same alignment between the camera and the surface, the same illumination conditions, etc., record the same information to within the measurement error of the camera, resulting in no net gain of information. With slightly differing alignments, however, the observed pixel values will be different, because the camera is observing slightly different patches on the surface. By relating these differences to locations on the surface, it is possible to reconstruct a model grid at a finer resolution than that of the observed pixel grid. This technique for combining overlapping information is closely related to deconvolution (e.g. radar imaging) and computed tomography (e.g. CAT scan), and is explained in more detail in section 3.. In particular, this information combining technique goes beyond the Nyquist limit for a single observed image. Fig. 1 shows schematically why subpixel resolution is possible.

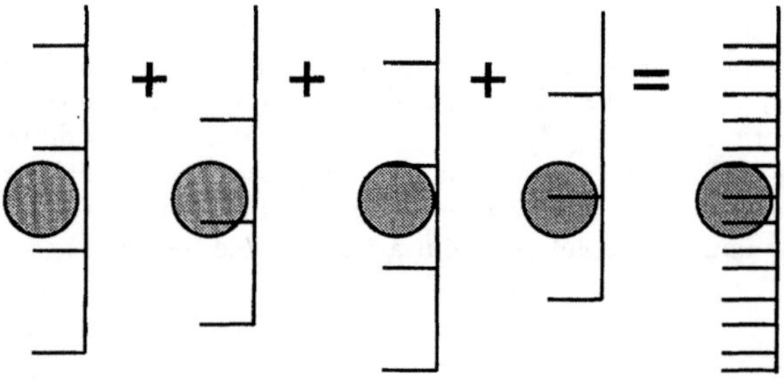

Figure 1: Because several sampling grids are used, offset randomly with respect to each other, resolution beyond the Nyquist limit of any one frame is possible.

We start by considering "flat" surface reconstruction. This is the best that can be achieved when the images are taken from essentially the same camera position and sun angle, but with slightly different registrations. This occurs with Landsat images, for example, where each location on Earth is imaged from essentially the same position in space. The reconstruction gives the "emittance" of the surface, which is a combination of the effects of surface albedo, illumination conditions and ground slope. We develop this theory in detail in section 2., and show its application to Viking Orbiter data of Mars. This theory includes the use of prior knowledge in the form of neighbor correlations. In section 5., we outline how to extend this approach to a 3-D surface reconstruction, where images from different directions allow us to separate variations in pixel values due to albedo from those due to ground slope.

2. 2-D Surface Reconstruction

Our approach is based on Bayesian probability theory. We use a likelihood function, defined to be the probability of the observed data given a model of how the data were generated. This model of the observation process is normally parameterized with respect to any variables that affect the process. For the current problem, these observational parameters include surface illumination, surface albedo, camera orientation, camera characteristics, optical distortions, and any data preprocessing. Computational considerations may require that some of these parameters be simplified or omitted, but doing so always entails some loss of precision.

We have made several such simplifications in the work described in this section. The first is that we model the "surface" as a plane lacking curvature and local relief — i.e., as a grid only of emittance values. Thus, the value of each mixel is simply a scalar — its emittance. The second is the substitution of a simple transformation (affine or quadratic) for the projective observation geometry and for any optical and electronic distortions of the camera system. A third lies in using a preprocessing step to deal with telemetry noise. For Mars images, we ignore ambient light contribution from a diffuse background and atmospheric attenuation, as they are negligible in the data sets we use.

In our approach, we begin by constructing a likelihood function that gives the probability of each pixel value, given the imaged surface and observation conditions. We take the likelihood of the entire image to be just the product of likelihoods of each pixel. This means we are assuming that the measurement error of a pixel is conditionally independent of the value of its neighbors. This conditional independence assumption is symbolically represented as:

$$P\,[\text{all pixel values} \,|\, \text{observation params, mixels}] = \prod_p P\,[(\text{pixel(p)} = \Phi_p \,|\, \text{observation params, mixels}]\,. \tag{1}$$

Here, pixel(p) is a location of a pixel on some image in the image set, and Φ_p is an observed energy value[1]. What we read off the camera is the radiant energy received by each pixel[9]. Note the split of parameters into two sets: observation parameters and mixels. We will explain the significance of this split below.

We assume that the probability of an observed pixel value is normally distributed, so that the likelihood of each pixel is given approximately by:

$$P\,[\text{pixel(p)} = \Phi_p \,|\, \text{observed params, surface model}] \approx N[\Phi_p \,|\, \hat{\Phi}_p, \sigma]\Delta\Phi_p. \tag{2}$$

Here $N[x \,|\, \mu, \sigma]$ is the standard normal (or Gaussian) distribution of x given a mean μ and standard deviation σ. The $\Delta\Phi_p$ term is the observed minimum gray-scale difference. The standard deviation σ of the observed pixels from their expected values is assumed to be the same for all pixels in an image. This deviation results from measurement error (especially quantization error[2]) and model errors of various kinds (e.g. slight mis-registration). If these many sources of error are largely independent, the central limit theorem leads us to expect the resulting error distribution to be close to normal. Experimental data confirm this expectation, as is discussed in section 3.2..The normal approximation in this case assumes that $\sigma >> \Delta\Phi_p$. This distribution is just the trapezoid approximation to the integral of a normal density over the interval from Φ_p to $\Phi_p + \Delta\Phi_p$.

[1]This is not physically correct, as the camera outputs joules. However, one can multiply the flux Φ by the exposure time and the pixel size to obtain joules Q. We will stick to flux values to keep Eqn.(1) camera independent.

[2]The quantization of continuous emittance values into integers

In Eqn. (2), the term $\hat{\Phi}_p$ represents the expected radiant energy[3]value for pixel p and is a function of the observation parameters and surface model. The parameters used in determining $\hat{\Phi}_p$, as used in likelihood Eqn. (2), are:

1. **Mixel Values:** This is the model of the reconstructed 2-D surface represented by an "emittance" value at each grid point (mixel);
2. **Registration Parameters:** These geometric parameters define how a pixel image maps onto the reconstructed mixel grid. Here, we use affine and quadratic transformations to define a 2-D (camera) to 2-D (mixel grid) function;
3. **Point Spread Function (PSF):** This function defines how points on the surface (mixels) contribute to the observed pixels through the camera optics, including any distortions produced by camera readout;
4. **Camera Shading:** These parameters are necessary for cameras, such as a vidicon[4], with a nonuniform readout gain across the image plane. These parameters define a scaling factor that varies depending on where on the image plane a particular pixel falls.

The contribution of these parameters to $\hat{\Phi}_p$ is shown diagrammatically in Fig. 2. Given

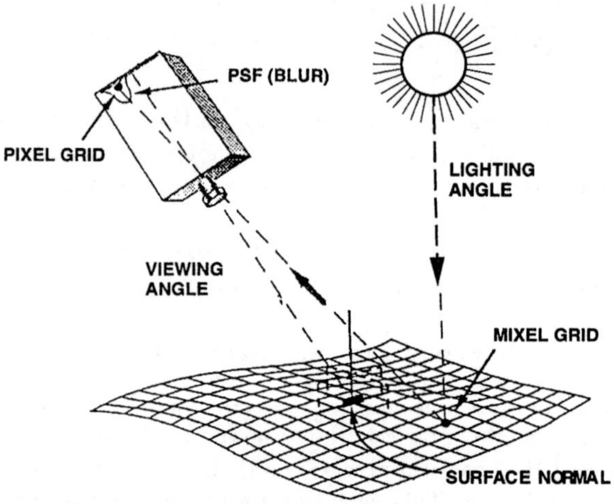

Figure 2: Parameters Relating Pixels to Mixels

values for the mixels and the parameters relating mixels to pixels, it is possible to calculate the expected value of a given pixel, $\hat{\Phi}_p$, by summing the contribution of each mixel, as weighted by the PSF. This is explained in detail in the next section. This pixel prediction process is just the "forward" graphics problem, shown in Fig. 2.

[3]See discussion below in section 3.0 for terminology.

[4]A vidicon camera is an obsolete electron beam readout camera, such as used in the Viking Orbiter images shown in this paper.

In a maximum likelihood (ML) approach, the goal is to find the set of parameter values that maximizes Eqn. (1)—in particular, the ML estimates of the mixels is a way of reconstructing an unknown surface from the images. Note that finding the ML mixel values is a way of solving the "inverse" graphics problem (i.e. finding a model from data) given the likelihood (i.e. probability of the data given a model). When the resolution chosen for the mixel grid is *over-determined* by the corresponding pixel values, the ML approach is reasonable. The mixels are over-determined by the pixels when there is no value for the mixels which can exactly predict all the pixel values. The over-determined situation means that the mixel grid is at a coarser spatial resolution than is otherwise achievable. If the ML approach is tried at too fine a resolution, the mixel values are *under-constrained*—i.e., there are many mixel grids that would predict the pixel values exactly, and there is no principled way of choosing among them.

The Bayesian approach used here is similar to the ML approach, but it uses additional (prior) knowledge in the form of expectations about correlations among neighboring mixels. This additional knowledge in the Bayesian maximum a posterior (MAP) estimate allows any scale mixel grid. If too coarse a mixel grid is used (i.e. the mixels are over-determined by the pixels), then the neighbor correlations have little effect, and the MAP estimate is essentially the same as the ML estimate. However, if a very fine mixel grid is used (i.e. the mixels are under-determined), then the effect of the neighbor correlations competes with the fit to the data to give a reasonable compromise result that uses all the information. The optimal mixel resolution is near the borderline between over-determined and under-determined, where the neighbor correlation information begins to suppress the affects of the noise in the data. Thus the prior term acts much in the same manner as a "regularization" term in related approaches.

3. MAP Reconstruction

Given pixel data and the parameters that specify the imaging model, we want to jointly estimate the mixel grid values together with other auxiliary model parameters. In a Bayesian approach one seeks a combination of all these parameters which has maximum posterior (MAP) probability, which is the same (up to a normalization factor) as seeking a maximum joint probability:

$$\begin{aligned} \text{Joint Probability} &= \text{Likelihood} \times \text{Prior Probability} \\ P[\text{Mixels, Pixels, Params}] &= P[\text{Pixels} \mid \text{Mixels, Params}] \\ &\quad \times P[\text{Mixels} \mid \text{Params}] \times P[\text{Params}], \end{aligned}$$

where "Mixels" refers to the set of all mixel values, "Pixels" refers to the set of all pixels in all images, and "Params" refers to the auxiliary observational parameters (registration parameters, PSF, etc.) listed above.

Repeating Eqn. (2), the likelihood term is:

$$P[\text{Pixels} \mid \text{Mixels, Params}] = \prod_p N[\Phi_p \mid \hat{\Phi}_p, \sigma_p]\Delta\Phi_p. \tag{3}$$

We now specify the mean for each pixel $\hat{\Phi}_p$ to be a linear combination of the emittances from mixels projected near the pixel location:

$$\hat{\Phi}_p = \sum_i \omega_{ip} m_i. \tag{4}$$

Here m_i is the emittance of the ith mixel and ω_{ip} is the mixel-pixel weight defined by the PSF and registration information. For images that are not significantly diffraction blurred, the radiant energy at a point in the image plane is a sum of contributing emittance values (not amplitudes) from the generating surface, as shown in equation (4).

So far, we have used the *physical* terminology appropriate for describing the quantities radiating from the mixel grid (emittance) and captured by the camera (radiant energy). While this convention is admittedly arbitrary, (we could, for example, equip the camera model with an exposure term E(Q) that turns radiant energy values Q to flux values ϕ) it is more precise than the alternate computer graphics convention of labeling a host of quantities with "intensity", regardless of its being a light source, a CRT raster, a planetary surface, etc. While in the present 2-D super-resolution case these distinctions may not seem useful, such precision is helpful when the model is extended to 3-D surface reconstruction. It is in this context of anticipated extension that the terminology in this paper is chosen.

3.1. Prior

The prior probability term $P[\text{Mixels} \mid \text{Params}]$ is the distinctly Bayesian contribution, and it embodies one's beliefs *before* seeing the data about the kinds of scenes or landscapes one might observe. The simple prior used in this paper describes how mixel intensities m_i relate to each other. The remaining model parameters — the point spread function coefficients, optical angles, etc.—are highly over-determined by the data, so we can reasonably neglect the priors $P[\text{Params}]$ on these parameters.

To gain insight into the appropriate prior over the mixel intensities, we analyzed Viking Orbiter imagery. This prior can be thought of as a means of preferring a given solution when many solutions fit the data equally. We choose a prior that makes a reconstruction more likely if its mixel values are highly correlated with their neighbors (i.e. there is emittance continuity). A simple, probabilistic model of continuity would be to estimate the value of a mixel m_i by a weighted sum of its neighbors:

$$\hat{m_i} = \sum_j \alpha_{ij} m_j. \tag{5}$$

Here, the α_{ij} are weights: mixel m_j contributes α_{ij} to mixel m_i. While this form is fairly general, we choose to start with a particularly simple relationship where $\alpha_{ii} = 0$, $\alpha_{ij} = \alpha_{ji} = 1/4$ if $|i - j| = 1$, and $\alpha_{ij} = 0$ otherwise. This just means that a mixel is directly correlated only with its four cardinal neighbors. Because the neighbors are correlated with *their* neighbors, etc., this also indirectly implies long range correlation.

Figure 3 shows the well-groundedness of the above continuity preference. First, the central peak at 0 shows that mixels are in fact correlated with the four cardinal neighbors. The moments of this distribution suggests that a multivariate normal form for the prior,

$$P[\text{Mixels} \mid \text{Params}] = N[m_i \mid \overline{m_i}, \Sigma_{ij}] \prod_i dm_i. \tag{6}$$

is appropriate. Here, the matrix Σ_{ij} collects the α_{ij} dependencies. The $\overline{m_i}$ in Eqn. (6) represents putative "mean" values for a mixel at position i. The model we implement only uses a constant value $\overline{m_i} = \overline{m}$, the mean value of all mixels. Possible extensions could make use of varying means $\overline{m_i}$ to capture "trends" in an image.

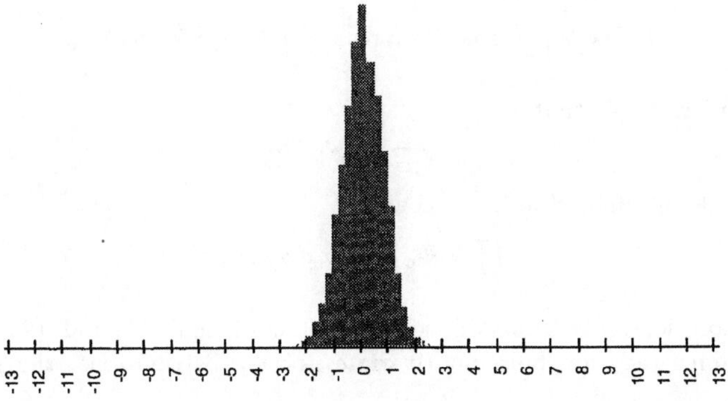

Figure 3: Distribution of $m_i - \hat{m}_i$ for a Viking Orbiter image of Mars. Horizontal axis is in 7-bit Data Numbers.

Appearances may be deceptive, however. Figure 4 shows the same calculation as in Figure 3; although the shape looks fairly Gaussian, examining the moments of the distribution reveals that there is much more energy in the tails than is the case for a normal model. This is because individual pixels in Earth imagery can differ substantially from their neighbors, *e.g.* a road traversing ground.

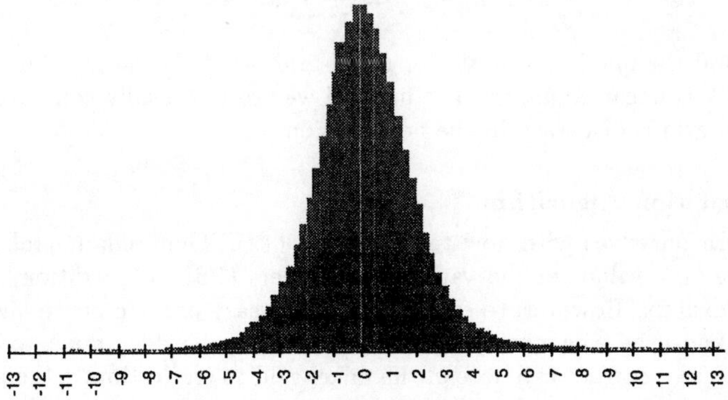

Figure 4: Distribution of $m_i - \hat{m}_i$ for a Landsat image of Kansas. Horizontal axis is in 8-bit Data Numbers.

3.2. MAP Equation

Now that we are equipped with the prior

$$P[\text{Mixels} \mid \text{Params}] = N[m_i \mid \overline{m}, \Sigma_{ij}] \prod_i \Delta dm_i \tag{7}$$

we combine it with our likelihood

$$P[\text{Pixels} \mid \text{Mixels, Params}] = \prod_p N[\Phi_p \mid \hat{\Phi}_p, \sigma] \Delta\Phi_p. \tag{8}$$

Keeping in mind Eqn. (4) that

$$\hat{\Phi}_p = \sum_i \omega_{ip} m_i,$$

we can rewrite the likelihood as

$$\prod_p N[\Phi_p \mid m_i, \omega_{ip}, \sigma] \Delta\Phi_p. \tag{9}$$

The MAP solution seeks to maximize the product of Eqns. (7) and (9), i.e. the "joint" distribution. Since they are both multivariate normal distributions, the joint is as well. Thus,

$$\begin{aligned} MAP &= Max(N[m_i \mid \overline{m}, \Sigma_{ij}] * \prod_p N[\Phi_p \mid \hat{\Phi}_p, \sigma]) \\ &= Max(N[m_i \mid \overline{m}, A_{ij}]) \end{aligned} \tag{10}$$

where

$$A_{ij} = \Sigma_{ij} + \frac{1}{\sigma^2} \sum_p \omega_{ip}\omega_{jp}. \tag{11}$$

The matrix A_{ij} in Eqn. (11) is calculated by standard methods in completing the square for multivariate distributions[10]. The peak of the distribution in Eqn. (10) are the m_i that satisfy

$$\sum_i A_{ij}(m_i - \overline{m}) = \frac{1}{\sigma^2} \sum_p \omega_{ip}(\Phi_p - \overline{m}). \tag{12}$$

We can thus find the maximum posterior mixel grid m_i given the auxiliary parameters by simply solving this linear equation. The method we use to actually compute the maximum posterior mixel grid is discussed in the next section.

4. Reconstruction Algorithm

We now concern ourselves with how to solve Eqn. (12). The fundamental "catch" is that *if* we knew the true values of the various parameters (PSF, registration, etc.), we could solve Eq. (12) exactly. However, to estimate the necessary parameters to high accuracy, we would have to know the true mixel grid! Our way around this dilemma is to iterate between two processes: use a current estimate of the mixel grid to (re-)estimate the parameters; use these new parameter estimates to re-calculate a better mixel grid, and so on. To start this process, we need either a nominal mixel grid or nominal set of parameters. Our data includes nominal values for parameters such as the camera location, the sun angle, etc., so we choose these as starting values. We begin our bootstrap process by estimating the registration parameters.

4.1. Registration

The registration parameters define the correspondence between points on the image plane and those on the modeled surface. The parameters can be thought of as coefficients to a function that projects surface points to the image plane; such a function depends on the planetary curvature, the imaging system's optics, and the camera's location and orientation relative to the surface. In the case of vidicon cameras, there is additional image distortion due to the read-out process that cannot easily be distinguished from the above geometric effects. Thus the projection function in principle varies for each point pair. The registration problem is to estimate the projection parameters for each image to the mixel grid that captures all of these components of the projection function.

As stated above, the strategy to estimate registration parameters first involves constructing a first guess at the mixel grid. To do this, we pick at random one of the pixel images and interpolate its pixel values onto a grid at the desired mixel resolution. Using this interpolated mixel grid as a reference image, we search for accurate relative registration parameters that map each image optimally onto the reference grid. But the quantization in the reference grid causes pixel-sized "jumps" in registration values. This in turn creates a hazardous search process! To avoid these jumps, we smooth the reference image with a Gaussian-like filter.

In theory, we could find a MAP estimate of the registration (or any other) parameters; instead, we seek a simpler ML estimate and ignore priors on the parameters. This is because of the large ratio of information (pixels) to the number of parameters that need be estimated. If we assume an independent Gaussian likelihood for each pixel relative to its projected value from the reference mixel grid, as in Eqn. (3), then finding the ML estimate of the registration parameters reduces to finding the registration with the smallest sum of squared pixel differences from their projected values (i.e., a minimum squared error). In other words, the optimal registration parameters for an image gives the minimum squared error when the mixel values projected through the PSF are compared to the corresponding pixels.

There is one difficulty: moving features of an image "off the edge" of the reference mixel grid during registration. Clearly, image pixels not matched with anything shouldn't contribute to the total error; however, pushing hard-to-register features out of the picture is a false minimum! In our data we had available a larger image that contained the entire image set as sub-images, and avoided the "edge" problem by processing the larger image as the reference.

Optimal registration parameters were determined by the Simplex algorithm [1], which searches for a minimum of the squared error by systematically varying the registration parameters, and then calculating the squared error for each such registration. (However, we assume that the error is a *smooth* function of the registration parameters. This is the reason for the Gaussian filter referred to above.) The algorithm stops when successive squared error values of the trial registrations are indistinguishable. We found that unless the registration search starts relatively close to the true registration (i.e., one has good nominal information), the search can get trapped in local minima. There are more efficient search algorithms than the Simplex algorithm, but they are not generally as robust. Note that standard methods for accurate relative image registration required locating "features" common to both images and finding a global mapping for all features to their counterparts in the other image [2]. The method described here uses *all* the information in both images, and this is part of the reason for the very high (subpixel) accuracy achieved by the method described here. However, feature based methods may be a good way of obtaining a close initial registration, when nominal registrations are not available or too inaccurate.

The affine transformation set is the parameter space in which the registration search is executed, and is sufficient for accurate registration provided that the principle nonlinear camera effects[3] are not severe. When these effects interfere, we have extended to a quadratic family of transformations.

4.2. PSF and Other Parameters

The point spread function (PSF) describes how the light energy from a point on the external surface is distributed over the image plane. The spreading of the surface point energy is usually due to the optical system's diffraction and aberration pattern. Typically, the PSF diameter is significantly smaller than the pixel dimensions, so that the images are not diffraction limited. With the scanning electron beam detector used in a vidicon, the PSF can be extended to model the diffuse *readout* spot as well. Since the PSF is a function of the imaging system, it does not depend on the particular image. In practice, the PSF can vary across the image plane, and with time. We have not attempted to model this variation, and work with an average PSF derived from the instrument's bench calibration [3].

"Shading" is the characteristic smooth variation in detector sensitivity across the image plane in vidicon tubes, equivalent to the variation of individual cell sensitivities in array detectors. The likelihood model must take shading into account, and can be learned from the data, given a rough idea of the registration: since all images contain the same subregion under similar lighting and viewing angles, any systematic differences in their appearance must be due to shading. We assume the shading function is a second order polynomial function of pixel position, and currently search for coefficients which make the subregions have the most similar mean intensities.

Defects in the optical system or on the image plane generate blemishes — e.g., dust particles and scratches — common to all images from that camera. A blemish map is used to identify suspect pixels. Rather than interpolating the missing values as is common practice [4], we ignore these pixels, so that the corresponding mixels may be influenced only by the other frames. Also, since spacecraft that are many light-minutes away cannot be asked to retransmit corrupted data packets, they do not implement a reliable transport protocol, and some pixels have incorrect values. Usually no more than two bits are affected; our preprocessor uses this to help detect corrupted pixels. In principle it could use it to recover the correct value, but this would make little practical difference in our case. We simply ignore all suspected corrupt pixels, as well as missing pixels and reseaux marks[5].

4.3. Initial Composite

Once the above methods are used to find good initial estimates of the basic parameters (PSF, registration parameters etc.), we next construct a composite mixel grid using information from *all* the pixel images. We construct the value of a composite mixel by calculating the "votes" from every pixel that could affect it from any frame, as weighted through a *compositing kernel.* These "votes" are accumulated to give a total mixel value

[5]These are permanent marks on the camera faceplate used for calibrating the optics in the Vidicon camera[3].

$$m_i = \frac{\sum_p \omega_{ip} \Phi_p}{\sum_p \omega_{ip}}$$

for each mixel, replacing the values of the reference grid. Clearly, those pixels that are nearest the projected position of a mixel have the strongest vote for that mixel. The compositing kernel functions algorithmically like a PSF, but needn't be the same function. For narrow kernels, the pixel-mixel "voting" is almost 1-to-1, but for diffuse kernels, each mixel value is the weighted combination of information from many pixels, leading to a "blurred" composite. In fact, if a small kernel is used that accurately models the actual PSF, and the noise content of the imagery is relatively small, this becomes a quick method for producing a super-resolved image.

4.4. Iterative Improvement

The composite is used as a starting point in a search for the MAP estimate of Eqn. (12). We use a standard iterative method (Jacobi's method) to solve the matrix equation. The Jacobi method solves an equation of the form $Ax = b$ by triangular decomposition

$$A = L + D + U$$

and updating

$$D \cdot x^{(r)} = -(L + U) \cdot x^{(r-1)} + b$$

which, if $\Delta x = x^{(r)} - x^{(r-1)}$ can be rewritten

$$\Delta x = D^{-1}(b - A \cdot x^{(r-1)}) \tag{13}$$

This can be shown to be equivalent to the method of "substitution", a useful fact for extensions of the model. To implement Eqn. (13), note that

$$D = \frac{1}{s^2}(1 + \sum_j \alpha_{ij}^2) + \frac{1}{\sigma^2} \sum_p \omega_{ip}^2 \tag{14}$$

is the denominator of Eqn. (13). Combining Eqn. (14) with Eqn. (13) for the numerator, one obtains the following iterative mixel re-estimation formula.

$$\Delta m_i = \lambda \frac{\frac{s^2}{\sigma^2} \sum_p \omega_{ip}(\Phi_p - \hat{\Phi}_p) - (m_i - \hat{m}_i) + \sum_j \alpha_{ij}(m_j - \hat{m}_j)}{\frac{s^2}{\sigma^2} \sum_p \omega_{ip}^2 + 1 + \sum_j \alpha_{ij}^2} \tag{15}$$

The results of applying this iterative formula to initial composite mixel grids is shown in Fig. 5; a noticeable sharpening of the composite is demonstrated. When the mixel grid resolution is too coarse, the mixels are over-determined by the pixels, so the MAP mixel estimate is essentially the same as the ML estimate. In Eqn. (15), this means that the first (data) term in the numerator dominates the other two (mixel neighbor correlation) terms. When the mixel grid resolution is large enough (under-constrained by the pixels), the two terms in the numerator balance each other—i.e. the data term tries to force the mixels to exactly agree with the data, while the mixel neighbor term tries to make all mixels look like their neighbors ("smoothing"). It is the tension between these two effects that leads to plausible images, even when the mixels are under-constrained by the data. This is the case where the prior term acts as a regularization term.

In Eqn. (15), all the necessary parameters (s, σ, and the registration, PSF, etc. parameters that go into ω_{ip}) are assumed known. The λ parameter regulates the amount that any mixel can change, and is there purely for purposes of numerical stability. Some of these parameters, such as the PSF, are often well known ahead of time. Other parameters, such as the registration, can be initially estimated from an interpolated version of a single image. Since we find a much more probable mixel grid as a result of compositing and iteration, we can then re-estimate these parameters, and even repeat this convergence cycle. Fortunately, this re-estimation is not needed in practice more than twice. The reason for this is that parameters, such as the registration parameters, are typically estimated from thousands of pixels in the interpolated initial mixel grid, and so are already very accurate.

The ratio s^2/σ^2 of mixel to pixel deviation is more difficult estimate, as the most probable value can be many orders of magnitude different from what one estimates from a composite. We initially intended to re-estimate these parameters *during* the iterative convergence cycle from the residual error in each new mixel grid. What we did not realize was that in some cases this dynamic re-estimation would result in *diverging* from the correct answer. So now when prior information is not enough to set these parameters, we must resort to an explicit search. We take a small but hopefully representative patch of an image and seek parameters values which maximize our quality measure, the determinant of the matrix A_{ij}.

4.5. Complexity

One may ask why an iterative method, like the one above, was chosen over an algebraic computation of Eqn. (12). Essentially, the former method has a lower computational complexity. To keep the comparison clear we will stick to the "just constrained" case, which is described as follows. If we have f frames, with p pixels per frame, the number of mixels N is set to the total number of pixels P:

$$N = P = fp$$

It is "just constrained" because the number of data P is equal to the number of parameters N. Let k denote the radius of the point spread function in mixels. Then each iteration step of Eqn. (15) is of complexity $\mathcal{O}(k^2)P$, whereas the complexity of an algebraic solution of Eqn. (12) is $\mathcal{O}(k^4)P$. As a point of comparison, the compositing routine is of order $\mathcal{O}(k)P$, suggesting that finding an optimal compositing routine would be a good strategy for obtaining quick (if improbable) results.

4.6. Results

Fig. 5 gives results for a U.S. postage stamp digitized at low resolution by a scanner, and for Viking Orbiter images of Mars [5]. The Viking reconstruction uses a series of 24 vidicon images of Mars; the data are from a high spacecraft altitude, with frames of very similar sun and camera angles. From these large images, which also include the edge of the polar cap, we extracted rather under exposed 128×128 pixel regions containing the same four prominent craters. These regions represent the same area to within a few pixels. The images were preprocessed using the techniques described in section 4.2.. Vidicon blemishes and

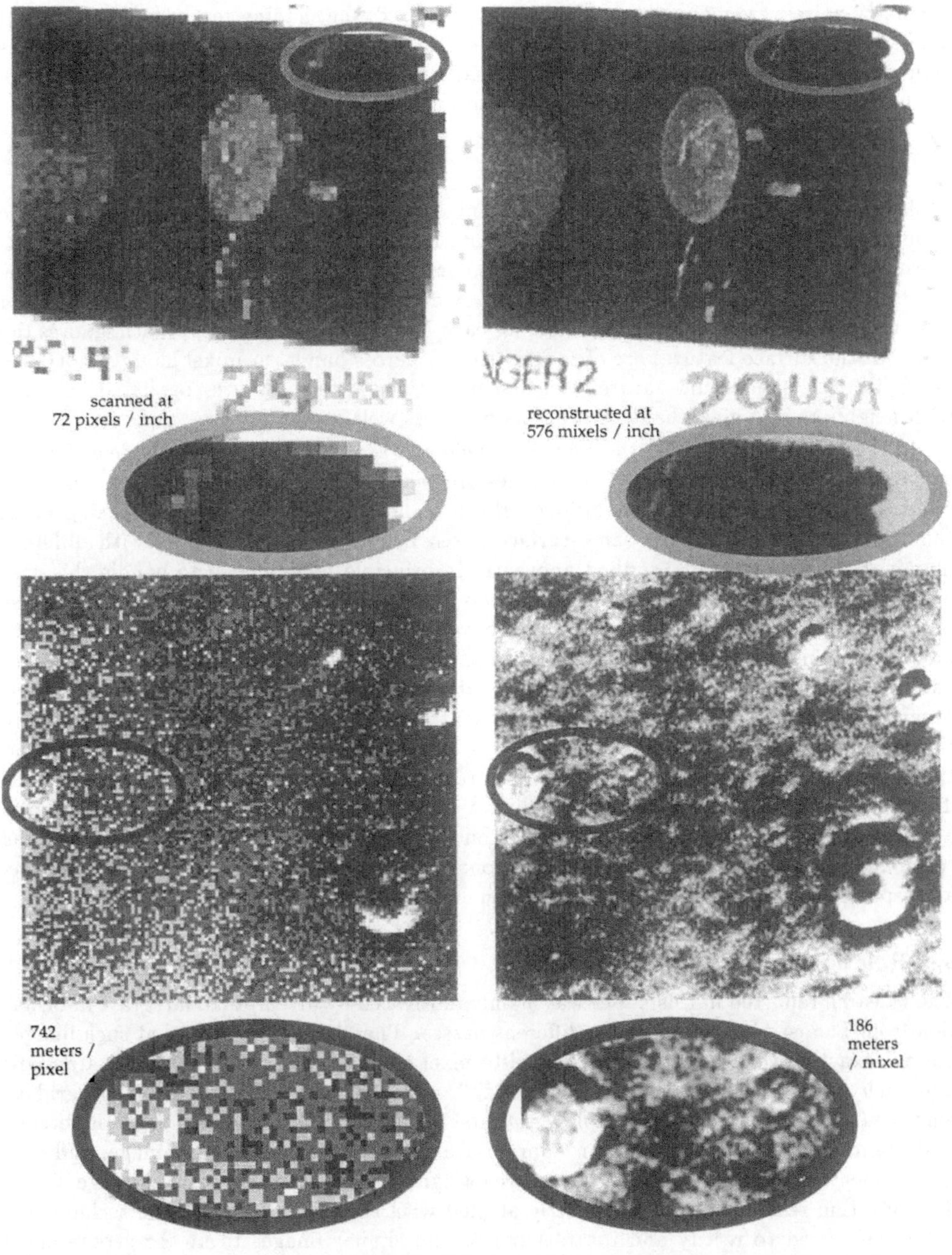

Figure 5: Surface Reconstruction

telemetry noise were mapped and subsequently ignored; the shading response was modeled; image registration used only affine (as opposed to quadratic) transforms. Restoration was done at a 1:4 mixel scale (1:16 area ratio), making the restoration slightly over-constrained. We leave it to the reader to judge the restoration's quality.

5. Extensions

In the above surface reconstruction, we gave mixels a single scalar emittance value. In some applications, color information is available; e.g., Landsat/TM records seven spectral bands in each exposure, and Viking Orbiter and Voyager took gray-scale pictures through various color filters. The 2-D reconstruction described above can be used on each spectral band separately, to get super-resolved surfaces for each band. However, this approach ignores the fact that the surface features are often very similar across bands. A mixel having emittance values in each band could ensure even higher resolution if mixels are correlated not only with neighbors in a given band, but across bands as well.

In the above we combined the effect of albedo and ground effects into a single emittance value, which is appropriate if all the images are taken from essentially the same direction under the same illumination conditions. However, for most of Voyager and Viking data, there are many views of the same surface taken from different directions with different illumination. The theory described above can in principle be extended to handle this case as well. It requires the mixels to have albedo and height values; the registration process is similar, but with more parameters. We have derived the ML equations for the surface model assuming all of the lighting differences in the images are due to either slope or albedo, and not to shadows or occlusions. The effects due to slope and albedo can be distinguished because the effects of parallax vary independently of effects due to surface albedo. Note that representing the surface emittance by a single scalar (albedo) is an approximation that assumes Lambertian scattering. Many real surfaces are not Lambertian. Using bi-directional reflectance parameters, including a specular reflectance component, would give a more accurate surface model. The priors on the surface may involve properties such as continuity, smoothness, and texture. Additionally, we would need to model effects such as atmospheric attenuation, clouds, and the camera "hot-spot" for Earth observation data.

6. Relation to Other Work

The research reported in this paper was mainly motivated by attempts to integrate information from Landsat images taken on different passes. The difficulty here is that such images did not exactly overlay each other, so pixel-to-pixel comparison is not possible. A standard approach to this problem is "rubber-sheeting", which attempts to fit one image grid to another (reference) grid by resampling the first image onto the reference grid. Reference grid points are mapped, through an appropriate transform, onto the new image, and new grid elements are computed by taking an area weighted average of the overlaid image pixels. The resulting resampled grid is perfectly aligned with the reference grid. The technique is extensively used to rectify and rotate Landsat and similar images to fit the geographical survey grid.

From the Bayesian perspective, the rubber-sheeting approach makes little sense, because the new averaged "pixels" are neither actual observations nor a surface model. Worse, the averaging process destroys information—it is impossible to recover the original image from the rubber-sheeted image. This information loss makes pixel-by-pixel comparison very dubious. The super-resolved surface modeling described in this paper does allow the integration and comparison of information from many images through the accumulated super-resolution surface model.

A related approach to the Bayesian 3-D surface reconstruction described above is called "Shape from Shading" [6]. This approach integrates observed surface intensity gradients from a single image to give a 3-D elevation model of the generating surface. It assumes a constant albedo, known illumination conditions, and surface continuity. Shape from shading can be extended to multiple images [7], and the result is greater detail in the elevation map because each grid point contains information from multiple images. However, the constant albedo assumption is a strong limitation on the ability to extract information from multiple images.

A Bayesian approach very similar to ours is described in [8]. This approach does surface reconstruction using images from different viewpoints, and a neighbor correlation prior with a Gaussian noise model. The surface is represented by planar patches joined to form a curved surface. Unlike our work, these authors assume smooth large scale surfaces that can be represented by large parameterized "surface patches". Because these patches are estimated from many pixels from many images, the parameters that describe them are accurately determined, and so the overall surface is accurately estimated. In our approach we achieve super-resolution, and there is no aggregation of surface mixels into large scale patches. Although our goals and assumptions are significantly different we use the same basic Bayesian approach.

A related area of study is in combining images from video [11]. Here, the registration of images passes from the discrete to the continuous, and thus the techniques of "optical flow" are used.

Acknowledgments

We gratefully acknowledge the fruitful contributions of Chris Wallace and Wray Buntine.

References

[1] W. Press, B. Flannery, S. Teukolsky, and W. Vetterling, *Numerical Recipes in C*, Cambridge University Press, 1988.

[2] R. W. Gaskell, "Digital Identification of Cartographic Control Points," *Photogrammetric Engineering and Remote Sensing*, Vol. 54, No. 6, Part 1, pp. 723-727, 1988.

[3] M. Benesh, and T. Thorpe, *Viking Orbiter 1975 visual imaging subsystem calibration report*, JPL Document 611–125, Jet Propulsion Laboratory, Pasadena, Ca., 1976.

[4] E. Eliason et. al., "Adaptive box filters for removal of random noise from digital images," *Photogrammetric Engineering and Remote Sensing*, **56**, 453-456, 1990.

[5] M. Carr et. al., *Archive of Digital Images from NASA's Viking Orbiter 1 and 2 Missions*, Planetary Data System, National Space Science Data Center, CD-ROM: USA_NASA_PDS_VO_1003, Frames VO217S42-88.

[6] B.K.P. Horn and M.J. Brooks, *Shape from Shading*, MIT Press, Cambridge, Massachusetts, 1989

[7] J. Thomas, W. Kober, and F. Leberl, "Multiple Image SAR Shape-from-Shading," *Photogrammetric Engineering and Remote Sensing*, Vol. 57, No. 1, pp. 51-59, 1991.

[8] Y. P. Hung and D. B. Cooper, "Maximum a posteriori probability 3D surface reconstruction using multiple intensity images directly," *SPIE Vol. 1260 Sensing and Reconstruction of Three-Dimensional Objects and Scenes*, pp. 36-48, 1990.

[9] C. Elachi,*Introduction to the Physics and Techniques of Remote Sensing*, Wiley and Sons, New York, 1987.

[10] K. Mardia, J. Kent, and J. Bibby, *Multivariate Analysis*, Academic Press, London, 1979.

[11] S. Mann and R. Picard, "Virtual Bellows: Constructing High Quality Stills from Video," *Proc. ICIP*, Austin, Texas, 1994.

BAYESIAN ANALYSIS OF LINEAR PHASED-ARRAY RADAR

Andrew G. Green* and David J.C. MacKay†
Cavendish Laboratory
Cambridge, CB3 0HE. United Kingdom.

ABSTRACT. A number of methods have been developed to analyze the response of the linear phased array radar. These perform remarkably well when the number of sources is known, but in cases where a determination of this number is required, problems are often encountered. These problems can be resolved by a Bayesian approach.

Here, a linear phased-array consisting of equally spaced elements is considered. Analytic expressions for the posterior probability distribution over source positions and amplitudes, and the corresponding Hessians are derived. These are integrated to give the evidence for each model order.

Tests using model data showed that performance at the second level of inference is critically determined by the accuracy of position estimation. If adequate parameter optimization is available, the Bayesian approach is demonstrated to work well, even in extreme circumstances. A commonly employed method of source location, noise subspace eigenanalysis of the correlation matrix, was tried and found to be inadequate. A Newton-Raphson optimization was then used starting from the positions predicted by eigenanalysis.

1 Introduction

We investigate the analysis of data from linear phased-array radar. Recent improvements in the speed of computers have made feasible the real-time use of more sophisticated methods than simple beam-sweep methods. One technique of interest is noise sub-space eigenanalysis of the correlation matrix [1]. This is one of a class of algorithms commonly known as super-resolution methods, because of their ability to resolve sources below the Rayleigh criterion [2, 3].

The merits of Bayesian inference have been demonstrated in many diverse fields of data analysis [4, 5, 6, 7]. Here, the improvements which may be made to position inference by eigenanalysis, with the application of Bayesian methods, is assessed. The locations of a finite number of sources are inferred, with error bars, by maximizing the integrated posterior probability. The number of sources is similarly inferred by evaluating the appropriate evidence.

2 Radar Configuration

The phased-array radar consists of a series of equally spaced elements with (ideally) isotropic far field responses. This arrangement is indicated in figure 1. For each source configuration, a number of data-sets, S, are collected in rapid succession. These are known as snapshots.

*andrewg@thphys.ox.ac.uk
†mackay@mrao.cam.ac.uk

G. R. Heidbreder (ed.), Maximum Entropy and Bayesian Methods, 309–318.

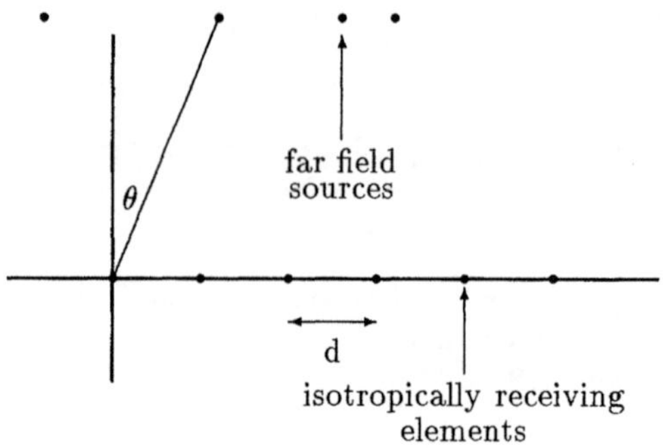

Figure 1: Antenna Configuration.

d: element separation; λ: wavelength; θ: source positions.

For a single snapshot the phase difference between the signal at adjacent elements of the antenna due to the jth source is given by:

$$\phi_j = (2\pi d/\lambda)\sin\theta_j \ . \tag{1}$$

The response of the kth element may be written as

$$x_k = \sum_j F_j \exp i(k\phi_j) + n_k \ . \tag{2}$$

This is more elegantly expressed in terms of matrices,

$$\underline{x} = \mathbf{V}\underline{F} + \underline{n} \ , \tag{3}$$

where $\underline{x}$ is the antenna response, $\underline{n}$, the noise vector, $\underline{F}$, the source amplitude vector and $\underline{\phi}$, the source position vector. The steering matrix $\mathbf{V}$ is given by

$$V_{kj} = \exp(ik\phi_j) \ . \tag{4}$$

The source locations are assumed not to change between snapshots, although their complex amplitudes may. We have, therefore, a single position vector $\underline{\phi}$ for all snapshots and a set of amplitude vectors $\underline{F}^s$, one for each snapshot.

3 Eigenanalysis of correlation matrix

There are a number of interrelated techniques in spectrum analysis based upon the eigenanalysis of the data correlation matrix [2, 3]. The method described in brief below, due to Reilly *et al.* [1], is one of several super-resolution algorithms known as "maximum likelihood". These can be shown to depend upon the orthogonality relationship given at the end of this section [2, 3].

The correlation matrix is formed by taking the outer product of the data vector with itself. Using equation (3) we find

$$\underline{x}\underline{x}^{\dagger} = \mathbf{V}\underline{F}(\mathbf{V}\underline{F})^{\dagger} + \underline{n}\underline{n}^{\dagger} + \mathbf{V}\underline{F}\underline{n}^{\dagger} + \underline{n}(\mathbf{V}\underline{F})^{\dagger} . \tag{5}$$

Averaging over snapshots we note that the final two terms will ideally go to zero, since the sources and noise are uncorrelated. The matrix $\mathbf{V}$ may be removed from the averaging since the source positions are the same for all snapshots. Equation (5) may then be written

$$\mathbf{R} = \mathbf{V}\,\mathbf{Rs}\,\mathbf{V}^{\dagger} + 2\sigma^2\mathbf{I} , \tag{6}$$

where $\mathbf{R} = \langle \underline{x}\underline{x}^{\dagger} \rangle$ is the correlation matrix, $\mathbf{Rs} = \langle \underline{F}\underline{F}^{\dagger} \rangle$ is the source correlation matrix, $\mathbf{I}$ is the identity matrix, σ^2 is the variance of real and imaginary parts of the noise, and $\langle\, \rangle$ indicates averaging over snapshots. Note that the noise power is assumed identical for each element of the antenna.

$\mathbf{R}$ is Hermitian and positive definite by construction. For an N element antenna in the presence of k uncorrelated sources, $\mathbf{R}$ will have a series of real positive eigenvalues of decreasing magnitude, that is,

$$\lambda_1 + 2\sigma^2 \geq \lambda_2 + 2\sigma^2 \geq \ldots \geq \lambda_k + 2\sigma^2 > 2\sigma^2 = \ldots = 2\sigma^2 . \tag{7}$$

Multiplying equation (6) by one of the $(N-k)$ noise subspace eigenvectors (*i.e.* an eigenvector $\underline{e}_{noise}$ of $\mathbf{R}$ corresponding to an eigenvalue $2\sigma^2$), one obtains the orthogonality relation

$$\mathbf{V}^{\dagger}\underline{e}_{noise} = 0 . \tag{8}$$

If an average correlation matrix of order $(k+1)$ is formed by averaging $(k+1) \times (k+1)$ minors along the leading diagonal of $\mathbf{R}$, there will be only one noise subspace eigenvector, corresponding to the lowest eigenvalue. In this case, taking the Z-transform of the noise eigenvector, one obtains a polynomial whose solutions are $z_k = \exp(i\phi_k)$ for each of the k sources present.

There exist more sophisticated ways of averaging $\mathbf{R}$ to form an order $(k+1)$ matrix [3]. These have the property of resolving correlated sources (*i.e.* where F_i^s is correlated with F_i^{s+1} or F_j^s) as well as uncorrelated sources. These are disregarded here for simplicity. An alternative method due to Burg et al. [8] involves finding the maximum likelihood Toeplitz structure matrix from the data.

4 Bayesian analysis

Single snapshot

The inferred parameters divide naturally into: the source amplitudes $\{\underline{F}^s\}$ (which are different for each snapshot), the source positions $\underline{\phi}$, and the number of sources k, (specified by the hypothesis H). Initially a single snapshot will be considered. This is extended to several snapshots in the following subsection.

We write Bayes' theorem for a series of levels of inference, as follows;

$$P(\underline{F}|H,D,\underline{\phi}) = \frac{P(D|H,\underline{F},\underline{\phi})P(\underline{F}|H,\underline{\phi})}{P(D|H,\underline{\phi})} , \tag{9}$$

$$P(\underline{\phi}|H,D) = \frac{P(D|H,\underline{\phi})P(\underline{\phi}|H)}{P(D|H)}, \tag{10}$$

$$P(H|D) \propto P(D|H)P(H). \tag{11}$$

We note that $P(\underline{F}|H,\underline{\phi}) = P(\underline{F}|H)$, since $\underline{F}$ and $\underline{\phi}$ are independent.

We assume that these distributions are strongly peaked and may be approximated by Gaussians about their peaks. In fact given the choice of likelihood and priors to be made later, equation (9) is exactly Gaussian. By making a Gaussian expansion of equation (9) and integrating, we obtain the expression for the evidence,

$$P(D|H,\underline{\phi}) = P(D|H,\underline{F}_M,\underline{\phi})P(\underline{F}_M|H)(2\pi)^k \det{}^{-1}\mathbf{A}(\underline{\phi}), \tag{12}$$

where $\underline{F}_M = \underline{F}_M(\underline{\phi})$ is the $\underline{F}$ that maximizes (9) and the Hessian $\mathbf{A}$ is given by

$$\mathbf{A}(\underline{\phi}) = -\nabla_{\underline{F}}\nabla_{\underline{F}} \ln P(\underline{F}|H,D,\underline{\phi}). \tag{13}$$

Substituting for $P(D|H,\underline{\phi})$ from equation (12) in equation (10), with the Hessian,

$$\mathbf{B} = -\nabla_{\underline{\phi}}\nabla_{\underline{\phi}} \ln P(\underline{\phi}|H,D), \tag{14}$$

gives upon integration,

$$P(H|D) \propto P(D|H,\underline{F}_M,\underline{\phi}_M)P(\underline{F}_M|H)P(\underline{\phi}_M|H)(2\pi)^{\frac{3k}{2}} \det{}^{-1}\mathbf{A} \det{}^{-\frac{1}{2}}\mathbf{B}, \tag{15}$$

where

$$\mathbf{A}(\underline{\phi}) = -\nabla_{\underline{F}}\nabla_{\underline{F}} \left[\ln P(D|H,\underline{F},\underline{\phi}) + \ln P(\underline{F}|H)\right] \tag{16}$$

and

$$\mathbf{B} = -\nabla_{\underline{\phi}}\nabla_{\underline{\phi}} \left[\ln P(D|H,\underline{F}_M,\underline{\phi}) + \ln P(\underline{F}_M|H) + \ln P(\underline{\phi}|H) - \ln \det{}^{-1}\mathbf{A}(\underline{\phi})\right]. \tag{17}$$

Equations (16) and (17) have been obtained from (13) and (14) by substitution from (9) and (10), noting that the normalizing factors, $P(D|H,\underline{\phi})$ and $P(D|H)$ are constant with respect to the differentiating variables, $\underline{F}$ and $\underline{\phi}$ respectively. The determinants of $\mathbf{A}$ and $\mathbf{B}$ appear to different powers in equation (15) because $\underline{F}$ is a complex vector and $\underline{\phi}$ is a real vector.

Extension to several snapshots

In the case of several data sets or snapshots the above theory must be modified. The positions of the sources, given by $\underline{\phi}$, are the same for all snapshots. Their inference is based upon the data from all the snapshots taken together. The complex source amplitudes, however, may be different for each snapshot, giving rise to S source amplitude vectors $\{\underline{F}^s\}$, where S is the number of snapshots.

It follows from the product rule of probability that one must take the product of the likelihood and priors over snapshots. The amplitude vectors, $\{\underline{F}^s\}$, are assumed, as a rather crude first approximation, to have independent priors between snapshots. Then the likelihood $P(D|H,\underline{F},\underline{\phi})$ is replaced by $P^{\pi}(D|H,\underline{F}^s,\underline{\phi}) = \prod_{s=1}^{S} P(D|H,\underline{F}^s,\underline{\phi})$, and the prior

$P(\underline{F}|H)$ is replaced by $P^{\pi}(\underline{F}^s|H) = \prod_{s=1}^{S} P(\underline{F}^s|H)$. The prior on the source positions $\underline{\phi}$ is unchanged.

Using these distributions with Bayes' theorem we obtain the multi-snapshot analogue of equation (9),

$$P^{\pi}(\underline{F}^s|H,D,\underline{\phi}) = \frac{P^{\pi}(D|H,\underline{F}^s,\underline{\phi})P^{\pi}(\underline{F}^s|H)}{P(D|H,\underline{\phi})} . \tag{18}$$

Equations (10) and (11) are used without alteration. Expanding the distributions as Gaussians about their maxima and integrating, as in section 4, we derive the final result,

$$P(D|H) \propto P^{\pi}(D|H,\underline{F}^s_M,\underline{\phi})P^{\pi}(\underline{F}^s_M|H)P(\underline{\phi}_M|H)(2\pi)^{k(S+1/2)}\det{}^{-S}\mathbf{A}\,\det{}^{-1/2}\mathbf{B}\ , \tag{19}$$

where

$$\mathbf{A}(\underline{\phi}) = -\nabla_{\underline{F}}\nabla_{\underline{F}}\left[\ln P(D|H,\underline{F}^s,\underline{\phi}) + \ln P(\underline{F}^s|H)\right]\ , \tag{20}$$

and

$$\mathbf{B} = -\nabla_{\underline{\phi}}\nabla_{\underline{\phi}}\sum_{s=1}^{S}[\ln P(D|H,\underline{F}^s_M,\underline{\phi}) + \ln P(\underline{F}^s_M|H) + \ln P(\underline{\phi}|H) - \ln\det\mathbf{A}(\underline{\phi})] . \tag{21}$$

Equation (20) is identical to equation (16), and equation (21) is simply the sum over snapshots of equation (17).

Application to Phased-Array radar

For a single snapshot, assuming Gaussian noise, the likelihood function is

$$P(D|H,\underline{F},\underline{\phi}) = \left(\frac{1}{2\pi\sigma^2}\right)^N \exp\left[-\frac{|\underline{n}|^2}{2\sigma^2}\right] . \tag{22}$$

The noise vector $\underline{n}$ is defined for each snapshot as the difference between the data and mock data, *i.e.*,

$$\underline{n} = \underline{x} - \mathbf{V}\underline{F} . \tag{23}$$

The noise variance σ^2 is assumed to be known for a particular antenna rather than included as a hyperparameter. Noise is generated in the antenna and can be measured. Taking the product over snapshots, the likelihood is given by

$$P^{\pi}(D|H,\underline{F}^s,\underline{\phi}) = \left(\frac{1}{2\pi\sigma^2}\right)^{SN} \exp\left[-\sum_{s=1}^{S}\frac{|\underline{n}|^2}{2\sigma^2}\right] . \tag{24}$$

The prior on positions is simply a uniform distribution between $\pm\frac{2\pi d}{\lambda}$, $P(\underline{\phi}|H) = \left(\frac{\lambda}{4\pi d}\right)^k$. The prior on source amplitudes is taken to be Gaussian with variance δ^2 on real and imaginary parts, where δ^2 is entered as a user defined parameter, *i.e.*,

$$P^{\pi}(\underline{F}^s|H) = \left(\frac{1}{2\pi\delta^2}\right)^{Sk} \exp\left[-\frac{1}{2\delta^2}\sum_{s=1}^{S}|\underline{F}^s|^2\right] . \tag{25}$$

Using the above priors, the Hessian matrices **A** and **B** defined in equations (20) and (21) are found to be:

$$\mathbf{A}(\underline{\phi}) = \frac{\mathbf{V}^\dagger\mathbf{V}}{\sigma^2} + \frac{\mathbf{I}}{\delta^2}, \tag{26}$$

and

$$\begin{aligned} B_{ij} = {} & \frac{1}{2\sigma^2}\sum_{s=1}^{S} \underline{x}^{s\dagger}\Big[\mathbf{B1}_{(ij)} + \mathbf{B1}^\dagger_{(ij)} - \mathbf{B2}_{(ij)} - \mathbf{B2}^\dagger_{(ij)} - \mathbf{B5}_{(ij)} \\ & - \mathbf{B2}_{(ji)} - \mathbf{B2}^\dagger_{(ji)} + \mathbf{B3}_{(ij)} + \mathbf{B3}^\dagger_{(ij)} + \mathbf{B4}_{(ij)} + \mathbf{B4}^\dagger_{(ij)}\Big]\underline{x}^s \\ & + \mathrm{Tr}\left[-\mathbf{A}^{-1}\frac{\partial\mathbf{A}}{\partial\phi_j}\mathbf{A}^{-1}\frac{\partial\mathbf{A}}{\partial\phi_i} + \mathbf{A}^{-1}\frac{\partial^2\mathbf{A}}{\partial\phi_i\partial\phi_j}\right] \end{aligned} \tag{27}$$

where

$$\begin{aligned} \mathbf{B1}_{(ij)} &= \frac{1}{\sigma^2}\left(\frac{\partial^2\mathbf{V}}{\partial\phi_i\partial\phi_j}\mathbf{A}^{-1}\mathbf{V}^\dagger\right) \\ \mathbf{B2}_{(ij)} &= \frac{1}{\sigma^2}\left(\frac{\partial\mathbf{V}}{\partial\phi_i}\mathbf{A}^{-1}\frac{\partial\mathbf{A}}{\partial\phi_j}\mathbf{A}^{-1}\mathbf{V}^\dagger\right) \\ \mathbf{B3}_{(ij)} &= \frac{1}{\sigma^2}\left(\frac{\partial\mathbf{V}}{\partial\phi_i}\mathbf{A}^{-1}\frac{\partial\mathbf{V}^\dagger}{\partial\phi_j}\right) \\ \mathbf{B4}_{(ij)} &= \frac{1}{\sigma^2}\left(\mathbf{V}\mathbf{A}^{-1}\frac{\partial\mathbf{A}}{\partial\phi_j}\mathbf{A}^{-1}\frac{\partial\mathbf{A}}{\partial\phi_i}\mathbf{A}^{-1}\mathbf{V}^\dagger\right) \\ \mathbf{B5}_{(ij)} &= \frac{1}{\sigma^2}\left(\mathbf{V}\mathbf{A}^{-1}\frac{\partial^2\mathbf{A}}{\partial\phi_i\partial\phi_j}\mathbf{A}^{-1}\mathbf{V}^\dagger\right) \end{aligned} \tag{28}$$

5 Results

Code was written to simulate the antenna response $\{\underline{x}^s\}$ for up to thirty-two snapshots of data. The simulated source environment consisted of up to five uncorrelated sources, with arbitrary position and amplitude, and Gaussian noise of arbitary amplitude. Using these data, the eigenanalysis procedure was tested, and the ability of equation (19) to evaluate the evidence for different model orders was determined.

Eigenanalysis

Without noise, and given the correct model order (*i.e.*, number of sources), the eigenanalysis predicted the position and amplitude of sources to within the computer accuracy as was expected. If the procedure was used assuming a model order greater than the actual, the extra sources were predicted to have zero amplitude to within the computer accuracy.

With Gaussian noise added to the data, eigenanalysis predicted positions and powers well, as long as the sources were well separated. Estimates of resolution for a range of noise powers were made by moving two unit amplitude sources together until the eigenanalysis predicted a single source of twice their amplitude at their average position. Resolution reduced with increasing noise power.

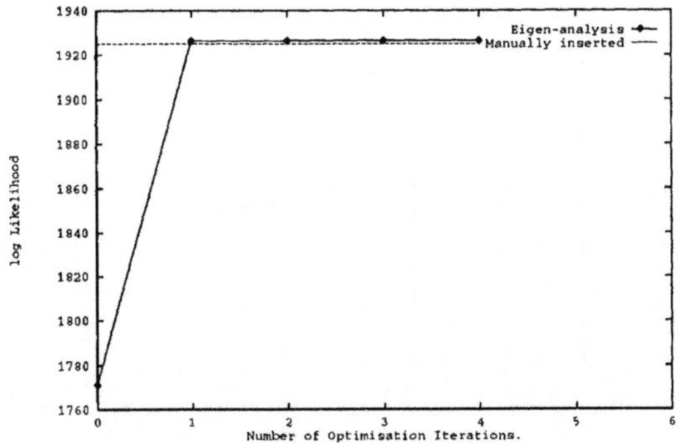

Figure 2: Variation of log Likelihood with Optimization iterations.

Eigenanalysis as described here depends upon the noise-subspace eigenvector, $\underline{e}_{\text{noise}}$, corresponding to the lowest eigenvalue of $\mathbf{R}$. For an S snapshot data-set with Gaussian noise, this eigenvalue is $\sim 2\sigma^2(1 \pm 1/\sqrt{S})$. For a source configuration containing two narrowly separated sources, the lowest source subspace eigenvalue, λ_k, reduces with separation and amplitude of the sources. It is evident from equation (7), that the resolution limit will occur when $\lambda_k \simeq \frac{2\sigma^2}{\sqrt{S}}$.

Does Eigenanalysis Maximize the Likelihood?

The parameters predicted by eigenanalysis do not maximize the likelihood. This conclusion was drawn for two reasons:

1. Manual insertion of the source parameters used to generate the data gave higher likelihoods than the parameters generated by eigenanalysis.
2. The log likelihood is expected to increase by ~ 0.5 for each additional parameter beyond the correct number. This was not the case. (In this case, an increase in k of 1 introduces $(1 + 2S)$ extra parameters giving an expected increase in log likelihood of about 32 between models.)

Although the predicted parameters do not maximize the likelihood, they do give a fair first approximation. A simple Newton-Raphson procedure was used to optimize the parameters predicted by eigenanalysis. (This was easily done, since the relevant Hessian matrix, $\mathbf{B}$, has already been evaluated using equation (27).) Initially, the integrated likelihood $P(D|H,\phi)$ was optimized by setting δ^2 to a very large value. Figure 2 shows typically how the likelihood increased with the number of iterations of the optimization routine. The likelihood increases above that for manual insertion of the parameters after only one iteration and remains fairly constant at this value through the subsequent iterations. Reassuringly, this indicates that the distributions are indeed Gaussian at their peaks.

Figure 3 shows the variation of log likelihood with model order. The expected increase is observed for each additional parameter beyond the correct model order.

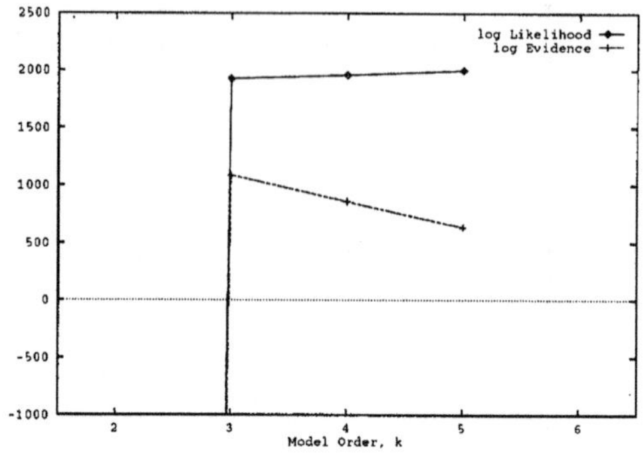

Figure 3: Variation of Likelihood and Evidence (Parameters Generated by Eigenanalysis and Newton-Raphson Optimization.)

$$\begin{aligned} \theta &= -0.2 \quad 0.1 \quad 0.3 \\ |F|^2 &= \quad 1 \quad\quad 1 \quad 1/4 \end{aligned}$$
$$S = 32,\ N = 32,\ \sigma = 0.1,\ \delta = 1.0$$

Estimated Evidence

Figure 3 shows the variation of log likelihood and log evidence with model order for a well resolved source environment with three sources. The evidence has a maximum at the correct model order. Figure 4 similarly shows the variation of log likelihood and log evidence, but for a situation where two of the sources have not been resolved by the eigenanalysis. Newton-Raphson optimization of the eigenanalysis parameters did not resolve these sources. The model order predicted by the evidence is correspondingly reduced.

Note that in cases such as that shown in figure 4, where the unresolved sources have large amplitude, the peak evidence is greatly reduced (as compared with cases where parameters are correctly evaluated). It is tempting to interpret this as an indication of error in the inferring of the parameters, however, it is not at all clear that such deductions may be drawn consistently.

Limitations of Model Comparison

In all cases discussed up to now, Bayesian model comparison has worked well. No severe test of this level of inference has been made, due to the limitations of the techniques used to determine the source positions. In lieu of a good optimization method the following test of the model comparison was made.

Several well separated sources of unit amplitude were generated and a source at the noise level introduced at a small angle from one of these. Below the correct model order eigenanalysis was used to seed the Newton-Raphson optimization of parameters. At and above the correct model order, optimization was seeded with actual source positions cou-

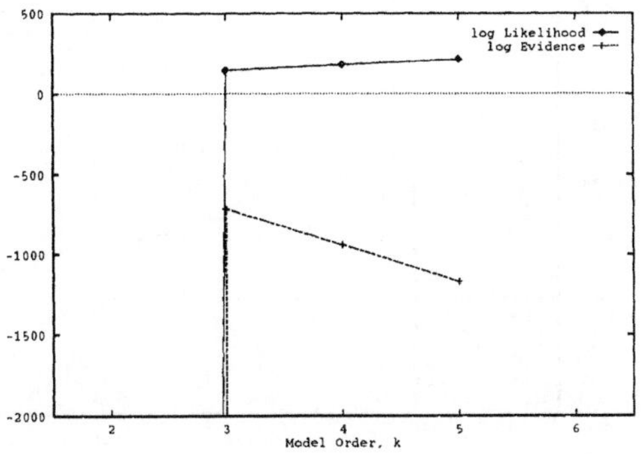

Figure 4: Variation of Likelihood and Estimated Evidence (Sources Unresolved by Eigenanalysis.)

$$\begin{array}{rclcccc} \theta & = & -0.21 & -0.2 & 0.1 & 0.3 \\ |F|^2 & = & 1 & 1 & 1 & 1/4 \end{array}$$
$$S = 32,\ N = 32,\ \sigma = 0.1,\ \delta = 1.0$$

pled with spurious source positions, predicted by eigenanalysis (note that this amounts to starting the optimization in the correct place, so this is not a demonstration of the entire system; it is only a test of the model comparison part). The variations of log likelihood and log evidence obtained in this way are shown in figure 5. The Bayesian analysis correctly predicts the number of sources present.

6 Conclusions

Limitations to the resolution of noise subspace eigenanalysis have been exposed. For real systems, where the number of snapshots is large, resolution will still be much better than the Rayleigh limit which restricts Fourier transform and beam-sweep methods.

The eigenanalysis technique as employed here does not give the maximum likelihood parameters. The predicted parameters are, however, a good approximation to the optimum. The success of Newton-Raphson optimization shows that assumptions of Gaussian probability distributions are well founded.

The application of Bayesian techniques has enabled the prediction of source positions to be given error bars. Bayesian model comparison has been shown to give consistent predictions even when positions are not well determined. In cases where parameters are well optimized, the Bayesian approach correctly infers the number of sources k.

Finally, it has been shown that the use of Bayesian techniques to make model comparisons is limited only by the standard of optimization routines employed.

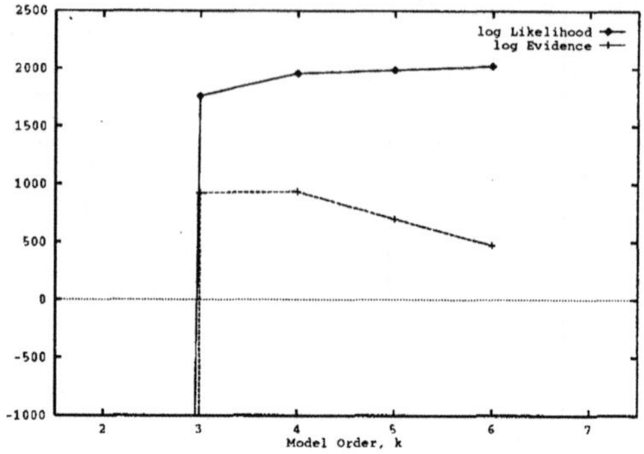

Figure 5: Variation of Likelihood and Evidence (Parameters Generated by Manually Seeded optimization)

$$\begin{array}{rcllll} \theta & = & -0.21 & -0.2 & 0.1 & 0.3 \\ |F|^2 & = & 0.01 & 1 & 1 & 1/4 \end{array}$$
$$S = 32,\ N = 32,\ \sigma = 0.1,\ \delta = 1.0$$

References

[1] J. Reilly, J. Litva, P. Bauman, "New Angle-of-Arrival Estimator : Comparative Evaluation Applied to the Low Angle Tracking Problem", *proc. IEEE* **135F**, 5, 1988.

[2] S.M. Kay, S.L. Marple, "Spectrum Analysis : A Modern Perspective", *proc. IEEE* **69**, 11, 1981.

[3] S. Alaykin, "Adaptive Filter Theory", Prentice Hall Information and System Sciences Series.

[4] G.L. Bretthorst, "Bayesian Spectrum Analysis and Parameter Estimation", Lecture notes in statistics, Springer-Verlag, 1990.

[5] S.F. Gull, "Bayesian Inductive Inference and Maximum Entropy", in *Maximum Entropy and Bayesian Methods in Science and Engineering, Vol. 1, Foundations*, ed. G.J. Erickson and C.R. Smith, pp. 53-74, Kluwer, 1988.

[6] E.T. Jaynes, "Bayesian Methods : General Background", in *Maximum Entropy and Bayesian methods in applied statistics*, ed. J.H. Justice. pp.1-25. C.U.P. 1986.

[7] D.J.C. MacKay, "Bayesian Interpolation", *Neural Computation* **4**, 415-447, 1992.

[8] J.P. Burg, D.G. Luenberger, D.L. Wenger, "Estimation of Structured Covariance Matrices", *proc. IEEE* **70**, 9, 1982.

NEURAL NETWORK IMAGE DECONVOLUTION

John E. Tansley, Martin J. Oldfield and David J.C. MacKay*
Cavendish Laboratory
Cambridge, CB3 0HE. United Kingdom.

ABSTRACT. We examine the problem of deconvolving blurred text. This is a task in which there is strong prior knowledge (*e.g.*, font characteristics) that is hard to express computationally. These priors are implicit, however, in mock data for which the true image is known. When trained on such mock data, a neural network is able to learn a solution to the image deconvolution problem which takes advantage of this implicit prior knowledge. Prior knowledge of image positivity can be hard–wired into the functional architecture of the network, but we leave it to the network to learn most of the parameters of the task from the data. We do not need to tell the network about the point spread function, the intrinsic correlation function, or the noise process.

Neural networks have been compared with the optimal linear filter, and with the Bayesian algorithm MemSys, on a variety of problems. The networks, once trained, were faster image reconstructors than MemSys, and had similar performance.

1 Traditional image reconstruction methods

Optimal linear filters

In many imaging problems, the data measurements $\{d_m\}$ are linearly related to the underlying image $\mathbf{f}$:

$$d_m = \sum_j R_{mj} f_j + \nu_m. \tag{1}$$

The vector ν denotes the inevitable noise which corrupts real data. In the case of a camera which produces a blurred picture, the vector $\mathbf{f}$ denotes the true image, $\mathbf{d}$ denotes the blurred and noisy picture, and the linear operator $\mathbf{R}$ is a convolution defined by the point spread function of the camera. In this special case, the true image and the data vector reside in the same space; but it is important to maintain a distinction between them. We will use the subscript $m = 1 \ldots N$ to run over data measurements, and the subscripts $i, j = 1 \ldots k$ to run over image pixels.

One might speculate that since the blur was created by a linear operation, then perhaps it might be deblurred by another linear operation. We derive the *optimal linear filter* in two ways.

Bayesian derivation

We assume that the linear operator $\mathbf{R}$ is known, and that the noise ν is Gaussian and independent, with a known standard deviation σ_ν.

$$P(\mathbf{d}|\mathbf{f}, \sigma_\nu, \mathcal{H}) = \frac{1}{(2\pi\sigma_\nu^2)^{N/2}} \exp\left(- \sum_m \left(d_m - \textstyle\sum_j R_{mj} f_j \right)^2 \Big/ (2\sigma_\nu^2) \right) \tag{2}$$

*Corresponding author. Email: mackay@mrao.cam.ac.uk.

G. R. Heidbreder (ed.), Maximum Entropy and Bayesian Methods, 319–325.

We assume that the prior probability of the image is also Gaussian, with a standard deviation σ_f.

$$P(\mathbf{f}|\sigma_f, \mathcal{H}) = \frac{\det^{-\frac{1}{2}} \mathbf{C}}{(2\pi\sigma_f^2)^{k/2}} \exp\left(-\sum_{i,j} f_i C_{ij} f_j / (2\sigma_f^2)\right) \tag{3}$$

If we assume no correlations among the pixels then the symmetric, full rank matrix $\mathbf{C}$ is equal to the identity matrix $\mathbf{I}$. The more sophisticated 'intrinsic correlation function' model uses $\mathbf{C} = [\mathbf{G}\mathbf{G}^{\mathrm{T}}]^{-1}$, where $\mathbf{G}$ is a convolution that takes us from an imaginary 'hidden' image, which is uncorrelated, to the real correlated image. The intrinsic correlation function should not be confused with the point spread function $\mathbf{R}$ which defines the image to data mapping. A zero-mean Gaussian prior is clearly a poor assumption if it is known that all elements of the image $\mathbf{f}$ are positive but let us proceed. We are now able to infer the posterior probability of an image $\mathbf{f}$ given the data $\mathbf{d}$.

$$P(\mathbf{f}|\mathbf{d}, \sigma_\nu, \sigma_f, \mathcal{H}) = \frac{P(\mathbf{d}|\mathbf{f}, \sigma_\nu, \mathcal{H}) P(\mathbf{f}|\sigma_f, \mathcal{H}))}{P(\mathbf{d}|\sigma_\nu, \sigma_f, \mathcal{H})} \tag{4}$$

In words,

$$\text{Posterior} = \frac{\text{Likelihood} \times \text{Prior}}{\text{Evidence}}. \tag{5}$$

The 'evidence' $P(\mathbf{d}|\sigma_\nu, \sigma_f, \mathcal{H})$ is the normalizing constant for this posterior distribution. Here it is unimportant, but it is used in a more sophisticated analysis to compare, for example, different values of σ_ν and σ_f, or different point spread functions $\mathbf{R}$.

Since the posterior distribution is the product of two Gaussian functions of $\mathbf{f}$, it is also a Gaussian, and can therefore be summarized by its mean, which is also the *most probable image*, $\mathbf{f}_{\mathrm{MP}}$, and its covariance matrix,

$$\Sigma_{\mathbf{f}|\mathbf{d}} \equiv [-\nabla\nabla \log P(\mathbf{f}|\mathbf{d}, \sigma_\nu, \sigma_f, \mathcal{H})]^{-1}, \tag{6}$$

which defines the joint error bars on $\mathbf{f}$. In this equation, the symbol ∇ denotes differentiation with respect to the parameters $\mathbf{f}$. We can find $\mathbf{f}_{\mathrm{MP}}$ by differentiating the log of the posterior, and solving for the derivative being zero. We obtain

$$\mathbf{f}_{\mathrm{MP}} = \left[\mathbf{R}^{\mathrm{T}}\mathbf{R} + \frac{\sigma_\nu^2}{\sigma_f^2}\mathbf{C}\right]^{-1} \mathbf{R}^{\mathrm{T}}\mathbf{d}. \tag{7}$$

The operator $\left[\mathbf{R}^{\mathrm{T}}\mathbf{R} + \frac{\sigma_\nu^2}{\sigma_f^2}\mathbf{C}\right]^{-1} \mathbf{R}^{\mathrm{T}}$ is called the optimal linear filter. When the term $\frac{\sigma_\nu^2}{\sigma_f^2}\mathbf{C}$ can be neglected, the optimal linear filter is the pseudoinverse "$\mathbf{R}^{-1}$" $= [\mathbf{R}^{\mathrm{T}}\mathbf{R}]^{-1}\mathbf{R}^{\mathrm{T}}$. The term $\frac{\sigma_\nu^2}{\sigma_f^2}\mathbf{C}$ 'regularizes' this ill-conditioned inverse.

The optimal linear filter can also be manipulated into the form:

$$\text{Optimal linear filter} = \mathbf{C}^{-1}\mathbf{R}^{\mathrm{T}}\left[\mathbf{R}\mathbf{C}^{-1}\mathbf{R}^{\mathrm{T}} + \frac{\sigma_\nu^2}{\sigma_f^2}\mathbf{I}\right]^{-1}. \tag{8}$$

MINIMUM SQUARE ERROR DERIVATION

The orthodox derivation of the optimal linear filter starts by assuming that we will 'estimate' the true image $\mathbf{f}$ by a linear function of the data:

$$\hat{\mathbf{f}} = \mathbf{W}\mathbf{d}. \tag{9}$$

The linear operator $\mathbf{W}$ is then 'optimized' by minimizing the expected sum-squared error between $\hat{\mathbf{f}}$ and the unknown true image . (Interestingly, any quadratic metric using a symmetric positive definite matrix gives the same optimal linear filter.) In the following equations, summations over repeated indices i, j, m are implicit. The expectation $\langle\cdot\rangle$ is over both the statistics of the random variables $\{\nu_m\}$, and the ensemble of images $\mathbf{f}$ which we expect to bump into. We assume that the noise is zero mean and uncorrelated to second order with itself and everything else, with $\langle \nu_m \nu_{m'} \rangle = \sigma_\nu^2 \delta_{mm'}$.

$$\begin{aligned} \langle E \rangle &= \frac{1}{2}\left\langle (W_{im} d_m - f_i)^2 \right\rangle & (10) \\ &= \frac{1}{2}\left\langle (W_{im} R_{mj} f_j - f_i)^2 \right\rangle + \frac{1}{2} W_{im} W_{im} \sigma_\nu^2. & (11) \end{aligned}$$

Differentiating, and introducing $\mathbf{F} \equiv \langle f_{j'} f_j \rangle$ (cf $\sigma_f^2 \mathbf{C}^{-1}$ in the Bayesian derivation above), we find that the optimal linear filter is

$$\mathbf{W}_{\rm opt} = \mathbf{F}\mathbf{R}^{\rm T} \left[\mathbf{R}\mathbf{F}\mathbf{R}^{\rm T} + \sigma_\nu^2 \mathbf{I} \right]^{-1}. \tag{12}$$

If we identify $\mathbf{F} = \sigma_f^2 \mathbf{C}^{-1}$, we obtain the optimal linear filter (8) of the Bayesian derivation. The ad hoc assumptions made in this derivation were the choice of a quadratic error measure, and the decision to use a linear estimator. It is interesting that without explicit assumptions of Gaussian distributions, this derivation has reproduced the same estimator as the Bayesian posterior mode, $\mathbf{f}_{\rm MP}$.

OTHER IMAGE MODELS

The better matched our model of images $P(\mathbf{f}|\mathcal{H})$ is to the real world, the better our image reconstructions will be, and the less data we will need to answer any given question. The Gaussian models which lead to the optimal linear filter fail to specify that all images are positive. This leads to the most pronounced problems where the image under observation has high contrast. Optimal linear filters applied to radio astronomical data give reconstructions with negative areas in them, corresponding to patches of sky that suck energy out of radio telescopes. The 'Maximum Entropy' model for image deconvolution [2] was a great success principally because this model forced the reconstructed image to be positive. The spurious negative areas and complementary spurious positive areas are eliminated, and the dynamic range of the reconstruction is greatly enhanced.

The 'Classic maximum entropy' model assigns an entropic prior $P(\mathbf{f}|\alpha, \mathbf{m}, \mathcal{H}_{\rm Classic}) = \exp(\alpha S(\mathbf{f}, \mathbf{m}))/Z$, where $S(\mathbf{f}, \mathbf{m}) = \sum_i (f_i \log(m_i/f_i) + f_i - m_i)$ [6]. This model enforces positivity; the parameter α defines a characteristic dynamic range by which the pixel values are expected to differ from the default image $\mathbf{m}$.

The 'ICF maximum entropy' model [1] introduces an expectation of spatial correlations into the prior on $\mathbf{f}$ by writing $\mathbf{f} = \mathbf{G}\mathbf{h}$, where $\mathbf{G}$ is a convolution with an intrinsic correlation function, and putting a classic maxent prior on $\mathbf{h}$.

The 'Fermi-Dirac' model generalizes the entropy function so as to enforce an upper bound on intensity as well as the lower bound of positivity. This model is appropriate where the underlying image is bounded between two grey levels, as in the case of printed text.

All these models are implemented in the MemSys package.

2 Supervised neural networks for image deconvolution

'Neural network' researchers often exploit the following strategy. Given a problem currently solved with a standard data modelling algorithm: interpret the computations performed by the algorithm as a parameterized mapping from an input to an output, and call this mapping a neural network; then adapt the parameters to examples of the desired mapping so as to produce another mapping that solves the task better. By construction, the neural network can reproduce the standard algorithm, so this data-driven adaptation can (one expects) only make the performance better.

There are several reasons why standard algorithms can be bettered in this way. (1) Algorithms are often not designed to minimize the real objective function. For example, in speech recognition, a hidden Markov model is designed to model the speech signal, whereas the real objective is to discriminate between different words. If an inadequate model is being used, the neural-net-style training of the model will focus the resources of the model on the aspects relevant to the discrimination task. Discriminative training of hidden Markov models for speech recognition does improve their performance. (2) The neural network can be more flexible than the standard model; some of the adaptive parameters might have been viewed as fixed features by the original designers. (3) The net can find properties in the data that were not included in the original model.

In this paper we apply this neural network attitude to a toy image reconstruction problem. The task is to reconstruct an image of a piece of text from blurred data. This is not viewed as a character recognition task; we perform the reconstruction on a pixel by pixel basis; the neural network is expected to learn general characteristics of the font, but not to memorize the alphabet. We start from the optimal linear filter. If the point spread function is a convolution, then the filter of equation (9) should also be a convolution. Such a filter can be viewed as the very simplest neural network — a single linear neuron that computes

$$\hat{f}_{(x,y)} = \sum_{(u,v)} w_{(u,v)} d_{(x+u,y+v)}. \tag{13}$$

where (x, y) label the coordinates of points in the image. The neuron has a two-dimensional input which might be about twice the size of the point spread function, and a single output corresponding to a single pixel in the image. The network receives a patch from a data image $\mathbf{d}$ as input, and its single output would be trained to produce the pixel value at the centre of that patch of data in the true image $\mathbf{f}$. As the trained network is scanned across a blurred image, its output produces a deconvolved image, pixel by pixel. The minimum square error derivation of the optimal linear filter in the previous section corresponds to training this neuron on an ensemble of examples $\{\mathbf{d}, \mathbf{f}\}$ where the original images $\mathbf{f}$ have correlations defined by the matrix $\mathbf{F}$.

The first advantage of training such a neuron on real data is that the neuron can

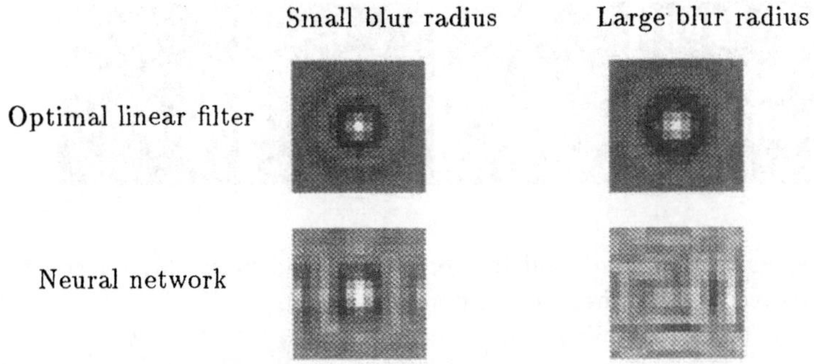

Figure 1: **Optimal linear filters and neural networks**

implicitly learn the correlations $\mathbf{F}$ from the data. One need not explicitly know the point spread function $\mathbf{R}$, the noise statistics σ_ν^2 or the correlation statistics $\mathbf{F}$; the optimization process implicitly learns all these for itself. This network could also learn the appropriate filter if the noise in the data were spatially correlated. Further advantages of a neural network approach arise when we imagine using a more sophisticated network than a linear one. By changing the function performed by the output unit, we can hard-wire prior knowledge into the net. For example, if we know that the true image is everywhere positive, then we can use a non-linear output function which only assumes positive values. In the toy problem studied here, we know that the true image has only two possible intensity levels (black and white, or $t = 0$ and $t = 1$), so we can make the network into a classifier which discriminates between these different possibilities. We define the output of the network to be

$$P(t = 1|\mathbf{d}, \mathbf{w}) = \frac{1}{1 + e^{-(\mathbf{d}\cdot\mathbf{w}+w_0)}}. \tag{14}$$

By introducing additional non-linear processing between the input and the output, one might allow the network to select from a richer space of non-linear filters. Such a network could implicitly learn a more complicated prior probability distribution for images, learn a more complicated noise model, and learn about non-linear detector responses. We do not go that far in this paper. Here we report the performance achievable using just a single neuron.

TRAINING WITH LIMITED AMOUNTS OF DATA

If our training set $\{\mathbf{d}, \mathbf{f}\}$ is small in size, a network trained to minimize the error on training data will 'overfit' the data. We cope with this by putting a standard Gaussian prior on the network parameters. We find parameters $\mathbf{w}$ that maximize the posterior probability, *i.e.*, the product of the likelihood (factors of the form (14)) and the prior. We optimize the variance of the prior (the 'weight decay constant') using approximate Bayesian methods [5, 4]. (Amusingly, these Bayesian regularization methods are descended from those developed in the Bayesian Maximum entropy method.)

Figure 2: From left to right: original image; blurred and noisy data; reconstruction by MemSys; reconstruction by trained network.

EXAMPLE

We created data sets from an image of text, using various degrees of blurring and adding various amounts of noise. The blur was spatially Gaussian, and noise was additively Gaussian. For each data set, an optimal linear filter was created, a MemSys deconvolution using the Fermi-Dirac prior was performed, and a neural network was trained on a small patch of the image. The network was trained on three hundred examples (*i.e.*, just a six letter word in the image).

We first contrast the properties of the network with the optimal linear filter. In all cases the trained network outperforms the optimal linear filter in terms of sum-squared error, the difference being greatest for the most difficult problems. In figure 1 we display examples of the parameters of the optimal linear filter and the network. In the case of a small blur radius (a standard deviation of one pixel), the network looks similar to the optimal linear filter except for a slight squareness produced by the font statistics. At a larger blur radius (two pixels standard deviation), the neural net's weights are completely unlike the optimal linear filter, and are also satisfyingly hard for a human to explain — a good sign that the network is doing something useful!

In table 1 we summarize the relative performance of MemSys and a neural network with a 13×13 input. In the cases with blurring and noise, the neural net's performance is slightly inferior to MemSys's. Where there is noise only, the net significantly outperforms MemSys. It is conjectured that the network would have done better had it been trained on more examples (the training set consisted of only 6 characters of text). Figure 2 shows patches of reconstructions given by the two methods for the case of the small blurring radius.

The comparison between these methods favours the network most strongly when we turn to the computational requirement. Once trained, a network can process an image in seconds (about 8 seconds for a well-programmed network with a 13×13 input on a 256×256 image). Whereas MemSys takes 10–15 minutes to process the same image.

3 Discussion

The neural network approach has proved a viable image reconstruction strategy in a problem where there are strong implicit priors in the data. Bayesian Maxent image reconstruction with MemSys depends on knowledge of the point spread function, and assumptions about the noise process and the prior on images. In contrast, the network approach requires examples of data for which the true image is known, but does not require explicit knowledge

Blur radius	Noise level	'Difficulty'	MemSys error	Net error
1	medium	15.7	7.4	8.8
2	medium	26.3	16.3	18.2
0	high	66.9	22.0	13.9

Table 1: **Performance of network relative to MemSys**
The 'difficulty' of a task is the sum-squared error between the data and the true image. The performance measure for reconstruction is the sum-squared error between the reconstruction and the true image. Both are in the same arbitrary units.

of the point spread function. noise level, or image statistics; these are 'learnt' implicitly from the data, so that our reconstruction ability is not limited by our inability to express a good prior over images. Once trained, a neural network is a much faster image reconstruction device.

It will be interesting to attempt more realistic problems, and investigate networks using more complex non-linear computations. A more sophisticated form of prior knowledge that could be incorporated is the spatial smoothness of the point spread function, which leads us to expect spatial smoothness in the deconvolving filter also. This prior expectation can be incorporated by changing the regularizer from $\alpha \sum W_m W_m/2$ to $\alpha \sum C_{mm'} W_m W_{m'}/2$, with appropriate cross terms between the parameters. Equivalently, one can retain the former regularizer, and blur the input data before feeding it to the network. This may sound surprising, but blurring the data even more can indeed enhance the performance of such networks [3].

References

[1] S.F. Gull. Developments in maximum entropy data analysis. In J. Skilling, editor, *Maximum Entropy and Bayesian Methods, Cambridge 1988*, pages 53–71, Dordrecht, 1989. Kluwer.

[2] S.F. Gull and G.J. Daniell. Image reconstruction from incomplete and noisy data. *Nature*, 272:686–690, 1978.

[3] I. Guyon, V.N. Vapnik, B.E. Boser, L.Y. Bottou, and S.A. Solla. Structural risk minimization for character recognition. In J.E. Moody, S.J. Hanson, and R.P. Lippmann, editors, *Advances in Neural Information Processing Systems 4*, pages 471–479, San Mateo, California, 1992. Morgan Kaufmann.

[4] D.J.C. MacKay. The evidence framework applied to classification networks. *Neural Computation*, 4(5):698–714, 1992.

[5] D.J.C. MacKay. A practical Bayesian framework for backpropagation networks. *Neural Computation*, 4(3):448–472, 1992.

[6] J. Skilling. Classic maximum entropy. In J. Skilling, editor, *Maximum Entropy and Bayesian Methods, Cambridge 1988*, Dordrecht, 1989. Kluwer.

BAYESIAN RESOLUTION OF CLOSELY SPACED OBJECTS

Nielson W. Schulenburg
Systems Engineering Division
The Aerospace Corporation
El Segundo, California

ABSTRACT. A technique is demonstrated for recovering positional and radiometric information on unresolved objects that are so closely spaced that their individual blur functions overlap. Emphasis is on point sources. A Bayesian spectral analysis method has been modified to two dimensions and applied to resolving "clumps" of objects for both simulated and real data. The method enables one to judge the amount of noise in the data and provide error bars in the individual pulse positions and amplitudes from a single data set rather than from the deviations observed after measuring many independent sets of data. The Bayesian technique also can estimate the number of discrete objects in a given clump. Noisy simulated data containing three sources were fitted by one-, two-, three-, and four-source models. By the way it formulates the model, the Bayesian approach naturally includes a factor which reflects the reduction in the number of degrees of freedom for a model with a greater number of sources. As a result, the algorithm gives a higher probability for the three-source model than for the four-source model while resoundingly rejecting the one- and two-source models. The estimated centroids and amplitudes are shown to agree with the truth within the derived error bars to the degree expected by Gaussian errors.

Studies of data taken during a flight test by a sensor that measured a scene simultaneously in the visible and long-wavelength regions show that positional information derived from visible-wavelength data can be "fused" with infrared images to derive the long-wavelength infrared (LWIR) intensities of individual objects in an unresolved clump. The estimated LWIR intensities using the visible assist are shown to be an improvement over working with the LWIR data alone.

The technique is also applied to real visible CCD data of observations of star clusters in NGC 6819 and is shown to be internally consistent in counting.

1. Introduction

Every optical system has a point response function (PRF) which is the image generated by a sensor from a point source located at infinity. An example of a PRF is shown in Figure 1 in one dimension for demonstration purposes. In general, the PRF depends on two spatial dimensions. Many objects viewed by optical sensors such as stars or distant space vehicles appear as point sources, because the geometrical angular subtense of the object is much less than the width of the PRF. The PRF width is due to diffraction of the input radiation through the system aperture and to the presence of aberrations in the optical system. The amplitude of a point source is measured by determining the height of the pulse response. The location of a point source is found by computing the centroid of the pulse response.

If two point sources are separated in object space by less than the width of the PRF, the sensor pulse responses from the sources will overlap. If many point sources are located within the resolution limit of the sensor, the optical system will produce an image which appears as a "clump." The objects in this case are referred to as a cluster of closely spaced

G. R. Heidbreder (ed.), Maximum Entropy and Bayesian Methods, 327–338.

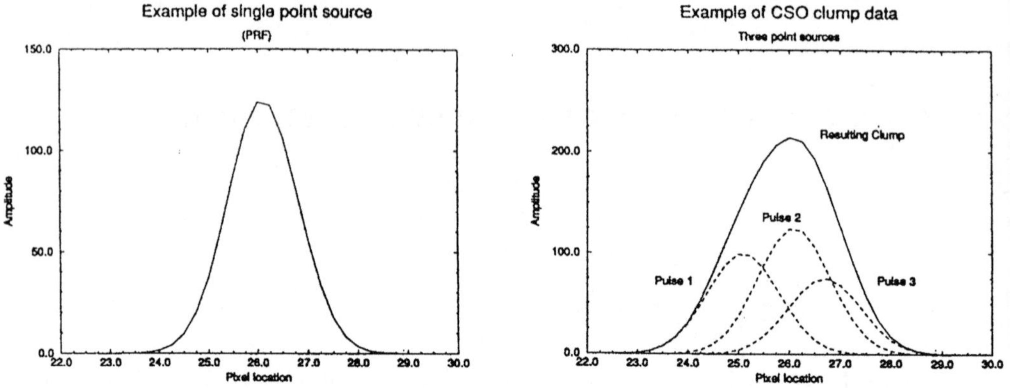

Figure 1: Example of single-point source and three-point source CSO clump. *Note: Figure 1a represents a two-dimensional single point source, i.e., a scaled PRF. The point source centroid is located at 26.1. The amplitude is 125. Pixel values correspond to quarter integers. The width of this PRF is 1.41 pixels. Figure 1b represents a three-source CSO clump. The pulse centroids are located at 25.1, 26.1, and 26.7. The amplitudes are 100, 125, and 75. The overlapping sources create a clump with a peak of 214.2 at pixel 26.0.*

objects (CSO). An example of a three-source CSO is shown in Figure 1. The individual pulses are represented by the dashed lines. Note that the individual pulse amplitudes and centroids determine the amount of overlap between the pulses and thus the shape of the clump, i.e., the solid line.

This report demonstrates a technique for counting and recovering positional and amplitude information of individual pulses from data consisting of a clump of CSO. The technique is a modification of the Bayesian spectral analysis method described by Bretthorst.[1] The Bayesian Probability Theory (BPT) uses probability as the measure of confidence or plausibility in a particular theory or hypothesis. The probability describes the level of likelihood that a hypothesis is true given the available data and prior information about the data. In terms of the CSO problem, we use the prior knowledge that a clump is of unresolved CSO, so the clump can be modeled with overlapping point sources. The sensor PRF is assumed known in either functional form or as a matrix of measured pixel samples.

The clump is modeled with m-point sources, each located at a hypothesized position with a particular amplitude. Because BPT computes the likelihood of each hypothesized model, a methodical procedure can be employed to determine the most likely placement and amplitudes of the m-point sources. Furthermore, the likelihoods allow one to compare models with different numbers of point sources. Thus, the most likely number of point sources in the clump can be discerned. In addition, the technique can quantify the noise power in the data and thereby provide error bars in the individual pulse positions and amplitudes from a single data set. Knowledge of the error bars can be valuable when fusing data from multiple sensors in real-time. Conventional least squares fitting techniques either require assumptions about the noise or calculate deviations observed after measuring many independent sets of data to determine error bars.

Because BPT was developed to incorporate a priori knowledge into the decisionmaking process, data fusion becomes simplified. For example, positional data from a spatially well-resolved spectral band can be utilized to enhance the derivation of radiometric data from a less well-resolved spectral band. We did exactly this with real data from a sensor taken during a flight test. Specifically, we used target positions from visible data to process long-wavelength infrared (LWIR) (10 μm) data and improve LWIR CSO radiometry. This is described in Section 3 of this report. The technique is applied to simulated data in Section 2 and to visible CCD data in Section 4. The theory will not be covered since the principles of BPT are explained in detail in References 1 and 5. We extended the BPT equations to two dimensions; these equations can be found in References 6 and 7.

2. Simulated data examples

We generated simulated clumps of data to test the theory. With a known number of pulses, centroid locations, and amplitudes, we investigated the effectiveness of the parameter estimation, the legitimacy of the error bar estimates, and the ability of the technique to decide on the most likely number of pulses in a clump.

In the following examples, clumps of three-point sources were created using the actual PRF from a sensor. The sensor is a scanner which simultaneously reflects radiation onto visible and LWIR focal plane arrays (FPA) at a 3-sec scan rate. The LWIR PRF and FPA were used to simulate the clumps. The LWIR PRF contains 80% of its energy within a 70 μrad blur. The LWIR pixels are 32.8 μrad square, and there are about four samples per dwell in the in-scan direction. The three pulses were placed at the centroids indicated in Table 1, scaled to the peak amplitudes in Table 1, and added to Gaussian noise having a standard deviation of 56.3 counts. The pulses had peak signal-to-noise ratios (SNR) of 7.6, 22.4, and 27.6. The centroids were taken from a scan of data during a flight test. This test is discussed in detail in Section 3. We want to emphasize that the simulated clump is made up of a real PRF and pulse locations from a real flight test. An 8 x 32-pixel window was isolated and used as the data. The resulting clump is shown in Figure 2.

	Pulse 1	Pulse 2	Pulse 3
Cross-scan centroid (pixels)	3.807	3.943	4.078
In-scan centroid (measurement)	13.105	15.354	17.339
Amplitude (counts)	430.0	1555.0	1264.0

Table 1: Simulated Clump

Pulse 1 is separated from Pulse 2 by less than 1/7 of a pixel in the cross-scan direction and separated from Pulse 3 by about a 1/4 pixel. In the in-scan direction, the pulses are separated by 1/2 and 1 pixels, respectively. The above separations are a factor of two smaller in terms of the PRF blur width.

We analyzed the data using one-, two-, three-, and four-point source models. The technique used a genetic algorithm search routine to find the set of centroids to maximize the posterior probability for each of the models. The posterior probability was derived by marginalizing over the amplitudes and noise terms. The error of each centroid was determined by finding the half-power point about the maximum of the posterior probability in each dimension independently. The amplitude, noise power, and amplitude error bars were

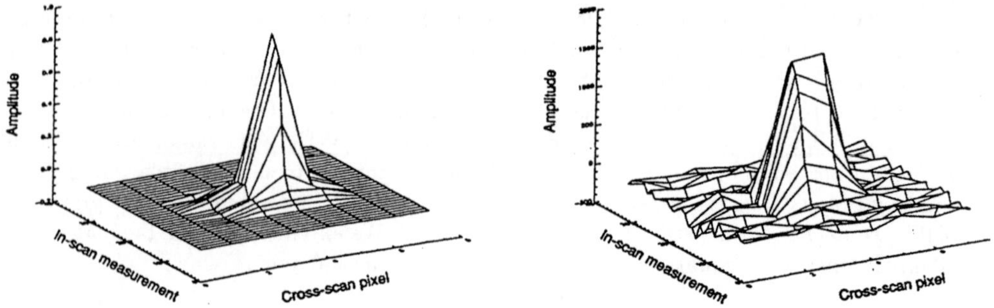

Figure 2: Sensor PRF and simulated three-source CSO clump. *Note: Figure 2a represents a real three-dimensional LWIR PRF from the sensor. It was used to create the simulated three-source clump in Figure 2b. The three sources were located at the centroids in Table 1, scaled to the amplitudes in Table 1, and added together with Gaussian noise. The pulses had SNR values of 7.6, 22.4, and 27.6.*

computed at the peak of the posterior probability. The pulse movement was accomplished using a two-dimensional bicubic spline interpolator, rather than fitting the PRF with a closed, analytical function.

The results for each model are contained in Table 2. The one- and two-source models both attempt to fit the clump midway between the actual points. The amplitude estimates are high in order to fit the width of the clump. The noise estimates are high because the signal within the data is not fit completely. As a result, the root mean square (r.m.s.) residuals are high. The BPT assigns these models virtually no probability of representing the data in comparison to the models which contain more sources.

The three- and four-source models represent very similar situations in that three of the pulses are at similar locations. The four-source model attempts to add a small pulse in addition to the three included in the simulated data. This fourth source reduces the r.m.s. residual and the noise estimate, since it is fitting the noise. However, BPT assigns a higher probability to the three-source model, even though the r.m.s. residuals are higher than the four-source model. The reduction in r.m.s. residual of the small fourth source is not enough to make up for the lower prior probability of the four-source model. Thus, the Occam factor in BPT has enabled the correct model selection.

The error bars in Table 2 are one standard deviation. Comparison of the three-source model parameters to the truth reveals that three of the six centroids and one of the amplitudes are out of the 1σ bound. Only the cross-scan location of Pulse 2 is out of the 2σ bound. We created five more clumps to have a larger sample size to evaluate the error bars. Thus, we had 54 estimated parameters: 36 centroids and 18 amplitudes. The numbers within 1, 2, and 3 standard deviations are shown in Table 3. The distribution of the 54 estimated parameter error bars follows that expected from a Gaussian parent distribution.

3. Sensor-flight test mission data

The flight test missile was launched from Wallops Island on 13 April 1992. The sensor viewed the flight test payloads from the Firepond test site in Massachusetts. Just after

	Position					
	cross scan	inscan	Amplitude	Noise	r.m.s.	
Actual Values	3.807 3.943 4.078	13.105 15.354 17.339	430.0 1555.0 1264.0	56.3		
One-Source Model	3.516 ± 0.029	16.823 ± 0.045	2798.9 ± 46.9	93.0	1491.6	$10^{-51.2}$
Two-Source Model	3.926 ± 0.017 4.043 ± 0.021	15.048 ± 0.064 17.289 ± 0.095	1789.7 ± 43.7 1330.2 ± 43.9	62.1	997.3	$10^{-10.3}$
Three-Source Model	3.758 ± 0.047 3.999 ± 0.020 4.000 ± 0.022	13.159 ± 0.197 15.485 ± 0.086 17.314 ± 0.091	488.4 ± 43.2 1536.7 ± 63.5 1214.6 ± 49.9	55.5	892.2	0.99
Four-Source Model	3.784 ± 0.050 3.958 ± 0.023 3.996 ± 0.025 5.008 ± 0.250	13.115 ± 0.207 15.474 ± 0.088 17.295 ± 0.092 15.639 ± 0.530	461.4 ± 43.1 1529.7 ± 63.9 1213.3 ± 50.3 91.1 ± 28.5	55.0	881.4	0.01

Table 2: Three-Source Clump Simulation Results

apogee, the post-boost vehicle (PBV) released two large balloons. The balloons' surface optical properties were designed to have large diffuse signatures in the visible and LWIR bands. Their deployment angle with respect to Firepond and delta velocities were such that the balloons resolved quickly in the visible but remained a clump in the LWIR for a number of scans before they became resolved.

The sensor data of the closely spaced objects allow clump processing in two ways. First, since there were simultaneous visible and LWIR measurements, the visible centroids can be fused with the LWIR to assist in the LWIR radiometry. Second, the clumps can be processed like the simulated clumps of the previous section. Both analyses are presented here.

3.1. Visible Assist

The balloons were released at 413 and 415 sec into the flight. Figure 3 shows the actual sensor simultaneous visible and LWIR data at 424 and 457 sec, respectively. The three-times improvement in spatial resolution of the visible over the LWIR is readily apparent. The visible FPA could resolve the balloons at 418 sec as seen in Figure 3. The LWIR FPA could not resolve the balloons from each other until 442 sec analytically and until 457 sec

	1σ	2σ	3σ
Amplitudes	9	16	17
Centroids	26	35	36
Total	35	51	53
Expected for Gaussian	37	52	54

Table 3: Number of Estimated Parameters within $n\sigma$ Error Bounds

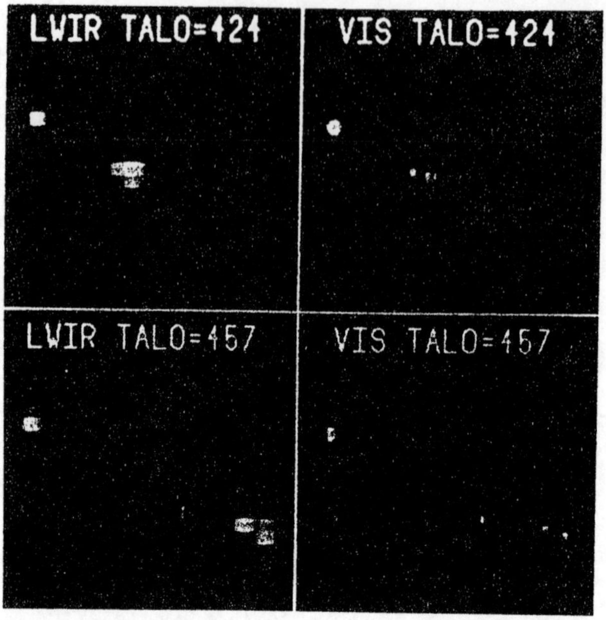

Figure 3: Simultaneous visible and LWIR sensor data at 424 and 457 sec. *Note: The sensor data taken during the flight test at 424 and 457 sec time after liftoff (TALO) are shown. Two balloons were released from a PBV at 413 and 415 sec. The three-times improvement in spatial resolution of the visible over the LWIR is readily apparent. The visible focal plane resolved the objects immediately as is apparent from the figure. The balloons were not resolved visually from each other in the LWIR until 457 sec. The fourth object in the data is the spent second-stage of the booster.*

visually. The LWIR pulses from the balloons overlapped until well past 470 sec. Thus, we used the visible centroids to enhance the LWIR radiometry from 418 to 448 sec. Specifically, the relative spacing between the pulses was fixed using the visible centroids. Then we used the posterior probability to fix the location of the clump, i.e., search over a two-dimensional space. We then solved for the amplitudes, noise estimates, and error bars.

The results are indicated in Figure 4. From 418 to 433 sec, the data contained three pulses (the PBV and two balloons), so a three-source Bayesian model was employed. After 433 sec, the pulse from the PBV did not overlap the balloons, so a two-source model was used from 433 to 448 sec. After 448 sec, the balloons were separated enough to be able to use a single-pulse pulse-matcher, which is an algorithm that estimates the amplitude and centroid of a well-resolved single. The SNR for the resolved balloons was approximately 50. As a point of comparison, the results from the single-pulse pulse-matcher are included for the whole measurement interval, even when clumps were present to show its performance on CSO. This figure dramatically reveals the earlier detection time and measurement improvement of individual pulse amplitudes by using the Bayesian m-pulse model.

Table 4 compares the estimated model amplitudes to the resolved pulse measurement amplitudes averaged over the observation times. The 1-sigma errors about the time averages

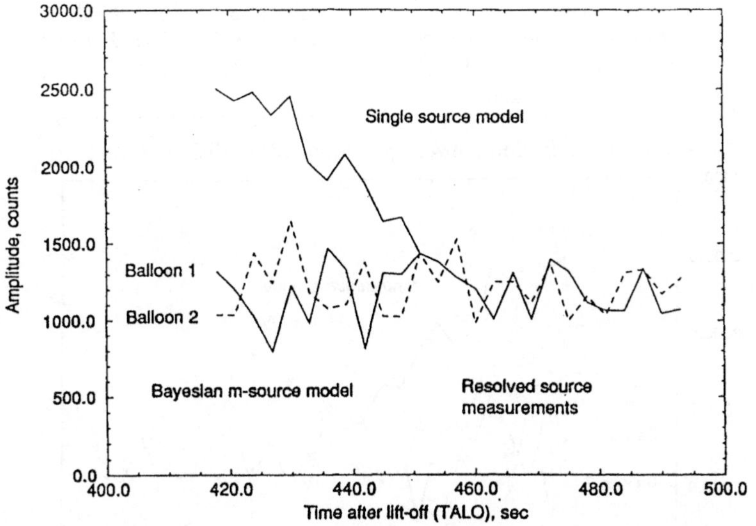

Figure 4: Bayesian CSO LWIR radiometry with visible assist. *Note: The sensor data taken during the flight test were used with the Bayesian m-source model from 418 to 448 sec. The visible centroids were combined with the LWIR data to improve the LWIR radiometry. After 448 sec the balloons were resolved in the LWIR, so resolved pulse measurements were obtained using a single-pulse pulse-match algorithm. From 418 to 433 sec, a three-source model was used. From 433 to 448 sec, a two-source model was used, because the PBV was resolved from the balloons. The figure shows that the Bayesian model produced amplitudes with scan-to-scan variations very close to the resolved pulse measurements. The results using the single-pulse pulse-match algorithm on the clumps reveals the improvement in detection time obtained by using the Bayesian model with visible assist.*

are indicated. On average, the visible-assist Bayesian pulse model amplitudes are within 4% of the resolved pulse measurements. Further, the model reflects the measurements in that the amplitude of Balloon 2 is about 3% higher than the amplitude of Balloon 1. The scan-to-scan precision, however, is about 30% greater for the model estimates than for the resolved measurements.

The Bayesian error bars on the amplitudes are about 40 counts, which is less than the scan-to-scan precision of the measurements. The small error bars are different from the results for the simulated data, where the derived error bars matched very well the difference between the estimated parameters and truth. We believe that atmospheric effects and scan mirror jitter resulted in a nonstationary PRF in the real data. For the simulated data, the same pulse used to create the clump was also used in the model. For the real data, this was not the case. The model pulse was a measured point source at the same elevation angle as the balloon/PBV clump but at a different time. The Firepond site essentially is at sea level, and atmospheric distortions in seeing and speckle were prominent. Furthermore, significant scan mirror jitter was observed during the measurements. The PRF modeling

	Balloon 1	Balloon 2
Resolved Pulse Measurements	1204 ± 155	1247 ± 142
Bayesian m-Pulse Model (with visible assist)	1162 ± 223	1199 ± 207

Table 4: Balloon Average Amplitudes, Visible Assist

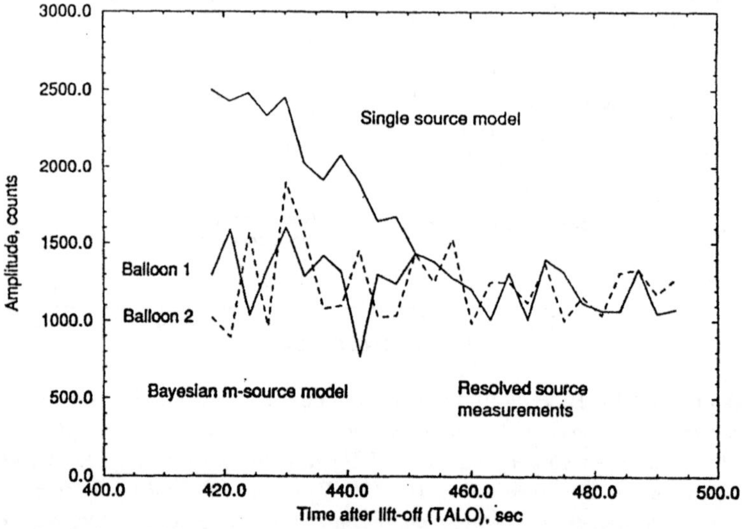

Figure 5: Bayesian CSO LWIR radiometry without visible assist. *Note: The Bayesian m-source models described in Figure 4 also were employed without using the visible centroids. Rather, the posterior probability was maximized by moving the three LWIR sources independently. Note that the scan-to-scan variation is greater than the visible assist case.*

mismatch serves to underestimate the noise power and thereby create error bars that are too small. This emphasizes the need for excellent PRF measurements for any algorithm designed to deconvolve CSO clumps.

3.2. No Visible Assist

For this analysis, we used prior knowledge only to specify the number of objects, m, in each clump. The search algorithm had to maximize the posterior probability over a 2m-dimensional space corresponding to the 2m model pulse centroids. The results are displayed in Figure 5 and summarized in Table 5.

Whereas the amplitudes derived using the methods agree well for Balloon 2, the amplitudes for Balloon 1 differ from one another by over 7%. Furthermore, the Bayesian model without visible assist assigns a greater amplitude to Balloon 1 rather than to Balloon 2. The scan-to-scan precision for the model without visible assist is greater than the model with visible assist. This illustrates the benefits of data fusion, i.e., using visible centroids to improve the LWIR amplitude estimates and the scan-to-scan amplitude variation. Furthermore, the BPT formulation easily allowed the use of the visible centroids in the LWIR model via the prior probability.

	Balloon 1	Balloon 2
Resolved Pulse Measurements	1204 ± 155	1247 ± 142
Bayesian m-Pulse Model (no visible assist)	1290 ± 232	1233 ± 326

Table 5: Balloon Average Amplitudes, No Visible Assist

4. Visible sensor data

We hypothesized in Section 3 that using the measured PRF in the model equations resulted in non-Gaussian noise processes due to the scan mirror jitter, non-integer samples per dwell, and atmospheric fluctuations. These problems caused the Bayesian amplitude error bars to be smaller than the scan-to-scan variation.

We thus desired sensor data with a more stationary PRF, i.e. a PRF which would not vary significantly across the field-of-view nor be adversely affected by atmospheric fluctuations. We acquired a set of data from a staring visible CCD sensor attached to a 24-inch telescope on Table Mountain, CA. The visible band was filtered to cover 450-550 nm. The PRF was reported to be stationary across the field. The PRF width was 6-7 pixels; each pixel was 2 μrad so the response was well-sampled. The array contained 512 x 512 pixels. The sensor took 60 second exposures so the atmospheric fluctuations of the point source response should theoretically have been smoothed. Since the background was very smooth it appeared to be true. Figure 6 shows the center 256 x 256 pixel scene of NGC 6819 measured September 19, 1992. To the eye, the single point source responses appear circularly symmetric and consistent across the scene. We chose four stars as model PRFs and ran 1-4 source models on 32 different clumps. The clumps were chosen to include single and multiple sources with a variety of amplitudes at locations all over the scene. Three cases will be discussed below. In these examples, the bright star at (250, 320) was used as the PRF. It produced the greatest posterior probability of the candidate PRFs. The three clumps are displayed in Figure 7 as detailed contour plots.

Star clump 423 at ~(300, 190) is commonly thought to be a single source and is used for calibration photometry. Our algorithm agreed with this hypothesis with an extremely high confidence of 99%. The estimated amplitude was 1198 counts with an error bar of only 3 counts. The location was 9.704 ± 0.006 pixels east and 11.493 ± 0.006 pixels north. The small error bars result from the high signal to noise ratio of ~150.

Star clump 416 at ~(200, 360) is also commonly taken to be a single source. Our technique, however, assigned virtually no probability to a one-source model compared to a two-source model. The two-source model was also preferred over the three-source model by a factor of 100. The two-source model put a source about 11 times dimmer than the other separated by 2.2 pixels to the east and 2.5 pixels to the south. The location error bars are greater than ~1/6 pixel for the dimmer source compared to 0.013 pixel for the brighter source. The amplitude error bars for the dimmer source were about 6% compared to 0.5% for the brighter source.

Star clump 414 at ~(340, 210) looked interesting because the bulge to the south and west of the doublet gives evidence for another source. The technique, in fact, strongly preferred a three-source model with a dim source located at 8.8 pixels east and 6.9 pixels

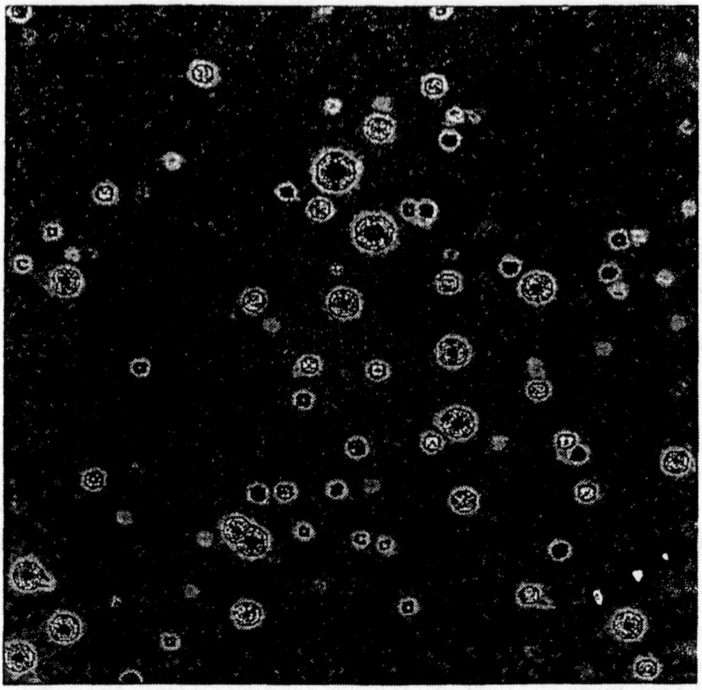

Figure 6: *Figure 6 shows real data from a staring visible CCD sensor attached to a 24-inch telescope. The visible band was filtered to cover 450-550 nm. The PRF width was 6-7 pixels; each pixel was 2 μrad so the response was well-sampled. The array contained 512 x 512 pixels. The sensor took 60 second exposures so the atmospheric fluctuations on the point source response should theoretically have been smoothed. Since the background was very smooth it appeared to be true.*

north. It was separated by 4.6 pixels east and 0.5 pixels south from one source and 0.6 pixels east and 5.2 pixels south from the other pulse.

We are currently searching for other data sources to verify the source counting of the technique. However, since it counted nine apparent single sources as singles we are confident in the technique. We found that choosing the PRF in the model is the key to success. The examples above used a PRF from the same region of the focal plane. PRFs chosen from other regions on these stars produced a much lower posterior probability which consequently led to miscounts and different amplitude estimates. This leads us to believe there are spatial distortions across the field of view. Furthermore, it was observed that the pulse response was nonlinear in amplitude. Thus if a star chosen as a PRF had a significantly lower amplitude than a particular star dataset, the posterior probability was less than if a brighter star was used as the PRF. The amplitude estimates invariably would be low; in effect, the model was not able to extrapolate well to higher amplitudes. We also tried to fit a Gaussian to a star and use the fit as the PRF. This resulted in posterior probabilities hundreds of orders of magnitude less than using the measurements themselves. We thus concluded that the

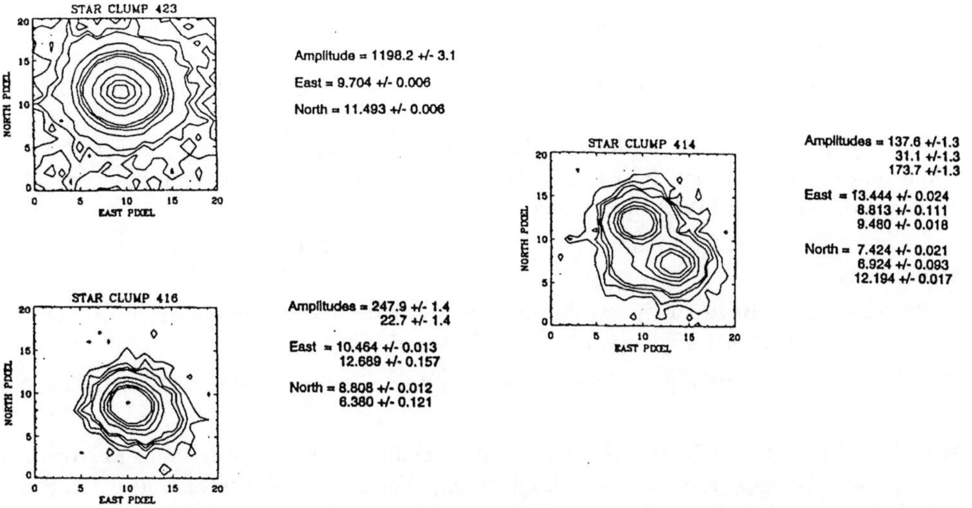

Figure 7: *Figure 7 displays the detailed contours of three star clumps which were analyzed. For the number of sources with the greatest posterior probability, the estimated amplitude(s) and centroids are stated with the corresponding derived error bars. Clumps 423 and 416 are commonly taken to be single stars.*

pulse responses were decidedly non-Gaussian.

In the results discussed above, the data were preprocessed to remove the background. However, because the model in the Bayesian formulation is completely general, we added a term to account for the background on unprocessed data. We added the equation for a plane. This created three new amplitudes in the formulation, namely a dc term and a linear term in both the x and y directions. The technique correctly solved for the background amplitude terms and resulted in nearly identical centroid and amplitude estimations for the stars. This has implications for how data from optical sensors can be processed. Typically, raw sensor data is put through a number of time-dependent processing algorithms to remove the background and identify regions of interest or detections. The detections are then individually sent to an object-dependent processor which performs the parameter estimation for the amplitudes and centroids. With the Bayesian formulation, it may be possible to combine these two steps into one calculation.

5. Summary

In this report we have used real and simulated data to show that the Bayesian Probability Theory can be employed to deconvolve clumps of closely spaced objects into individual pulse measurements. Pulse centroids and amplitudes with error bars can be determined from a single scan of data, since the sensor noise power can be estimated. This has implications for the way sensors can work together. Instead of having to communicate all of the data from one sensor to another so that the receiving sensor can compute relative confidences in the measurements, only the covariances for all estimated parameters need to be sent because they are available.

We also showed that a decision can be made on the most likely number of objects in a clump. The decision comes directly from using maximum entropy in the Bayes theorem and the rules of probability theory without relying on contrived penalty functions or fudge factors.

Finally, we showed how easily the Bayesian Probability Theory can be used to fuse visible centroid information into LWIR data to obtain better LWIR amplitude estimates compared to using the LWIR data alone.

References

[1] Bretthorst, L. "Bayesian spectrum analysis and parameter estimation," Ph.D. Dissertation, Washington University, St. Louis, Missouri, 1987.

[2] Jeffreys, H. "Theory of probability," Third Edition, Clarendon Press, Oxford, England, 1961.

[3] Gull, S. "Bayesian inductive inference and maximum entropy," Maximum-Entropy and Bayesian Methods in Science and Engineering, Vol. 1, G. J. Erickson and C. R. Smith (eds.), Kluwer Academic Publishers, Boston, Massachusetts, 1988.

[4] Jaynes, E. "Detection of extra-solar system planets," Maximum-Entropy and Bayesian Methods in Science and Engineering, Vol. 1, G. J. Erickson and C. R. Smith (eds.), Kluwer Academic Publishers, Boston, Massachusetts, 1988.

[5] Bretthorst, G. and Smith, C. "Bayesian analysis of signals from closely-spaced objects," Infrared Systems and Components III, R. L. Caswell (ed.), Proc. SPIE, Vol. 1050, 1989.

[6] Schulenburg, N. and Hackwell, J. "Bayesian approach to image recovery of closely spaced objects," Proc. SPIE, Vol. 1954, 1993.

[7] Lillo, W. and Schulenburg, N. "A Bayesian closely spaced object resolution technique," Proc. SPIE, Vol. 2235, 1994.

ULTRASONIC IMAGE IMPROVEMENT THROUGH THE USE OF BAYESIAN PRIORS WHICH ARE BASED ON ADJACENT SCANNED TRACES

Louis Roemer and Jianmin Zhang
Electrical Engineering Department
Louisiana Tech University
Ruston, Louisiana 71272 USA

ABSTRACT. Improvement in an ultrasonic image can be obtained by using the echo amplitude in adjacent traces of the ultrasonic signal to form a Bayesian prior. The location of an echo in a trace is usually accompanied by an echo in the adjacent traces, as the beam width is greater than the distance between traces. A scaled version of the sum of the adjacent traces is used as a prior. By keeping the area under the curve at a constant value (as the probability must sum to unity), a consistent criterion is maintained. Adjacent traces with no echoes would result in a uniform prior. An example is presented in which the known shape of an object is more closely recovered than the shape inferred when the adjacent trace information is not used. The technique might be regarded as a Markoff process, as only the adjacent trace (channel) is used in forming the prior.

1 Background

Ultrasonic beams, particularly unfocused beams, sweep out a volume of space. Due to the finite beam diameter, each trace contains some fraction of the adjacent trace information. If we model the signal as an autoregressive model, following [1], a smoothly changing estimate of the echo source versus depth of beam penetration is obtained [2, 3].

By using adjacent traces to form a Bayesian prior, information from adjacent regions is incorporated in the estimation of the validity of the current trace's echoes. If the adjacent traces contained no echoes, then they would form a uniform prior. All priors, of course, would have the area under their curves normalized, as the probability must sum to unity.

The example illustrated, scanning a piece of spaghetti, produced a more circular cross section than the cross section described by the raw data. This improvement in resolving the known target, as well as the sharpness of the image, show promise in more complicated images. Though spaghetti is of little clinical interest, the example provides encouragement to apply the method in clinical trials on biological tissues.

Some criterion of goodness must be chosen to determine when to stop calculating the autoregressive filter coefficients. For this very simple example, a low order filter was sufficient by any criterion [5, 6]. Low order gives a slowly changing prior, which is intuitively pleasing.

2 Example

The method of echo identification using the phase information has proved useful [3, 4]. The highly overlapped echoes are shown in Figure 1, the original time domain signal. The signal source and receiver are a conventional pulsed medical ultrasound instrument

G. R. Heidbreder (ed.), Maximum Entropy and Bayesian Methods, 339–342.

whose transducer is centered at 3.5MHz. The object (a piece of boiled spaghetti) was mechanically scanned. By processing the signal phase using maximum entropy methods [2, 3], the localized echo can be identified. This localized echo is used as the raw signal for further processing. Figure 2 is the localized echo from the data of Figure 1. Conventional

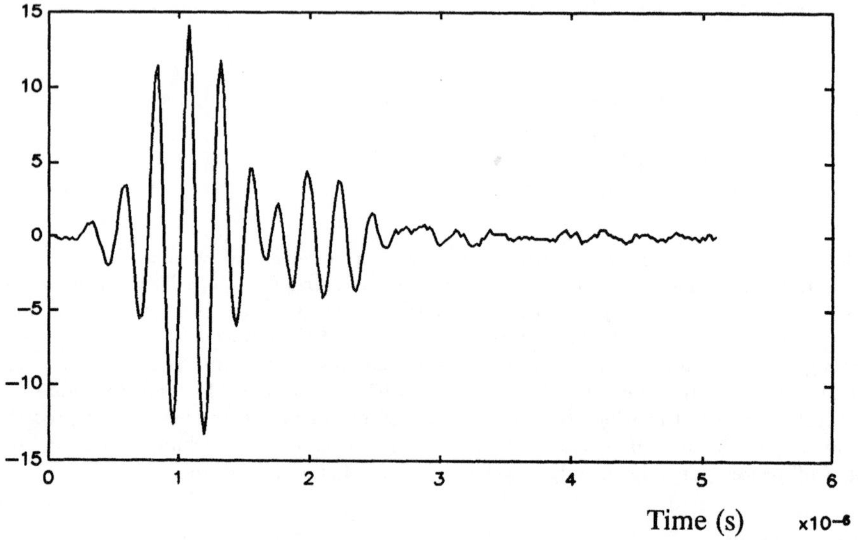

Figure 1: The Received Signal in Trace 4 (Volts)

signal processing would display each trace, independent of the adjacent traces. In this example, however, the two adjacent traces of raw signal are summed; then the area under the curve normalized. The resultant summed curve is used as the Bayesian prior. The justification is that each adjacent curve contains some information about the center curve due to beam overlap. The method is illustrated in Figure 3. Biological structures of interest are usually of larger extent than the ultrasound beam width. This would be an intended application area. Finally, Figure 4 provides a reconstruction of the object, giving some confidence in the method by both its shape and sharpness.

3 Conclusion

The example shows a relatively simple method of incorporating adjacent traces of a broad beam sensor into the image analysis.

References

[1] J.P. Burg, "Maximum Entropy Spectral Analysis", Modern Spectrum Analysis, D.G. Childers ed, IEEE Press, New York, pp. 34-41, 1978.

[2] J. Yao, L. Roemer, N. Ida, Ke-Sheng Huo, "Delay Estimation Using Maximum Entropy Method Derived Phase Information," Maximum Entropy and Bayesian Methods, 1989,

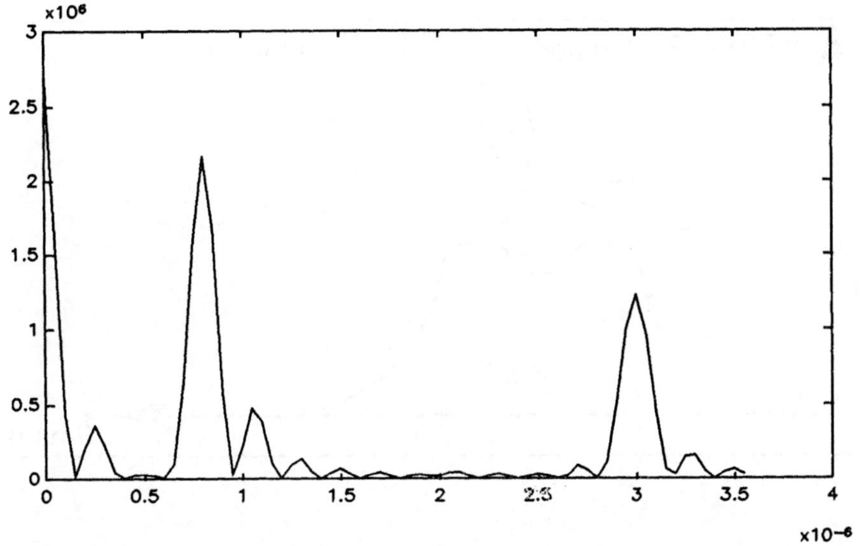

Figure 2: Echo (raw signal, unscaled) for Trace 4 versus Time (s)

P. Fougere ed., Kluwer, a division of D. Reidel Publishers, Dordrecht, Holland.

[3] Jie Hu, Maximum Entropy Estimation Method Applied to Ultrasound Overlapping Echoes Identification, M.S. Thesis, Louisiana Tech University, November 1991.

[4] J. Zhang, Ultrasound Detecting The Shape Of An Object By Maximum Entropy Method M.S.Thesis, Louisiana Tech University, May 1993.

[5] H. Akaiki, "Power Spectrum Estimation through Autoregression Model Fitting," Ann. Inst. Stat. Math., vol 21, pp 407-419, 1969.

[6] W.J. Fitzgerald and M. Niranjan, "Speech Processing using Bayesian Inference", Maximum Entropy and Bayesian Methods, 1992, A. Mohammad-Djafari ed., Kluwer, a division of D. Reidel Publishers, Dordrecht, Holland.

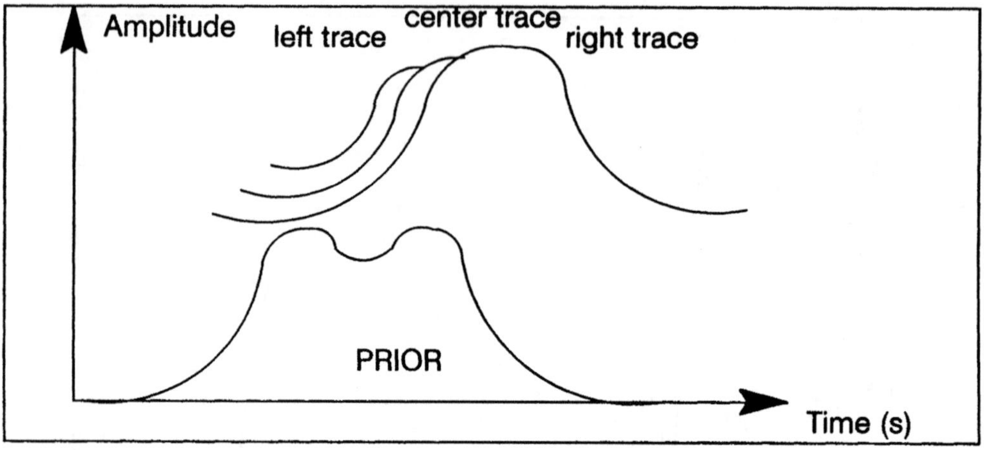

Figure 3: Schematic of Formation of Bayesian Prior

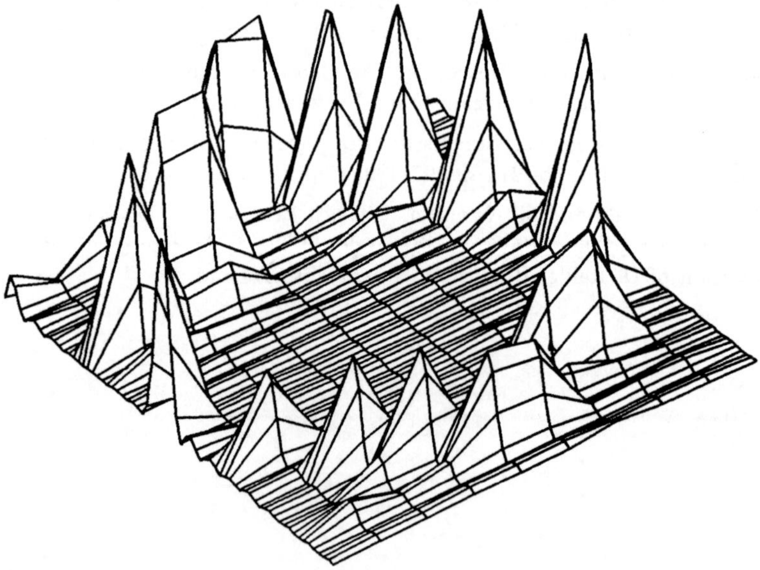

Figure 4: A 3-Dimensional Perspective Plot of the Time Delay Versus Depth

APPLICATION OF MAXENT TO INVERSE PHOTOEMISSION SPECTROSCOPY

W. von der Linden, M. Donath, and V. Dose
Max-Planck-Institut für Plasmaphysik, EURATOM Association
D-85740 Garching b. München, Germany
e-mail: wvl@ibmop5.ipp-garching.mpg.de

ABSTRACT. Information about the spectral density gained by inverse photoemission spectroscopy is distorted by the Fermi distribution and the apparatus function. In many cases recovery of the desired physical quantities is hampered by an ill-posed inversion problem.

It is shown, based on the spin- and temperature dependent quasiparticle spectrum of Ni, that the maximum entropy method yields unbiased access to the spectral density independent of model assumptions. The effective energy resolution is thereby improved by a factor of 5 and structures below the Fermi level E_F, which are generally lost in inverse photoemission, are recovered.

1. Introduction

In this article we will show that the maximum entropy method [1] is an ideal data-analysis tool for recovering "hidden information" from experimental data without making any model-assumptions. We will address a longstanding problem in the field of itinerant magnetism.

For the microscopic understanding of collective magnetism of itinerant electrons, as in transition metals, the temperature and spin dependence of the spectral density $A_{k\sigma}(\omega)$ close to the Fermi level E_F play a vital role. In an homogeneous magnetic field the electronic spectral density consists of a single δ-function $\delta(\omega - \omega_{k\sigma})$ for given spin direction σ and momentum k. The quasiparticle energy, $\omega_{k\sigma} = \varepsilon_k + \alpha B\sigma$, depends on the free electron dispersion ε_k and the Zeeman term which splits spin-up and spin-down energies proportional to the external magnetic field. Inside a transition metal, below the Curie temperature T_C, there exists an effective magnetic field which is proportional to the net magnetization of all electrons, $B_{\text{eff}} \propto \langle S_z \rangle$. Within the mean field approximation, B_{eff} acts like an external field and one expects $A_{k\sigma}(\omega)$ to show a pair of peaks, one for each spin direction. More elaborate approximations to the many-body problem allow for changes in the electronic spin due to electron-electron interactions. Thus an electron with initial spin σ experiences also states of opposite spin. One would therefore expect a "multiband structure" with temperature dependent pole strength and quasiparticle energies. With increasing temperature correlation effects are expected to lead to a mixing of spin-up and spin-down states resulting in "extraordinary" peaks. Above T_C the spin asymmetry disappears, as the rotational symmetry is restored. The multiband structure is, however, retained at and above T_C, owing to short-range ferromagnetic spin-correlations. These ideas are underpinned by, for example, the fluctuating band theory [2, 3, 4] or approximate many-body calculations based on Hubbard-type model Hamiltonians [5, 6] and cluster-calculations [7]. At present, however, there is no generally accepted theory for band magnetism. Even the more fundamental question,

G. R. Heidbreder (ed.), Maximum Entropy and Bayesian Methods, 343–350.

whether model Hamiltonians such as the Hubbard model describe ferromagnetism at all is not settled [8]. It is therefore important to have accurate and conclusive experimental data to test the various theories.

Experimentally the situation cannot be solved rigorously either, since $A_{k\sigma}(\omega)$ cannot be measured directly. The deconvolution of the experimental IPE data is hampered by an ill-posed inversion problem. There exists an infinity of possible solutions consistent with the experimental data within the error bars. Experimental efforts have been made to reveal the detailed spin and temperature dependence of the electronic states. A number of photoemission and inverse photoemission (IPE) studies on Fe and Ni have been performed [9, 10, 11, 12, 13, 14]. While for Fe clear evidence has been found for non-collapsing band behavior at specific points in **k**-space [9, 10], the situation is more subtle for Ni and direct conclusions from the raw experimental data are not possible.

2. Formalism

To reveal such detailed features of the spectral density, as the quasiparticle energy and lifetime, particularly for states lying below the Fermi energy, which are buried under the Fermi distribution, a more subtle analysis of the experimental data is required. To this end we invoke the Maximum Entropy (MaxEnt) method which is based on Bayesian probability theory, the importance of which has been emphasized recently by P.W. Anderson. [16].

The experimental IPE intensities for 100% spin-polarized electrons of spin σ are proportional to

$$I^\sigma(\omega, T, \mu) = j^\sigma \int A^\sigma(\omega')\{1 - f(\omega', T, \mu)\}\, g(\omega' - \omega) d\omega' \quad . \tag{1}$$

Here $A^\sigma(\omega)$ is the required spectral density of quasiparticles with energy ω, spin σ and wavevector **k**. j^σ represents the current density of incoming electrons of spin σ. To derive (1) standard approximations have been made, in particular ignoring matrix element- and relaxation-effects. The information about the electronic structure is contained entirely in the electronic spectral density. Dependence on temperature T and chemical potential μ enters via the Fermi distribution $f(\omega, T, \mu) = 1/(1+\exp((\omega-\mu)/kT)$. In (1) $g(\omega'-\omega)$ stands for the apparatus function, which is a convolution of the energy distribution of the incoming electrons and the energy window for the detected photons. The apparatus function can be estimated quite accurately from a comparison of image-potential surface states on Ni(111) measured by IPE and two-photon photoemission. We find that the apparatus function can be approximated fairly well by a Gaussian with standard deviation 195 meV [17].

For numerical purposes we evaluate the spectral density $A_n^\sigma = A^\sigma(\omega_n)$ at discrete energies ω_n with $n = 1, 2, \ldots, N_{\text{var}}$ and interpolate it linearly between these. The integrals of the piece-wise linear $A^\sigma(\omega)$ in (1) and the exact Fermi function $f(\omega, T, \mu)$ are computed numerically. Eq. 1 transforms into a set of linear equations

$$I_l^\sigma = I^\sigma(\tilde{\omega}_l, T, \mu) = j^\sigma \sum_{i=1}^{N_{\text{var}}} M_{li}\, A_i^\sigma \qquad l = 1, 2, \ldots, N_{\text{eq}} \quad , \tag{2}$$

where $\tilde{\omega}_l$ represents the (coarser) mesh on which the experimental data are available. The spin polarization of the incoming electron beam has been estimated as $p = \frac{N_\uparrow - N_\downarrow}{N_\uparrow + N_\downarrow} \approx 0.33$

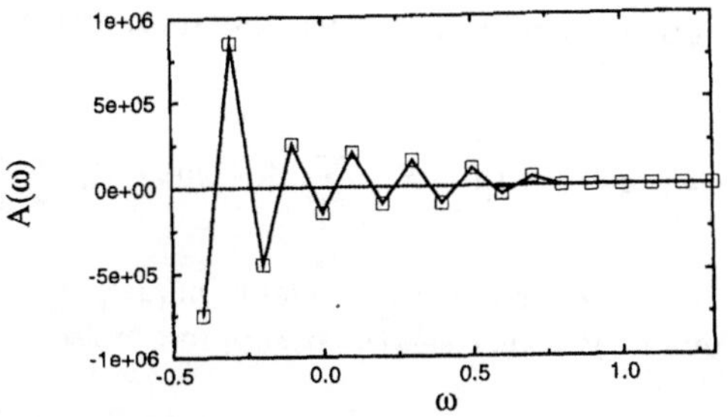

Figure 1: *Direct inversion of Eq.2 with* $N_{\text{eq}} = N_{\text{var}} = 20$

[15]. The incoming beam of predominantly spin-σ electrons contains, therefore, a proportion $n_{\sigma,\sigma} = (1+p)/2$ of spin-σ electrons, while the remaining proportion of electrons $n_{\sigma,-\sigma} = (1-p)/2$ has opposite spin. Therefore the measured intensity for a σ-polarized beam of incident electrons is

$$g_l^\sigma(\vec{A}) = \sum_{\sigma',i} M_{li} n_{\sigma,\sigma'} A_i^{\sigma'} \quad . \tag{3}$$

To recover the spectral density, Eq. (3) has to be inverted. At first sight this inversion appears to be utterly ill-posed. The kernel M_{li} is almost singular due to the Fermi function, which suppresses structures below the chemical potential ($|\omega - E_F| > k_B T$) exponentially. Therefore the inverse matrix has very large eigenvalues and the experimental errors are strongly amplified. The scatter of solutions compatible with the experimental data, is therefore enormous. A direct inversion of (3) as depicted in fig. 1 (with $N_{\text{var}} = N_{\text{eq}}$) leads to results fluctuating between $+10^5$ to -10^5, while the real values for $A(\omega)$ are positive and of order 1. Only if the experimental data have a relative accuracy of better than 10^{-6} is direct inversion of (3) feasible. Moreover, this direct approach is restricted to $N_{\text{var}} \leq N_{\text{eq}}$.

We use Bayesian probability theory to determine the *posterior* probability $P(\vec{A}|\vec{g}^e, \lambda)$ for a particular solution $\vec{A}$ given the experimental data $\vec{g}^e$ and additional experimental parameters λ, such as the scale of the error bars, the chemical potential or the width of experimental resolution:

$$P(\vec{A}|\vec{g}^e, \lambda) = P(\vec{g}^e|\vec{A}, \lambda)\,\frac{P(\vec{A}|\lambda)}{P(\vec{g}^e|\lambda)} \quad . \tag{4}$$

$P(\vec{g}|\vec{A}, \lambda)$ is the *likelihood* function which contains the new information provided by the experiment. In IPE experiments the data are independent and normally distributed with error σ_i. The likelihood function is therefore

$$P(\vec{g}^e|\vec{A}, \lambda) \;=\; e^{-\frac{1}{2}\chi^2} \qquad \text{with} \quad \chi^2 = \sum_{l=1}^{N_{\text{eq}}} \Big(\frac{g_l^e - g_l(\vec{A})}{\sigma_l}\Big)^2 \quad .$$

Here $g_l(\vec{A})$ is the theoretically predicted result for given $\vec{A}$. The spectral density is a positive, additive distribution function, for which the appropriate uninformative entropic *prior* is invoked [18].

$$P(\vec{A}|\lambda) = e^{\alpha S} \quad \text{with} \quad S = \sum_i A_i - m_i - A_i \ln(\frac{A_i}{m_i}) \quad ,$$

the information theory entropy relative to a default model m_i. We have chosen $m_i = \varepsilon$, where ε is a small quantity which serves to suppress noise in regions of insufficient information.

The MaxEnt solution for $\vec{A}$ is obtained by maximizing the posterior probability, or equivalently $\alpha S - \frac{1}{2}\chi^2$, with respect to A. The regularization parameter α is determined self consistently as elaborated by Skilling [18] upon maximizing the evidence $P(\alpha|\vec{g}^e)$ for α, given the experimental data. Other parameters, like chemical potential, width of the Gaussian resolution, and degree of polarization, can likewise be determined.

3. Discussion and Results

We have applied the MaxEnt deconvolution to temperature-dependent spin- and angle-resolved IPE data of Ni(110) for the $Z_4 \rightarrow Z_2$ transition [15]. The experimental data are taken for 20 energies per spin direction and $A_\sigma(\omega)$ is reconstructed for 80 energies. The inversion problem of Eq. (3) is therefore highly underdetermined. Experimental data are available for temperatures $T/T_C = 0.48, 0.64, 0.72, 0.82, 0.95$ and 1.02, covering the range from almost perfect ferromagnetic order out into the paramagnetic regime.

Before discussing the physical conclusions we will address characteristic parameters of the experiment. The statistical errors of the IPE data are known and fairly small ($\leq$ 2%). MaxEnt analysis consistently leads to a confirmation of these values. The same holds for the chemical potential, for which only slight deviations $|\Delta\mu| \lesssim 0.04$ eV from the experimentally determined values were found. A further convincing result of MaxEnt concerns the apparatus function. We allowed for more flexibility by supposing that the "Gaussian" can fall off at rates corresponding to different standard deviations d_l, d_r on the left and right flank of the peak. We find that the evidence is sharply peaked at a value $d_l = 183$ meV and $d_r = 195$ meV, with an uncertainty of ± 1 meV. These values are in good agreement with the estimate based on the comparison with two-photon photoemission data [17]. As the MaxEnt values for σ_l, σ_r have only very small uncertainty, this approach is very useful for determining the apparatus function whenever it is not accessible by experimental means.

Due to the incomplete spin polarization of the incoming beam one always observes two peaks in the experimental raw data. The polarization of the incoming beam had been set experimentally to make the extraordinary peak vanish for the $T/T_C = 0.48$ data [15]. Using this experimentally determined value $p \approx 33\%$ for all temperatures we find almost negligible and temperature independent "extraordinary" peaks. As MaxEnt is not a linear method it is expedient to use it on the full experimental information in the form of Eq. 3 using p as an adjustable parameter. Since the polarization of the incoming beam is independent of the sample temperature, the same polarization p is used for all temperatures and the combined

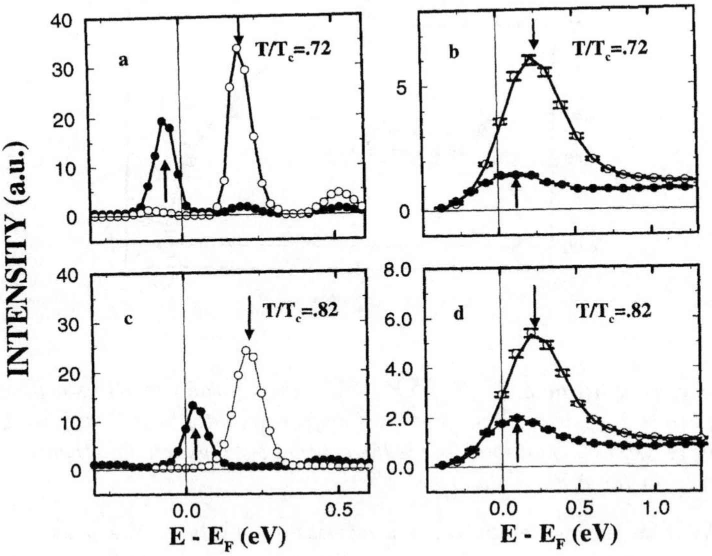

Figure 2: *Spin-dependent quasiparticle spectral density (a,c) and experimental IPE data (b,d) of the $Z_4 \rightarrow Z_2$-transition in Ni for two temperatures $T/T_C = 0.72$ (a,b) and 0.82 (c,d).*

evidence $\prod_i P(p, T_i)$ has to be maximized simultaneously. The maximum evidence is obtained for $p = 0.32$, which is in good agreement with the experimentally determined value of $p = 0.33 \pm 0.03$. At this value of p extraordinary peaks disappear for all temperatures.

Typical results obtained by the MaxEnt deconvolution are given in fig. 2(a), (c) for $T/T_C = 0.72$ and 0.82. For comparison we also depict the experimental data. In the experimental data (fig. 2(b), (d)), both spin-up and spin-down features appear above E_F [15]. The reconstructed spectral densities, however, reveal the spin-up peak clearly below (above) E_F for $T/T_C = 0.72$ (0.82) with a line-width of about 80 meV independent of temperature. The resolution of IPE+MaxEnt is better then 40 meV, which is an improvement by at least a factor of 5 over the raw experimental resolution. The explanation is that the experimental resolution is due to a convolution with a smooth function which can be characterized extremely accurately by a few parameters, independent of temperature. There is no significant indication of "extraordinary" peaks at all temperatures. To quantify this statement, we have determined the posterior probability of a two-peak structure, where we have mixed in a proportion q of the minority peak to the majority structure. It appears that the posterior probability falls like $P(q)/P(0) \approx e^{-aq^2}$, were the constant a depends on temperature. Remarkably, in the $T/T_C = 0.72$ data of fig.2 the posterior probability falls to $1/e$ already for $q = 0.009$ even though the results are still within the experimental error bars. This observation demonstrates emphatically that extraordinary peaks can be

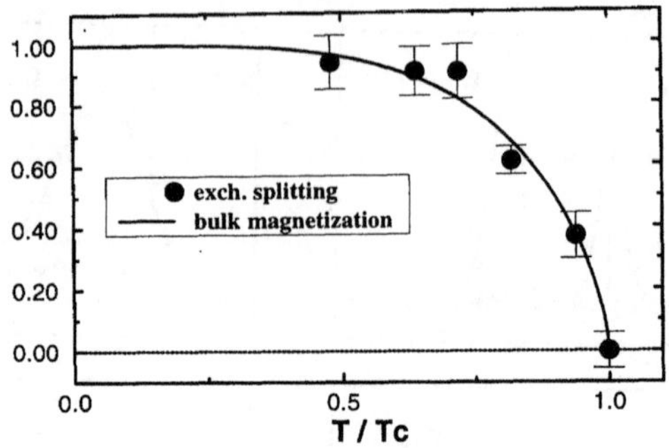

Figure 3: *Exchange splitting $\Delta E_{ex}(T)/\Delta E_{ex}(0)$ of the Z_2-band in Ni as a function of temperature (full circles). Errorbars are obtained selfconsistently from MaxEnt. Extrapolation yields $\Delta E_{ex}(0) \approx 0.28eV$. The solid line is the experimental bulk magnetization $M(T)/M(0)$ of Ni rescaled to fit the ΔE_{ex} data [19].*

ruled out. With increasing temperature a decreases slightly as the peaks approach each other. For $T/T_C = 0.95$, the posterior probability drops below $1/e$ at $q = 0.02$. The slight structures visible in fig.2 for $T/T_C = 0.72$ are attributed to noise and are completely absent for $T/T_C \geq 0.82$. The minor structure at ≈ 0.5 eV stems from irregularities in the experimental data a few eV above E_F and has no physical relevance.

It is instructive to compare the experimental data with those obtained by using the MaxEnt result for $A(\omega)$ in Eq. 3 (solid lines through the data points in fig. 2(b),(d)). The agreement is perfect, but for ill-posed inversion problems this is not surprising. It is likewise a completely useless test for theories to compare the theoretical and experimental values of g_l. The different heights of the peaks above and below the chemical potential should not be taken too seriously for the following reason. With a uniform model, MaxEnt reduces structures where the signal-to-noise ratio is poor, which is the case below μ due to the exponential decay of the Fermi function. The same argument leads also to a slight shift of structures in the direction in which the kernel of the transformation increases. In the present case we therefore expect that structures below μ are actually somewhat lower in energies. This effect is, however, accounted for in the error bars given by MaxEnt. The temperature dependence of exchange splitting $\Delta E_{ex}(T)/\Delta E_{ex}(0)$ for the Z_2 band in nickel is given in fig.3. The zero temperature value is estimated by extrapolation as $\Delta E_{ex}(0) \approx 0.28eV$. The data follow nicely the rescaled experimental bulk magnetization curve [19] which yields strong support for a Stoner-like band behavior. The extrapolated ground-state exchange splitting of the magnetic Z_2-band is 0.28 ± 0.05 eV. Similar values, ranging from 0.17 to 0.33 eV, have been reported for occupied d-bands in Ni [20, 21].

In conclusion, we have shown that the maximum entropy method gives spin-dependent quasiparticle spectral densities from IPE data. In the present case the resolution is improved by a factor of 5 and structures below E_F, which are generally lost in inverse photoemission, are recovered. This is important for the study of electronic structures in general and high temperature superconductors in particular, where the detailed behavior of quasiparticle energies and lifetimes is important for theoretical understanding.

We found that the quasiparticle spectral density of Ni consists of only one peak per spin direction for all temperatures. The exchange splitting ΔE_{ex} decreases with increasing temperature and vanishes at T_C. Hence, it appears that the influence of transverse spin fluctuations is negligible for the electronic bands in Ni in the energy regime under consideration. The Maximum Entropy concept is clearly very useful to deconvolve experimental data, and can be applied immediately to other types of spectroscopy.

References

[1] "Maximum Entropy in Action", ed. B. Buck and V.A. Macaulay, *Oxford Science Publications, Oxford*, 1990.

[2] V. Korenman and R. E. Prange, "Local-band theory analysis of spin-polarized photoemission spectroscopy", *Phys. Rev. Lett.*, **53**, 186, 1984.

[3] V. Korenman, "The local band theory", in *Metallic Magnetism*, H. Capellmann, ed., p. 109, Springer, Berlin, 1987.

[4] D. M. Edwards, "Itinerant magnetism", *J. Magn. Mat.*, **45**, 151, 1984.

[5] W. Borgiel, W. Nolting, and M. Donath, "On the temperature dependent exchange-splitting in the quasi-particle band structure of Ni", *Solid State Commun.*, **72**, 825, 1989; J. Braun, G. Borstel, and W. Nolting, *Phys. Rev.*, B46, 3510, 1992.

[6] J. Braun, G. Borstel, and W. Nolting, "Theory of temperature-dependent photoemission in 3D-band ferromagnetism", *Phys. Rev. B*, **46**, 3510, 1992.

[7] H. Gollisch and R. Feder, "Temperature-dependent spin resolved photoemission from Ni", *Solid State Commun.*, **76**, 237, 1990.

[8] W. von der Linden and D.M. Edwards, "Ferromagnetism in the Hubbard model", *J.Phys.* **C3**, 4917, 1991.

[9] E. Kisker, K. Schröder, M. Campagna, and W. Gudat, "Spin polarized angle-resolving photoemission of the electronic structure of Fe", *Phys. Rev. Lett.*, **52**, 2285, 1984.

[10] J. Kirschner, M. Globl, V. Dose, and H. Scheidt, "Wave-vector-dependent temperature behavior of empty bands in ferromagnetic iron", *Phys. Rev. Lett.*, **53**, 612, 1984.

[11] D. E. Eastman, F. J. Himpsel, and J. A. Knapp, "Experimental bandstructure and temperature-dependent magnetic exchange-splitting of Ni", *Phys. Rev. Lett.*, **40**, 1514, 1978.

[12] C. J. Maetz, U. Gerhardt, E. Dietz, A. Ziegler, and R. J. Zelitto, "Evidence for short-range magnetic order in Ni above T_C", *Phys. Rev. Lett.*, **48**, 1686, 1982.

[13] H. Hopster, R. Raue, G. Guntherodt, E. Kisker, R. Clauberg, and M. Campagna, "Temperature-dependence of the exchange-splitting in Ni", *Phys. Rev. Lett.*, **51**, 829, 1983.

[14] K. P. Kamper, W. Schmitt, and G. Gunterhodt, "Temperature and wave-vector dependence of the spin-split band-structure of Ni", *Phys.. Rev. B*, **42**, 10696, 1990.

[15] M. Donath and V. Dose, "Temperature behavior of a magnetic band in Ni", *Europhys. Lett.*, **9**, 821, 1989.

[16] P. W. Anderson, "The Reverend Thomas Bayes, needles in haystacks, and the fifth force", *Physics Today*, **45**, 9, 1992.

[17] W. von der Linden, M. Donath, and V. Dose, "Unbiased access to exchange splitting of magnetic bands using the maximum entropy method", *Phys. Rev. Lett.*, **71**, 899, 1993.

[18] J. Skilling, "Quantified maximum entropy" in Maximum Entropy and Bayesian Methods, ed. P. F. Fougere, Kluwer Academic Publishers, 1990.

[19] P. Weiss and R. Forrer, "Ferromagnetism", *Ann. Phys.*, **5**, 153, 1926.

[20] F.J. Himpsel, J.A. Knapp, and D.E. Eastman, Experimental energy-band dispersion and exchange-splitting in Ni", *Phys.Rev.*, **B19**, 2919, 1979;

[21] P. Heimann, F. J. Himpsel, and D. E. Eastman, "Experimental energy-bands, exchange-splitting, and lifetime for Ni", *Solid State Commun.*, **39**, 219, 1981.

AN ENTROPY ESTIMATOR ALGORITHM AND TELECOMMUNICATIONS APPLICATIONS

Nina T. Plotkin
Electrical Engineering Department
U.C. Berkeley

Abraham J. Wyner
Statistics Department
Stanford University

ABSTRACT. In this paper we present a novel technique for calculating the entropy of a dataset. We then apply this technique to measure the entropy of traffic streams in high-speed telecommunications networks. The entropy estimation technique is motivated by the Lempel-Ziv universal data compression algorithm, which is well known to asymptotically compress sequences to their entropy. However an LZ based "compression ratio" test has no characterizable statistical signifigance. We utilize a string matching technique, based on the LZ algorithm, but modified to estimate entropy rather than to compress data. This technique provides an estimate in a framework where statistical analysis is possible, and retains the universal properties of Lempel-Ziv. Such a tool is useful because the traffic streams in high-speed networks have distributional properties that are analytically intractable. We represent a network path by $G/D/1/\infty$-FCFS queues connected in series, and use the entropy estimator to calculate the entropy of the queue output processes. We consider queues with two inputs and two outputs, and we examine the entropy of a single output class. Our results show that the entropy can either increase or decrease depending upon the type of input traffic. We show that even with large amounts of data it is not possible to confirm the hypothesis that the entropy converges as the number of queues grows.

1. Introduction

High-speed telecommunications networks will support voice, video, data, fax and other traffic simultaneously. Since resources are shared in these networks, heterogeneous traffic streams will be merged and split apart as they traverse various devices inside the network. In order for the network to guarantee a specific level of performance, the network must understand the nature of the traffic traversing its devices. We use the term *through traffic* to refer to a traffic stream which travels through the network from a specific source to a specific destination. The expression *cross traffic* refers to other traffic streams which may have different sources and destinations. As a through traffic stream travels across the network, it traverses multiple network buffering devices such as switches and multiplexers. The cross traffic will interact with the through traffic *temporarily*, that is at a few devices and then continue on a path distinct from the path of the through traffic. See Figure 1. Thus both the network device and the cross traffic will affect the statistical nature of the through traffic.

A connection from one end of the network to the other, usually consists of a number of

G. R. Heidbreder (ed.), Maximum Entropy and Bayesian Methods, 351–363.

these network buffering devices in tandem. In order to understand the nature of through traffic as it traverses multiple network devices, we study the behavior of traffic in a tandem queueing system. We are thus interested in properties of queue departure processes. Typically when one uses analytical methods to model end-to-end network paths, one is forced to use a model containing a single queue, due to the intractable nature of most tandem queueing systems. Aside from a few exceptions, including for example quasi-reversible queues, it is not known how to find the distribution of a queue output process. Therefore our main method of study is simulation.

Upon requesting a connection through a network, a user must describe the traffic it will send with a few simple descriptors. The network then uses this information to setup a connection and to allocate sufficient resources (eg: bandwidth) to that connection. The user's description is of its traffic at the *input* to the network, and does not reflect the statistical nature of the traffic *inside* the network, once it has been mixed with other heterogeneous traffic streams. Common parameters to describe traffic are average, variance and peak size of clusters of cells. The interaction of the heterogeneous traffic traversing network devices causes cell scattering and clustering. In order to describe this scattering and clustering phenomenon, we propose the usage of entropy as a new traffic descriptor for high-speed networks. In this work we examine the entropy of successive departure processes.

Since we do not know the distributions of the queue departure processes, we cannot compute the entropy analytically. Instead, we use an entropy estimation technique that is motivated by the Lempel-Ziv (LZ) universal data compression algorithm, which is well known to asymptotically compress sequences to their entropy. However, an LZ based "compression ratio" has few known statistical properties; and those that are known indicate that the compression ratio converges to the entropy only very slowly [7]. In [7], A. J. Wyner develops a new entropy estimation method which avoids these problems. The entropy estimator utilizes a string matching technique, based on the LZ algorithm, but modifies the LZ method to estimate entropy rather than to compress data. This technique provides a more efficient estimate, in a framework where statistical analysis is possible, and retains the universal properties of Lempel-Ziv.

The high-speed Asynchronous Transfer Mode (ATM) network has received considerable attention in the last few years and is the expected network of the future. We study a queueing system which is useful for modeling ATM networks. The ATM network carries small fixed sized packets (53 bytes in length), usually called *cells*, hence the transmission (service) time is deterministic. Therefore we use $G/D/1/\infty$-FCFS queues to represent network devices. These queues can represent either ATM statistical multiplexers or non-blocking output-buffered ATM switches. Each queue has two input classes. Input class 1 carries through traffic and input class 2 carries cross traffic. Each cross traffic stream enters a particular queue and departs the system after service by that queue. In the case of simultaneous arrivals we use a discipline which always serves the cell from input class 1 first. The system we consider, with Q queues in tandem, is depicted in Figure 1. To capture the heterogeneous traffic, we consider traffic streams with varying amounts of correlation. We calculate the entropy of the through traffic at each of the queue outputs.

In Section 2., we present the basic traffic models and discuss the scattering and clustering phenomenon. The entropy estimator is given in section 3. and a description of its accuracy is given in subsection 3.1.. The implementation of the estimator algorithm is discussed

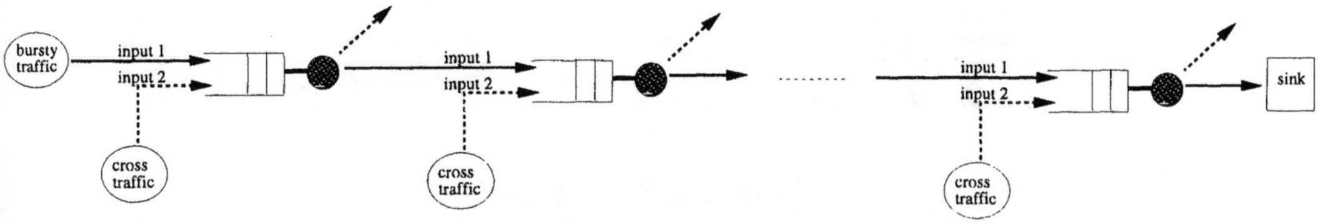

Figure 1: Tandem Queueing System

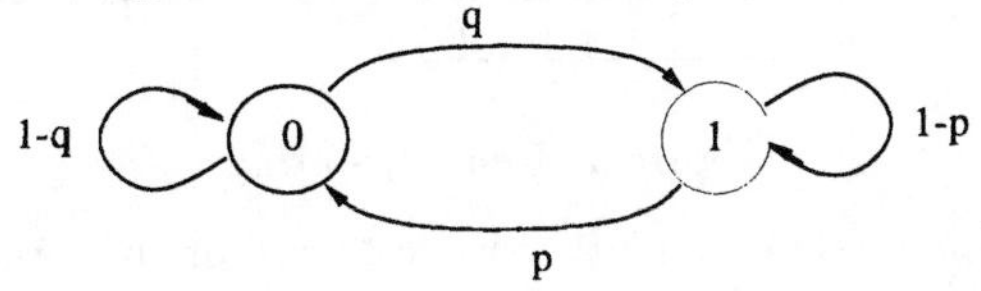

Figure 2: On-Off Traffic Model

in subsection 3.2. We designed a simulator, written in the C programming language, to simulate multiple queues and to carry out the estimator algorithm. The simulation results are given in section 4.

2. Traffic

We consider both an "on-off" bursty traffic generator and and a Bernoulli traffic generator. The former is a two-state discrete time Markov Chain, as depicted in Figure 2. In state 1 we generate a cell and in state 0 we generate an empty slot. A cluster of back-to-back cells, with no spacing between them is called a *burst*. Similarly a sequence of back-to-back empty slots is called an *idle* period. This on-off model generates an alternating burst/idle renewal process. Let B be the random variable which denotes the length of a burst, and I be the random variable which denotes the length of an idle period. B and I are geometric random variables with parameters p and q respectively; $E(B) = \frac{1}{p}$ and $E(I) = \frac{1}{q}$. Bursty traffic streams generated this way exhibit correlations over successive slots. We generate traffic streams with varying amounts of correlation by adjusting p and q. Note that the correlations here are only among the cells of a given burst; there is no correlation between different bursts. In a Bernoulli traffic stream, the probability that a slot contains a cell is p and the probability that a slot is empty is $1-p$. The Bernoulli traffic is a special case of the on-off model when $q = 1 - p$.

Because we consider a system with fixed sized time slots which may or may not contain a cell, a traffic stream can be represented by a binary sequence of 0's and 1's, where a '1' indicates the presence of a cell and a '0' indicates an empty slot. This observation provides us with a representation of traffic streams for which it is then possible to apply entropy estimation techniques. Define the random variable X_i as

$$X_i = \begin{cases} 1 & \exists \text{ cell in slot } i \\ 0 & \text{no cell is slot } i \end{cases}$$

Thus $\{X_i\}_0^\infty$ denotes the stochastic traffic process.

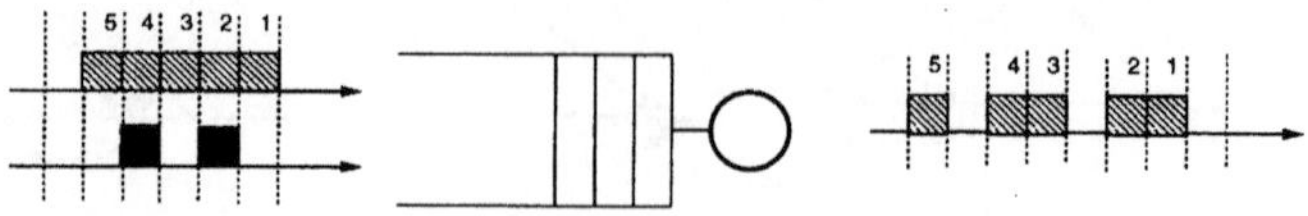

Figure 3: Cell Scattering

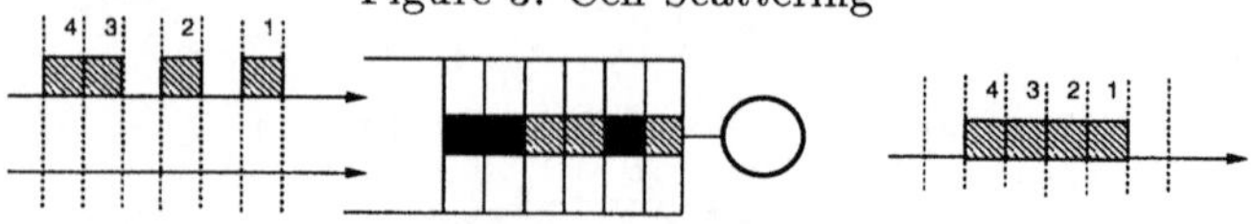

Figure 4: Cell Clustering

The cells in the through traffic can either be scattered or clustered as they traverse each queue. We say that a group of cells become more *scattered* (*clustered*) when the distances, measured in time slots, between individual cells increase (decrease), respectively. Examples of cell scattering and clustering are given in Figures 3 and 4. The figures depict snapshots in time. The input streams and queue depict the state of the system at some time t. The output streams depict the through traffic at a later time, after all the cells in the example have been served. The cells in the through traffic are numbered, at both the input and the output, to show the change in their relative positions. In the cell scattering example, the queue is empty immediately before the arrival of cell #1. Since the cross traffic cells leave the system after service, they effectively create *holes* in the original input stream, scattering the initial group of back-to-back cells. In the cell clustering example, the queue has 7 cells in it immediately before the arrival of cell #1. During the next 6 time slots, cells numbered #1-#4 will arrive and be positioned back to back in the queue. After they have all been served they will be clustered together, with no holes between them.

The entropy descriptor captures the behavior of cell scattering and clustering by detecting frequently occurring patterns, in this case binary patterns. The entropy descriptor is one dimensional, which makes it a feasible descriptor to use in high-speed networks. Certainly the entropy measure is only a partial traffic descriptor since different underlying traffic distributions can generate the same entropy. [1] We discuss the entropy of traffic streams more formally in the next section.

3. Entropy

The entropy of a Bernoulli traffic source with load p is given by the familiar equation

$$H(p) = -p\log p - (1-p)\log(1-p) \tag{1}$$

since such a traffic stream is equivalent to coin tossing. Recall that for a stationary Markov chain, $\{X_i\}$, with stationary distribution μ and transition matrix P, the entropy rate [2] is

$$H(X_2/X_1) = -\sum_{ij} \mu_i P_{ij} \log P_{ij} \tag{2}$$

[1] The most informative traffic descriptor would be the entire distribution function, but this infinite dimensional quantity is unlikely to be available even to an approximation.

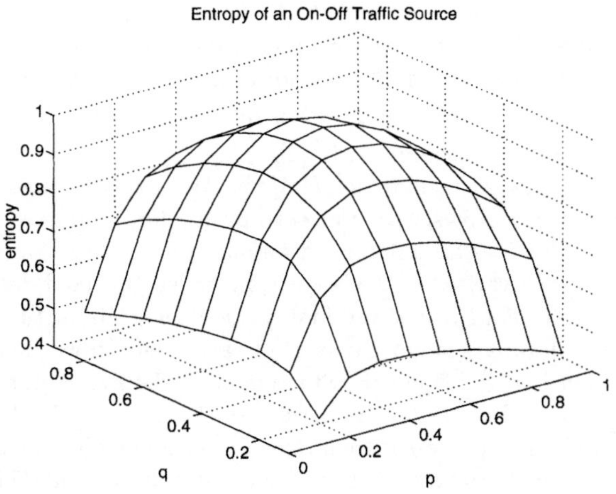

Figure 5: Entropy of an On-Off Markov Traffic Model

Using this equation (2) the entropy of our bursty on-off traffic source is

$$\frac{p}{p+q}H(q) + \frac{q}{p+q}H(p) \tag{3}$$

A plot of the entropy of this traffic process for various p and q is given in Figure 5. Although we can find the entropy of the input traffic processes analytically, recall we do not know the distribution of the underlying stochastic process $\{X_i\}$ at queue outputs. Therefore we estimate the entropy of these processes from empirical data produced in simulations, using the following entropy estimation method.

The LZ universal data compression algorithm [2] can compress a sequence derived from any finite alphabet. We consider binary alphabets, $\Theta = \{0,1\}$, and thus represent each sample traffic stream as a sequence of 0's and 1's. The *compression ratio* is given by the ratio of the length of the compressed string over the length of the original string. Ziv proved that as the length of the data string goes to infinity, the compression ratio approaches the entropy. One might then consider finding the entropy of a dataset (or a traffic stream in this case) by compressing the dataset and observing the compression ratio. We now discuss the problems with this and related methods in order to explain why we chose the method used here.

The basic LZ data compression algorithm is as follows. An input sequence is processed sequentially starting from the first bit. The sequence is parsed into strings that have not yet been seen, and a comma is placed after every new string. For example, the string 01000111011 would be parsed as 0,1,00,01,11,011. After every comma, we look along the sequence until we come to the shortest string that has not been marked off before. Since this is the shortest such string, all its prefixes must have occurred earlier. Actually the string consisting of all but the last bit of this string must have occurred earlier. Each sequence between two commas is called a phrase. Each phrase can be coded by a tuple (a,b) where

[2] LZ compression has become the Unix standard for file compression.

'a' indicates the location of the prefix and 'b' $\in \{0,1\}$ indicates the value of the last bit. Let $c(m)$ be the number of phrases after an input sequence of length m has been parsed. Then $\log c(m)$ bits are needed to describe the location of the prefix and 1 bit is needed to describe the last bit. Therefore the compression ratio is given by $\frac{c(m)(\log c(m)+1)}{m}$. A proof that $\frac{c(m)(\log c(m)+1)}{m} \rightarrow H(\Theta)$ can be found in [2].

We call the set of strings seen so far the *database.* With this LZ algorithm the database is not bounded, but rather grows with the entropy of the input sequence. Clearly any program which potentially uses an unbounded amount of memory cannot be implemented on a computer. Therefore a variant of this algorithm, called the fixed database Lempel-Ziv, FD-LZ, algorithm was developed. Let the first n, $(n < m)$, bits of the input sequence be the database. Now the database is not parsed, but rather any subsequence in the database starting at any position is a valid phrase. Starting with the first input bit, we look along the input sequence until we find the longest string which also exists in the database. This is called the *longest match.* The process is then repeated starting with the next sequential bit, not in the previous match. For example, consider the database 01000111011 and the input sequence 00101011011. The sequence of matches found is 001,010,11011. Now each match is coded into tuples (a,b) where 'a' denotes the position in the database of the first bit in the match, and 'b' denotes the length of the match. The code for this last example would be (4,3),(1,3),(7,5). It has been proved in [6] that FD-LZ compression also asymptotically approaches the entropy.

The "compression ratio" of the FD-LZ algorithm has almost no known statistical properties, and thus cannot be used in statistical analysis. The one known property states that the compression ratio yields $H * (1 + \frac{\log\log m}{\log m})$ [7]. In other words, a FD-LZ compression ratio test, on an input sequence of size m, would have an error term of $\frac{\log\log m}{\log m}$. For $m = 1,000,000$, this yields an error term of approximately 1/4. Since for a binary dataset, $0 \leq H \leq 1$, this is unacceptably large.

From a recent result of A. J. Wyner in [7], one can develop an entropy estimator that uses a fixed size database and that has a small error term. Consider the random process $X^{\infty}_{\{-n+1\}}$, where $X_i \in \{0,1\}$. For any fixed positive n, let $D_n = X^{0}_{\{-n+1\}}$ be the database. We look along the input sequence, $X_1, X_2, \ldots$ sequentially, looking for the longest match in the database. Let L_i denote the ith longest match found. The result states

$$\left| E(L_i) - \frac{log n}{H} \right| \leq O(1) \tag{4}$$

where n denotes the database size. This implies that we can calculate the entropy of a dataset by computing the longest matches. It is required that the database be formed from the first n elements of the random process, because the theorem only holds if the distribution of the database is the same as the distribution of the input sequence (other theorems apply in the general case, see [7]). In addition, the result requires that the $\{X_i\}$ process be stationary, ergodic and Markov. Within this large class of distributions (larger still considering that Markov processes can approximate general stationary ergodic sources) the theorem is universally applicable. Although the error term, $\frac{O(1)}{\log n}$, can be reduced by increasing the size of the database, we cannot increase n arbitrarily. We are constrained, as we will see in §3.2., in our choice of n by the memory required and the running time of the estimator algorithm. At the time of this writing, nothing is known about the $O(1)$ error constant; however based on experience as described in the following section, it is believed to be very small.

3.1. Accuracy of the Entropy Estimator

To examine the accuracy of this entropy estimator, we observe its performance on two processes, namely the Bernoulli and the on-off traffic processes, whose entropies are known analytically (equations (1) and (3), respectively). First we empirically generated a Bernoulli traffic stream with parameter p. In Figure 6, we plot both the estimated entropy ($\hat{H}$) and the true entropy (H) versus p. Since computer random number generators are not perfect, we plotted $\hat{p}$ on the abscissa, which is the actual coin bias produced by the random number generator. This figure shows that the error is largest for the maximum entropy case, $p = 0.5$. For $0.2 < p < 0.8$, the entropy is biased on the low side, and for $p < 0.2$ and $p > 0.8$, the estimator is slightly biased on the high side.

For the datum in Figure 6 a database of 20,000 bits and an input of 30,000 bits was used. For a given process, the error is a function of the database size n, the input size m and the process parameter(s). We must choose m and n in such a way as to minimize the error in our estimate. We need m large enough to generate a sufficient amount of data to achieve a 98% confidence interval in our estimate. (Note that rather than generate k realizations of traffic streams, each of length m, we can generate one traffic stream of length km since the process is ergodic.) We also want n large because, as we will see, our estimator has an inherent bias, which can be partially reduced by increasing n. Intuitively, this occurs because the database must contain sufficiently many bit patterns to represent the true process entropy and generate a meaningful $E(L_i)$. However we also want n and m small enough so that the algorithm will run in minimal time and utilize minimal space.

Let $\hat{H}(n,m)$ denote the random variable which gives an entropy estimate using a database of size n and an input of size m. Then $\hat{H}_i(n,m)$ denotes a single realization of the random variable. Let $\bar{H}(n,m) = \frac{1}{k}\sum_{i=1}^{k} \hat{H}_i(n,m)$ denote the sample mean. The root mean squared error (RMSE) is given by

$$\sqrt{\frac{1}{k-1}\sum_{i=1}^{k}(\hat{H}_i(n,m) - \bar{H}(n,m))^2}$$

We choose the smallest n and m which are large enough to minimize the RMSE for the Bernoulli process which generated the worst case error, namely $p = 0.5$. With $n = 20,000$ and $km = 30,000$ the $RMSE = 0.001$. Increasing either n or km did not reduce the RMSE any further. For this n and km we obtained $H - \bar{H}(n,m) = 0.021$, which indicates that the estimator itself has an inherent bias.

We performed the same test for on-off traffic streams. The error now is a function of m, n, p and q. For small p and q (we consider $p < 0.1$ and $q < 0.1$ since our bursts sizes are typically larger than 10), the errors were slightly larger than in the case of Bernoulli traffic. In order to bring the RMSE down to 0.001 we needed to increase km to $200,000$. In this case the inherent bias (for worst case p and q) was 0.027.

We should not be confounded by this bias for several reasons. First, we are estimating entropy using a universal algorithm that clearly cannot compete with a variety of simpler techniques that exploit *known* properties of the distribution. Secondly, the worst case error occured in the Bernoulli process with $p = .5$. Since it is known that this distribution is

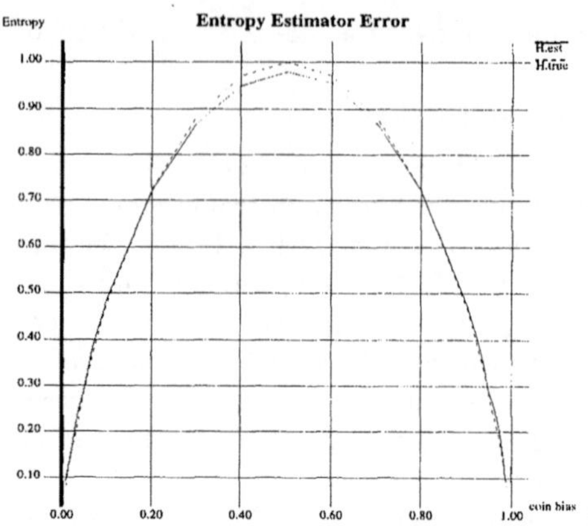

Figure 6: Entropy Estimator Error for Coin Tossing

pathological, in the sense that the distributions of the longest match converge weakly to a different limit than the limit for any other stochastic source [7], we have some hope to believe that the worst bias occurs in this example as well. In point of fact, the worst case bias is still quite reasonable.

We chose to use $n = 20,000$ and $m = 200,000$ in all our simulation runs. For a $RMSE = 0.001$ we can be 98% confident that the true mean of $\hat{H}(m,n)$ lies within ± 0.0022 of the sample mean. Even though the traffic inside the virtual circuit is neither Bernoulli nor an on-off stream, these tests give us some confidence that the error in our entropy estimator will be small.

3.2. Implementation of the Entropy Estimator

The implementation of the entropy estimator requires careful consideration of the tradeoffs between memory and speed. A brute force implementation of the estimator would require only n bits to hold the database, but would have a running time of $O(nm)$ comparison operations. We sometimes process hundreds of queues and are thus more concerned with speed than memory; however we cannot use so much memory that the speed is reduced due to memory swapping.

Our algorithm has two stages. In the first stage the database is preprocessed to form a binary tree. In the second stage we find the longest matches. The tree is constructed such that all possible subsequences of strings in the database can be found by tracing a path which starts at the root. Each subsequence has a unique path in the tree. Traversing the left (right) edge from a node at a depth k indicates a '0' ('1') in the kth position of a string, respectively. To build the tree we start at the first bit in the database, and build the tree for all subsequences starting with that bit, of which there are n. The process is repeated starting at the second bit (for which there are $n-1$ subsequences), and so on. If a particular subsequence already exists in the tree, it is not added again. The total number of subsequences in the database is $\sum_{i=1}^{n} i = \frac{n(n-1)}{2}$, which is therefore an upper bound on the total number of *unique* subsequences. Because we have a binary alphabet, and exactly two edges emanating from each node, we do not need to label the edges in the tree. The amount of memory required to store the tree is $O(\frac{n^2}{2})$ node units, and the running time of this stage is also $O(\frac{n^2}{2})$. A node unit requires 8 bytes to hold 2 pointers.

We find the longest matches of the input sequence in the database as follows. The input sequence index starts at X_0. The tree index starts at the root. If the next bit in the input sequence is a '0' ('1'), then traverse the left (right) edge of the current tree node. Repeat this process until a bit in the input sequence is encountered for which no corresponding edge exists in the tree. Then the current depth of the tree gives the length of the match. In this algorithm the string matching time is linear in the size of the input sequence; i.e. it uses $O(m)$ comparison operations.

We are motivated to use large databases in order to reduce the error term. With the above method a database of size 3000 can require up to 4.5×10^6 nodes units, or 36 Mbytes of storage just for the matching tree. We cannot use much more memory than this because then even a good Sparc station with 32M RAM will spend most of its time swapping to disk, which drastically reduces the speed of the simulator. In order to increase the database size by an order of magnitude, and not increase the amount of memory needed, we modified the algorithm as follows. Our observations showed that typical match lengths range from 2 to 200. In the modified algorithm we do not allow the tree to grow beyond a depth of 250. In the rare event that there is a match of length greater than 250, we do a brute force search in the database for that longest match. Such matches occur typically less than 0.5% of the time. With this method the time to construct the tree is now $O(n)$, where the constant hidden in the big-O notation is 250. This method allows us to use a database of size 20,000, which requires at most 40 Mbytes of storage. Increasing the database from 3000 to 20,000 cut the bias by 30%. In summary, our algorithm has a running time of $O(n + m)$ where n is the size of the database and m is the size of the input sequence.

Some implementations of data compression algorithms are discussed in [3, 4, 5]. Our algorithm is slower and uses more memory than the compression algorithms used in practice because we compute entropy rather than compress data. This necessitates a database size on the order of 10^5. The type of algorithms used in practice are in the FD-LZ class and many implementations use a database size as small as 12. In many computer systems, a database for compression and decompression is maintained as part of the system (i.e. it exists before compression starts). Recall that our method requires that the database have the same distribution as the input sequence, and thus cannot use a pre-existing database. Instead we have to generate a new database and preprocess it for each traffic stream.

4. Simulation Results

4.1. Basic Properties

Recall that the *through* traffic corresponds to the traffic on input stream 1 and departure class 1. The *cross* traffic corresponds to the traffic on input stream 2 and departure class 2. Since we are only interested in the properties of the through traffic we use the term *input* to denote input stream 1 and the term *output* to denote departure class 1. We can view the cross traffic as part of the queuing system, and together they transform the input into the output.

Intuitively, the cross traffic is the predominant cause of cell scattering, which leads to entropy increases, and the queue is the predominant cause of cell clustering, which leads to entropy reduction. Which of the cell scattering or cell clustering effects is predominant

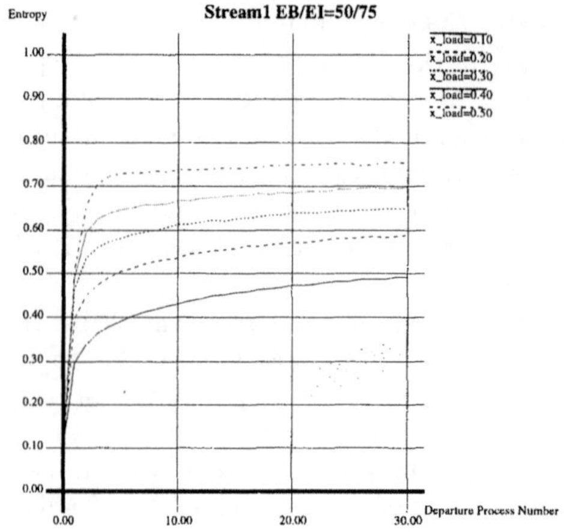

Figure 7:

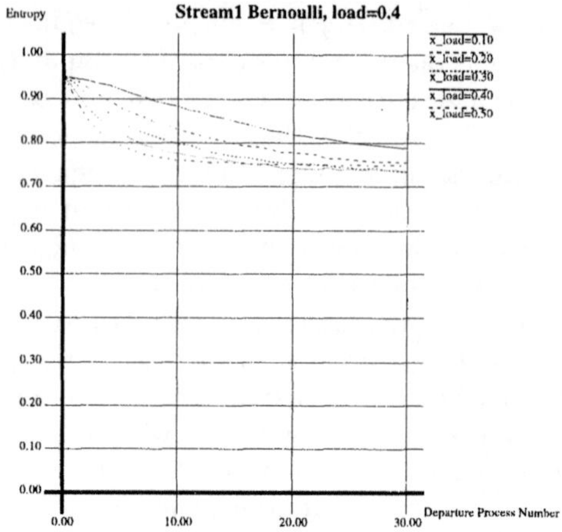

Figure 8:

will depend upon the traffic types and the loading conditions. We examine these effects, in terms of entropy, in the figures below.

The notation $EB/EI = x/y$ indicates the average burst size and average idle period size, which specify the parameters in the on-off input model. The load for the on-off model is thus given by $\frac{EB}{EB+EI}$. The notation x_load indicates that the cross traffic is a Bernoulli process with parameter equal to x_load. In Figures 7 and 8 we consider Bernoulli cross traffic and in Figures 9 and 10 we consider correlated cross traffic. The abscissa label "Departure Process Number" i correspondes to the traffic process at the output of queue i. (Departure process number 0 specifies the input to the whole system.)

In Figure 7 we consider correlated input traffic with 0.4 load, and we vary the load of the Bernoulli cross traffic. As expected, the entropy of the output is greater when the cross traffic load is larger. We see that the entropy grows rapidly during approximately the first six queues, and from there on grows slowly. In Figure 8 we consider Bernoulli inputs on stream 1 as well as Bernoulli cross traffic. In this case the entropy of the output decreases. It is reasonable that under these traffic conditions, cell clustering is more predominant than cell scattering since the i.i.d inputs are completely random to begin with. The entropy is reduced more quickly at higher cross traffic loads, although not significantly. Notice that these first two graphs together imply that the entropy of a single departure class can either increase or decrease depending upon the input types.

In Figure 9 we consider correlated traffic on both the through and cross traffic streams. Here the total load and cross traffic load, 0.8 and 0.4 respectively, are fixed. The cross traffic streams represent streams of increasing correlation. The entropy remains rather low (less than 0.5 for example), even after 20 queues, when both traffic streams have average initial burst sizes greater than 40. We can compare this figure to the $x\_load = 0.4$ case in Figure 7. The entropy of the through traffic grows substantially slower when the cross traffic is correlated (Figure 9) than when the cross traffic is Bernoulli (Figure 7). We saw in Figure 7 that the higher cross traffic load, the higher the entropy of the output. However in comparing Figures 7 and 9 we see that a lightly loaded Bernoulli cross traffic

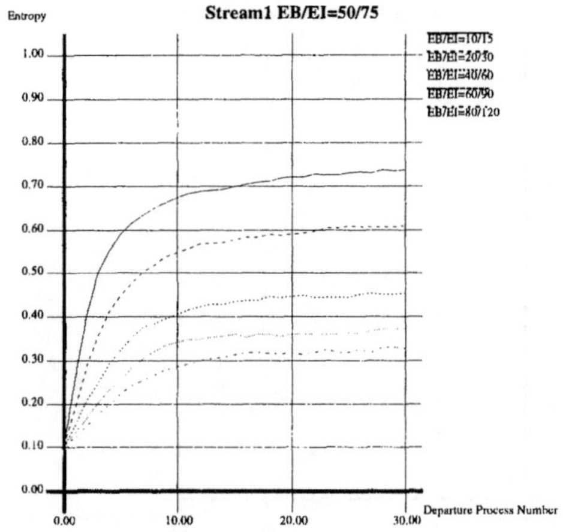

Figure 9:

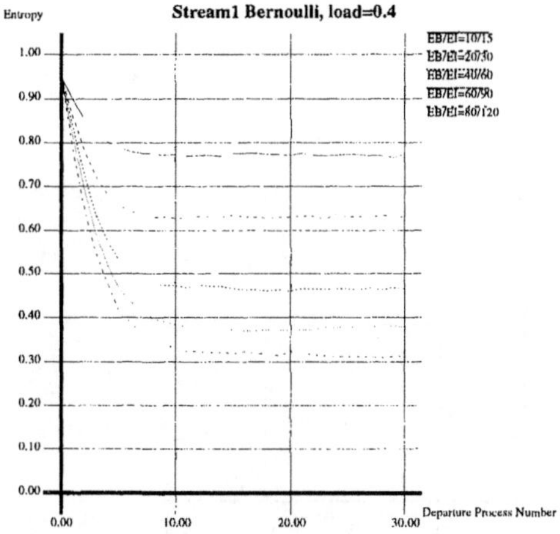

Figure 10:

stream increases the entropy of the ouput more than a highly loaded correlated cross traffic stream. For example, consider $x\_load = 0.2$ in Figure 7 and $EB \geq 60$ in Figure 9.

Figure 10 demonstrates the entropy behavior for Bernoulli through traffic and correlated cross traffic. Here again the total (cross) load remains the same 0.8 (0.4). We vary the amount of correlation. In comparing Figures 8 and 10, (i.e. Bernoulli inputs) we see that the output entropy is sensitive to increasing amounts of correlation, but insensitive to increasing Bernoulli traffic loads.

4.2. Convergence

Next we consider the asymptotic behavior of the entropy of the through traffic as the number of queues grows. We consider only Bernoulli cross traffic in this section. Let $\{X_i\}^j$, $i = 0, 1, \ldots$, denote the input to a queue $j + 1$, and the output process of queue j. We can view the behavior of the queue and cross traffic as applying some function $\mathcal{F}$ which is an input-output mapping and has parameter p (cross traffic load). In other words $P(\mathcal{X}^j) = \mathcal{F}(P(\mathcal{X}^{j-1}), p)$ where $P(\mathcal{X}^j)$ denotes the joint probability distribution of the $\{X_i\}^j$ process. Passing the input through many queues in tandem is equivalent to repeatedly applying this function to each successive output process, i.e. $P(\mathcal{X}^j) = \mathcal{F}^j(P(\mathcal{X}^0), p)$. We can hypothesize that as j grows, the traffic process might reach an invariant distribution, and hence the entropy would no longer change. A weaker hypothesis would be that just the entropy converges, without requiring that the distribution converge. We distinguish two separate convergence hypotheses. The first hypothesis is that the entropy of a traffic stream *converges* as the number of queues grows, for Bernoulli cross traffic. The second hypothesis is that *all* types of through traffic streams, at a specific load, converge to the *same* limit, for a given load of Bernoulli cross traffic.

In Figures 11 and 12 we examine the entropy through 2000 queues, calculating the entropy at the output of every 50th queue. The cross traffic load was 0.3 in Figure 11 and 0.4 in Figure 12. Both Bernoulli and correlated through traffic were considered in each case. Although the entropy values appear to be leveling off, a close examination of the data reveals a very slow downward trend. This means that if the entropy converges it does

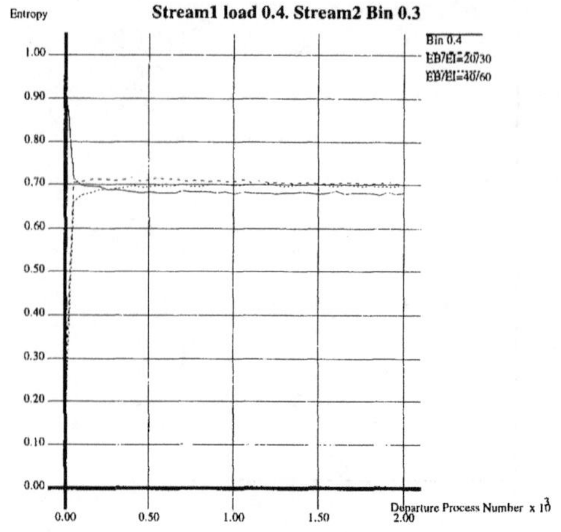

Figure 11:

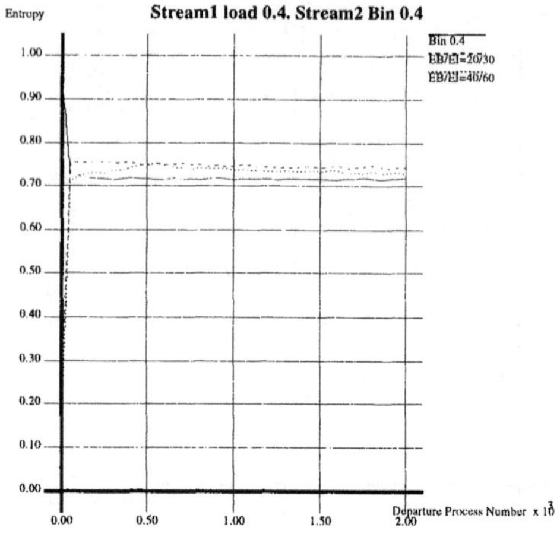

Figure 12:

so extremely slowly (not even by the 2000th queue). Both these figures exhibit *bunching* behavior; i.e. regardless of the initial input type, by the 100th queue, all the entropies are *close*, but not precisely the same. Although this bunching behavior intuitively supports the second hypothesis, the different convergent values cast doubt on this hypothesis. The data here is insufficient to validate these hypotheses. The difficulty in validation may be due to very slow convergence, in which case simulation is not a suitable method of study.

5. Summary

In this paper we first presented some basic trends in high-speed telecommunications networks today. We described the motivation for studying tandem queueing. We then presented two traffic models and discussed entropy as a traffic descriptor that can be used by network designers. We considered networks in which traffic streams can be represented by a binary sequence. We compute the entropy of sample traffic streams, generated via simulation, using a novel entropy estimation algorithm. The algorithm is a modified version of Lempel-Ziv data compression schemes designed to compute entropy rather than compress data. We describe an implementation of the algorithm whose running time is linear in the size of the data sequence.

We developed a fast simulator to simulate multiple queues in tandem, and looked at the entropy of successive queue departure processes. These departure processes represent internal network traffic. Our results show that the entropy of the queue output stream can either be larger or smaller than the entropy of the queue input stream, depending upon the types of input traffic. The amount of change in the entropy - from the input to the output - is greater at higher loads. We show that even with large amounts of data it is not possible to confirm the hypothesis that the entropy converges as the number of queues grows.

References

[1] T.C. Bell, J.G. Cleary, and I.H. Witten. *Text Compression*. Prentice Hall, 1990.

[2] T. Cover and J. Thomas. *Elements of Information Theory.* John Wiley & Sons, Inc., 1991.

[3] Edward McCreight. A Space-Economical Suffix Tree Construction Algorithm. *Journal of the ACM*, April 1976.

[4] M. Rodeh, V. Pratt, and S. Even. Linear Algorithm for Data Compression via String Matching. *Journal of the ACM*, January 1991.

[5] Terry Welch. A Technique for High-Performance Data Compression. *Computer*, 17, 1984.

[6] A. D. Wyner and J. Ziv. Fixed Data Base Version of the Lempel-Ziv Data Compression Algorithm. *IEEE Trans. Information Theory*, 1991.

[7] Abraham J. Wyner. *String Matching Theorems and Applications to Data Compression and Statistics.* PhD thesis, Stanford University, 1993.

A COMMON BAYESIAN APPROACH TO MULTIUSER DETECTION AND CHANNEL EQUALIZATION

Laurence Mailaender and Ronald A. Iltis *
Center for Information Processing Research
Department of Electrical and Computer Engineering
University of California
Santa Barbara, CA 93106
e-mail: iltis@ece.ucsb.edu

ABSTRACT. Multiuser detection is the problem of detecting the data sequence from several simultaneous code division multiple access (CDMA) users and cancelling the interference between users. Equalization of intersymbol interference (ISI) is the separate problem of data detection in the presence of interference generated by the channel's effect on the user's own data stream. A commonality in these two problems arises from the finite memory associated with each. Both problems can be solved by a maximum likelihood sequence estimation (MLSE) approach, for example by the Viterbi algorithm (VA). The VA is driven by likelihoods conditioned on every possible data sequence with length equal to the channel memory. As is well known, the VA incurs a probabilistic decoding delay, and is burdened with the need to maintain track of all survivor sequences through the trellis. An alternative "symbol-by-symbol" approach to this problem offers a fixed decoding delay and no need to maintain survivor sequences. This approach is based on a Bayesian recursion of posterior probabilities due to Abend and Fritchman (AF). In the AF algorithm, maximum a-posteriori (MAP) symbol decisions are made (based on the entire history of past measurements) at each symbol time. In this paper, we provide a common framework that can be applied to both the multiuser detection and ISI equalization problems, and explicitly develop the Bayesian recursions for each. We emphasize the distinction between the MAP symbol detection and MLSE approaches, and discuss the trade-offs in performance and complexity.

1 Introduction

The code division multiple access (CDMA) communication system, in which many users modulated with special "signature waveforms" share the same transmission bandwidth, is being considered for use in future radio-based data networks. A well-known limitation of the basic system is the self- interference, or near-far effect, in which excessive bit-error-rate (BER) degradation can occur due to reception of "strong" signals from other users. The recent interest in multiuser detection stems from the result that the ideal multiuser detector is free of any such near-far effect, and thus does away with the need for power control or other ad hoc remedies. The multiuser detector theoretically outperforms the conventional matched filter detector simply because the matched filter is optimal for the Gaussian channel, whereas the CDMA interference is non-Gaussian [1].

The form of the optimal multiuser detector is known [1] [2] [3], and consists of sampling the matched filter outputs and applying either a maximum-likelihood or maximum

*This work was supported in part by Rockwell International Co.

G. R. Heidbreder (ed.), Maximum Entropy and Bayesian Methods, 365–374.

a-posteriori criteria. Such implementations tend to be highly complex, with complexity exponential in the number of users. The optimal linear multiuser detector is also known [4], which offers satisfactory performance and greatly reduced complexity (linear in the number of users). Implementation of the optimal multiuser detector requires that the amplitudes and delays of each signal be known a priori, or jointly estimated. Our previous work describes a non-linear estimator/detector employing extended Kalman filter (EKF) estimation of amplitudes and time delays [5]. We have also described an importance sampling approach for simulating the bit error rates of such systems [6].

In this paper we give a discussion of the related topic of intersymbol interference (ISI) channel equalization. Examination of the ISI problem will reveal that it is closely related to multiuser detection, and in particular, that multiuser detection is simply ISI equalization with M-ary[1] composite symbols, and a "channel memory" of one (i.e. the current and previous composite symbols affect the current data vector). This close connection led us to consider solving the multiuser detection problem using the Abend and Fritchman [7] algorithm; a technique that was originally developed for the ISI problem. We develop here the Bayesian recursion in posterior metrics that has been used in our prior work, and demonstrate that this technique provides a common framework for solving both the multiuser detection and channel equalization problems. While the AF algorithm has not enjoyed the widespread popularity of the VA, in some applications the AF approach may be preferable. For example, metric pruning provides a natural way of reducing the complexity of the algorithm [8], and some applications may benefit from the fixed-delay property of the algorithm [9].

Both the multiuser detection and ISI channel equalization problems have been solved previously by a MLSE approach (i.e. by a Viterbi algorithm) [10] [11] [12]. We consider here the differences between the MAP symbol detection (AF) and MLSE techniques. The two algorithms give similar performance and the increased computations required for AF are not very significant given current technology.

2 Bayesian Equalizer

We consider the problem of optimal data detection over a channel containing intersymbol interference (ISI). ISI may result from a bandlimited channel, and/or multipath conditions. We employ the conventional discrete-time channel model established in [10]. The received samples at time k are given by

$$r(k) = \sum_{n=0}^{D-1} b(n)d(k-n) + z(k). \tag{1}$$

where D denotes the effective memory length of the channel in bit durations, $b(n)$ denotes the equivalent channel tap weights, $d(n)$ the transmitted binary data sequence, and $z(k)$ denotes a circular white Gaussian noise sample with

$$E\{z(k)z^*(j)\} = \sigma^2\delta_{k,j}. \tag{2}$$

The tap coefficients in the above model represent the combined impulse response due to baseband pulse shaping, channel transfer function, matched filter, and (if necessary)

[1] M, the alphabet size, is herein assumed to equal 2.

whitening filter. For our purposes here we will consider this to be a fixed, known sequence. Baseband samples are produced at the rate of $1/T_b$. Define the following sequences:

$$d^{k,D} \equiv [d(k), d(k-1), \ldots, d(k-D+2), d(k-D+1)], \tag{3}$$

$$d^k \equiv d^{k,k}, \tag{4}$$

$$r^k \equiv [r(k), r(k-1), \ldots, r(2), r(1)]. \tag{5}$$

The likelihood function for a single sample is given by

$$p(r(k)/d^{k,D}) = \frac{1}{\pi\sigma^2} \exp\left(-\frac{1}{\sigma^2}|r(k) - \sum_{l=0}^{D-1} b(l)d(k-l)|^2\right), \tag{6}$$

and by the i.i.d. property of the noise, the likelihood for a sequence of samples is

$$p(r^k/d^k) = \prod_{j=1}^{k} p(r(j)/d^{j,D}). \tag{7}$$

The optimal MAP symbol decision for symbol k is

$$\hat{d}_k = \arg \max_{d(k)} \; p(d(k)/r^{k+D-1}), \tag{8}$$

which utilizes the entire received sequence up to time $k+D-1$. This rule implicitly takes a "symbol-by-symbol" approach; a decision is made on each symbol $d(k)$ at time $k+D-1$. This is in contrast to the sequence detection strategy employed in the VA solution, to be described later.

We now proceed to determine a Bayesian recursion for the posterior probabilities, $p(d(k)/r^{k+D-1})$, according to [7]. The development closely follows [13]. First, applying Bayes' rule to (8),

$$\hat{d}_k = \arg \max_{d(k)} \; p(r^{k+D-1}/d(k))\frac{p(d(k))}{p(r^{k+D-1})}, \tag{9}$$

$$\hat{d}_k = \arg\max_{d(k)} \frac{1}{c} p(r^{k+D-1}/d(k))p(d(k)), \tag{10}$$

where (10) follows because the denominator does not depend on the argument $d(k)$ and is thus a normalization constant. This constant will be carried through the remaining analysis because the re-normalization will minimize numerical underflows in the resulting recursive formula. Note that if all symbols are equiprobable, the MAP detector is equivalent to a ML symbol detector.

Consider the detection of the first symbol ($k = 1$), which is written as

$$\hat{d}_1 = \arg\max_{d(1)} p(r^D/d(1))\frac{p(d(1))}{c_1}. \tag{11}$$

Note that the required likelihood is conditioned on a single symbol, and may be found from

$$p(r^D/d(1)) = \sum_{\forall j : d(1) \in d_j^{D,D}} p(r^D/d_j^{D,D})p(d_j^{D,D-1}). \tag{12}$$

The notation $\forall j : d(1) \in d_j^{D,D}$ denotes a sum over all sequences of length D whose first element is equal to $d(1)$. This notation will be used extensively throughout this paper, and is simply a more compact way of writing the sum that occurs in the following variation of the law of total probability

$$p(r^D/d(1)) = \sum_{d(D)} \sum_{d(D-1)} \cdots \sum_{d(2)} p(r^D/d(D), d(D-1), \ldots, d(2), d(1)) \cdot \tag{13}$$

$$p(d(D), \ldots, d(2), d(1)/d(1)). \tag{14}$$

Define the posterior metric at time $k = 1$ as

$$m_1(d^{D,D}) = \frac{1}{c_1} p(r^D/d_j^{D,D})p(d_j^{D,D}), \tag{15}$$

where we note there are M^D possible sequences, and thus M^D metrics at time $k = 1$. The decision for symbol 1 is

$$\hat{d}_1 = \arg\max_{d(1)} \sum_{\forall j : d(1) \in d_j^{D,D}} m_1(d_j^{D,D}). \tag{16}$$

Now consider the MAP symbol decision at time $k = 2$

$$\hat{d}_2 = \arg\max_{d(2)} p(d(2)/r^{1+D}) = \arg\max_{d(2)} p(r^{1+D}/d(2)) \frac{p(d(2))}{c_2}. \tag{17}$$

By total probability we get,

$$\hat{d}_2 = \arg\max_{d(2)} \sum_{\forall j : d(2) \in d_j^{1+D,D}} p(r^{1+D}/d_j^{1+D,D}) \frac{p(d_j^{1+D,D})}{c_2}, \tag{18}$$

or,

$$\hat{d}_2 = \arg\max_{d(2)} \sum_{\forall j : d(2) \in d_j^{1+D,D}} m_2(d_j^{1+D,D}). \tag{19}$$

Now consider the relationship between $m_2(d^{1+D,D})$ and $m_1(d^{D,D})$. By the i.i.d. property of the noise we may break up the likelihood expression,

$$m_2(d_j^{1+D,D}) = p(r(1+D)/d_j^{1+D,D})p(r^D/d_j^{1+D,D})p(d_j(1+D)) \frac{p(d_j^{D,D-1})}{c_2}, \tag{20}$$

and then, because r^D is independent of $d(1+D)$,

$$m_2(d_j^{1+D,D}) = p(r(1+D)/d_j^{1+D,D}) \frac{p(d_j(1+D))}{c_2} p(r^D/d_j^{D,D-1})p(d_j^{D,D-1}). \tag{21}$$

By total probability,

$$m_2(d_j^{1+D,D}) = p(r(1+D)/d_j^{1+D,D})c_1 \frac{p(d_j(1+D))}{c_2} \sum_{\forall i: d_i^{D,D} \in d_j^{1+D,D}} m_1(d_i^{D,D}), \tag{22}$$

and consolidating constants,

$$m_2(d_j^{1+D,D}) = \frac{1}{c} p(r(1+D)/d_j^{1+D,D}) \sum_{\forall i: d_i^{D,D} \in d_j^{1+D,D}} m_1(d_i^{D,D}). \tag{23}$$

or, in general,

$$m_{k+1}(d_j^{k+D,D}) = \frac{1}{c} p(r(k+D)/d_j^{k+D,D}) \sum_{\forall i: d_i^{k+D-1,D} \in d_j^{k+D,D}} m_k(d_i^{k+D-1,D}). \tag{24}$$

This last equation gives the recursion of Bayesian posterior probabilities. It states that the updated metric is obtained via a sum over the previous metric, multiplied by the likelihood of the newest sample. The normalization constant c ensures that $m_k(d^{k+D,k})$ sums to unity (i.e. is a valid probability density).

The recursive algorithm requires computation of M^D metrics at each step, or kM^D metrics for k symbols. We observe that if $p(d^k/r^{k+D})$ were calculated directly, it would require calculation of M^k metrics. Thus, the recursive algorithm is highly efficient, and implicitly collapses an exponentially growing sequence tree into a fixed-length trellis.

3 Multiuser Detection

The multiuser detection problem refers to the simultaneous detection of several users, each employing a unique signature waveform. The waveforms are traditionally pseudo-noise (PN) sequences, which are quasi-orthogonal (i.e. have low cross-correlation). However, since the cross-correlation is not zero, a large number of simultaneous users, or a single strong user could substantially degrade the BER performance of a traditional correlation-based detector. In contrast, the multiuser detector uses knowledge of all the signature sequences, including their amplitudes and time delays, to effectively cancel the interference.

The received signal is downconverted, low-pass filtered with bandwidth T_c, and sampled at the Nyquist rate. Samples are grouped into vectors of length N_s, where $N_sT_s = T_b$. The received samples due to N simultaneous users have the form

$$r(mN_s + k) = \sum_{l=0}^{1} \sum_{n=1}^{N} d_n(m-l) a_n s_n(kT_s + lT_b - T_n) + z(k) \tag{25}$$
$$for\ k = 0, 1, \ldots, N_s - 1,$$

where T_b refers to the bit period, T_s is the sampling period, $T_n \in [0, T_b]$ is the time delay of the n-th user, a_n is the (complex) amplitude of the n-th user, $d_n(k)$ is the binary data bit of the n-th user at time k, and $s_n(\cdot)$ denotes the output of an ideal low-pass (anti-aliasing) filter of bandwidth T_c whose input is the signature sequence of user n. The index over l indicates that two bits from each user may contribute to any given received vector (because the vector is not time synchronous with the bit transitions). The noise samples, $z(k)$, are circular Gaussian with

$$E\{z(k)z^*(j)\} = \sigma^2\delta_{k,j}. \tag{26}$$

For the purposes of this paper, we will treat a_n and T_n as fixed, known quantities. Previous work has established that it is also possible to use EKF's to estimate these parameters. We now consider (25) as a variation of equation (1) which specified the received samples under ISI. It is readily apparent that the multiuser waveform is equivalent to N simultaneous ISI signals of memory length 2 ($l = 0, 1$). The sum over $a_n s_n(\cdot)$ acts as a generalized $b(\cdot)$. If we use likelihoods conditioned on all possible length-2 sequences from N users (2^{2N} possibilities) then the problems are essentially equivalent.

We now develop a recursion for posterior probabilities as was done for the Bayesian equalizer. Define the sequences:

$$\mathbf{d}_i^{m,2} = [\mathbf{d}_i(m), \mathbf{d}_i(m-1)], \tag{27}$$

$$\mathbf{r}(m) = [r(mN_s), r(mN_s+1), \ldots, r((m+1)N_s - 1)], \tag{28}$$

$$\mathbf{r}^m = [\mathbf{r}(m), \mathbf{r}(m-1), \ldots, \mathbf{r}(1)]. \tag{29}$$

Note that $\mathbf{d}_i(m)$ is now an N-dimensional vector of binary data bits ("composite symbol"), and that $\mathbf{d}_i^{m,2}$ is a length-2 sequence of vectors. Likewise, $\mathbf{r}(m)$ denotes a vector of complex samples, and $\mathbf{r}^m$ a sequence of such vectors. The likelihood for a single sample is written,

$$p(r(mN_s+k)/\mathbf{d}^{m,2}) = \frac{1}{\pi\sigma^2}\exp(-\frac{1}{\sigma^2}|r(mN_s+k) - \sum_{l=0}^{1}\sum_{n=1}^{N} d_n(m-l)a_n s_n(kT_s + lT_b - T_n)|^2), \tag{30}$$

and the likelihood of a sequence can be written as a product of such terms, as in (7). The MAP symbol decision for $k = 1$ is

$$\hat{d}_1 = \arg\max_{d(1)} p(d(1)/\mathbf{r}^2) \tag{31}$$

$$\hat{d}_1 = \arg\max_{d(1)} \sum_{\forall j: d(1)\in\mathbf{d}_j(1)} p(\mathbf{d}_j(1)/\mathbf{r}^2) \tag{32}$$

$$\hat{d}_1 = \arg\max_{d(1)} \sum_{\forall j: d(1)\in\mathbf{d}_j^{1,2}} \sum_{\forall i: \mathbf{d}_j^{1,2}\in\mathbf{d}_i^{2,2}} p(\mathbf{d}_i^{2,2}/\mathbf{r}^2) \tag{33}$$

$$\hat{d}_1 = \arg\max_{d(1)} \sum_{\forall j: d(1)\in\mathbf{d}_j^{1,2}} \sum_{\forall i: \mathbf{d}_j^{1,2}\in\mathbf{d}_i^{2,2}} p(\mathbf{r}^2/\mathbf{d}_i^{2,2})\frac{p(\mathbf{d}_i^{2,2})}{c_1} \tag{34}$$

$$\hat{d}_1 = \arg\max_{d(1)} \sum_{\forall j: d(1)\in\mathbf{d}_j^{1,2}} \sum_{\forall i: \mathbf{d}_j^{1,2}\in\mathbf{d}_i^{2,2}} m_1(\mathbf{d}_i^{2,2}) \tag{35}$$

The MAP symbol decision at time k=2 is

$$\hat{d}_2 = \arg\max_{d(2)} p(d(2)/\mathbf{r}^3) \tag{36}$$

$$\hat{d}_2 = \arg\max_{d(2)} \sum_{\forall j:\mathbf{d}(2)\in\mathbf{d}_j^2} p(\mathbf{d}_j(2)/\mathbf{r}^3) \tag{37}$$

$$\hat{d}_2 = \arg\max_{d(2)} \sum_{\forall j:d(2)\in\mathbf{d}_j^2} \sum_{\forall i:\mathbf{d}_j^{2,2}\in\mathbf{d}_i^{3,2}} p(\mathbf{d}_i^{3,2}/\mathbf{r}^3) \tag{38}$$

$$p(\mathbf{d}_i^{3,2}/\mathbf{r}^3) = p(\mathbf{r}^3/\mathbf{d}_i^{3,2})\frac{p(\mathbf{d}_i^{3,2})}{c_2} \tag{39}$$

$$p(\mathbf{d}_i^{3,2}/\mathbf{r}^3) = p(\mathbf{r}(3)/\mathbf{d}_i^{3,2})p(\mathbf{r}^2/\mathbf{d}_i^{3,2})\frac{p(\mathbf{d}_i^{3,2})}{c_2} \tag{40}$$

$$p(\mathbf{r}^2/\mathbf{d}_i^{3,2}) = \sum_{\forall i:\mathbf{d}_l^{2,2}\in\mathbf{d}_i^{3,2}} p(\mathbf{r}^2/\mathbf{d}_l^{2,2})p(\mathbf{d}_l^{2,2}) \tag{41}$$

$$m_2(\mathbf{d}_i^{3,2}) = p(\mathbf{r}(3)/\mathbf{d}_i^{3,2})\frac{1}{c} \sum_{\forall i:\mathbf{d}_l^{2,2}\in\mathbf{d}_i^{3,2}} m_1(\mathbf{d}_l^{2,2}) \tag{42}$$

Finally, in terms of general time k,

$$m_{k+1}(\mathbf{d}_j^{k+2,2}) = \frac{1}{c}p(\mathbf{r}(k+2)/\mathbf{d}_j^{k+2,2}) \sum_{\forall i:\mathbf{d}_i^{k+1,2}\in\mathbf{d}_j^{k+2,2}} m_k(\mathbf{d}_i^{k+1,2}) \tag{43}$$

We see that this metric update is essentially identical to (24) with a channel memory length of $D = 2$.

4 Comparison With Maximum Likelihood Sequence Estimation

The previous sections have developed Bayesian recursions that can be used as the basis for optimal data detection in both the multiuser detection and ISI channel equalization problems. The decisions are optimal in the sense of maximizing the posterior probabilities for an individual symbol.

In this section, we follow a different strategy and derive a recursion in likelihoods according to the VA. The decisions are optimal in the sense of maximizing the likelihood of a given *sequence*. Consider again the equalization problem. The MLSE of the entire received sequence is,

$$\hat{d^k} = \arg\max_{d^k} p(r^k/d^k) \tag{44}$$

$$p(r^k/d^k) = p(r(k)/d^k)p(r(k-1)/d^k)\dots p(r(1)/d^k) \tag{45}$$

$$p(r^k/d^k) = p(r(k)/d^{k,D})p(r(k-1)/d^{k-1,D})\dots p(r(1)/d(1)) \tag{46}$$

$$p(r^k/d^k) = \prod_{i=1}^{k} p(r(i)/d^{i,D}) \tag{47}$$

At time $k+1$

$$p(r^{k+1}/d^{k+1}) = p(r(k+1)/d^{k+1})p(r^k/d^{k+1}) \tag{48}$$

$$= p(r(k+1)/d^{k+1,D})p(r^k/d^k). \tag{49}$$

This last line gives a recursion in likelihoods, however the number of metrics to compute is increasing exponentially with k. The Viterbi algorithm collapses this expanding tree of sequences into a fixed-size trellis as follows. At time k=D-1,

$$p(r^{D-1}/d^{D-1}) = \prod_{i=1}^{D-1} p(r(i)/d^{i,D}). \tag{50}$$

At time $k = D$ we can compute all M^D metrics

$$p(r^D/d^D) = \prod_{i=1}^{D} p(r(i)/d^{i,D}), \tag{51}$$

but "prune" according to

$$p(r^{D,D-1}/d^{D,D-1}) = \max_{d(1)} \prod_{i=1}^{D} p(r(i)/d^{i,D}), \tag{52}$$

which keeps the number of retained "survivor" metrics fixed at M^{D-1}. The general recursion is,

$$p(r^{k,D-1}/d^{k,D-1}) = \max_{d(k-D+1)} \prod_{i=1}^{k} p(r(i)/d^{i,D}). \tag{53}$$

The AF algorithm also calculates M^D metrics, but retains M^D metrics at each step, according to

$$p(r^{k+1,D}/d^{k+1,D}) = \frac{1}{c} p(r(k+1+D)/d_j^{k+1+D,D}) \sum_{\forall j: d_i^{k+D,D} \in d_j^{k+D+1,D}} p(r^{k,D}/d^{k,D}). \tag{54}$$

Comparison of equations (53) and (54) reveals the essential difference between the AF and VA procedures. Calculation of (53) is especially simple because taking natural logarithms turns the product of exponentials into a sum of arguments followed by the "max" operation. For the AF procedure, no such simplification occurs; we must calculate the argument, take the exponential, perform the sum, and multiply. Although these calculations are trivial, this extra burden is undoubtedly the reason why the VA is used far more extensively than the AF algorithm.

As mentioned previously, the VA procedure in general does not make a decision on any given symbol until the entire sequence is received. In practice, any implementation will have finite memory and a release depth of five times the constraint length is a commonly used rule of thumb. A subtle point is that if a "merge" has not occurred when memory is exhausted, the algorithm is forced to make a decision on the oldest symbol. It does this on the basis of which sequence has the greatest likelihood. This is of course a non-optimal symbol decision (in the sense of MAP symbol detection). Consider the following simple example. Let the decoder have a memory of 2. Thus we have four possible sequences and their likelihoods, for example:

d_2 d_1	$p(d(2), d(1)/r^2)$
0 0	0.4
0 1	0.3
1 0	0.0
1 1	0.3

Forced to make a symbol decision for d_1 the decoder will choose sequence 0, 0 (since it has the maximum metric), and hence decide $\hat{d}_1 = 0$. The AF strategy, on the other hand, is to sum all sequences stemming from the same $d(1)$ value. Hence the probability in favor of $d(1) = 1$ is $0.3 + 0.3 = 0.6$, and for $d(1) = 0$ is $0.0 + 0.4 = 0.4$. Thus, the AF algorithm decides $\hat{d}_1 = 1$.

5 Summary

The problem of multiuser detection has been shown to be closely related to the problem of channel equalization. Because of this close relation, the Abend and Fritchman (AF) algorithm, which was originally developed for the equalization problem, may also be applied to the multiuser detection problem. The AF algorithm calculates the maximum a-posteriori probabilities of each possible symbol in a recursive manner. This accords nicely with the Bayesian view of probability, as we decide in favor of the symbol in which we have the greatest "belief" in being correct. We have considered the fundamental differences between the AF algorithm and the better-known Viterbi algorithm (VA). The AF algorithm is optimal in the sense of MAP symbol detection, while the VA is optimal in the sense of maximum-likelihood sequence detection. Examining the algorithms closely shows that while both must compute M^D metrics, the VA requires only addition to update its metrics, while the AF algorithm requires calculation of exponentials, sums, and products. Given the state of modern digital signal processing technology, the additional operations are no longer very significant, and the common selection of the VA over the AF algorithm needs to be reconsidered. We believe the AF algorithm could be preferable in some applications, for example in blind equalization, where it leads to an efficient parallel adaptive filtering structure [9].

References

[1] S. Verdu, "Recent Progress in Multiuser Detection," in *Advances in Communication and Signal Processing* (A. Porter and S. Kak, eds.), Springer-Verlag, 1989.

[2] S. Verdu, "Minimum Probability of Error for Asynchronous Gaussian Multiple-Access Channels," *IEEE Trans. on Information Theory*, vol. IT-32, No. 1, Jan. 1986.

[3] R. Lupus and S. Verdu, "Near-Far Resistance of Multiuser Detection in Asynchronous Channels," *IEEE Trans. on Communications Theory*, vol. Vol 38 No. 4, April 1990.

[4] R. Lupus and S. Verdu, "Linear Multiuser Detectors for Synchronous Code- Division Multiple-Access Channels," *IEEE Trans. on Information Theory*, vol. IT-35, no.1, pp. 123–136, Jan. 1989.

[5] R. Iltis and L. Mailaender, "A Symbol-by-Symbol Multiuser Detector with Joint Amplitude and Delay Estimation," in *Twenty-Sixth Annual Asilomar Conference on Signals, Systems, and Computers*, (Pacific Grove, California), 1992.

[6] L. Mailaender and R. Iltis, "Importance Sampling Simulation of a Symbol-by-Symbol Multiuser Detector," in *Twenty-Seventh Annual Asilomar Conference on Signals, Systems, and Computers (to appear)*, (Pacific Grove, California), 1993.

[7] K. Abend and B. Fritchman, "Statistical Detection for Communication Channel with Intersymbol Interference," *Proceedings of the IEEE*, vol. 58, pp. 779–785, May 1970.

[8] R. Iltis and L. Mailaender, "An Adaptive Multiuser Detector with Joint Amplitude and Delay Estimation," *Submitted to the IEEE Journal on Selected Areas in Communications*, 1993.

[9] K. Giridhar, J. Shynk, and R. Iltis, "Bayesian/Decision-Feedback Algorithm for Blind Adaptive Equalization," *Optical Engineering*, vol. 31 no. 6, pp. 1211–1227, June 1992.

[10] G. D. Forney, "Maximum Likelihood Sequence Estimation of Digital Sequences in the Presence of Intersymbol Interference," *IEEE Trans. on Information Theory*, vol. IT-28, May 1972.

[11] R. Iltis, "A Digital Receiver for Demodulation of CDMA Waveforms with A-Priori Unknown Delays and Amplitudes," in *Proceedings of the IEEE Military Communication Conference (MILCOM)*, (McLean, Va.), Nov. 1990.

[12] R. Iltis, "A Bayesian Maximum-Likelikhood Sequence Estimation Algorithm for A-Priori Unknown Channels and Symbol Timing," *IEEE Journal on Selected Areas in Commnication*, vol. 10, pp. 579–588, April 1992.

[13] J. G. Proakis, *Digital Communications.* McGraw-Hill, 1989.

THERMOSTATICS IN FINANCIAL ECONOMICS

Michael Stutzer
Department of Finance, Carlson School of Management
University of Minnesota, 271 19th Ave. S., Minneapolis, MN 55455

ABSTRACT. At a recent MaxEnt conference, the Bayesian econometrician Arnold Zellner noted that "much more empirical and theoretical work needs to be done to get a 'maximum entropy, thermodynamic model' that performs well in explaining and predicting the behavior of economic systems". This paper reports modest progress toward such a model in financial economics. Arbitrage-free financial models imply the existence of risk-neutral probability measures, which are used in the prediction of asset prices. This paper uses MaxEnt to select a risk-neutral probability measure, and develops a few applications of it. In this way, the connection between Jaynes' formulation of statistical mechanics and important problems in financial economics are made clear.

1. Introduction

The constrained maximization of entropy (MaxEnt) has been widely and successfully used to select probability measures for a myriad of applications. Subsequent to Josiah Willard Gibbs' pioneering use of expectations taken with respect to the canonical measure, which maximizes entropy subject to linear constraint(s), one of the best known successes of MaxEnt has been Jaynes' [8] parametric sensitivity analysis of this MaxEnt problem, providing a foundation for a "generalized statistical mechanics" and thermostatics.

Economists call the parametric sensitivity analysis of *general* constrained optimization problems "comparative statics". It has been universally adopted as the foundation for much of neoclassical economics, since the publication of Paul Samuelson's influential Foundations of Economic Analysis [13] in 1947. For example, consumer behavior is modelled by assuming that consumers act as if they maximized a concave utility function subject to a linear budget constraint. Using this formulation, one can study the parametric sensitivity of consumption choices when, say, the consumer's income is increased.

Because Samuelson's thinking was heavily influenced by the lectures of Edwin Bidwell Wilson, J.W. Gibbs' last protegé [14], it is puzzling that MaxEnt – in particular, the canonical measure – hasn't been utilized more in conventional economic theory. Examining some past efforts to do so [1] prompted Arnold Zellner to note at a recent MaxEnt conference [2] that "much more empirical and theoretical work needs to be done to get a 'maximum entropy, thermodynamic model' that performs well in explaining and predicting the behavior of economic systems."[21, p.21]

---

[1] Of course, there have been other papers in economic theory which employ MaxEnt. For a survey, see Kapur [chaps. 13,19][10]. But these results are seldom, if ever, included in economics textbooks, and are accordingly out of the mainstream–rightly or wrongly.

[2] At this point, MaxEnt has only found a measure of acceptance in some parts of econometrics described by Zellner (also see the survey article by Maasoumi [11]).

G. R. Heidbreder (ed.), Maximum Entropy and Bayesian Methods, 375–390.

This paper reports some modest progress toward such a model. The economic system modelled will be the financial markets. Both academicians and many financial market participants have adopted the assumption that competition for financial gains will eventually eliminate any *arbitrage opportunities*, i.e. investment strategies which cost nothing yet risklessly earn investment income. It is argued that attempts to exploit such "free lunches" will eventually become self-defeating, by changing demand and supply for financial assets in ways which move the assets' prices away from those at which the lunches *were* free. As an approximation to reality, it is thus not unreasonable to develop an *equilibrium* theory which rules them out. When coupled with other assumptions about the stochastic processes governing the movements of asset prices, a fundamental duality theorem in financial economics shows that this *no arbitrage assumption* is equivalent to the existence of a set of *risk neutral probability measures*, which satisfy a set of linear contraints. These risk neutral measures are widely used to predict the prices of *contingent claims*, such as stock options, as well as to test and interpret more elaborate behavioral theories of asset prices.

Under the conditions of incomplete information that analysts typically work under, this paper uses MaxEnt to select a *canonical risk neutral probability measure.* The computation of expected values taken with respect to the canonical distribution, a procedure familiar from statistical mechanics, provides an alternative, simple derivation of a well-known asset pricing prediction called a *multi-beta, approximate arbitrage pricing model.* Its "market prices of risk" are provided by the canonical distribution's parameter vector (the Lagrange multipliers from the MaxEnt problem), which is also the vector of risky asset weights in a *canonical mean-variance efficient portfolio.* Because a similar procedure has previously been used to derive the justly celebrated Black-Scholes formula for predicting the price of a stock option [18], it is hoped that MaxEnt may eventually be used to provide a foundation for a general theory of arbitrage-free, contingent claim pricing under conditions of incomplete information.

Having established that some conventional asset pricing results can be derived by taking expectations with respect to the canonical risk neutral measure, we turn to an exploratory investigation of the financial analog of thermostatics. In particular, we will explore both similarities and differences between the thermostatic analysis of weakly interacting physical systems brought into thermal contact, and an analysis of the integration of formerly segmented financial markets. We will argue that a combination of thermostatic modeling and empirical investigation is a promising route to sharp, testable predictions about the effects of financial market integration [3].

2. A Standard Model of Arbitrage-Free Asset Prices

We utilize an important special case of the standard, finite dimensional securities market model with a finite number of states and a single consumption good [4]. At the beginning of any period, an uncertain state of nature is drawn independently from a set of possible states of nature $\Omega = \{\omega_1, \ldots, \omega_K\}$. The particular state drawn determines the end-of-period price

[3] It is highly unlikely that more traditional structural economic models of market integration will imply anything analogous to, say, the Ideal Gas Law. To obtain even qualitative comparative statics predictions, the framework must be augmented by auxiliary hypotheses, e.g. specific functional forms for agents' optimization problems and specific values of their free parameters. Because no one would ever accuse thermostatics of failing to make sharp predictions, it is reasonable to consider thermostatic reasoning as an alternative predictive framework in this setting.

[4] For more detail about the standard securities market model, see Dothan [1, chaps. 1-2], or Willinger and Taquu [20]

and/or dividend paid for each of the $N+1$ primary assets available for trade. Formally, asset i pays $X_i(\omega_j)$ of the consumption good to the buyer from the seller when state j occurs at the period's end, with (objective) probability $\pi_j > 0$. Because there are a large number of events which could affect the future prices and dividends of financial assets, we will assume that the number of states K is quite large. Denoting the price paid for asset i at the beginning of the period by P_i, the asset's *total (gross) return* $R_i(\omega_j)$ over the period is $R_i(\omega_j) \equiv X_i(\omega_j)/P_i$ when state j occurs. In other words, a dollar invested in asset i returns $R_i(\omega_j)$ at the end of a period, after adjusting for inflation. If state j occurs at the end of some period in which an investor owned asset i, and $R_i(\omega_j) > 1$, the investment appreciated at a rate in excess of the rate of inflation. Of course, the expected (gross) return per period is $E_\pi[R_i] = \sum_{j=1}^{K} R_i(\omega_j)\pi_j$.

A myriad of financial assets can be represented in this way. For example, we will assume that asset 0 is a *riskless asset*, i.e. its gross return per period is not random. This is represented by

$$R_0(\omega_j) = r, \quad j = 1, \ldots, K. \tag{1}$$

A riskless asset plays an important role in many asset pricing theories, serving as a perfect hedge against inflation. The constant r is then the *(gross) real rate of interest.* In some applications, a riskless asset is approximated by a government treasury bill with maturity equal to the period length.

2.1. The Fundamental Duality Principle

Suppose some asset $i \neq 0$ has the random return $R_i(\omega_j) > r$, $j = 1, \ldots, K$. In other words, the asset returns a variable amount which will *always* be in excess of the rate of interest. If an investor sold a dollar's worth of the riskless asset and invested that dollar in asset i, one would earn $R_i(\omega_j) - r > 0$ when state j occurs at the end of the period. That is, for a net investment of 0 dollars, one would always earn something net of inflation. Effectively, the investor financed her purchase of asset i by borrowing at the interest rate r, knowing that she would always be left with some purchasing power after selling the asset at the end of the period to repay her loan. Clearly, this is a good deal, for by borrowing a large number of dollars, one would always obtain a larger number back. It is a "free lunch", so why not eat to your heart's content? In fact, investors should have no trouble convincing lenders to lend them unboundedly large amounts to do this, because they will never have any trouble repaying them out of the investment proceeds.

It is this sort of opportunity that economists dub an *arbitrage opportunity.* An arbitrage opportunity allows the investor to produce something out of nothing. Of course, arbitrage opportunities might be much more complex, involving the simultaneous purchase and sale of many assets, perhaps in some complex way over time. But the result of an arbitrage opportunity is always the same: purchasing power in the future is produced at no cost now.

But how likely are these opportunities to arise and persist? Investors attempting to exploit the above opportunity would drive up the demand for asset i, raising its price P_i and lowering its return $R_i(\omega_j) \equiv X_i(\omega_j)/P_i$. And their attempts to borrow large amounts will drive up the interest rate r. Once the interest rate rises above $\max_j R_i(\omega_j)$, the free lunch vanishes, for the investor would lose $r - \max_j R_i(\omega_j)$ if state $\arg\max_j R_i(\omega_j)$ occurred, which it does with positive probability. The fear of this occuring would curb lenders' appetites to give the investors unlimited funds, and fear of losses in those states where the asset's return is lower than r would curb some investors' desire to do so. In other words, attempts to exploit arbitrage opportunities sow their own seeds of destruction in the financial markets.

As such, it is reasonable to assume that the assets' random returns and the riskless rate r can not assume values permitting arbitrage opportunities. But exactly what values does this assumption rule out? Note that in our example, because $R_i(\omega_j) > r$ for all states $j = 1, \ldots K$, it is also true that $E_Q[R_i] \equiv \sum_j R_i(\omega_j)Q_j > r$ for *any* probability distribution Q over the possible states of nature, including the actual state probability distribution π. Note that if for some asset m, $R_m(\omega_j) < r$ for all states $j = 1, \ldots K$, an arbitrage opportunity would arise by selling a dollar's worth of asset m and investing the proceeds in the riskless asset. This would produce $r - R_m(\omega_j) > 0$ for any state j at no cost to the investor. In that case, $E_Q[R_i(\omega_j)] < r$ for any probability distribution Q over states of nature. Furthermore, any portfolio formed by buying some assets and selling some others will admit an arbitrage opportunity when it *always* returns a random amount $R(\omega)$ in excess of r, in which case $E_Q[R] > r$, and the sense of the inequality would be reversed if the portfolio always returned less than r.

The discussion above linked the *presence* of arbitrage opportunities to the *nonexistence* of a probability measure under which the expected returns of some assets would equal the riskless rate of interest. Furthermore, it would be reasonable to conjecture that the converse is true, i.e. that the *absence* of arbitrage opportunities requires the *existence* of a probability measure Q under which the Q-expected returns of all risky assets must equal the riskless rate r. Of course, we didn't examine other potential arbitrage opportunities involving portfolios whose returns are not *always* greater or *always* less than r, nor did we examine conceivable multiperiod investment strategies. But in the special model described in this section, it turns out that the conjecture and its converse are true. While we won't formally state and prove it here [5], a fundamental duality theorem of financial economics shows that the assumption of no arbitrage opportunities is equivalent to the existence of a probability distribution Q under which all assets have expected return equal to the riskless rate r. That is:

Theorem 2..1 *Under the assumptions of this section, there are no arbitrage opportunities possible if and only if there exists a strictly positive probability measure Q, called a* **risk neutral probability measure,** *satisfying:*

$$E_Q[R_i] \equiv \sum_{j=1}^{K} R_i(\omega_j)Q_j = r, \quad i = 1, \ldots N \tag{2}$$

Theorem 2.1 provides the restriction on the N random asset returns and the riskless rate of interest which is equivalent to the no arbitrage assumption. Theorem 2.1 also places restrictions on the returns from a *portfolio* of these assets. Let θ_i denote the *share* of the portfolio's cost tied up in asset i (so $\theta_i < 0$ for an asset which is sold, rather than bought). The return from the portfolio in state j is then $R(\omega_j) = \sum_{i=0}^{N} \theta_i R_i$, where $\sum_{i=1}^{N} \theta_i = 1$. Due to linearity of the expectations operator, the following is a corollary of theorem 2.1.

[5]For a more formal definition of arbitrage opportunities and a proof of this result, see Stutzer [18].

Corollary 2..1.1 *For any portfolio of the $N+1$ assets with random return $R(\omega)$, and any risk-neutral measure Q, $E_Q[R] = r$.*

3. The Canonical Risk Neutral Probability Measure

Any modern graduate text in financial economics contains numerous applications of risk neutral measures [6]. Here, we develop an application of the MaxEnt choice of risk neutral measure. The *canonical risk neutral probability measure* solves the following MaxEnt problem:

Definition 3..1 *The* **canonical risk neutral probability measure** $\hat{Q}$ *solves the constrained maximum entropy problem:*

$$\max_Q S \equiv -\sum_{j=1}^{K} Q_j \log Q_j \tag{3}$$

subject to:

$$E_Q[R_i] = r,\ i = 0, \ldots, N \tag{4}$$

Note that due to the existence of the riskless asset with return $R_0(\omega_j) \equiv r$, the constraint in (4) corresponding to $i = 0$ is just the normalization constraint on probabilities. If we obtain data on $N + 1 = K$ linearly independent assets [7], the no arbitrage assumption and Theorem 2.1 guarantee that there is a unique feasible point in problem (3), in which case the probabilities are also called *normalized Arrow-Debreu state prices.* In our applications, however, $N + 1 < K$, so there is a convex polytope of risk neutral measures, and MaxEnt selects the familiar Gibbs' canonical measure:

Theorem 3..1 *The* **canonical risk neutral measure** *has the form:*

$$\hat{Q}_j = e^{\sum_{i=1}^{N} \hat{\gamma}_i R_i(\omega_j)}/Z,\ j = 1, \ldots, K \tag{5}$$

where the **partition function** *Z is the normalizing constant:*

$$Z \equiv \sum_{j=1}^{K} e^{\sum_{i=1}^{N} \hat{\gamma}_i R_i(\omega_j)}. \tag{6}$$

The parameter vector $\hat{\gamma}$ may be computed by finding a stationary point of the **free energy** *function F:*

$$\hat{\gamma} = \arg\min_{\gamma_1,\ldots,\gamma_N} F(\gamma; r) \equiv \sum_{j=1}^{K} e^{\sum_{i=1}^{N} \hat{\gamma}_i (R_i(\omega_j) - r)} \tag{7}$$

and the maximum entropy attained is

$$S_{\max} = \log F(\hat{\gamma}; r) \tag{8}$$

[6] Perhaps the best known application is the discrete time binomial option pricing model. Textbook presentations can be found in, e.g. Huang and Litzenberger [pp.248-54][6] or Jarrow and Rudd [chap.13] [7].

[7] Linear independence means that no asset payoff $X_i(\omega)$ can be written as a linear combination of the other N assets' payoffs. If this were not true, the return $R_i(\omega)$ could also be produced by a portfolio of the other assets, and investors would find asset i to be redundant.

3.1. Isomorphism with Jaynes' Generalized Statistical Mechanics

The canonical measure $\hat{Q}$ is the tool Jaynes [8] used to develop the Gibbsian statistical mechanics of systems subject to conservation laws [8]. Each of the N no arbitrage constraints in (4) is identified with one of Jaynes' *conservation laws.* Because it is generally presumed that the financial analyst is studying a much smaller number of risky assets than is necessary to complete the market, $N < K - 1$, which is isomorphic to Jaynes' assumption that the number of conservation laws is less than the number of states. In the most common application of statistical mechanics, there is only $N = 1$ conservation law, where $R_1(\omega_j)$ is identified with the energy level of a system (e.g. a gas) in state j, and the riskless rate r is identified with the expected value of the energy [19, p.158]. But unlike Jaynes' analysis of the general case of $N > 1$ conservation laws, here the Q-expected value of each conserved quantity must equal the *same* constant, r. Financial counterparts of temperature, heat, work, and forces will later be identified with the MaxEnt solution and its comparative statics, and used to interpret financial data.

3.2. An Example

Of course the size of the rate of interest r, as well as other assets' returns, will depend on the period length chosen. To model this, we let $T = 1$ represent the total length of time over which we wish to model asset prices, and assume that asset trading may occur during all of the n periods of length $\Delta t = 1/n$ within it. We assume that

$$\begin{aligned} R_i(\omega_j) &= 1 + \mu_i \Delta t + \sum_{l=1}^{K-1} e_{jl}\sigma_{il}\sqrt{\Delta t} \\ r &= 1 + \iota \Delta t \end{aligned} \tag{9}$$

where e_{jl} is element (j, l) of a $K \times K - 1$ matrix $\mathbf{e} = (e_1, \ldots, e_{K-1})$, whose columns form an orthonormal basis for the $K - 1$ dimensional linear space which is orthogonal to the K-vector of ones. Thus when state j occurs, the $e_{jl} = \mathbf{e}_l(\omega_j)$, $l = 1, \ldots, K - 1$, additively affect all the risky assets' returns, through the coefficients in the matrix σ with element σ_{il} in row i and column l. *We assume that there is a uniform actual probability distribution of states, i.e.* $\pi_j = 1/K$, $j = 1, \ldots, K$. It is then easy to show that the columns of $\mathbf{e}$ are uncorrelated random variables with zero mean and unit variance. The returns are contemporaneously correlated, and we assume that the symmetric matrix $\sigma\sigma'$, which determines their covariances, is positive definite (and hence invertible).

In the *continuous time limit* with the number of trading periods $n \to \infty$, so the period length $\Delta t = 1/n \to 0$, the method of Hua He [4] proves that the process of net rates of return $R_i - 1$ from (9) converges weakly to the continuous time, correlated geometric Brownian process,

$$\begin{aligned} \frac{dP_i}{P_i} &= \mu_i dt + \sum_{l=1}^{K-1} \sigma_{il} dW_l \\ \frac{dP_0}{P_0} &= \iota dt \end{aligned} \tag{10}$$

[8] See also the textbooks of Tribus [19], Hobson [5] and Haken [3].

for the $i = 1, \ldots, N$ risky assets. Here, W_l denotes component l of a $K - 1$-dimensional standard Brownian motion $\mathbf{W}$, μ_i is a drift parameter for asset i, and ι is the instantaneous riskless rate of interest. Thus, column $\mathbf{e}_l$ in (9) is used to approximate the increment dW_l, and the instantaneous covariance matrix of returns is $\sigma\sigma'$.

To obtain the canonical risk neutral probabilities, we must solve (7). Of course, the solution will depend on the number of trading periods n as well as on the matrix $\mathbf{e}$ whose columns model the Brownian shocks. To see what is likely to result when the number of trading periods n is large, we study the solution in the continuous time limit as $n \to \infty$. To do so, it is useful to multiply (7) by $\pi_j = 1/K$ and sum to produce the equivalent minimization

$$\hat{\gamma} = \arg \min_{\gamma_1, \ldots, \gamma_N} E_\pi \left[e^{\sum_i \gamma_i (R_i(\omega) - r)} \right] \tag{11}$$

and to note that, in this case where π is uniform, the canonical measure (5) may also be written

$$\hat{Q}_j = \pi_j e^{\sum_{i=1}^N \hat{\gamma}_i R_i(\omega_j)} / Z, \; j = 1, \ldots, K \tag{12}$$

where the **partition function** Z is now the normalizing constant [9]

$$Z \equiv E_\pi \left[e^{\sum_{i=1}^N \hat{\gamma}_i R_i(\omega_j)} \right]. \tag{13}$$

Take a Taylor series expansion of (11) about the N risky assets' mean gross returns $1 + \mu_i/n$ and take the logarithm to obtain the following equivalent minimization problem:

$$\min_{\gamma_1, \ldots, \gamma_N} \sum_{l=1}^N \gamma_l (\mu_l - \iota)/n + \log \left[1 + \gamma' \sigma \sigma' \gamma / 2n + o(1/n) \right]. \tag{14}$$

For each n, minimization of (14) is equivalent to the minimization of $n \times (14)$. Doing so, the series expansion for e^x identifies the second term to be $\gamma'\sigma\sigma'\gamma/2$ as $n \to \infty$, yielding a quadratic minimization having the solution:

$$\lim_{n \to \infty} \hat{\gamma} = -\left(\sigma\sigma'\right)^{-1} \mathbf{x} \tag{15}$$

where $\mathbf{x} \equiv (\mu_1 - \iota, \ldots, \mu_N - \iota)'$ is the vector of the excess mean returns over the instantaneous riskless rate ι.

By way of analogy to the thermodynamic limit employed in physics, the continuous time limit also provides sharp predictions, for calculations using it are independent of the matrix $\mathbf{e}$ used to model the random shocks.

We now use (15) to develop a general asset pricing result relating the N risky asset returns to another asset's return R. If an asset having return R is a portfolio of the N assets, then we know that R is a weighted average of the assets' returns, and that $E_{\hat{Q}}[R] = r$.

---

[9] For nonuniform state probabilities π, it turns out that (11) produces the parameter vector for the "exponentially twisted" measure (12) which minimizes the Kullback-Leibler Information Criterion (KLIC) "distance" $D(Q \mid \pi) \equiv \sum_j Q_j \log(Q_j/\pi_j)$ subject to (4). Thus, by adopting the KLIC to interpret our findings, rather than the Shannon Entropy, all our calculations using (11)-(13) are valid for the general case of nonuniform π.

But suppose it is not a portfolio of the assets, so that it is possible that $E_{\hat{Q}}[R] \neq r$. It may be possible to discover additional assets, which when added to the N assets, will enable R to be written as a weighted average of the returns from the larger set. Adding the additional no arbitrage constraints to (4) and solving for the different canonical probabilities Q^* would then permit us to investigate the consequences of $E_{Q^*}[R] = r$.

To test for the importance of searching for this potentially larger set of assets, we derive a testable relationship among asset returns which follows by assuming that the arbitrage-free relationship holds only approximately, i.e. that $E_{\hat{Q}}[R] \approx r$. We interpret the approximate arbitrage-free relationship using the Bayesian, MaxEnt perspective. In the absence of explicit knowledge about which additional assets are needed (if any), one views $\hat{Q}$ as the Bayesian estimate of the actual, but unknown risk neutral probabilities (Q^*) consistent with the larger set of assets [10]. Under this interpretation, we are using estimated risk neutral probabilities rather than actual ones, so one can only expect that $E_{\hat{Q}}[R] \approx r$, rather than the exact relationship. Substitute (12) and (13) and take logarithms to transform this into the equivalent statement $\log[E_\pi[R/r \exp[\sum_i \hat{\gamma}_i R_i]]] \approx \log[Z]$. As before, substitute (9) and expand both sides in a Taylor series about the assets' mean returns, multiply both sides by n, and take the continuous time limit $n \to \infty$ to yield the following result :

Theorem 3..2 (Canonical Pricing Theory) *Assume that only N asset returns with processes (9) are used to price all other assets. In the continuous time limit $n \to \infty$,*

$$\mu_R - \iota \approx -\sum_l \hat{\gamma}_l cov(R, R_l). \tag{16}$$

That is, MaxEnt predicts that an asset's mean excess return approximates a $-\hat{\gamma} \equiv (\sigma\sigma')^{-1}\mathbf{x}$-weighted sum of its covariances with the N risky assets.

The "market price vector of risks" $-\hat{\gamma}$ is also the vector of risky asset portfolio weights in the **canonical mean-variance efficient portfolio** formed from the factors and the riskless asset. This is the mean-variance efficient portfolio which has standard deviation equal to the maximal Sharpe performance measure $\sqrt{H} \equiv \sqrt{\mathbf{x}'(\sigma\sigma')^{-1}\mathbf{x}}$ attained by the tangency portfolio of risky assets, and which has expected return $H + \iota$ [6, p.76-7], [9, p.435]. Substitute (15) into (16) to obtain an approximate arbitrage pricing theory:

$$\mu_R - \iota \approx \beta_R \mathbf{x} \tag{17}$$

where $\beta_R \equiv (cov(R, R_1), \ldots, cov(R, R_N))(\sigma\sigma')^{-1}$

EXAMPLE:

Suppose there are $N = 3$ risky assets, with drift vector $\mu = (.062, .146, .128)'$, and a riskless rate $\iota = .05$, so $\mathbf{x} = (.012, .096, .078)'$. Suppose the positive definite covariance matrix $\sigma\sigma'$ of the returns is:

[10] Unlike the MaxEnt analysis in Grandy [2], we assume here that all investors agree about the actual state probabilities $\pi(\omega_j)$=1/K, $j = 1, \ldots, K$. Grandy assumes that investors don't know these probabilities, and use MaxEnt to form subjective probabilities. Here, we use MaxEnt to estimate risk neutral probabilities, rather than actual state probabilities.

$$\sigma\sigma' = \begin{pmatrix} .0146 & .0187 & .0145 \\ .0187 & .0854 & .0104 \\ .0145 & .0104 & .0289 \end{pmatrix}$$

used by Markowitz[p.176] [12]. The weights in the canonical mean variance efficient portfolio are $-\hat{\gamma} \equiv (\sigma\sigma')^{-1}\mathbf{x} = (-7.52124, 2.07369, 5.72635)'$. Because $\sum_i -\hat{\gamma}_i = .2788$, the riskless asset's weight in the canonical portfolio is $.7212 = 1 + \sum_i \hat{\gamma}_i$, while the tangency portfolio of risky assets is $-\hat{\gamma}/.2788$. Suppose another asset's covariances with the factors are known to be .01, .02, and .015, respectively. Then, $\beta_R = (.0123909, .176775, .4492)$ and (16) predicts that the asset's mean excess return will be close to $-(.01, .02, .015)\hat{\gamma} = .0522$, or $\beta_R(.012, .096, .078)'$ from (17).

Note that the canonical pricing theory does not presume that there *are* N assets which make (16) valid for other assets. Rather, it is a test of "factorhood", by testing restrictions on other assets' excess returns and covariances with the N assets'. If the priced asset is in fact a portfolio of the N assets, a simple computation shows that β_R just equals the vector of portfolio weights, so the relation is exact. If it isn't a portfolio of the factors, the multi-beta relation (17) is a Bayesian, MaxEnt inference obtained *solely* from the returns data on the N assets with returns process (9) and the hypothesis of no arbitrage. As in other applications of MaxEnt, failure of the relationship to explain asset returns of interest indicates that other contraints are needed in (4), i.e. that other assets are required to "span" it.

4. The Thermostatics of Financial Market Integration

The previous section illustrated one use of the canonical measure: expectations taken with respect to it can be used to predict pricing relationships among a group of assets. And in [18], the celebrated Black-Scholes model for predicting the price of a stock option was rederived using canonical expectations. So one ubiquitous procedure in Gibbsian statistical mechanics is useful in conventional financial economics. We now begin to explore the more provocative prospect that thermostatic reasoning may lead to a predictive theory of the effects of financial market integration. Consider two distinct groups of assets, $\mathbf{R}^a$ and $\mathbf{R}^b$, which originally were not linked by the absence of arbitrage opportunities. For example, the two groups might be used to represent assets in separate countries, in which currency controls and/or other regulations impeded the ability of investors to realize international arbitrage opportunities. Solnick [15, pp.iii-iv] noted that international integration of financial markets proceeded rapidly in the 1980's, spurred by recognition of the need for international diversification, by deregulation, and by technological innovation in trading technologies. The 1990's will bring further integration of formerly segmented markets. What effects will international integration have on the financial markets?

The conceptual framework explored here is to make these predictions by the same method used to predict the outcome of bringing two formerly isolated physical systems into thermal contact. The two countries' "isolated" financial systems could formerly sustain two separate riskless rates of interest, r^a and r^b. But once their markets are integrated, i.e. linked by the absence of international arbitrage opportunities, this can no longer occur.

Investors would rush to borrow at the lower of the two rates, taken without loss of generality to be country b, causing it to rise by Δr^b, and would use the loans to invest at the higher rate in country a, causing its riskless rate to fall by Δr^a, until the two rates were equilibrated at a common value r. Thus, unlike conventional thermostatics, it is not the *sum* of the two subsystems' internal (average) energies which need be conserved [19, p.118], but rather the right hand side of the "conservation constraints" (4) (i.e. the riskless interest rate) for each subsystem must adjust to attain a common value. Furthermore, changing investment flows may result in changes to the parameters of the two countries' price processes, i.e. their drift vectors μ^a and μ^b and volatility matrices σ^a and σ^b. In fact, adjustment of the riskless rates and the absence of arbitrage may necessitate some parametric change, for if all these parameters remained the same following integration, it is possible that intra-country arbitrage opportunities would arise with the new riskless rate r that were absent at the old rates r^a and r^b. Investor's attempts to realize the arbitrage profits would radically change demands for risky assets involved in the arbitrage strategies, changing their drift and/or volatility parameters. *Such a temporary arbitrage opportunity would be analogous to a type of phase transition, associated with parameters for which the constraints (4) fail to have a solution.*

The assumption that parameters change in such a way as to maintain the absence of arbitrage opportunities is analogous to the "quasi-static" assumption in thermostatics. Following Jaynes [8], we let α denote a parameter which affects the riskless rates (say, the flow of investment), and which possibly also affects the risky asset returns, now written $R_i(\omega_j;\alpha)$.

The highest conceivable value of $S_{\max}$ in (3) would be $\log K$, attained by the uniform canonical measure $\hat{Q}_j \equiv 1/K = \pi_j$. But from (4), that would require $E_\pi[R_i] = r$, $i = 1,\dots N$. That is, all assets' actual expected returns would have to be the riskless rate of interest. Of course, this is highly unlikely. Financial economists attribute this to the influence of *risk*, calling the nonzero difference $E_\pi[R_i] - r$ the *risk premium* for asset i. *So the analog of "heat death" is a "risk neutral" world with no risk premia.* The *degree of risk*, denoted $D_{\min}$, may be measured by a decrease of entropy from its maximum possible value. Because of our assumption that the actual state probabilities $\pi_j \equiv 1/K$, straightforward algebra shows that the degree of risk $D_{\min}$ is just the Kullback-Leibler Information Criteria (KLIC) "distance " between the canonical risk neutral measure and the actual state probability measure.

$$\begin{aligned} D_{\min} &\equiv \log K - S_{\max} \\ &= D(\hat{Q} \mid \pi) \\ &\equiv \sum_{j=1}^{K} \hat{Q}_j \log(\hat{Q}_j/\pi_j) \end{aligned} \tag{18}$$

Direct substitution of (12) and (13) into (18) verifies that

$$D_{\min} = -\log E_\pi\left[e^{\sum_i \hat{\gamma}_i (R_i(\omega;\alpha)-r)}\right] \tag{19}$$

Because the degree of risk $D_{\min}$ is a scalar index of the vector of risk premia [11], a theory of its temporal change would be useful. Following the Jaynesian treatment of thermostatics, we decompose changes in the countries' values of $D_{\min}$ into "thermal" and "mechanical" parts. Applying the envelope theorem to (19) yields the following decomposition of the changes in the countries' degrees of risk:

$$\begin{aligned} dD^a_{\min} &= \sum_i \hat{\gamma}_i{}^a \left(dr^a - E_{\hat{Q}^a}[dR^a_i]\right) \\ dD^b_{\min} &= \sum_i \hat{\gamma}_i{}^b \left(dr^b - E_{\hat{Q}^b}[dR^b_i]\right). \end{aligned} \tag{20}$$

In the "generalized statistical mechanics" of Jaynes [8, p.627], equation (20) decomposes the changes of countries' degrees of risk into the sum of N terms, term i being the product of the "integrating factor" $\hat{\gamma}_i$ and the "ith type of heat". Each expectational term represents a "generalized force" contributing to the summed "work effects" [19, chap. 6].

After financial market integration, it is possible that *only* the riskless rates of interest will change, leaving other asset price process parameters, e.g. the drift and volatility matrices, unchanged. This corresponds to the special case of strictly "thermal interaction". From (20) we then have

$$\begin{aligned} dD^a_{\min} &= dr^a \sum_i \hat{\gamma}^a_i \\ dD^b_{\min} &= dr^b \sum_i \hat{\gamma}^b_i. \end{aligned} \tag{21}$$

Thus, when the interaction is strictly thermal, $\sum_i \hat{\gamma}_i$ is the slope of its $(r, D_{\min})$ curve. This is analogous to the relationship between internal energy and thermostatic entropy, so *it is natural to think of the slope $\sum_i \hat{\gamma}_i$ as the "conjugate" [8] concept to temperature in thermostatics.*

It is easy to show that $S_{\max}$ is concave in the riskless rate, so that the curves $(r^a, D^a_{\min})$ and $(r^b, D^b_{\min})$ must be convex in the plane, each having a global minimum at a riskless rate where its slope is zero [12]. To test this prediction requires empirical estimates of the curves.

In regimes of constant parameters, it is possible to use vector time series of returns to empirically estimate $\hat{\gamma}^a$ and $\hat{\gamma}^b$ for given values of r^a and r^b, producing estimates of $D^a_{\min}$ and $D^b_{\min}$. To do so, one first uses a law of large of numbers (i.e.ergodicity) to identify the phase average in (19) with the time average. Formally,

$$E_\pi\left[e^{\sum_i \hat{\gamma}_i (R_i(\omega;\alpha)-r)}\right] = \lim_{T\to\infty} \frac{1}{T}\sum_{t=1}^{T} e^{\sum_i \hat{\gamma}_i (R_i(\omega(t);\alpha)-r)}. \tag{22}$$

Upon substituting a time series of returns of length T into the right hand side of (22), the Newton-Raphson method can be used to numerically minimize it. This produces an

[11] See [16] or [17] for additional reasons to interpret $D_{\min}$ as a preference-free index of the influence of risk.

[12] This property is also true when π is not uniform.

estimate of $\hat{\gamma}$. Because a finite length time series is used, the estimate is subject to sampling error, which can be quantified as well [17]. Using monthly indices of U.S. stock and long term government bond returns, Figure 1 reports the estimated $(r, D_{\min})$ curve for the country over a reasonably long period. Note that the predicted strict convexity is confirmed empirically, with a global minimum occuring when $\sum_{i=1}^{2} \hat{\gamma}_i = 0$, at a monthly gross real interest rate around 1.002 (i.e. an effective annual net real rate of 2.4%)[13].

Because the $(r, D_{\min})$ curve is strictly convex, we may replace the differential relations (21) with inequalities valid for discrete changes in interest rates, and add to obtain the discrete change in the sum of the two countries' degrees of risk:

$$\Delta(D^a_{\min} + D^b_{\min}) > \Delta r^a \sum_i \hat{\gamma}_i{}^a + \Delta r^b \sum_i \hat{\gamma}_i^b \tag{23}$$

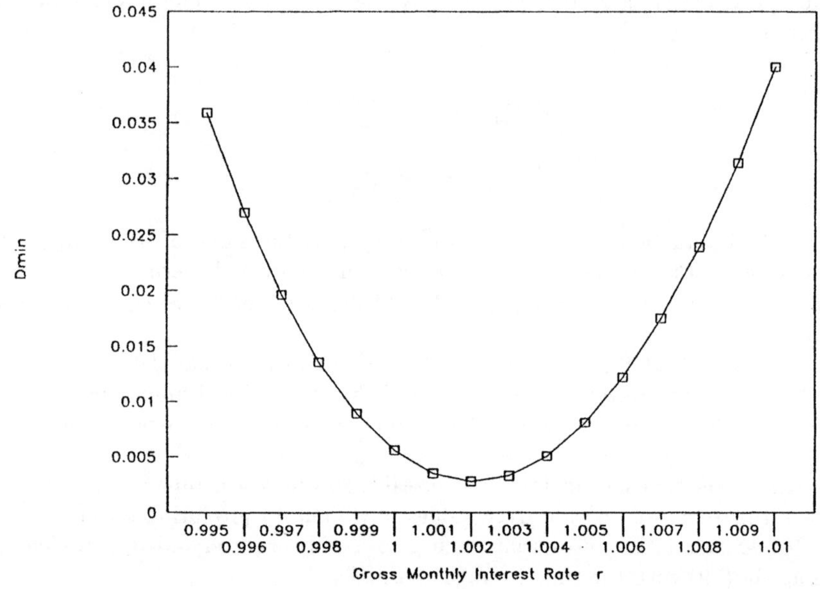

Figure 1: Convexity of $D_{\min}$
U.S. Stocks and Bonds: 1959:7–1986:12

[13]In [16], this curve was used to provide a diagnostic aid for interpreting estimated consumption-based asset pricing models.

The financial interpretation of (23) is straightforward. We first interpret the left hand side of (23). To do so, *we additionally assume that* $e^{\sum_i \gamma_i^a R_i^a}$ *is uncorrelated with* $e^{\sum_i \gamma_i^b R_i^b}$, so that the expectation in (19) for the combined markets after integration factors into the product of the country-specific expectations: [14]

$$E_\pi\left[e^{\sum_i \gamma_i^a(R_i^a(\omega;\alpha)-r)+\sum_i \gamma_i^b(R_i^b(\omega;\alpha)-r)}\right] = E_\pi\left[e^{\sum_i \gamma_i^a(R_i^a(\omega;\alpha)-r)}\right] E_\pi\left[e^{\sum_i \gamma_i^b(R_i^b(\omega;\alpha)-r)}\right]. \tag{24}$$

By (19), this assumption insures that $D_{\text{min}} = D^a_{\text{min}} + D^b_{\text{min}}$ (or $S_{\text{max}} = S^a_{\text{max}} + S^b_{\text{max}}$) *after* the financial markets are integrated. So this assumption is the financial analog of the "weak interaction", or "additivity of entropy" postulate in thermostatics. Thus, the sum of the two countries' degrees of risk at the common, post-integration riskless rate is in fact the integrated market's degree of risk. The right hand side of (23) is a lower bound on the sum of the countries' separate degrees of risk. Because both countries must have a common interest rate r after integration, the discrete changes in the countries' interest rates must satisfy the following equation:

$$r = r^a + \Delta r^a = r^b + \Delta r^b. \tag{25}$$

Solving (25) for Δr^b and substituting in (23) yields

$$\Delta(D^a_{\text{min}} + D^b_{\text{min}}) > \Delta r^a\Big(\sum_i \hat{\gamma}_i^a + \sum_i \hat{\gamma}_i^b\Big) + \sum_i \hat{\gamma}_i^b(r^a - r^b) \tag{26}$$

By way of analogy to the Second Law of thermostatics, one might conjecture that the total entropy should increase, i.e. that the total degree of risk will decrease, so the left hand side of (26) will be less than zero. Unlike the Second Law of thermostatics, there is no obvious theoretical reason why this should be true in our context. But it is not implausible either, and would have to be verified empirically. Adopting this assumption, (26) becomes the following testable restriction on the change in a country's interest rate following integration:

$$\Delta r^a\left(\sum_i \hat{\gamma}_i^a + \sum_i \hat{\gamma}_i^b\right) < \sum_i \hat{\gamma}_i^b(r^b - r^a) \tag{27}$$

Thus, if the asset price process parameters remain invariant following integration, empirical validation of the analog of the Second Law would enable a prediction (27) about the size of the change in a countries' riskless rate following integration. This is a testable prediction, requiring estimates of the countries' pre-integration interest rates and "tempers" (obtained by the aforementioned time series methods). It is hard to imagine that a sharper and/or more easily tested prediction about Δr^a would arise from a structural general equilibrium model of market integration.

The general relationship incorporating "mechanical interaction", i.e. changes in asset price processes induced by financial market integration, is similarly found from (20) to be:

[14] Alternatively, the partition function for the system factors into the product of the separate countries' partition functions.

$$
\begin{aligned}
d(D^a_{\min} + D^b_{\min}) &= dr^a \sum_i \hat{\gamma}_i{}^a + dr^b \sum_i \hat{\gamma}_i^b \\
&+ \sum_i \hat{\gamma}_i^a E_{\hat{\mathbf{Q}}^a}[dR_i^a] + \sum_i \hat{\gamma}_i^b E_{\hat{\mathbf{Q}}^b}[dR_i^b]
\end{aligned} \tag{28}
$$

The first term in (28) is the effect just analyzed (i.e."thermal" interaction). The second term in (28) corresponds to the "generalized forces" associated with the flux of investment. The expectational terms play a role like pressure does in thermostatics, the flow of investment is like a change in volume [15], and the product of the two is a "work effect". Once again, it is not implausible that the left hand side of (28) will be negative. International diversification of assets which would accompany financial market integration might very well result in smaller risk premia than existed in the segmented markets, thereby lowering the degree of risk in one or both countries. In thermostatics, relations like (28) are very useful, because they lead to results which were known and used long before their micro foundations were provided by statistical mechanics. But in much of economics, widely accepted and used macro relations of this nature are few and far in between, in part because of the inability to do controlled experiments. While volume is an experimentally controllable quantity in repeatable experiments, the flow of investment induced by financial integration is not controllable. Furthermore, observed instances of financial market integration are not as clearly demarcated as, say, a supernova. And it is not as simple to compare one instance of financial market integration to another as it is to compare repeatable experiments differing by one controlled variable. Econometric methods are a poor substitute for rigorous experimental control of parameters.

5. Conclusions

The absence of arbitrage opportunities in a popular model of financial economics is equivalent to a set of linear constraints on probability measures. Using MaxEnt to select a measure from this set produces Gibbs' canonical distribution. Computation of expectations taken with respect to the canonical distribution – one of the hallmarks of Gibbsian statistical mechanics – provides an alternative means of deriving some asset pricing formulae.

The connection between Gibbs' canonical distribution and arbitrage-free asset pricing permits us to explore financial uses of thermostatics, which is based on a parametric sensitivity analysis of the MaxEnt problem. The first step was taken toward a thermostatic-like theory of effects caused by the integration of once segmented international financial markets. A lack of adequate empirical evidence helped ensure that this wasn't an earth-shaking step.

[15]Much of thermostatics is based on the analysis of changes in "extensive parameters'. An extensive parameter is one which adds to a constant across the two subsystems, thereby playing a role analogous to internal energy or volume in thermostatics. Using them, one may develop differential relations linking changes in subsystem entropy (or degree of risk) to "forces and fluxes" associated with extensive parameters, as in Haken [sec.3.5][3]. Because there is no guarantee that $\Delta r^a + \Delta r^b = 0$, the riskless rate itself cannot be treated as an extensive parameter. But a conceivable extensive parameter would be the total investment in the two countries' financial assets. Under the plausible (but not inevitable!) assumption that total investment will remain unchanged after financial market integration, any changes in riskless rates and/or other asset returns will be associated with a "flux" of investment between the two countries.

Still, the method has the potential to generate sharp, testable predictions about interest rate and asset price process movements following integration.

We thus see that the two hallmarks of generalized statistical mechanics, i.e. expectations taken with respect to Gibbs' canonical measure and the parametric sensitivity of the MaxEnt problem, both provide useful insights into important problems in financial economics. This extends Gibbs' influence on economic theory beyond the fundamental contributions related to Samuelsons' (and others' subsequent) stress on the solution and parametric sensitivity of *general* constrained optimization problems.

Economic theorists' current understanding of its mainstream conceptual framework has taken over 200 years. The centenary of the publication of Gibbs' Elementary Principles of Statistical Mechanics is less than a decade away. Perhaps future research will bring MaxEnt into the mainstream economic theorists' liturgy by its bicentennial.

References

[1] Michael U. Dothan. *Prices in Financial Markets.* Oxford University Press, 1990.

[2] Christopher Grandy. The principle of maximum entropy and the difference between risk and uncertainty. In W. T. Grandy and L. H. Schick, editors, *Maximum Entropy and Bayesian Methods*, pages 39–47, Kluwer, 1991.

[3] H. Haken. *Synergetics: An Introduction.* Springer-Verlag, 1977.

[4] Hua He. Convergence from discrete to continuous-time contingent claim prices. *Review of Financial Studies*, 3(4):523–546, 1990.

[5] Arthur Hobson. *Concepts in Statistical Mechanics.* Gordon and Breach, 1971.

[6] Chi-fu Huang and Robert H. Litzenberger. *Foundations for Financial Economics.* North-Holland, 1988.

[7] Robert Jarrow and Andrew Rudd. *Option Pricing.* Irwin, 1983.

[8] Edward Jaynes. Information theory and statistical mechanics. *Physics Review*, 106:620–630, 1957.

[9] J.D. Jobson and Bob Korkie. Potential performance and tests of portfolio efficiency. *Journal of Financial Economics*, 10:433–466, 1982.

[10] K.N. Kapur. *Maximum-Entropy Models in Science and Engineering.* Wiley Eastern, Ltd., 1989.

[11] Esfandiar Maasoumi. Information theory. In *The New Palgrave: Econometrics*, Norton, 1990.

[12] Harry Markowitz. *Portfolio Selection.* Basil Blackwell, 1991.

[13] Paul A. Samuelson. *Foundations of Economic Analysis.* Harvard University Press, enlarged edition, 1983.

[14] Paul A. Samuelson. Gibbs in economics. In G. Caldi and G.D. Mostow, editors, *Proceedings of the Gibbs Symposium*, pages 255–267, American Mathematical Society, 1990.

[15] Bruno Solnick. *International Investments.* Addison-Wesley, 1991.

[16] Michael Stutzer. *A Bayesian Approach to Diagnosis of Asset Pricing Models.* Technical Report, Dept. of Finance, CSOM, University of Minnesota, 1992.

[17] Michael Stutzer. *An Information Theoretic Index of Risk in Financial Markets.* Technical Report, Dept. of Finance, CSOM, University of Minnesota, 1993.

[18] Michael Stutzer. The statistical mechanics of asset prices. In K.D. Elworthy, W.N. Everett, and E.B. Lee, editors, *Differential Equations, Dynamical Systems, and Control Science*, Marcel Dekker, 1993 (forthcoming).

[19] Myron Tribus. *Thermostatics and Thermodynamics.* Van Nostrand, 1961.

[20] Walter Willinger and Murad Taquu. The analysis of finite security markets using martingales. *Advances in Applied Probability*, 19:1–25, 1987.

[21] Arnold Zellner. Bayesian methods and entropy in economics and econometrics. In W. T. Grandy and L. H. Schick, editors, *Maximum Entropy and Bayesian Methods*, pages 17–31, Kluwer, 1991.

LESSONS FROM THE NEW EVIDENCE SCHOLARSHIP

G A Vignaux
Institute of Statistics and Operations Research
Victoria University, PO Box 600, Wellington, New Zealand,
Tony.Vignaux@vuw.ac.nz

Bernard Robertson
Department of Business Law
Massey University, PO Box 11222, Palmerston North, New Zealand,
B.W.Robertson@massey.ac.nz

ABSTRACT. There has been much debate on whether Bayesian probabilistic analysis of legal disputes can improve court decision-making. In this paper we ask what Bayesians can learn from problems faced by the courts.

Though debate continues, the Bayesian approach is clearly right for analysing those clearly definable and quantifiable problems which arise in forensic science. When we attempt to generalise and apply these techniques to other forms of evidence some fundamental difficulties arise. Typically there are two responses. The statisticians respond by redefining the question so that it can be answered using orthodox frequentist techniques. Alternatively, some lawyers respond that the evidence should be treated 'holistically'. The problems are difficult and are general to real life decision-making but only Bayesian probability theory offers an approach for analysing and eventually overcoming them.

1. Introduction

There is vigorous debate and substantial literature on the use of Bayesian methods and techniques such as inference charts to analyse evidential problems in court cases. This movement is commonly termed the New Evidence Scholarship as descibed by Eggleston (1983), Tillers and Green (1988) and Watson (1991). Most of this literature is written on the assumption, at least by those arguing in favour of Bayesian methods, that the application of well established techniques will inevitably improve court decision-making. Others argue that practical courtroom problems are much too difficult ever to be captured by analytical techniques.

There has been less thought about whether we can improve and develop Bayesian methods using lessons learned from analysing problems which, in size and complexity, usually dwarf the problems used to expound Bayesian methods. Here we consider a few points that have arisen from our attempts to analyse legal problems from the Bayesian view.

2. Scientific evidence

Even opponents of the general applicability of probabilistic methods in legal problems will usually agree that forensic scientific evidence can often be properly analysed and presented

G. R. Heidbreder (ed.), Maximum Entropy and Bayesian Methods, 391–399.

in this way. Despite this general approval there are vigorous arguments within the forensic science community between the proponents of orthodox and Bayesian methods.

The question the court must answer is "how much does the evidence presented tend to prove or disprove the defendant's liability?" In our view this is best answered by the value of the likelihood ratio of the evidence for two competing positive and specific hypotheses (Vignaux and Robertson, 1993). Currently prevailing orthodox methods, in contrast, answer pre-data questions such as "what is the probability of obtaining a 'match' by chance using the procedure I am about to use?"

Orthodox statistical techniques may have survived for so long because in one special case they give the same answer as the Bayesian techniques. This special case is when:

1. The characteristic concerned is either present or absent (i.e. the samples either 'match' or do not 'match').

2. A single mark is being examined, e.g. only one bloodstain at the scene of the crime is to be matched or there is only one group of glass fragments on the accused's clothing.

3. The population from which any frequency is derived is homogeneous. This means that if we are considering the human population it must be in Hardy-Weinberg equilibrium, ie, randomly mating. We must not have sub-populations where the distribution of the characteristic differs from the population as a whole.

4. Only one comparison is made (either between the accused and a mark at the crime scene or between the crime scene and a mark on the accused), we are not screening large number of suspects or searching through databases for a match.

Suppose a single bloodstain at the scene of a crime is analysed by conventional blood typing methods and found to have a combination of characteristics shared by only 1 in 1000 of the population. The accused, after arrest, is found to share these characteristics. Assume there is no other evidence about the perpetrator. The Bayesian report would state that the blood evidence was 1000 times more probable if the accused were the perpetrator than if a 'randomly selected' member of the population were. The orthodox statistical report would be that the probability of a 'match by chance' is 1 in 1000. In this case, then, the jury would probably see little difference in meaning between the two reports.

If any of these conditions is violated the orthodox techniques produce a wrong answer. Firstly, where the characteristic is continuously variable (like glass refractive index or DNA band position) we meet the problems inherent in significance testing. A very early paper by Lindley (1977) pointed out the consequences of this "fall off a cliff" problem but most scientists, under the influence of orthodox statistics teaching, failed to change their practices and the courts are used to receiving evidence in this form. The introduction of DNA evidence makes the problem more acute. DNA bands separated by 2.99 standard deviations (sd) may be 'declared a match' while bands separated by 3.01 sd are considered 'not to match'. Authorities disagree on appropriate 'match criteria' and argument has become diverted into this question and into whether, as in the famous Castro case (1989), the chosen arbitrary criteria have been adhered to.

Secondly, where more than one mark is found the orthodox techniques gives the same answer *regardless of which mark the accused matches* even when the characteristics of one mark are common and the others rare. Evett (1987) showed that the value for the Likelihood Ratio where there are n marks (such as n different bloodstains) is $1/nf$ where f is the frequency of the characteristic shared by the accused.

Thirdly, orthodox techniques lead to considering the frequency of the accused's characteristics within *his* sub-group of the population whereas the correct alternative hypothesis, assuming no other information, is that the perpetrator was some unknown person. As Walsh, Buckleton and Evett (1991) pointed out, the relevant sub-population is correctly defined by what is known of the perpetrator, not the accused. This point has generated most of the misunderstanding and argument about the value of DNA evidence (Lewontin and Hartl, 1991). It is often argued that, as the accused comes from some peculiar sub-population, the evidence is of questionable value if there is no database for that sub-population. This would arise, for example, in New Zealand if the accused is from a small Pacific island population such as Niue and the database contains no samples from that sub-group. This argument ignores two matters: (a) the correct alternative hypothesis usually relates to a much larger population and (b) evidence, if it exists, identifying the perpetrator as a member of a small sub-group also affects the prior probability that should be used. This can easily be handled in the Bayesian framework.

Fourthly where a large number of people are screened or a database is searched the probability of finding a 'match by chance' is obviously increased. Orthodox techniques therefore mandate that this is found (approximately for small frequencies) by multiplying the proportion of the characteristic in the population by the number of comparisons carried out. Logical analysis shows that the value of the evidence is not in fact affected by databases size if the whole database is searched. A characteristic with a frequency of 0.001 yields a likelihood ratio of 1,000 however many comparisons are carried out.

There are two points which need to be made about databases. First we may be using the database to assess the frequency of a characteristic about which we have no initial data such as that of facial tattoos. In this case for any given number of 'matches' the value of the evidence actually increases with the size of the database, the extreme example being where only one 'match' is found in a database of the entire population. The idea that to an orthodox statistician an increase in the size of a database makes evidence weaker while to a Bayesian it makes the evidence stronger starkly illustrates the difference between pre-data and post-data approaches.

Again, where a database has been screened, the prior odds and the strength of the case as a whole need to be carefully examined. We suggest that the prior odds should be 1: the size of the population from which the database is drawn. Thus if the database is the criminal records drawn from the entire population of the USA the prior odds should be 1:250 million. Detection of a matching characteristic with a frequency of 1 in one million will therefore yield posterior odds of 1:250 (ie 250 to 1 that the accused is not the perpetrator). The Bayesian scientist must ensure that the jurors do not invert the conditional here; but that should be done by careful explanation, not by doctoring the evidence.

Application of probability theory forces us to identify the right questions and to realise that we can only produce a likelihood ratio for competing hypotheses. If the alternative hypothesis is altered the value of the evidence will be affected. The dependence of the value of the evidence on all the circumstances is dramatically illustrated by the example of 'associative evidence' which can *reduce* the probability that the accused was the perpetrator.

Evett (1987 showed that this may occur where n bloodstains at the scene of a crime are examined and the characteristics of the stain which the accused matches have a frequency greater than $1/n$.

Presentation of the evidence in likelihood ratio form leads to two difficulties that do not appear in the orthodox approach:

1. If we ask a question like "What are the probabilities this accused would have this glass on him if he were or alternatively were not the person who broke the window?" we have to assess the probability that, given everything we know about the incident, glass would be transferred from the window to the person who broke it (the transfer probability) and remain until the forensic scientist observed it (persistence). Before the use of Bayesian techniques for forensic work neither these probabilities of transfer and persistence nor the question of what proportion of people generally have glass on their clothing was addressed (Abadom, 1983, Evett and Buckleton, 1990, and Evett, 1986). We have identified a factor which not only affects the value of the evidence but also increases the complexity of the problem. Thus even if we limit ourselves to analysing forensic scientific evidence a Bayesian approach may increase complexity. However this is a minor problem compared with the question to which this analysis ineluctably leads:

2. If evidence is to be presented as a likelihood ratio which is to be applied to prior odds where do these prior odds come from? In other words, is the other evidence in the case susceptible to the same analysis?

3. Evidence Generally

When we try to apply Bayesian methods to the analysis of evidence other than forensic evidence we meet more severe problems. First there is the complexity of the problems and the interactions between different pieces of evidence and inferences. Diagrams have been suggested for dealing with this but such techniques seem to be unable to accommodate the richness of courtroom problems. Then there is the need to deal explicitly with the differences in background information available to different players in the courtroom.

3.1. Diagrams of Evidence

Diagrammatic methods have been proposed in the past for analysing the structure of complex legal problems by Wigmore(1913) and presented by Vignaux and Robertson (1993a,c). Such diagrams enable the human mind to grapple with complex problems by allowing one to examine the problem one part at a time while at the same time documenting the connections between the parts. Bayes networks (or belief nets or influence diagrams), as described in Oliver and Smith (1990), set out in an acyclic network of nodes connected by directed arcs the propositions (hypotheses) to be proved and the evidence, in the form of propositions, on which we expect the proof to be based. These form the nodes of the network; each node represents all the possible values that a proposition can have.

An arc between two nodes represent conditional dependence between those nodes; lack of an arc represents the assumption of conditional independence between them.

Having defined the problem, identified the influences, and drawn the diagram, conditional probability tables are established for each proposition (node). The first version of the diagram represents the problem prior to the receipt of evidence, taking into account all the possible values that the evidence could take. Evidence is then gathered to give the values of some of the lower-level propositions. These values propagate through the network to establish probabilities for the various propositions conditional on the given evidence.

Standard probability calculations are represented on the network by changing the directions of the arcs (corresponding to the Bayes Rule calculation of $P(H|EI)$ from $P(E|HI)$) and updating the conditional probability tables. Additional links appear between the different pieces of evidence. These links correspond to the predictability of unknown evidence from known evidence. The diagram quickly becomes very dense.

Though such a graphical presentation can be helpful it has a number of limitations which mean that it still does not capture the messy and dynamic nature of real-life inferential problems.

First, the standard exposition of Bayes networks assumes a defined problem, a predetermined set of influences, and a process clearly divisible into problem definition and data gathering. But knowledge of the problem is required in order to identify relevant influences and as one acquires more knowledge one may discover new effects which may have to be investigated. The stages of the process are thus more interactive than the model can describe.

Second, a pre-condition for the admissibility of any evidence into court is that it is "relevant". Relevance is defined in the US Federal Rules of Evidence, Rule 401: ' "Relevant evidence" means evidence having any tendency to make the existence of any fact that is of consequence to the determination of the action more probable or less probable than it would be without the evidence.'

Leaving aside the question of whether the existence of facts can be made more probable by evidence, this definition clearly implies that proposition B is relevant to proposition A whenever the likelihood ratio is other than 1. However, for any A and B it is highly improbable that the likelihood ratio is precisely 1. In other words, any proposition is potentially relevant to any other. This means that there is no clear legal limit to what should be included in any Bayes network nor a clear distinction between nodes that should be connected by arcs and those that are not.

Another Federal Rule, Number 403, states: "Although relevant, evidence may be excluded if its probative value is substantially outweighed by the danger of unfair prejudice, confusion of the issues, or misleading the jury, or by considerations of undue delay, waste of time or needless presentation of cumulative evidence."

There is a cost to introducing any item of evidence and so in every case the application of Rule 403 involves balancing probative value (ie how much the likelihood ratio differs from 1) against the increasing complexity. This is also the judgement that has to be made in deciding whether to connect two nodes by an arc. Each decision to draw or not to draw an arc is a matter of balance and judgement, in contrast to the standard view that propositions are either dependent or conditionally independent.

The same difficulty arises in defining the boundary of the diagram. Why are some matters included and others left out? Some matters, such as the law of gravity, are so obvious and pervasive, that they are automatically included in every assessment and therefore do not appear as a node. Other matters we may know, because of our background knowledge, will have no influence. For example, until recently it was not possible to obtain a DNA analysis from cut hair. An investigator who knew this would not have included a node for a DNA match in a Bayes network of the problem, indeed would not even have collected the hair sample. In other instances a factor's influence may be so tenuous so as not to justify consideration. In the end judgement has to be made about what factors to include in a Bayes net and that judgement will depend on knowledge of the world and about the problem. This further emphasises that problem definition and data gathering are not distinct phases.

The parallel legal concept is "judicial notice". Matters such as "Mr Clinton is currently the President of the USA" may be "judicially noticed" if it is important in the case. This is conventionally expressed as an exception to the rule that cases are decided only on the evidence presented in court. Seen in a Bayesian light, however, judicial notice is not an exception but represents the grey area between those matters which must clearly appear in the Bayes network and the very large body of background knowledge (that will be absorbed into I).

3.2. Different Background Information

In the odds form, Bayes rule tells us that, given evidence E_1

$$\frac{P(H|E_1,I)}{P(\bar{H}|E_1,I)} = \frac{P(H|I)}{P(\bar{H}|I)} \frac{P(E_1|H,I)}{P(E_1|\bar{H},I)}, \tag{1}$$

where H is the hypothesis to be proved and I is our background knowledge. It would then conventionally be said that the posterior odds in (1) provide the prior odds when a new piece of evidence, E_2 is considered. The effect of the new piece of evidence would then be shown as:

$$\frac{P(H|E_1,E_2,I)}{P(\bar{H}|E_1,E_2,I)} = \frac{P(H|E_1,I)}{P(\bar{H}|E_1,I)} \frac{P(E_2|H,E_1,I)}{P(E_2|\bar{H},E_1,I)} \tag{2}$$

We discuss two problems that arise from this exposition.

The rule is obviously only formally valid if the I, represents the same body of knowledge wherever it appears. If the same piece of evidence is examined against different background knowledge, different conclusions may result (Jaynes, draft, Chapter 5).

However, where a forensic scientist expresses a likelihood ratio for an item of evidence it will be informed by the scientist's own I. the prior odds being determined by the jury will be informed by the jury's own I. This leads to a number of observations.

1. Each juror's I will also be different. The jury may, *inter alia*, be a device for ensuring that decisions are taken against a background of a number of different Is in the hope that distortions in perception are evened out. During jury vetting procedures in the USA parties attempt to obtain a jury whose composite I is likely to lead to a favourable result. In England and New Zealand such procedures are not permitted, the theory being that the jury should be representative of the population.

2. Strategies are necessary to make the value of evidence insensitive to differences in the I. Chief among these is the Opinion Evidence Rule. Whilst a strict distinction between fact and opinion cannot be sustained, the principle is that witnesses should only testify as to what they have perceived with one of the five senses. Thus ordinary witnesses must not give their opinions (their inferences). They are allowed only to give observations. Inference should, so far as possible, be conducted only by the jury. There is however

an irreducible element of interpretation in all evidence so the problem of different Is for different items of evidence provided by different witnesses cannot be avoided; it can only be minimised.

This problem will arise whenever evidence from different sources is considered. The Opinion Evidence Rule tells us that, as far as possible, one should go back to "raw" data. This enables one to interpret the data not only in the light of the hypotheses in which one is interested but also one's own consistent I. The techniques proposed by Garrett and Fisher(1992) in dealing with evidence from different sources achieve the same object. Even if the different investigators were all Bayesians and provided likelihood ratios, combination of their results still requires resort to the original data to eliminate the effect of different Is.

3. This point will be missed if one is not rigorous about including the Is in the notation, a point made frequently by Jaynes (draft, Chapter 15).

4. Responses

Attempts to analyse evidential problems in court in Bayesian fashion are frequently met with the objection that these methods do not capture human thought processes even when those processes are rational. We have previously shown (Vignaux and Robertson, 1993b) that many of these objections are based on a purely frequentist model of probability. However it may be true that some such objections identify ways in which Bayesian techniques, as conventionally explained, are not as helpful as they might be. Certainly, Bayesian analysis of legal problems is a complex and subtle process. There have been two stock responses to this complexity.

The first is that of the orthodox statistician, to redefine the problem so that it can be tackled by orthodox statistical techniques. This has two important drawbacks: first we do not advance our understanding of the problems of inference since we simply define them away; secondly, we saw in dealing with the forensic scientific evidence that the questions addressed by this approach are actually quite different from the questions the court wants to answer. This means, as Pratt (1961) puts it, "the problems it solves, however precisely it may solve them, are not even simplified theoretical counterparts of the real problems to which it is applied".

The lawyer's response has been to assert that jurors can or should assess the evidence "holistically" using some non-quantitative technique. There are a number of problems to this approach. First it is inaccessible. It is unclear how evidence is to be assessed holistically. It is even unclear what it means. Secondly, as Jaynes demonstrates (draft, Appendix A), any system of analysis must either be reducible to quantitative form or violate the requirements of rationality. Thirdly "it is hard to imagine how we can imbibe the evidence we 'see' without performing *some* sort of mental analysis, which, by definition, seems to involve some sort of dissection."(Tillers,1989).

Neither of the above approaches enables us to proceed. We must recognise as Friedman(1992) said that "the world is a complex place but that is not the fault of Bayesian analysis". At their present stage of development Bayesian techniques are based upon models which, while constituting a vast improvement on those used in classical statistical analysis, still do not capture the complexity of the reasoning process. Thus we cannot pretend today to have all the answers to every objection raised to probabilistic analysis of legal decisions but the only hope of analysing and eventually overcoming these problems is offered by Bayesian probability theory.

References

[1] Case. R v Abadom. *All England Reports*, 1:364–369, 1983.

[2] K A J Walsh J S Buckleton and Ian Evett. Who is 'random man'? *Journal of the Forensic Science Society*, 31(4):463–468, 1991.

[3] Mary P Watson (ed). Decision and inference in litigation. *Cardozo Law Review*, 13(2-3):253–1079, November 1991.

[4] Case. People v Castro *545 New York Supplement* 985 (SC/NY).

[5] Richard Eggleston. *Evidence, Proof and Probability.* Law in Context. Weidenfeld and Nicolson, London, 2nd edition, 1983.

[6] Ian Evett. A Bayesian approach to the problem of interpreting glass evidence in forensic science casework. *Journal of the Forensic Science Society*, 26:3–18, 1986.

[7] Ian Evett. On meaningful questions : a two-trace transfer problem. *Journal of the Forensic Science Society*, 27:375–381, 1987.

[8] Ian Evett and John Buckleton. The interpretation of glass evidence. A practical approach. *Journal of the Forensic Science Society*, 30:215–223, 1990.

[9] R D Friedman. Infinite strands, infinitesimally thin: Storytelling, Bayesianism, Hearsay and other evidence. *Cardozo Law Review*, 14:79 – 101, 1992.

[10] A J M Garrett and D J Fisher. Combining data from different experiments: Bayesian and Meta-analysis. In C.R. Smith et al (eds), *Maximum Entropy and Bayesian Methods*, 273–286. Kluwer Academic Publishers., 1992.

[11] E T Jaynes. *Probability theory - the logic of science.* in draft, 1993.

[12] R C Lewontin and Daniel L Hartl. Population Genetics in Forensic DNA typing. *Science*, 254:1745–1750, December 1991.

[13] Dennis Lindley. A problem in forensic science. *Biometrika*, 64:207–213, 1977.

[14] Robert M Oliver and James Q Smith (eds). *Influence Diagrams, Belief nets and Decision Analysis.* John Wiley and Sons, Chichester, 1989.

[15] John W Pratt. Book review of Testing Statistical Hypotheses by E L Lehmann. *JASA*, 56:163–166, March 1961.

[16] Bernard Robertson and G A Vignaux. Taking Fact Analysis seriously. *Michigan Law Review*, 92(6):1442–1464, 1993a.

[17] David A Schum and Peter Tillers. Marshalling evidence throughout the process of fact-investigation : a simulation - part iii. investigating fact investigation. *unpublished report 89-03*, Oct 1989.

[18] Peter Tillers. Webs of things in the mind : A new science of evidence. *Michigan Law Review*, 87:1225–1258, May 1989.

[19] Peter Tillers and Eric D. Green. *Probability and Inference in the Law of Evidence.* Kluwer, 1st edition, 1988.

[20] G A Vignaux and Bernard Robertson. Bayesian methods and Court Decision-making. In A Mohammad-Djafari, editor, *Maximum Entropy and Bayesian Methods, Paris, 1992 (Proc 12th MaxEnt Workshop)*, 85–92. Kluwer Academic Publishers, 1993b.

[21] G A Vignaux and Bernard Robertson. Wigmore Diagrams and Bayes networks. In A Mohammad-Djafari, editor, *Maximum Entropy and Bayesian Methods, Paris, 1992 (Proc 12th MaxEnt Workshop)*, 93–98. Kluwer Academic Publishers, 1993c.

[22] John Henry Wigmore. *Principles of judicial proof.* Little, Brown and Co, Boston, 2nd edition, 1931.

HOW GOOD WERE THOSE PROBABILITY PREDICTIONS? THE EXPECTED RECOMMENDATION LOSS (ERL) SCORING RULE

David B. Rosen
Center for Biomedical Modeling Research
University of Nevada, Reno
Present address:
Department of Medicine
New York Medical College
Valhalla, NY 10595 USA
Internet: `d.rosen@ieee.org` (or `rosen@unr.edu`)

ABSTRACT. We present a new way to understand and characterize the choice of scoring rule (probability loss function) for evaluating the performance of a supplier of probabilistic predictions after the outcomes (true classes) are known. The ultimate value of a prediction (estimate) lies in the actual utility (loss reduction) accruing to one who *uses* this information to make some decision(s). Often we cannot specify with certainty that the prediction will be used in a particular decision problem, characterized by a particular loss matrix (indexed by outcome and decision), and thus having a particular decision threshold. Instead, we consider the more general case of a distribution over such matrices. The proposed scoring rule is the *expectation*, with respect to this distribution, of the loss that is actually incurred when following the decision *recommendation*, the latter being the decision that would be considered optimal *if* we were to *assume* the predicted probabilities. Logarithmic and quadratic scoring rules arise from specific examples of these distributions, and even common single-threshold measures such as the ordinary misclassification score obtain from degenerate special cases.

1. Introduction

1.1. Purpose

One of two outcomes (events or classes) will occur in an observation or experiment. We consider a forecaster providing an assessment, i.e. estimate, opinion, or *prediction*, of the probability that one of them (say outcome 1) will occur. We use the term *forecaster* broadly: this could be a human expert, maximum-likelihood fit of some parametric model, classifier / learning machine, or Bayesian inference procedure given a particular prior, to name some possibilities. We are not concerned here with how this prediction was or should have been generated from some set of available information (such as training sample data and prior knowledge), but rather with the question of what figure of merit, i.e. scoring rule or *probability loss function*, we should assign to the (probabilistic) prediction in hindsight once the true outcome is known.

1.2. Probability Loss Function

As an example, consider a weather forecaster who states that "the probability of *rain* today is $\hat{p}$", and of course, perhaps implicitly, "the probability of *no rain* today is $(1 - \hat{p})$".

We wish to choose a function L in order to assign a score to today's prediction by the forecaster as $L(i, \hat{p})$, where i is the actual outcome: 0 for no rain or 1 for rain. Examples of such a function include the logarithmic loss $L(i, \hat{p}) = -i \log(\hat{p}) - [1 - i] \log(1 - \hat{p})$ [6], the quadratic loss (squared

G. R. Heidbreder (ed.), Maximum Entropy and Bayesian Methods, 401–408.

error) $L(i, \hat{p}) = [i - \hat{p}]^2$ [3, 15, 5], and of course the binary misclassification loss, which is zero or one depending merely on whether $\hat{p}$ is on the appropriate side of $\frac{1}{2}$.

1.3. Importance of the Probability Loss Function

To place this problem into a particular practical setting for concreteness, suppose that some *user* of weather forecasts wishes to hire one forecasting consultant (or purchase one computer-based weather forecasting system) from among several available. The user has access to the forecasters' respective predictions and the true weather every day for the past year, and wishes to decide which one to hire (or buy) based on performance on this set of test data[1]. The relevant measure of performance is the expected benefit (i.e. utility) that *would have* accrued to the user during the test period if he had relied entirely on a given forecaster's predictions.

Then the user might hire the forecaster providing the best performance (after subtracting off the consulting fee each charges, if these differ).

We consider the predictions of a single forecaster. Since (expected) utilities are additive, it suffices to consider each test point (this forecaster's prediction and the actual outcome, for a single day) separately, and then sum these later. Thus, we are back to the problem of choosing a loss function for a single probability prediction, but now with the idea that it should perhaps represent the actual loss to the user of a prediction $\hat{p}$ when the ith outcome occurs.

1.4. Overview of Paper

Section 2. explains how a probability prediction can be viewed as a mapping from possible decision problems the user may face, each characterized by a decision loss (utility or regret) matrix, to corresponding *decision recommendations*. For any one such decision problem, the loss actually incurred after making the recommended decision defines the quality of the probability prediction. Then in Section 3. we consider the case in which we might use the probability prediction in any of a continuum of decision problems, described by a *distribution* of decision loss matrices. We show that the recommendation loss approach to scoring or loss functions can be generalized by using as our scoring function the *expected* recommendation loss (ERL), where the expectation is over (only) the decision loss matrix distribution. We explain how the commonly used scoring functions mentioned above arise in this recommendation loss approach. Table 1 summarizes the quantities and notation used in this paper.

Section 4. briefly discusses some of the literature on probability loss functions based on truth- or honesty-rewarding properties, and relates this to the ERL results of the present paper.

2. Decision, Prediction, and Recommendation

A decision problem concerns the choice of a course of action having real-world consequences (costs) that depend on both the action (decision) and an outcome event (true class). In the case of weather, examples would include deciding whether to water the lawn, bring an umbrella, or cancel some outdoor social event. The "cost" of cancelling a social event unnecessarily (vs. having the event rained on) may be quite different from the "cost" of the lawn being neither watered nor rained on (vs. being both watered and rained on), so you might make a different decision for each based on the same probability prediction.

---

[1]This is an oversimplification since the user should be interested in expected future (generalization) performance rather than empirical past (test set) performance. But we certainly cannot determine the expectation (over outcomes) of performance if we cannot determine performance even when the true outcome is known.

Table 1: Quantities and functions appearing in this paper.

i	=	observed outcome event (true class)
$\hat{p}$	=	prediction (judgment or estimate) of probability of outcome $i = 1$
$L(i, \hat{p})$	=	probability loss function (prediction scoring rule)
j	=	decision index (in some decision problem where i is relevant)
$C = \{c_{ij}\}$	=	decision loss (regret or cost matrix) characterizing a decision problem
t	=	decision threshold $= \frac{c_{01}}{c_{01}+c_{10}}$
s	=	stakes $= c_{01} + c_{10}$
$\hat{j}$	=	decision recommendation $= I(\hat{p} > t)$
$L_C(i, \hat{p})$	=	recommendation loss $= c_{i\hat{j}} = c_{i,I(\hat{p}>t)}$
$L_{\mathsf{ERL}}(i, \hat{p})$	=	expected recommendation loss $= E^C L_C(i, \hat{p})$
$h(t)$	=	threshold importance density $= \mathrm{pr}(t) E^s\{s\|t\}$
p	=	"true" outcome probability or that believed by expert (Sec. 4.)

Table 1a: Notation.

I(inequal.)	=	1 if inequality is true; 0 otherwise	
$E^z\{g(z)\|A\}$	=	expectation over z of $g(z)$ given A	$= \int_Z dz\, \mathrm{pr}(z\|A) g(z)$

Probability predictions per se have no real-world consequences—until used to make decisions. Thus the true measure of a system's performance is of course the actual (or expected) gains or losses to those who use its predictions to make one or more decisions.

2.1. Decision Loss (Cost Matrix)

A decision problem is characterized by the *decision loss* c_{ij} for each observed outcome i and decision j; these c_{ij} can be said to form the elements of a cost matrix C. [2] In general, the number of decision alternatives need not be equal to the number of outcomes (classes), but we assume the two-outcome (i.e. two-class) two-alternative case for simplicity, giving

	$j = 0$	$j = 1$
$i = 0$	0	$c_{01} > 0$
$i = 1$	$c_{10} > 0$	0

,

where we have defined decision $j = 0$ as that most favorable when outcome $i = 0$, and ignored nonzero c_{00} or c_{11} (diagonal elements) as merely leading to overall offsets (Section 3.3.).

2.2. Decision Recommendation Implicit in Prediction

Presumably, if our system were designed for (or a human expert were apprised of) a particular known decision loss matrix C, it could simply plug this and its prediction $\hat{p}$ into elementary decision theory, and thus recommend the course of action $j = \hat{j}$ that minimizes the expected decision loss

$$E^i\{c_{ij}|\hat{p}\} = \hat{p}c_{1j} + (1 - \hat{p})c_{0j} = \begin{cases} \hat{p}c_{10} & \text{if } j = 0 \\ (1 - \hat{p})c_{01} & \text{if } j = 1 \end{cases} .$$

[2]Note that the decision loss is a function of (i.e. is indexed by) outcome and decision made, while a probability loss is a function of outcome and probability prediction.

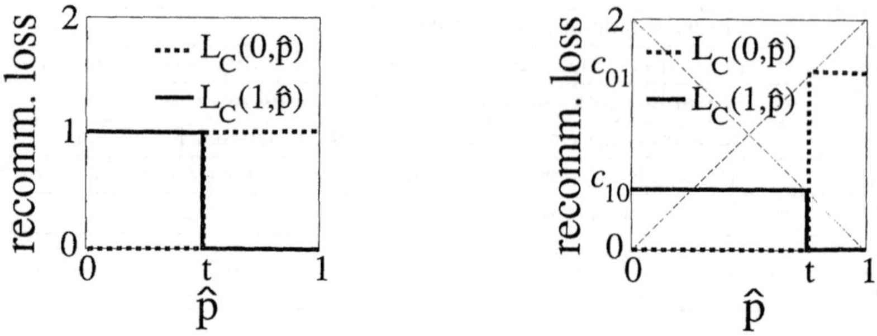

(a): $c_{10} = c_{01} = 1 \Leftrightarrow t = .5, s = 2$ (b): $c_{10} = .5, c_{01} = 1.5 \Leftrightarrow t = .75, s = 2$

Figure 1: Recommendation loss $L_C(i, \hat{p}) = c_{i,I(\hat{p}>t)}$ vs. prediction $\hat{p}$ for two fixed decision problems indicated. In (a) this gives the ordinary "0–1" misclassification loss.

The solution is given by

$$\begin{aligned} \hat{\jmath} &= \begin{cases} 1 & \text{if } (1-\hat{p})c_{01} < \hat{p}c_{10} \\ 0 & \text{otherwise} \end{cases} \\ &\equiv I((1-\hat{p})c_{01} < \hat{p}c_{10}) \\ &= I(\hat{p} > t), \end{aligned}$$

where *decision threshold* t is defined as $\frac{c_{01}}{c_{01}+c_{10}}$ and lies between 0 and 1 (inclusively). Of course this recommendation may be poor if the prediction is poor.

In the work of Thomas Bayes, one can interpret a personal probability as merely a convenient summary of one's decision rule, which is a function mapping the cost matrix to a decision preference. Similarly, we consider a probability *prediction* $\hat{p}$ to be merely a convenient summary of the function mapping cost matrix C to decision *recommendation* $\hat{\jmath}$.

2.3. Decision Recommendation Loss

We rewrite the cost matrix in terms of threshold t (Section 2.2.) and overall *stakes* $s = c_{01} + c_{10}$ as

$$c_{01} = ts, \qquad c_{10} = (1-t)s.$$

Since we consider a probability prediction to represent an implicit decision recommendation to the user, this user's question "how much would the prediction $\hat{p}$ (by itself) be worth to me?" naturally becomes equivalent to "what would be my losses if I followed the corresponding recommendation $\hat{\jmath} = I(\hat{p} > t)$". The user need not make this recommended decision, but then the user's decision loss would not be a measure of the value of the prediction $\hat{p}$ by itself in decision problem C, since we presume then that the decision was based (at least in part) on different believed/assumed outcome probabilities or other information.

The actual decision loss in a given decision problem, when following the recommendation implicit in $\hat{p}$ and the actual outcome is i, is given by the *recommendation loss*

$$\begin{aligned} L_C(i,\hat{p}) = c_{i\hat{\jmath}} = c_{i,I(\hat{p}>t)} &= \begin{cases} tsI(\hat{p} > t) & \text{if } i = 0 \\ (1-t)sI(\hat{p} < t) & \text{if } i = 1 \end{cases} \\ &= [1-i]tsI(\hat{p} > t) + i[1-t]sI(\hat{p} < t), \end{aligned} \tag{1}$$

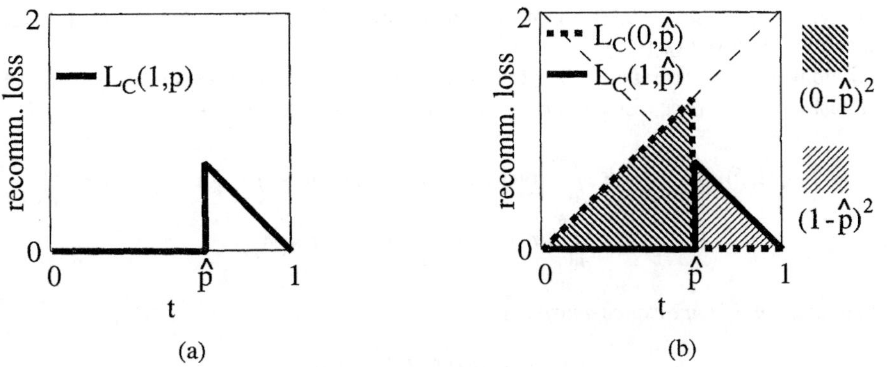

Figure 2: Recommendation loss (with constant stakes $s = 2$) vs. decision threshold t for fixed prediction $\hat{p}$ and (a) $i = 1$; (b) both values of i. In (b), the shaded areas represent $L(i, \hat{p})$ = quadratic loss (squared error), which is the expected recommendation loss (ERL) when importance $h(t) = 2$ (Section 3.1.).

which is plotted as a function of the prediction in figure 1, for two particular decision problems. *If* we are given a single decision problem of interest (with a specific cost matrix C determining a specific threshold), then our final performance measure (probability loss) should simply be given by this recommendation loss. We call this type of probability loss function *single-decision* or *single-threshold*, since it depends only on whether $\hat{p}$ is above or below a particular t.

2.4. Cost Matrix Unknown

If we don't know what the cost matrix, and thus the decision threshold, will be, we can plot the recommendation loss *as a function of this unknown threshold* for a given prediction and outcome, as in figure 2. By summing or averaging such curves over a data set consisting of many prediction-outcome pairs, we can completely characterize the performance on this data. We call this a(n empirical) *recommendation loss characteristic* (RLC)[12] curve, as it is an alternative to the widely-used *receiver operating characteristic* (ROC)[8] curve.

To measure and compare the overall performance of forecasters, we need in general a probability loss function to assign a single numeric score to each. A single-threshold loss is crude and not appropriate unless we are certain that the predictions are never to be used in any other decision problem. Other probability loss functions have been proposed and used historically, but which should we use, and what relationship (if any) will it have to the recommendation loss in decision problems?

3. Expected Recommendation Loss (ERL)

When it is uncertain in which decision problem a prediction will be used, we can describe the situation by a probability distribution over possible decision problems. Suppose pr(C) gives the probability density that our actual decision problem will be described by cost matrix C [3] The natural

[3] C will be known exactly by the decision-maker; we simply do not know it (and thus the decision threshold) now. In contrast, if pr(C) were to remain when the decision itself were made, the decision problem would simply be characterized

probability loss function is then the *expectation over cost matrices* of the recommendation loss, i.e. $E^C L_C(i, \hat{p}) = \int dC \, \text{pr}(C) L_C(i, \hat{p})$. We call this choice of probability loss function the *expected recommendation loss* $L_{\mathsf{ERL}}(i, \hat{p})$.

It is convenient to express the distribution of cost matrices as a joint probability density $\text{pr}(t, s)$ over threshold and stakes, instead of over c_{01} and c_{10}. From (1), we have

$$\begin{aligned} L_{\mathsf{ERL}}(i, \hat{p}) &= \int_0^1 dt \int_0^\infty ds \, \text{pr}(t, s) \, \{[1-i]tsI(\hat{p} > t) + i[1-t]sI(\hat{p} < t)\} \\ &= [1-i] \int_0^{\hat{p}} dt \, th(t) + i \int_{\hat{p}}^1 dt \, [1-t]h(t), \end{aligned} \quad (2)$$

where the *threshold importance density* is

$$h(t) = \text{pr}(t) E^s\{s|t\} \quad \geq 0,$$

i.e. the probability that a decision problem will use this threshold, times the expected stakes of decision problems having such a threshold.

The ERL probability loss function (2) can be described as an average of the decision recommendation loss per stakes (or recommendation loss for unit stakes), *weighted* by the importance $h(t)$ of each decision threshold. It reduces to a single-threshold loss (for example the misclassification loss) when $h(t)$ is concentrated at a single t (for example $\frac{1}{2}$).

3.1. Quadratic Loss

For uniform threshold importance density $h(t) = 2$ (for $t \in [0, 1]$), the ERL is given by the area under the recommendation loss curve, which is the quadratic loss $L_{\mathsf{ERL}}(i, \hat{p}) = [i - \hat{p}]^2$, as indicated by the shaded areas in figure 2b.

3.2. Logarithmic Loss

Let $h(t) = \{t[1-t]\}^{-1}$, sometimes called the Haldane density. From eqn. (2), $L_{\mathsf{ERL}}(i, \hat{p}) = -i \log(\hat{p}) - [1-i] \log(1-\hat{p})$, which is simply the logarithmic loss mentioned earlier. This assigns an unbounded penalty when the prediction $\hat{p}$ is near 0 (1) when the true outcome $i = 1$ ($i = 0$). This is due to the unbounded importance given to thresholds near zero or one by this improper (non-normalizable) density. It can be argued[12] that the Haldane density may be an appropriate noninformative prior when all that is known about the decision problem is that it is nontrivial, thus leading to the choice of the logarithmic loss as a performance measure in such a situation.

The logarithmic and quadratic probability loss functions are members of a particular one-parameter family[7] of ERL loss functions. Even the misclassification loss (recommendation loss when $c_{01} = c_{10} = 1$) is obtained [11] in a limiting case of this parameter.

3.3. Arbitrary offsets in loss functions

If we add to the costs in a decision problem an amount depending on the outcome i but not on the decision j, i.e. $ai + b[1-i]$ with arbitrary real a and b, we then have nonzero diagonal elements c_{00} and c_{11}, but the decision analysis does not change. Similarly, if we add such arbitrary offsets to a probability loss function, the comparison of two forecasters is never affected, even after summing over a data set or taking expectations[13]. If we had considered a distribution over a full cost matrix without assuming diagonal elements of zero, the effect in Section 3. would have been to add to the

---

by the expected cost matrix $E\{C\}$, with a *single* resulting decision threshold.

resulting probability loss function (2) terms that could simply be incorporated into its own arbitrary offsets. In addition, of course, multiplying all loss functions by a constant has no substantive effect.

The logarithmic loss, also called empirical cross-entropy, is related to Kullback-Liebler distance and mutual information by such arbitrary offsets and overall scale[5].

4. Truth-Rewarding Loss Functions

A *strictly proper* probability loss function $L(i, \hat{p})$ can be defined as one that is *truth-rewarding*, i.e. its expectation (over i) is minimized when and only when $\hat{p}$ is equal to the true probability p (input-conditional class probability[4] $\mathrm{pr}(i = 1|x)$ for classification from input features/predictors).

Equivalently, a strictly proper loss function is *honesty-rewarding*: if we dock an expert's pay by $L(i, \hat{p})$ for prediction $\hat{p}$ and outcome i, then her expected pay (conditional on her *belief*) will be maximized if and only if she gives us $\hat{p}$ equal to the probability p implied by her belief[13].

The form (2) of our ERL loss functions (or variants of it, or generalizations to continuous outcomes [2] or their expectation [13, 9], or to more than two discrete outcomes[5]) has previously been proposed as either an objective function to be optimized in parameter estimation [7, 9] or as a device to elicit honest predictions from an expert [1, 13, 10, 4], in contrast to our interpretation as the expected recommendation loss to the user. Also in those treatments, $h(t)$ is an arbitrary non-negative function, in contrast to our interpretation in terms of a probability distribution over cost matrices. For greater than two outcomes (classes), the logarithmic loss is often advocated based on its *locality*, i.e. that it depends only on the predicted probability of the outcome that did in fact occur, rather than on the entire predicted probability distribution. The characterization of this assumption as a kind of "likelihood principle" for probability loss functions is attributed by Bernardo [2] to one of his manuscript's referees. Locality would not necessarily seem appropriate in the ERL context, since the decisions one makes, and thus the value of predictions, would *in general* depend on the entire predicted distribution. Locality alone also leaves the choice of $h()$ completely arbitrary in the two-discrete-outcome case, which is precisely the case considered in the present paper.

Single-threshold loss functions are not strictly proper, since they have the same expected loss for any $\hat{p}$ on the same side of the threshold, not just the true p. They are however *loosely proper*, meaning that the true p does indeed minimize the expected loss, even if other values do as well. A loosely proper loss function never rewards an expert for lying, but it may not always *penalize* the expert for doing so.

It follows from the work cited above that all ERL loss functions are (at least loosely) proper. An ERL with $h(t) > 0$ almost everywhere (on $[0, 1]$) is strictly proper. In addition, at least one of those authors [13] showed that any absolutely continuous strictly proper loss function can be written in the form (2), or to restate this in our present framework, there exists some importance $h()$ that generates such a loss function as an ERL.

5. Conclusion

It has been said [14][6] that

---

[4] Often called a posterior probability in the pattern recognition literature, where typically Bayes' Theorem is used to calculate it from a class prior and the input features' class-conditional probabilities.

[5] In some cases the results are given in differential form instead of integral form.

[6] Parentheses added and clauses rearranged; "[strictly]" added; references can be found in the cited paper.

> ...in most situations, rewarding the assessor according to the value of his forecasts with respect to some decision-making problem (if such a value can be determined) (Murphy, 1968) would conflict with the desire to use a [strictly] proper scoring rule (Roberts, 1968).

The present paper can be viewed as a way around this conflict, where we replace "some decision-making problem" by "some *generalized* decision-making problem", the latter meaning the situation described by a distribution over ordinary decision problems.

Acknowledgments

Thanks to David Wolpert for several useful discussions and criticism, and to José Bernardo, Chris Fuchs, John Miller, Padhraic Smyth, Michael Stutzer for useful discussions and references. Supported in part by the (U.S.) Agency for Health Care Policy and Research (HS06830), and by the American College of Surgeons, AJCC prognostic systems project.

References

[1] J. Aczél. Remarks on the measurement of subjective probability and information. *Metrika*, 11(2):91–105, 1966.

[2] José M. Bernardo. Expected information as expected utility. *Ann. Stat.*, 7(3):686–691, 1979.

[3] G. W. Brier. Verification of forecasts expressed in terms of probability. *Monthly Weather Review*, 78(1):1–3, 1950.

[4] P. Fischer. On the inequality $\sum p_i f(p_i) \geq \sum p_i f(q_i)$. *Metrika*, 18:199–208, 1972.

[5] H. Gish. A probabilistic approach to the understanding and training of neural network classifiers. In *IEEE Int'l Conf. on Acoustics, Speech and Signal Processing*, pages 1361–1364, April 1990.

[6] I. J. Good. Rational decisions. *J. of the Royal Stat. Soc. B*, 14:107–114, 1952.

[7] John B. Hampshire II and Barak Pearlmutter. Equivalence proofs for multi-layer perceptron classifiers and the Bayesian discriminant function. In *Connectionist Models: Proc. of the 1990 Summer School*, pages 159–172. San Mateo, CA: Morgan Kaufmann Publishers, 1991.

[8] Charles E. Metz. Basic principles of ROC analysis. *Seminars Nuclear Med.*, 8(4):283–298, 1978.

[9] John W. Miller, Rod Goodman, and Padhraic Smyth. On loss functions which minimize to conditional expected values and posterior probabilities. *IEEE Tr. Information Theory*, 39(4):1404–1408, 1993.

[10] Gy. Muszély. On continuous solutions of a functional inequality. *Metrika*, 19:65–69, 1973.

[11] David B. Rosen. Cross-entropy vs. squared error vs. misclassification: On the relationship among loss functions. Submitted. For preprint info., e-mail d.rosen@ieee.org with Subject: QUERY PAPER CSM .

[12] David B. Rosen. Issues in selecting empirical performance measures for probabilistic classifiers. In Kenneth Hanson and Richard Silver, editors, *Maximum Entropy and Bayesian Methods (Proceedings of the Fifteenth International Workshop, July 1995)*. Kluwer, Dordrecht, The Netherlands, 1996. Paper title subject to revision. To appear. For preprint info., e-mail d.rosen@ieee.org with Subject: QUERY PAPER ISEPM .

[13] Leonard J. Savage. Elicitation of personal probabilities and expectations. *J. of the American Stat. Assoc.*, 66(336):783–801, 1971.

[14] Robert L. Winkler. Scoring rules and the evaluation of probability assessors. *J. of the American Stat. Assoc.*, 64:1073–1078, 1969.

[15] J. Frank Yates. External correspondence: Decompositions of the mean probability score. *Organizational Behavior and Human Performance*, 30:132–156, 1982.

Index

Fundamental Theories of Physics

22. A.O. Barut and A. van der Merwe (eds.): *Selected Scientific Papers of Alfred Landé. [1888-1975]*. 1988 ISBN 90-277-2594-2
23. W.T. Grandy, Jr.: *Foundations of Statistical Mechanics.* Vol. II: *Nonequilibrium Phenomena.* 1988 ISBN 90-277-2649-3
24. E.I. Bitsakis and C.A. Nicolaides (eds.): *The Concept of Probability.* Proceedings of the Delphi Conference (Delphi, Greece, 1987). 1989 ISBN 90-277-2679-5
25. A. van der Merwe, F. Selleri and G. Tarozzi (eds.): *Microphysical Reality and Quantum Formalism, Vol. 1.* Proceedings of the International Conference (Urbino, Italy, 1985). 1988 ISBN 90-277-2683-3
26. A. van der Merwe, F. Selleri and G. Tarozzi (eds.): *Microphysical Reality and Quantum Formalism, Vol. 2.* Proceedings of the International Conference (Urbino, Italy, 1985). 1988 ISBN 90-277-2684-1
27. I.D. Novikov and V.P. Frolov: *Physics of Black Holes.* 1989 ISBN 90-277-2685-X
28. G. Tarozzi and A. van der Merwe (eds.): *The Nature of Quantum Paradoxes.* Italian Studies in the Foundations and Philosophy of Modern Physics. 1988 ISBN 90-277-2703-1
29. B.R. Iyer, N. Mukunda and C.V. Vishveshwara (eds.): *Gravitation, Gauge Theories and the Early Universe.* 1989 ISBN 90-277-2710-4
30. H. Mark and L. Wood (eds.): *Energy in Physics, War and Peace.* A Festschrift celebrating Edward Teller's 80th Birthday. 1988 ISBN 90-277-2775-9
31. G.J. Erickson and C.R. Smith (eds.): *Maximum-Entropy and Bayesian Methods in Science and Engineering.* Vol. I: *Foundations.* 1988 ISBN 90-277-2793-7
32. G.J. Erickson and C.R. Smith (eds.): *Maximum-Entropy and Bayesian Methods in Science and Engineering.* Vol. II: *Applications.* 1988 ISBN 90-277-2794-5
33. M.E. Noz and Y.S. Kim (eds.): *Special Relativity and Quantum Theory.* A Collection of Papers on the Poincaré Group. 1988 ISBN 90-277-2799-6
34. I.Yu. Kobzarev and Yu.I. Manin: *Elementary Particles. Mathematics, Physics and Philosophy.* 1989 ISBN 0-7923-0098-X
35. F. Selleri: *Quantum Paradoxes and Physical Reality.* 1990 ISBN 0-7923-0253-2
36. J. Skilling (ed.): *Maximum-Entropy and Bayesian Methods.* Proceedings of the 8th International Workshop (Cambridge, UK, 1988). 1989 ISBN 0-7923-0224-9
37. M. Kafatos (ed.): *Bell's Theorem, Quantum Theory and Conceptions of the Universe.* 1989 ISBN 0-7923-0496-9
38. Yu.A. Izyumov and V.N. Syromyatnikov: *Phase Transitions and Crystal Symmetry.* 1990 ISBN 0-7923-0542-6
39. P.F. Fougère (ed.): *Maximum-Entropy and Bayesian Methods.* Proceedings of the 9th International Workshop (Dartmouth, Massachusetts, USA, 1989). 1990 ISBN 0-7923-0928-6
40. L. de Broglie: *Heisenberg's Uncertainties and the Probabilistic Interpretation of Wave Mechanics.* With Critical Notes of the Author. 1990 ISBN 0-7923-0929-4
41. W.T. Grandy, Jr.: *Relativistic Quantum Mechanics of Leptons and Fields.* 1991 ISBN 0-7923-1049-7
42. Yu.L. Klimontovich: *Turbulent Motion and the Structure of Chaos.* A New Approach to the Statistical Theory of Open Systems. 1991 ISBN 0-7923-1114-0

Fundamental Theories of Physics

43. W.T. Grandy, Jr. and L.H. Schick (eds.): *Maximum-Entropy and Bayesian Methods.* Proceedings of the 10th International Workshop (Laramie, Wyoming, USA, 1990). 1991 ISBN 0-7923-1140-X
44. P.Pták and S. Pulmannová: *Orthomodular Structures as Quantum Logics.* Intrinsic Properties, State Space and Probabilistic Topics. 1991 ISBN 0-7923-1207-4
45. D. Hestenes and A. Weingartshofer (eds.): *The Electron.* New Theory and Experiment. 1991 ISBN 0-7923-1356-9
46. P.P.J.M. Schram: *Kinetic Theory of Gases and Plasmas.* 1991 ISBN 0-7923-1392-5
47. A. Micali, R. Boudet and J. Helmstetter (eds.): *Clifford Algebras and their Applications in Mathematical Physics.* 1992 ISBN 0-7923-1623-1
48. E. Prugovečki: *Quantum Geometry.* A Framework for Quantum General Relativity. 1992 ISBN 0-7923-1640-1
49. M.H. Mac Gregor: *The Enigmatic Electron.* 1992 ISBN 0-7923-1982-6
50. C.R. Smith, G.J. Erickson and P.O. Neudorfer (eds.): *Maximum Entropy and Bayesian Methods.* Proceedings of the 11th International Workshop (Seattle, 1991). 1993 ISBN 0-7923-2031-X
51. D.J. Hoekzema: *The Quantum Labyrinth.* 1993 ISBN 0-7923-2066-2
52. Z. Oziewicz, B. Jancewicz and A. Borowiec (eds.): *Spinors, Twistors, Clifford Algebras and Quantum Deformations.* Proceedings of the Second Max Born Symposium (Wrocław, Poland, 1992). 1993 ISBN 0-7923-2251-7
53. A. Mohammad-Djafari and G. Demoment (eds.): *Maximum Entropy and Bayesian Methods.* Proceedings of the 12th International Workshop (Paris, France, 1992). 1993 ISBN 0-7923-2280-0
54. M. Riesz: *Clifford Numbers and Spinors* with Riesz' Private Lectures to E. Folke Bolinder and a Historical Review by Pertti Lounesto. E.F. Bolinder and P. Lounesto (eds.). 1993 ISBN 0-7923-2299-1
55. F. Brackx, R. Delanghe and H. Serras (eds.): *Clifford Algebras and their Applications in Mathematical Physics.* Proceedings of the Third Conference (Deinze, 1993) 1993 ISBN 0-7923-2347-5
56. J.R. Fanchi: *Parametrized Relativistic Quantum Theory.* 1993 ISBN 0-7923-2376-9
57. A. Peres: *Quantum Theory: Concepts and Methods.* 1993 ISBN 0-7923-2549-4
58. P.L. Antonelli, R.S. Ingarden and M. Matsumoto: *The Theory of Sprays and Finsler Spaces with Applications in Physics and Biology.* 1993 ISBN 0-7923-2577-X
59. R. Miron and M. Anastasiei: *The Geometry of Lagrange Spaces: Theory and Applications.* 1994 ISBN 0-7923-2591-5
60. G. Adomian: *Solving Frontier Problems of Physics: The Decomposition Method.* 1994 ISBN 0-7923-2644-X
61 B.S. Kerner and V.V. Osipov: *Autosolitons.* A New Approach to Problems of Self-Organization and Turbulence. 1994 ISBN 0-7923-2816-7
62. G.R. Heidbreder (ed.): *Maximum Entropy and Bayesian Methods.* Proceedings of the 13th International Workshop (Santa Barbara, USA, 1993) 1996 ISBN 0-7923-2851-5
63. J. Peřina, Z. Hradil and B. Jurčo: *Quantum Optics and Fundamentals of Physics.* 1994 ISBN 0-7923-3000-5

Fundamental Theories of Physics

64. M. Evans and J.-P. Vigier: *The Enigmatic Photon.* Volume 1: The Field $\boldsymbol{B}^{(3)}$. 1994
ISBN 0-7923-3049-8
65. C.K. Raju: *Time: Towards a Constistent Theory.* 1994 ISBN 0-7923-3103-6
66. A.K.T. Assis: *Weber's Electrodynamics.* 1994 ISBN 0-7923-3137-0
67. Yu. L. Klimontovich: *Statistical Theory of Open Systems.* Volume 1: A Unified Approach to Kinetic Description of Processes in Active Systems. 1995
ISBN 0-7923-3199-0; Pb: ISBN 0-7923-3242-3
68. M. Evans and J.-P. Vigier: *The Enigmatic Photon.* Volume 2: Non-Abelian Electrodynamics. 1995 ISBN 0-7923-3288-1
69. G. Esposito: *Complex General Relativity.* 1995 ISBN 0-7923-3340-3
70. *Forthcoming*
71. C. Garola and A. Rossi (eds.): *The Foundations of Quantum Mechanics – Historical Analysis and Open Questions.* 1995 ISBN 0-7923-3480-9
72. A. Peres: *Quantum Theory: Concepts and Methods.* 1995 (see for hardback edition, Vol. 57) ISBN Pb 0-7923-3632-1
73. M. Ferrero and A. van der Merwe (eds.): *Fundamental Problems in Quantum Physics.* 1995 ISBN 0-7923-3670-4
74. F.E. Schroeck, Jr.: *Quantum Mechanics on Phase Space.* 1996 ISBN 0-7923-3794-8
75. L. de la Peña and A.M. Cetto: *The Quantum Dice.* An Introduction to Stochastic Electrodynamics. 1996 ISBN 0-7923-3818-9
76. P.L. Antonelli and R. Miron (eds.): *Lagrange and Finsler Geometry.* Applications to Physics and Biology. 1996 ISBN 0-7923-3873-1
77. M.W. Evans, J-P. Vigier, S. Roy and S. Jeffers: *The Enigmatic Photon.* Volume 3: Theory and Practice of the $\boldsymbol{B}^{(3)}$ Field. 1996 ISBN 0-7923-4044-2

KLUWER ACADEMIC PUBLISHERS – DORDRECHT / BOSTON / LONDON

GPSR Compliance

The European Union's (EU) General Product Safety Regulation (GPSR) is a set of rules that requires consumer products to be safe and our obligations to ensure this.

If you have any concerns about our products, you can contact us on ProductSafety@springernature.com

In case Publisher is established outside the EU, the EU authorized representative is:

Springer Nature Customer Service Center GmbH
Europaplatz 3
69115 Heidelberg, Germany

Batch number: 10815974

Printed by Printforce, the Netherlands